Encyclopedia of Rainforests

Encyclopedia of
Rainforests

Diane Jukofsky
for the Rainforest Alliance

The rare Arabian Oryx is believed to have inspired the myth of the unicorn. This desert antelope became virtually extinct in the early 1960s. At that time several groups of international conservationists arranged to have 9 animals sent to the Phoenix Zoo to be the nucleus of a captive breeding herd. Today the Oryx population is over 1,000 and nearly 500 have been returned to reserves in the Middle East.

Library of Congress Cataloging-in-Publication Data

Jukofsky, Diane.
    Encyclopedia of rainforests / Diane Jukofsky for the Rainforest Alliance.
      p.   cm.
    Includes bibliographical references (p.) and index.
    ISBN 1–57356–259–9 (alk. paper)
    1. Rain forests—Encyclopedias.   I. Rainforest Alliance.   II. Title.
    QH86.J85   2002
    578.734'03—dc21   2001032154

British Library Cataloguing in Publication Data is available.

Library of Congress Catalog Card Number: 2001032154
ISBN: 1–57356–259–9

First published in 2002

Oryx Press, 88 Post Road West, Westport, CT 06881
An imprint of Greenwood Publishing Group, Inc.
www.oryxpress.com

Printed in the United States of America

The paper used in this book complies with the Permanent Paper Standard issued by the National Information Standards Organization (Z39.48–1984).

10 9 8 7 6 5 4 3 2 1

Photos in frontispiece and on part title pages courtesy Chris M. Wille.

The paper used for the text of this book is a 60-pound Lynx, made by Weyerhaeuser, which contains 30% post-consumer fiber.

Contents

A photo essay follows p. 184.

v

Preface

Writing an encyclopedia about a place on Earth that I love—tropical forests—seemed like a wonderful assignment. But after the Rainforest Alliance signed the contract with Oryx Press and handed me the task, I was momentarily daunted. How could I even begin? Tropical forests hold 50 percent of the world's known plants and animals, or more than 700,000 species. A book of manageable size could scarcely hold 0.2 percent of that number. How could I possibly go about choosing which plants and animals to include?

I set limitations. I focused only on the land area found between the Tropic of Cancer and the Tropic of Capricorn, officially known as "the Tropics," and within this area, only on plants and animals that inhabit forests. But I still faced a perplexing task of selection.

Luckily, minds wiser than mine had already done some helpful selecting. As director of the Rainforest Alliance's Conservation Media Center, my files and library are filled with books and reports by outstanding scientists, naturalists, and journalists. If they thought it worthwhile to include particular species in their own books, or focus an entire magazine article on one species of bird or tree, that was good enough for me. From these pre-selections, I got personal—I chose the plants and animals that seemed the most fascinating to me and also represented the astounding complexity of the tropical forest. The end result has been that I have described 261 plant families and species and 818 animal classes, orders, families and species.

It's that amazement I have at the incredible diversity of life in tropical forests, which hold so many species that are completely distinct from one another and yet so utterly connected, that I hope that readers of this book will share. This book's opening chapter, "An Introduction to Tropical Forests: Understanding a Mysterious Place on the Planet," serves as an introduction to these forests—where they are found, what they look like, what's found inside them and why, how they function as an ecosystem, and what are the principal causes of deforestation.

A chart on page xxix, "The World's Tropical Forests: Where They Are; How Much Is Left; How Fast They Are Vanishing," lists every tropical country, its size, how many square miles of forest stood in 1980, how much remained in 1995, and the percentage change, which will give you a good idea of the deforestation rate in each country. One thing to remember when reviewing these statistics—nearly every country in the tropics was once completely covered with trees.

Parts One through Three of the book provide individual descriptions of wildlife, plants, and people. Each description also provides one to three suggestions for further reading. In Part One, the wildlife section, you will find—in alphabetical order except for the section on birds—general descriptions of animal classes (e.g., *Mammalia*), orders (*Carnivora*), and families (*Felidae*), which note the habits, appearances, and ranges that each member species has in common. Descriptions of each animal class include the number of families in that class; descriptions of families include the number of species. The individual entries in the wildlife sections always describe what each animal looks like, in which tropical countries it can be found, and if it is endangered or threatened with extinction. Beyond that, descriptions include attributes and habits that are particularly interesting or unusual. Cataloging species in their proper classes, orders, and families helps us understand how animals are related to one another. The scientists who do this cataloging—the taxonomists—are still debating how to classify some species. Part One includes descriptions of 69 amphibians, 20 arachnids, 191 birds, 68 fish, 116 insects, 251 mammals, and 103 reptiles.

Entries in the plant section in Part Two include 261 plant families and species and describe each species' size, color, flowers and fruits, range, and status (endangered or threatened). If people use the plant—and a majority of those included in this book do have attributes valued by human beings—this is noted, often with historic detail.

Part Three, the section on people, begins with descriptions of 32 tribes of indigenous people who, against all odds, continue to dwell in the world's tropical forests. Entries include very rough population estimates, where and how they live, and what threats they face from non-indigenous people. This is followed by an alphabetical list of descriptions of 54 scientists, explorers, and activists who had—or have—an important role in our understanding of tropical forests and in their conservation. I made an effort to include people from many different countries and backgrounds. Regrettably, space limitations forced me to leave out scores of dedicated people who have contributed greatly to tropical biodiversity research and conservation.

In Part Four, "Saving Tropical Forests," you'll find essays on what people are doing to try to stop trees from falling, divided by topic, from "agriculture" to "wildlife research and management." This section concludes with ideas of how individuals can play a role in slowing the worldwide loss of tropical forests.

Part Five contains lists of resources on the rainforest. Check the table of contents to see the range this book offers—you'll find lists of reference publications, including books, articles, and Web sites; videos; names and addresses of organizations and embassies; and more.

Books and magazines listed after Further Reading, at the end of each description in Parts Two through Four, are in alphabetical order by the author's last name. For the full reference, check the bibliography, where they are also listed alphabetically by the author's last name and in four different categories: tropical forests in general, wildlife, plants, and people.

Web sites that offer helpful information are also listed under Further Reading. Since Web site addresses change so frequently, if you find that a particular URL is no longer valid, check the site's homepage. For example, if <http://www.baldeagleinfo.com/eagle/eagle6.html> no longer exists or no longer holds information about the New Guinea eagle, try <http://www.baldeagleinfo.com>. The Internet offers a wealth of information about many of the species described here, but the quality of the information varies widely. Web sites of established nonprofit organizations, research or zoological institutions, universities, and museums are usually the most reliable. If you cite information taken from a Web site, it's a good idea to verify it elsewhere.

You will also note that names of organizations with the country where they are headquartered precede many Web site URLs listed under Further Reading. If you need to contact these organizations for more information, you will find their addresses, telephone numbers, fax numbers and email addresses (when available) in the chapters "International Conservation Groups and Organizations" and "Organizations and Government Agencies Based in the United States and Canada."

Acknowledgments

The scientists and writers to whom I am indebted are all listed in this book's bibliography, but special acknowledgment goes to photographer Art Wolfe and botanist Ghillean Prance, whose beautiful book, *Rainforests of the World*, provided inspiration and guidance. *Tropical Nature*, by Adrian Forsyth and Kenneth Miyata; *Amazon: The Flooded Forest*, by Michael Goulding; *Costa Rican Natural History*, edited by Daniel Janzen; *A Neotropical Companion* by John Kricher; *The Primary Source* by Norman Myers; and *Tropical Rainforests* by Arnold Newman, were also of great help. I would have been lost without the books I borrowed, for many months, from the personal library of conservation biologist George Powell.

Because I live in Costa Rica, I had limited resources available, in particular when it came to information about flora and fauna of Africa and Asia. Luckily, I was able to depend upon a small corps of excellent research and writing assistants. My heartfelt thanks to journalist David Dudenhoefer, who provided most of the chapters on fish and spiders—I'm particularly grateful for his bravery in tackling the latter, since I have arachnophobia—and many of the frog and toad descriptions.

This is a Rainforest Alliance book and a few words should be said about that organization. The Rainforest Alliance is an international nonprofit organization that works to protect endangered ecosystems and the people, wildlife, and plants that live within them by transforming land use, business practices, and consumer behavior. The Alliance designs and implements environmentally and socially responsible management practices for natural resource–based activities such as logging, farming, and tourism. It focuses its efforts in the tropics, where most of the world's endangered ecosystems and biodiversity are located, and plays a unique role in ensuring environmentally responsible land use around parks and protected areas. The Alliance creates alliances with farmers, scientists, businesses, governments, and local peoples.

My colleagues in the Alliance office in New York—in particular Daniel Katz and Karin Kreider—backed me completely whenever I cried out for help. Thanks to Helena Albuquerque, I could count on extremely helpful interns and assistants, who kept office telephones and computers busy as they gathered information.

Sarah O'Braitis began working for the Alliance at the height of this book's creation and didn't need more than a day to figure out what needed to be done. She researched and wrote a good portion of the sections on people and ferns and provided steady guidance, bright ideas, and a never-ending supply of encouraging words. Lily Pashall was a tireless source of fast, thorough, and professional research from start to finish, from deforestation data to turtles to medicinal plants. If Lily couldn't find the answer, it was not to be found. Maria Ghiso's insightful assistance with research on indigenous people, videos, fungi, and embassies and conservation groups worldwide helped make this book a solid source of information on these topics. Ina Chaudhury helped launch this book in the first months of research and writing and established a good working rhythm for us all. Erika Diamond, Zeina Haurani, Christine Pinzon, and Asma Syed also provided important data and vital research. So did Nuria Bolaños, who also created the bibliography and helped to organize the manuscript.

Thanks to Tom Divney for carefully reviewing the entire plant section and to Oliver Bach, for his knowledgeable review of the reptile section. Chris Wille not only carefully read and edited every word of this book, but also helped shape it and kept its author encouraged and somewhat sane. While the comments of these experts were invaluable, any errors are entirely my fault.

It would be difficult to publish a book on tropical forests, plants, and wildlife without including color photographs that can do a far better job than I in depicting their striking beauty. Labeeb Abboud, Jerry Bauer, Gary Braasch, Gerry Ellis, Michael Goulding, David Julian, Chris Wille, and Art Wolfe are not only superb photographers, but also are generous in their dedication to the cause of tropical forest conservation.

Thanks for the mentoring go primarily, of course, to S.L. and E.A. Jukofsky, with boosts along the way from Charles Roberts, Fred Jerome, and especially Alan McGowan.

I offer my final words of gratitude to Elizabeth Z. Cushing, for first showing me wooded wildlands, E.O. Wilson for explaining to me why tropical forests can bring tears to my eyes, and especially to C.M. Wille for taking me there and forever sharing in the astonishment.

An Introduction to Tropical Forests: Understanding a Mysterious Place on the Planet

The unsolved mysteries of the rainforest are formless and seductive. They are like unnamed islands hidden in the blank spaces of old maps, like dark shapes glimpsed descending the far wall of a reef into the abyss. They draw us forward and stir strange apprehensions.
—E.O. Wilson, *The Diversity of Life*

Why do rainforests captivate human visitors? It may be the sight of so many intertwining shades of green, broken by sudden and startling flashes of brilliant colors. Or it might be the heady smell that simultaneously signals new life and recent death, sweet perfume and dusty musk. Or the sounds—a hushed silence at first, and then, as the ear adjusts, a steady, subtle buzzing and rustling punctuated by sharp calls, cracks, and high-pitched songs. It may be the feeling of refuge provided by a high ceiling of verdant green and the surprising spaciousness underneath that allows—and even beckons—visitors to enter and explore, in spite of a vague unease provoked by the unfamiliar and nearly overwhelming surroundings.

RAINFOREST BIODIVERSITY

Luckily, explorers have stepped back out of the rainforests and shared their discoveries. What they have found—and continue to find—is astounding to anyone who thought human beings had a reasonable grasp on what life on Earth is all about, who thought that the nearest unexplored frontier lay in the heavens.

As recently as the early 1980s, most biologists agreed that Earth had about two million species, based on the 1.5 million species already identified plus another 500,000 species that they believed probably existed. Then, in 1982, Terry Erwin, a U.S. entomologist with the Smithsonian Institution, opened a new frontier. He invented a way to fumigate the tropical forest treetops, or canopy, in Panama so that

bugs and beetles overcome by the chemical fog tumbled down to plastic sheets that he had spread on the ground. Based on the number of species he counted on the sheets—1,200 kinds of insects from 19 trees—he estimated that the number of species on Earth was probably closer to 30 million. Although many of Erwin's colleagues agreed that a serious readjustment needed to be made, they were more conservative in their estimates, concluding that Earth probably holds about 10 million species.

Now some experts, including Erwin, will go as high as 100 million. What's notable is that the best minds in biology are so uncertain about how many different forms of life share the globe with us human beings. Scientists have identified more insect species than any other group of plants or animals, and certainly many more exist than the 751,000 known kinds. But we know least about algae, fungi, protozoa, monera, and viruses, and the number of species within these groups is unquestionably many times higher than what scientists have described to date.

The word that indicates the variety of all life on Earth, from the tiniest viruses to the largest mammals and tallest trees, is biodiversity. Species is the unit most often used when measuring biodiversity, but the term means more than just numbers of plants and animals. Biodiversity is also all the ways that these species are genetically different from each other, and all the ways that they interact. When one ancient tree crashes to the ground in a mountaintop forest, biodiversity is affected by more than the loss of just a single species of tree. In this example, the biodiversity in the forest will also be affected by the fungi and beetles that thrive as the tree rots, the animals that feed on the fungi and beetles, the plants that sprout from the rich matter that was once a majestic tree, the afternoon rains that soak all this life, the warmth of the morning sun, and the enshrouding mists—all of this is biodiversity. The term, a shortened version of biological diversity, was first introduced to the public in 1986 at a symposium sponsored by the National Academy of Sciences and the Smithsonian Institution. It became popular during the United Nations Conference on the Environment and Development, held in 1992 in Brazil, when U.S. President George Bush's decision not to sign the Biodiversity Convention caused much controversy.

Biodiversity is not evenly distributed on Earth. It is found in greatest concentrations in tropical forests, which comprise just 7 percent of the planet's dry land surface but hold more than 50 percent of all species. It's easiest to grasp what that means with examples and comparisons (see table). Conservation International, an environmental group based in the United States, determined the 17 countries in the world that have the most biodiversity, based on the number of different species plus the number of species that are found nowhere else, which are called endemics. On the list are Australia, Brazil, China, Colombia, Congo, Ecuador, India, Indonesia, Madagascar, Malaysia, Mexico, Papua New Guinea, Peru, Philippines, South Africa, United States, and Venezuela. Every one of these nations has rainforests.

IN THE TROPICS YOU CAN FIND:	OUTSIDE THE TROPICS YOU CAN FIND:
700 tree species in 25 acres of forest in Borneo.	700 tree species in all of North America.
1,525 bird species in Colombia.	195 bird species in New York State, which is roughly the same size as Colombia.
43 species of ants on a single tree in Tambopata Reserve, Peru.	43 species of ants in the entire British Isles.
In the same 13,590-acre Peruvian reserve, 1,209 species of butterflies.	380 butterfly species in all of Europe and the Mediterranean coast of North Africa combined.
18,000 beetle species in just 2.5 acres of forest in Panama.	24,000 species of beetles in the United States and Canada combined.
80 percent of all the world's insects.	The other 20 percent.
2,000 species of fish in the Amazon River system.	250 species of fish in the Mississippi River system.
Hiding in the fur of one sloth in Panama, 3 species of beetles, 6 species of mites, 3 species of moths, and 3 species of algae.	No sloths.

WHERE AND WHAT IN THE WORLD ARE RAINFORESTS?

The epicenter of rainforests is at the equator, the navigational line that divides the Earth in two. From the equator, rainforests are found north to the Tropic of Cancer and south to the Tropic of Capricorn. These are lines on a map, and nature doesn't always stay within the lines. But in between these two borders lie nearly all the world's tropical forests.

In total, the humid tropics cover approximately 3.7 billion acres of land area, has about 2 billion human inhabitants distributed throughout some 60 countries, and contains 2.5 billion acres of the tropical rainforest that remains in the world. Forty-five percent of the tropics are in Latin America, Africa has 30 percent, and Asia 25 percent.

Originally, forests draped virtually all the land in the tropics, encircling Earth like a great green cummerbund, some 6 billion unbroken acres (9,460,000 square miles). Today only about 2.5 billion acres (3,865,000 square miles) remain in three principal areas of Latin America, Africa, and Asia. Latin America's forests are found in parts of Mexico and south to Brazil and Paraguay. The largest intact piece is in Amazonia, named for the mighty river system, the world's largest. About 1.37 million acres of Latin American tropical forests remain. Most lie in Peru, Ecuador, Colombia, Venezuela, Brazil, Bolivia, Guyana, Suriname, and French Guiana, which are the countries of the Amazon River basin. The Amazon also holds the greatest biological diversity on Earth, including many endemic species. Of the 100,000 known plant species in Latin America, 30,000 are found *only* in the Amazon. About 2,500 species of trees, or one-third of identified tropical trees, are found *only* in the Amazon. The region also has the highest diversity of freshwater fish, birds, and butterflies on Earth.

The Caribbean islands hold small pieces of rainforests; Puerto Rico and Trinidad have the most. Another relatively large amount—a total of 610 million acres—of tropical forest lies in Asia, particularly in India, Burma, Thailand, Malaysia, Indonesia, and Papua New Guinea. Congo has more forest remaining than other nations on the African continent, and relatively extensive stands are found in Cameroon, the Central African Republic, and Gabon. About 464 million acres of African tropical forest remain.

Throughout the year, the sun beats down on this verdant belt around Earth from a position nearly directly overhead. In temperate zones, the sun's rays come from a position closer to the horizon, producing much less heat and light. In the tropics, 1 square centimeter (.16 square inch) of ground receives more than twice as much solar energy as does 1 square centimeter of land at a latitude 50 degrees north (for example, Vancouver, Canada) or south (the tip of Argentina) of the equator. The length of the days in tropical countries is more or less the same year-round. At the equator, daylight lasts exactly 12 hours. As you move north, days are longer during part of the year, but that may merely mean the tropical sun sets at 6:15 P.M. rather than 6 P.M. It never gets cold in the tropics. The average temperature is 75 degrees Fahrenheit, with little fluctuation.

Precipitation is the other factor that defines tropical forests. Although the terms "rainforests" and "tropical forests" are frequently used interchangeably, rainfall is what makes a true rainforest. If a forest receives at least 100 inches of precipitation annually, never less than 6 inches of rain in any one month, it can be called a rainforest. These wet woods are generally evergreen—it's always the growing season—and lie below altitudes of 4,300 feet.

Oceans and atmospheric currents, wind currents, and mountains also play a role in the amount of rain that falls on tropical forests. The air over oceans is generally warmer than the air above the land just offshore. If winds move the moisture-laden air from the ocean to the cooler air above land, a rainshower results. Even more rain will fall if the warm ocean air moves over mountains, since air cools when it rises. The wettest forests in Latin America, and perhaps the world, are along the Pacific coast of Colombia. There, a warm weather pattern called El Niño ("The Child" in Spanish; the weather pattern usually coincides with Christmas) lies offshore, and that particularly warm air sweeps over the lowland forests until it hits the steep Andes Mountains. The combination of warm air hitting cooler air over land, and still cooler mountain air, results in prodigious rainfall, more than 360 inches per year.

Summer and winter in the tropics translate into the dry season and the rainy season. During the dry season, the sun's rays are more direct, and it rains only sporadically, while the rainy season brings more cloud cover and usually a shower every day, often a torrential downpour. Plants and wildlife naturally adjust to the change in weather. People carry around umbrellas and try not to schedule any errands during the afternoons, when downpours can make a dash from store to bus a drenching endeavor.

Tropical trees more commonly flower during the dry season, since pollinating insects are more active when the sun shines. Pollination during the dry season means that seeds will sprout at the start of the rainy season, when they most need moisture to grow. Tree leaves also sprout with the first rains, and leaf predation is highest then, as new leaves are less likely to have developed any chemical or structural defenses against munching. Many nectar-sipping and fruit-eating bird species breed during the start of the rainy season, when high-energy performances require adequate nutrition.

The tamandua (*Tamandua mexicana*), a rainforest anteater, switches from ants in the rainy season to termites in the dry season for a good thirst-quenching reason: Termites are juicier than ants. A small rodent called the agouti (*Dasyprocta*) buries seeds during the rainy season, then must locate them again to get through the seedless dry season months, much as a squirrel digs up the nuts it has buried to survive winters in the northern, or temperate, region. Depending on such factors as altitude, precipitation, weather, and location, tropical forests are made up of different ecosystems, or particular environments and all the organisms living within them. One ecosystem may overlap with another. Although some plants and animals are found in many different tropical forest ecosystems, others inhabit only one niche, unable to flourish anywhere else on Earth.

Dry Forests

As you move away from the equator, rains slacken. In some tropical forests called "dry" or "seasonal," rains are quite scarce during a definite period, lasting four to seven months each year. Seasonal forests are found in Indonesia, Thailand, Myanmar, India, and Sri Lanka and also in Africa; they are less common in Latin America, where the climate is generally wetter.

When the rains cease in a dry forest, trees draw upon water still in the soil, then drop their leaves. Some tree species bloom with flowers in vibrant or soft pastel shades, and then drop their leaves in reaction to the drought. Others keep their leaves, which fade from green to brown or gray. These seasonal forests appear much less luxurious and lush than rainforests, and indeed, they contain far fewer plants. But once the rains begin, they turn verdant once again. Dry forests do not lack for biodiversity—for example, they abound in insects, particularly butterflies and moths, bees, wasps, and ants. For insects and many species of mammals, the rainless period's flowering and fruiting trees provide rich bounty. Monkeys, wild cats, and iguanas abandon the dry forests when leaves drop, moving to moister, riverside forests.

Dry forests are the world's most threatened. Their lack of humidity means they burn more easily than rainforests, so beginning centuries ago, people set them on fire to clear land for farming and cattle grazing.

Flooded Forests

The Amazon River and its hundreds of tributaries cannot contain the downpour during the rainy season in South America. Water overflows into surrounding floodplains, and the river becomes in many places a wide lake, whose waters splash against the upper branches of tall tropical trees and fill valleys and meadows. The Amazon's flooded forest, which is about 2 percent of the entire rainforest in the region, consists of at least 25 million acres, an area larger than England. The rains inundate the flooded forest for an average of four to seven months each year. The river basin contains 20 percent of the world's fresh water.

Any keeper of houseplants knows that overwatering can be fatal. So how do the tremendous trees that poke out of the floodwaters survive half the year half submerged? Scientists are unsure of the answer to this rainforest mystery. But flooded forest trees hold their leaves and may even flower or fruit. Tree branches lie close enough to the water's surface so that some Amazon fish are able to pluck off fruits merely by raising their powerful jaws slightly out of the muddy water. A fish species called the arowhana (*Osteoglossum bicirrhosum*) can actually hurl itself out of the water—up to three feet in the air—to feed on the masses of bugs and spiders that crowd the tree tops to avoid drowning.

Though none is as large as the Amazon's, flooded forests are also found in the Mekong River basin in Vietnam, the Malay River in Borneo, and the Sepik River in New Guinea.

Mangrove Forests

Along coasts and lagoons lies another kind of tropical forest—mangroves. Worldwide, mangroves contain about 60 species of trees and shrubs and provide habitat for more than 2,000 species of fish, invertebrates, and epiphytic plants. Mangrove trees range in size from the giant forests found in parts of Brazil, Colombia, Ecuador, and Venezuela, where species of red mangrove (*Rhizophora mangle*) and black mangrove (*Avicennia germinans*) reach heights of 150 feet, to the barely three-foot-tall mangrove systems common in Asia and the Pacific islands. Along with red and black mangrove, white mangrove (*Laguncularia racemosa*) and buttonwood (*Conocarpus erecta*) are commonly found in the Neotropics. All four species can grow in saltwater, as their succulent leaves can secrete excess sodium and chloride. Arching prop roots support the trunks of red mangrove trees, providing a firm anchor in the mucky substrate. Black mangroves have developed a root system equally helpful for survival in water-logged soil: Their straight, shallow roots send up vertical shoots, called "pneumatophores," that emerge from the mud like snorkels and absorb oxygen.

The world's largest mangrove forest is in the Ganges Delta on the border of India and Bangladesh, while other large stands border Borneo's Malay Peninsula and New Guinea. Mangrove forests protect coastal areas from storms, build up soils along coastlines, and are the natural nurseries for sea creatures like shrimp, oysters, crab, and tarpon, which support and feed millions of people worldwide. Nonetheless, relatively few mangrove forests survive. They are felled to make way for coastal development, rice paddies, and shrimp farms, and for firewood and the tannins in their bark.

Cloud Forests

The woods that flourish at mid-elevations of tropical mountain ranges and volcanoes, usually on the windward side, are called cloud forests. The World Conservation Monitoring Centre has identified 605 cloud forests in 41 countries, with 280 sites in Latin America, the majority in Mexico, Colombia, Ecuador, and Venezuela. In Southeast Asia, 228 sites have been identified in 14 countries, mainly Indonesia and Malaysia. Africa holds 97 identified cloud forests in 21 countries. The higher the forest, the cooler the air; temperature decreases about 1 degree Fahrenheit every 330 feet. Clouds frequently enshroud these high-altitude forests, creating an ever-misty, dripping ecosystem. Since the nearly constant cloud cover makes photosynthesis more difficult, plants grow more slowly, and fewer species of trees are present than in lower altitude tropical forests. But a multitude of mosses, ferns, orchids, and bromeliads do flourish in cloud forests. These plants that are adapted to grow on other plants and trees are called epiphytes. In cloud forests, epiphytes make up 40 percent of the total plant biomass. Because temperatures are cooler, trees commonly found in northern temperate zones, like oaks, also grow in tropical cloud forests. Cloud forests hold moisture like sponges and often are the source of drinking water for people who live downslope. One result of a ravished cloud forest is the loss of potable water for millions of people.

At the highest altitudes where trees can still grow—around 11,500 feet—lack of sunlight and brisk temperatures make trees spindly, twisted, and stunted. Because trees are only a few feet high, this ecosystem is called "elfin forest." Moss and lichens grow in thick, wet carpets.

Few large mammals find their way up to the cloud forests because high altitudes result in slower plant growth, and thus, less food. But in the Virunga Volcanoes in eastern Congo, gorillas (*Gorilla gorilla*) and elephants (*Loxodonta africana*) roam lowlands and climb as high as 9,000 feet into the clouds. Both species eat a wide variety of plants, but are particularly fond of young and tender bamboo, which flourishes at higher altitudes. In Malaysia, plant eaters like gibbons (*Hylobates*) and tapirs (*Tapirdae*) are found in cloud forests, as well as in forests at lower altitudes. Some of the world's most beautiful birds are at home in cloud forests. In Latin America dozens of species of dainty, darting, and jewel-colored hummingbirds play an important role in spreading genes as plant pollinators. The emerald green plumage of the crimson-breasted resplendent quetzal (*Pharomachrus mocinno*) is a flash of brilliant color in the mists of Central America's cloud forests. The cloud forests of New Guinea are home to exuberantly plumaged birds of paradise (*Paradisaeidae*).

FROM THE GROUND UP: How Tropical Forests Work

It's thrilling to stand in a tropical forest and know—even if you can't spot them all—that before you are hundreds of different species of plants and animals. Harder to see are all the natural processes in full throttle. Just as the cells, nerves, and tissues that make up your body are all at work even though you are standing perfectly still, the forest is continually throbbing with life, growth, death, and decomposition, though it may seem that not even a leaf is stirring.

Poor Soils and Rich Litter

It starts with earth, and surely soil that supports so much life must be rich and deep. Surprisingly, however, tropical soils are thin and poor. Unlike temperate forests, where the ground is laden with layers of litter, in a tropical forest you can brush away the dead and rotting debris with merely a sweep of your hiking shoe. Then you'll unearth a tangled mass of thin, white threads, which are fungi and tree roots. The feathery roots are indeed connected to nearby towering trees, which have—like virtually everything in a rainforest—a particular and consequential relationship with the filaments of fungus. The trees provide the minute fungi with energy, and the fungi provide minerals vital to the forest giants. The topsoil beneath the leaf litter is usually no more than an inch or two deep. Beneath this is a red clay that's nearly devoid of minerals. Just 20 percent of the soils in the tropics are good for growing crops, and virtually all that land has indeed been developed. Fertile soils are found in the volcanic highlands of Central America, the Philippines, Indonesia, and Cameroon and the riverside floodplains of Asia's Ganges and Mekong.

Although they can't support food crops, the rainforest's shallow topsoil and leaf litter are teeming with nutrients, which are absorbed as quickly as they develop by the very trees that drop the leaves, fruit, and branches that will decompose and provide more nutrients. Key to the rainforest's lushness is this nutrient cycle, which is fed by heat, moisture and the all-important fungi.

Fungi: First in the Food Chain

Fungi lack roots, seeds, leaves, and chlorophyll and get their food from plants and animals, living or dead. Mycorrhizal fungi are those found in soils, pumping nutrients into tree roots and helping trees resist root diseases and withstand drought. Thus, the tallest, mightiest trees in the rainforest are dependent on one of the tiniest forms of life, and vice versa. When a forest is cleared, the fungi die, and there's little chance of reestablishing the forest unless the fungi too can be restored.

Parasitic fungi attack living plants and animals, often killing them in the process. Saprophytic fungi are decomposers that feed off dead plants or animal matter. They range in size from the miniscule threads in the soil to mushrooms that can be as large as 24 inches across. Bracket fungi protrude from dead trees and stumps like semicircular plates. They penetrate the wood, breaking it down. Another kind of fungus is the puffball. When rain or a falling twig strikes the membrane-thin center of a puffball, it bursts, hurling its spores into the air.

Fungi are also important sources of protein to animals, including human forest inhabitants, such as the Yanomami indigenous people of the Amazon. Mites and millipedes feed on soil fungi, and then larger insects in turn consume them. The bigger bugs are eaten by spiders, frogs and snakes, which are eaten by the larger animals at the top of the food chain. Without fungi to break it all down for easy absorption, rainforests would quickly overflow with dead organic matter.

Ground-Dwellers

The vast majority of animals in the tropical forest are insects and other arthropods. Thousands of species of bugs creep and crawl along the ground. A team of scientists in Gunung Mulu Park in Sarawak counted 8,000 species of beetles in only 131 acres. Another scientist examined 2.5 acres of forest in Brazil and calculated that they held 727 million mites, 120 million springtail insects, 40 million aphids and scale insects, and 8.6 million ants. Like fungi, the termites, ants, mites, and nematodes play an important role in decomposition and nutrient recycling. For example, most termite species eat dead wood, but they can't digest it. This is done by minute organisms called protozoans or by bacteria that live in termites' digestion systems. Since few earthworms are found in tropical soils, termites are the most important decomposing insect. Of the approximately 2,200 species of termites, nearly all live in the tropics. Their numbers are estimated to total 250,000 billion, which translates to nearly 45,000 termites per living human being.

The rainforest floor is also home to frogs, lizards, and snakes. Their coloring is often so well matched to their surroundings that they are nearly invisible, but many species sport brilliant hues that stand out like traffic lights, serving as a dazzling warning to predators that they are poisonous.

A large percentage of the mammals that inhabit rainforests are small rodents (*Rodentia*), mainly rats and also agoutis (*Dasyprocta*). Among the larger mammals are many species of wild pigs (*Suidae*) and wild cats (*Felidae*), including jaguars, ocelots, and leopards. In all major rainforest areas, the wild cats are at the top of the food chain. Human beings are their only enemies, and they have significantly damaged their populations by hunting them to near extinction and destroying most of their habitat.

Several species of deer and antelope roam tropical forests, including a smaller cousin of the white-tailed deer that's common in North American temperate forests. The smallest deer in the Neotropics, which includes Mexico, Central and South America, and the Caribbean, is the red brocket deer (*Mazama americana*), which weighs about 40 pounds. Of similar dainty size are the duikers (*Cephalophus monticola*), small antelope found in Africa, where forest ungulates, or hoofed animals, are more common than in the Neotropics.

Although most rainforest birds are strong flyers, some forage along the forest floor, making only short flights to low-hanging branches. They include the tinnamous, common in the Neotropics, the hornbills (*Bucerotidae*) of Asia, and the cassowaries (*Casuariidae*) of Australia and New Guinea. The oscillated turkey (*Meleagris ocellata*), related to the Thanksgiving variety but much more colorful, with iridescent blue and green feathers, inhabits forests of Guatemala and Belize. Perhaps the most gorgeous ground-dwelling bird is the argus pheasant found in Asian forests. Like peacocks, the male argus pheasant erects his magnificent tail and huge wing feathers during a mating dance to impress and attract a passing female.

A Fight for Light

Only 1 to 3 percent of the sun's light can permeate the rainforest canopy, and all the vegetation beneath that canopy, and reach the dim forest floor. A large variety of plants grow well in the faint light, including ferns, mosses, and liverworts. The light that flickers through the layers of foliage is sufficient for their photosynthesis and survival. Most tree seedlings that take root in the shaded forest grow exceedingly slowly or lie dormant in the soil for as long as 10 years. They may die unless a falling tree or limb creates a break in the forest canopy that allows sunlight to filter through. Every time a tree dies from old age or disease or is toppled by wind or a storm, or a limb or two breaks off and tumbles to the ground, it creates an opening in the canopy through which sunlight can seep—the breach is called a light gap. Tree seedlings not crushed in the fall now have a chance to reach for the sky.

When a light gap occurs, some seeds that were buried as deep as eight inches in the soil quite suddenly germinate, like a Sleeping Beauty awakened by the sun's kiss. Another source of plant growth in canopy gaps are "seed rains." Seeds dispersed by the wind get funneled into light gaps, while birds that fly over or hop about gaps defecate seeds from the fruit they've eaten. Animals that helpfully distribute seeds in this way are called, appropriately enough, seed dispersers, and they play a particularly vital role in the health of tropical forests.

Because young trees are at risk of being shaded by other species, their energy is devoted to growing. Once they have grown tall enough to receive a survivable share of sunlight, they can spare energy for growing fruit and seeds. For this reason, most tropical trees do not reach sexual maturity, bloom, or bear fruit until they are several years old. Plant and tree species that grow most rapidly when in light gaps are called pioneers or colonists. Neotropical trees such as *Cecropia*, *Jacaranda*, and balsa are common fast-growing colonists. One reason balsa wood is so light is that all the tree's energy is spent on growing, not developing dense wood. In Africa, *Musanga cecropiodes*, which resembles the *Cecropia*, is a familiar col-onizer, while *Trema micrantha*, a species of the elm family, is a pioneer found in Latin American, Asian, and African tropical forests. Pioneer tree species often produce abundant small seeds that are dispersed by birds and bats that fly over light gaps.

Until recently, scientists believed that light gaps accounted for the impressive number of tree species in tropical forests. The accepted theory was that the number of species in a forest depended on the number and size of light gaps. In early 1999, biologist Stephen Hubbell, a professor at Princeton University, reported that his studies in Panama revealed that there was no such relationship, that light gaps have no more species than areas where the sun barely breaks through the canopy. His research is helping conservationists understand how forest ecosystems recover after people have burned or logged trees.

Secondary Forests

After a tropical forest has been cleared, the herbaceous plants and the fast-growing pioneer tree species that compose what becomes a secondary forest provide a protective umbrella over newly exposed soils in just a few weeks. Within three years, they may be 40 feet tall. A secondary tropical forest may look much like a primary one, but it is different in an important way. Biodiversity in a secondary forest is far less, and once a primary forest has been cleared, hundreds of years may pass before the diversity of plant and animal life in a secondary forest comes near its original numbers, if ever. In Angor Wat, Cambodia, large pieces of forests that were cleared some six hundred years ago still do not equal the biodiversity of surrounding patches of undisturbed forests. The forests around the ancient temples in the Petén, Guatemala, have also not completely recovered from their clearing by the Maya more than a thousand years ago.

The plants that succeed in secondary tropical forests are much more tolerant of the sun and grow in a tangle—they look like what we think of as impenetrable "jungles." Climbing plants cling to their trunks, seeking a free ride to the sun. Ironically, the seeds of pioneer species won't germinate in their own shade, so the slower-growing shade-tolerant species eventually, over decades, replace them.

If a light gap is quite large—created by powerfully high winds, or by people armed with matches or chainsaws—a secondary tropical forest will be able to recover to an ecosystem that's healthy enough to be called "primary" only if there are nearby sources of seeds of the original towering trees. Because the survival of so many plant and animal species in a rainforest can be dependent on the health of other species, a secondary forest may never fully be able to resemble its original state—it is likely that extinguished in the ashes, vanquished in the clearing of trees and trampling of plants, were entire species of flora and fauna, gone forever.

The Understory: Middle Management

How plants respond to light or the lack thereof is what determines a tropical forest's vertical structure, or its strata. Some scientists see the structure of a forest as orderly, with specific kinds of plants and animals inhabiting each layer. Paul W. Richards was an advocate of this idea of organized growth, which he described in an influential book on tropical forests published in 1952. Other biologists disagreed strongly with the theory, seeing tropical forest growth as chaotic. The debate continues today. So much about how the rainforest engine runs remains a mystery.

The tropical forest ecosystem seems at its most exuberant in the understory, among the weave of shorter trees, vines, and plants. Palms often dominate the scene, particularly in the Neotropics, which has more than 1,000 palm species. Other shade-tolerant species seem familiar because we've adopted them as house and office plants—philodendrons, dracaenas, and dieffenbachia, which, in their natural environments, dwarf their pot-bound cousins.

Lianas and other vines compound the tangle of understory growth. These climbing plants hitch a ride on tree trunks, looking for a way up to the light. Some species coil tightly around trees, distorting their growth. A liana called the "monkey ladder" (*Entada gigas*) grows in half-inch-thick plaits with open windows down the middle, a perfect space for primate hand- and foot-holds. When a liana twists, turns, and climbs its way to its host tree's uppermost branches, called the tree's "crown," it then grows horizontally, looping from tree to tree, entangling itself in the canopy. If there's a gap between tree crowns, the liana sends several experimental shoots outward to try to make a connection with another crown branch.

Hardy lianas can outlive the trees that brought them to the light in the first place. When a forest giant falls, it often pulls down neighboring trees that are connected to it by lianas. The fallen tree dies, but the liana usually lives on, growing outward and up again—some species gain half-a-foot each day.

Emerging Behemoths

When we fly over a rainforest, we see the intertwining branches of a variety of slowly but steadily growing trees—they make up the rainforest canopy, which appears from above as a bumpy carpet of all possible shades of green. Occasionally, towering trees stick out of the choppy green sea like umbrellas. These forest Goliaths are called emergents. In the Amazon, emergent species include *Dinizia excelsa*, which can grow 200 feet high, the Brazil nut tree (*Bertholletia excelsa*), and the kapok tree (*Ceiba pentandra*). Emergents are widely scattered in rainforests. Their tree crowns are quite broad, since their leaves need to soak up as much sunlight as they can get to provide adequate energy to such a massive enterprise.

Among the most impressive emergents are the *Dipterocarpacae*, found in Southeast Asia and particularly in Borneo. An average dipterocarp tree trunk measures 16 feet around, and its uppermost tree branches are some 20 stories above ground, or 164 feet. In a dipterocarp forest, an occasional emergent may be twice that tall. Dipterocarps dominate the canopy in the Southeast Asian rainforest, but in the swampy lowland forests of Malaysia and Indonesia, the legume *Koompassia malaccenis* may soar even higher.

The straight, dense trunks of dipterocarps, whose name in Latin refers to their twin-winged seeds that twirl like helicopters during their groundward descents, provide some of the most handsome furniture and other wood products in the world. Not surprisingly, these majestic forests are heavily logged.

Life at the Top: The Forest Canopy

Until recently, we knew very little about life in the forest treetops, for an obvious reason: getting up there was nearly impossible. Early explorers ascended using ropes and pulleys or ladders carved into tree trunks, then camped out on platforms. They brought back stories of an amazing world, an unexplored frontier full of never-before-seen animals, harsh climates, and astounding beauty. Today, biologists explore the canopy via towers, suspension bridges, rafts lowered gently onto treetops by dirigibles, and even construction cranes. As a result, we know a great deal more about life in the canopy, but also more clearly understand that there's much to discover and understand.

Because the world's forest treetops are so extensive and so complex, ecologists estimate that they are the airy homes of more than half of the world's total biodiversity. The sun that barely reaches the forest floor strikes the canopy with full force, fueling the photosynthesis that results in leaves, fruit, and seeds. Since there's so much good food way up there, animals abound in the canopy, perfectly adapted in a world that human beings must struggle to explore, much as marine life prospers in a realm that only human beings loaded down with complicated scuba gear can venture to visit.

Bats and birds, of course, are canopy inhabitants, though many species also fly beneath the forest ceiling, in the understory. Insects are everywhere in the rainforest; the canopy is no exception. When added together, the surface area of tree leaves is 10 times that of the forest floor, and insects inhabit much of that space. Some are canopy residents; others are just passing through.

During the blooming season, the canopy is alive, buzzing, and humming with these winged and all-important pollinators. Because canopy trees have such expansive crowns and are so diverse, two like species may be many yards apart. As a result, the territory of canopy songbirds is four to five times larger than understory birds, where similar tree species are likely to be much closer together.

Arboreal animals, or those that live in trees, may have prehensile tails that they can use like a fifth appendage to grab onto branches as they jump and swing from limb to limb. These include monkeys and lemurs. Orangutans are the largest arboreal animal. Canopy dwellers also include many animal species you might assume stick to the ground floor: cockroaches, mice, earthworms, and scorpions.

Other arboreal fauna are good gliders, such as flying squirrels, colugos (*Cynocephalidae*), *Draco* lizards, and flying frogs (*Rhacophoridae*). These animals don't actually fly, but rather control their falls by spreading their limbs and flattening their bellies, maximizing their undersurfaces to slow their descents. Often their flights are helped by thin membranes that they can open and spread at will to catch the wind. Another canopy animal, the sloth (*Bradypodidae* and *Megalonychidae*), doesn't leap or glide; it barely moves at all, but clings to tree trunks and branches with razor-sharp claws.

The blooms in a rainforest canopy's garden are flowers of the trees themselves and of the plants that climb up tree trunks. But the most numerous and most spectacular canopy flowers are of the epiphytes, which have no need at all for the soil down below. Epiphytes festoon tree branches, sometimes in so many numbers that the entire branch gives way from

their weight. But even though they risk loss of limb, trees can benefit from providing a home to epiphytes. Organic matter accumulates around the roots of the hitchhiking plants, creating a mass of rich soil. U.S. ecologist Nalini Nadkarni found that trees send out roots from their branches that reach through the fertile soil, just another way that forests seek out nourishment in a nutrient-scarce environment.

The epiphytes, which are far more numerous in Latin American and Southeast Asian forests than in Africa, include species of orchids, bromeliads, cacti, ferns, mosses, lichens, and liverworts, totaling some 30,000 species in 850 genera and 65 plant families. One tree may hold hundreds of species of orchids and more than a thousand species of epiphytes. Biologists can pick out the paths of canopy wildlife along certain tree branches that are stripped clean of epiphytes, signaling heavy traffic along a principal aerial highway.

The complex structure of the rainforest canopy allows a huge range of plant species to exist, while providing different microhabitats to wildlife, both vertebrates and invertebrates. Ninety percent of all organisms in a rainforest are found in the canopy.

SURVIVAL IN THE RAINFOREST: Relationships and Adaptations

Scientists are unsure exactly why tropical regions contain so much more biodiversity than do temperate ones, but it seems likely that this richness of life is related to the ecological complexity of rainforests, not only in the canopy, but throughout the forest strata. The prodigious amounts of sun and rain and stability of temperature, resulting in ample food supplies year-round, also encouraged the evolution of thousands of species.

Unlike other parts of the planet, rainforest ecosystems have undergone few geological changes over the past millions of years. Fossil deposits in Southeast Asia show that the region's forests have existed, virtually unchanged, for at least 70 million years. This stability has allowed evolution to proceed with few disruptions until now, with the overwhelming dominance of human beings.

The immutable and fecund tropical-forest environment has led to countless micro-habitats, or niches, into which plant and animal species of common ancestry have spread in an evolutionary process called adaptive radiation. Earth's archipelagoes, such as Hawaii, Indonesia, and Ecuador's Galapagos Islands, offer the best example of this concept. As waters rose and divided land masses into independent islands, the suddenly separated populations of the same species evolved, through adaptive radiation, into distinct species that adapted to a changed environment. Each time waters rose and receded, over millions of years, new species split off from the ancestral line.

Hawaii has 10,000 known endemic species of insects, which likely evolved from 400 species that first arrived millions of years ago. The cardueline finch also was an original island colonizer, whose first flight to the isolated islands was probably aided by fierce storm winds. Over time, a pair or small flock of this one species evolved into some 50 species of Hawaiian honeycreepers (*Drepanididae*), the common name for a diverse group of small tropical birds that share tongue and bill adaptations for feeding on flower nectar, but are separate species. Most of these ancient descendents are gone now, recent victims of deforestation, overhunting, rats unwittingly brought to the islands by boat, and diseases spread by non-native birds that human immigrants also brought to Hawaiian shores.

One reason human intervention in tropical forests can be so devastating is that life in these ecosystems has evolved into a supremely complicated system of interactions between plants and animals and their habitats. Because so much life exists in small niches, the smallest change can have a serious impact on many different species. The examples below describe some of the myriad ways in which plants and animals depend on one another to survive and have evolved to enhance their own survival in an extremely competitive realm.

Spreading Genes

The most obvious relationship between plants and animals in tropical forests involves food and germination. Plants put a good deal of energy into producing nectar and fruit to entice hungry animals, but the payoff for the flora is that by feeding on their offerings, the animals perpetuate the plant species. Few tropical plants rely on the wind to spread their seeds. Of the 760 tree species in 100 acres of forest in Brunei, scientists found that the seeds of just one were spread by the wind. The rest depend on animals, particularly insects, bats, and birds.

Plants have evolved flowers with a variety of shapes, scents, colors, oils, and nectars to attract an-

imals and spread their pollen to another of the same species. Certain plants and animals have co-evolved in an intimate, often exclusive, relationship.

Birds pollinate flowers everywhere but in Europe and in Asia north of the Himalayas, but most bird pollination occurs in the tropics. Flowers pollinated by birds are usually bright orange or red—birds have excellent vision—loom at a distance from the plant's leaves, produce large quantities of watery nectar, and may be hardily built to support the weight of a bird. Since birds do not have a keen sense of smell, most bird-pollinated flowers do not have a scent. Many bird-pollinated blossoms are shaped like long tubes, a perfect fit for the long, thin bills of hummingbirds (*Trochilidae*). The elongated, needle-shaped bill of the swordbill hummingbird (*Ensifera ensifera*) slips easily into the oboe-shaped blossom of passion flowers (*Passiflora*), while heliconia flowers are a perfect fit for the downward curving bill of the white-tipped sicklebill (*Eutoxeres aquila*). Hummingbirds often do more than pollinate when they sip on flower nectar. Minute mites lie in flowers favored by hummers. When the last blossom blooms, the mites dash into a feeding hummingbird's nostrils. There they wait until they smell the familiar scent of a new host plant and abandon the nostrils for their new home and food source. Each plant species visited by hummingbirds shelters a different species of mite.

Bats are also important tropical flower pollinators. Bat-pollinated plants usually open only at night, are luminous white, and have a sour or fetid scent. Bats have excellent senses of smell, but most have poor eyesight. The bell- or ball-shaped blossoms dangle or are held aloft on stems, at a distance from branches and foliage, so they are in easy reach of bats. Durian trees (*Durio zibethinus*) of the Malay Peninsula are pollinated by just one species of bat, *Eonycteris spelaea*. The foul-smelling but delicious (according to aficionados) fruit of the durian is widely sold, so an important market is dependent on the bat, whose populations are threatened by deforestation.

When we think of cacti, we think of dry deserts, but in the Neotropics, cacti are also common epiphytes. The windy canopy in the dry season is actually quite an arid environment, so it's not surprising that some cacti species have adapted to this ecosystem. Their succulent stems allow them to store water and lose it slowly, so they can survive long, rainless periods. Cacti of the genus *Rhipsalis* dangle like long, green fingers from tree branches. Their seeds are particularly sticky, so they adhere to the beaks of birds that feed on them. The birds rub their beaks on branches to scrape off the clinging seeds, thereby spreading *Rhipsalis* seeds. Some species of *Epiphyllum* cactus have long, narrow, tubular flowers, with nectar at their base and petals at their tips. The petals open at night, when hovering hawk moths (*Sphingidae*) visit them. The long tongues of the moths slip perfectly into the cactus flower's shaft to reach the nectar.

The most widespread plant genera in the tropics is *ficus*, the figs, with more than 900 species. Each species is pollinated by a different species of wasp. When a pollen-laden female enters a fig to lay eggs, she also pollinates the fig flowers. The wasp larvae and fig seeds develop inside the fruit. The female wasps emerge, become dusted with fig pollen, mate with the wingless male wasps (who then die), and fly off to another fig tree of the same species, and so repeat the cycle.

The Neotropical bucket orchid (*Coryanthes*) has developed a convoluted, Rube Goldberg–like structure that promotes pollination by the orchid bee (*Euglossini*). The orchid's scent attracts the bees, but many slip on the waxy surface around the nectar-producing gland and tumble into the orchid's well. To escape the collected fluid, they struggle onto a raised area at the entrance of a tunnel on the side of the orchid's bucket shape. As the bee attempts to fly out of the passageway, the orchid is able to snap shut the tunnel, trapping the bee inside and, at the same time, coating its back with pollen. The orchid then releases the bee, which, in spite of its ordeal, obligingly carries the pollen to another sweet-smelling bucket orchid.

Fruit Eaters

Biologists are discovering that tropical plant species depend as much on the animals that eat their fruits as the animals depend on the plants for daily nutrition. More than 75 percent of tree species in Neotropical forests produce fruits adapted for consumption by vertebrates, mainly bats and birds.

In a forest in southern Mexico, many of the larger animals that disperse seeds, such as tapir, deer, and monkeys, are also sources of food for people. A sign that these animals have been overhunted is a pile of rotting fruit under the tree that bore them—fruits are produced to be eaten by animals, digested with the sweet pulp that surrounds them, and eventually defecated in some other part of the forest. Trees may sprout from the fruit that falls from branches and rots, but the seedlings are unlikely to survive under the shade of the parent tree. So the survival of the fruit-bearing tree is jeopardized by the decline of the fruit-eating mammal population.

Many tropical trees depend on bats to eat their fruits and spread their seeds. Bats eat as much as twice their body weight in fruit each night and may fly 25 miles throughout the forest, eliminating seeds as they go, thus spreading them over wider areas than any other animal. The scientist Donald Thomas studied straw-colored fruit bats in West Africa and found that one large bat colony scattered the seeds of almost half-a-million pounds of fruit in just one night. The feces eliminated with the seeds also provide fertilizer for germination. Thomas found that just 10 percent of the seeds he planted directly from the fruit grew, while all the seeds collected from bat feces sprouted.

The Brazil nut tree (*Bertholletia excelsa*), basis of a huge and profitable industry, depends on several shiny bee species for pollination and on one mammal, the agouti, to disperse its seeds. The agouti (*Dasyprocta*), a rodent with strong and sharp teeth, chews on the Brazil nuts and softens the seed coat, so that a seedling can emerge. Without the bees, without the agoutis, Brazil nut trees would disappear, and so would an industry upon which thousands of people depend.

Ants and Plants

A fascinating and mutualistic relationship exists between some ant and plant species. Some plants have developed cavities that ants use as nest sites and also produce nectar and nutritious material that ants consume. In return, the ants fend off herbivores, distribute the host plant's seeds, and provide nutrient-rich soils. Some plant and ant species appear to have coevolved so that each is specialized to take advantage of the other's services.

Plants that have developed structures to house ant colonies, known as domatia, are called myrmecophyte, from the Greek word for ant. The bull's-horn acacia of the Neotropics has hollow thorns that house ants of the genus *Pseudomyrmex*. In addition to a home, the acacia provides sweet nectar to adult ants and protein-rich leaf tips that the ants feed their larvae. In return, the ants, which have a nasty, burning sting, ward off insect predators. Kiskadees (*Pitangus sulphuratus*) take advantage of this ant and plant relationship. The birds commonly build nests in ant acacias, and although the ants may attack the bird at first, eventually they learn to live with the interloper. Even though the kiskadee nests are fairly low to the ground, and thus more vulnerable to egg and nestling predators, many of the birds' fledglings survive because the ants do ward off other animals that might prey on kiskadee eggs or newborns.

One of the most successful and fastest-growing trees in the Neotropics is the cecropia, one of the first to fill a light gap. The tree has a hollow-noded trunk and branches that provide a home to belligerent Aztec ants (*Azteca*). The cecropias also provide the ants with food, in the form of a nodule at the base of each leaf that contains a substance called glycogen. No other tree produces glycogen. It's obviously in the Aztec ant's interest that its cecropia home remain healthy, so it delivers a fierce bite to any trespasser. Even a vine that dares to rest a tendril against a cecropia trunk is quickly gnawed away. As nature would have it, there is one animal undeterred by the aggressive aztecs. Cecropia leaves are the favorite food of sloths (*Bradypodidae* and *Megalonychidae*), whose thick and coarse coats adequately protect them against ant bites.

In the forest canopy, certain ant species turn to gardening, to the benefit of a variety of epiphyte plants. The ants make their homes in the root ball of an epiphyte, and the detritus they add to the root ball nourishes the plant. The ants also add seeds of other epiphyte species to their home, and as they grow, these new plant roots add to the structure of the ant nest. The plants, then, always have nourishment; the ants have a home structure, and feed on the plant's fruit pulp.

Fitting In

The fascinating and satisfactory results of evolution are everywhere evident in tropical forests. Virtually all living things grow, eat, and multiply in a particular, often singular, way for a distinctive reason. What they leave behind in death serves a multitude of living things. The leaves of many tree and shrub species in rainforests have similar shapes—oblong and with a pointed tip. The tapering leaf ends are called "drip tips" because they allow the copious rain to run off, like the corner of a roof. Fast-drying leaves have several advantages. One, a film of water on the surface would reflect the already dim light, reducing the already limited opportunities for photosynthesis. Two, wet leaves reduce transpiration, limiting the uptake of essential minerals from the soil. Finally, leaves that are dry are less likely to be colonized by mosses and fungi. Plants in the understory are more likely to exhibit drip tips, since canopy vegetation is exposed to drying sunlight. Also common in tropical tree leaves are joints, called pulvini, on the leaf stem, which allow the leaf to adjust its position to expose itself to as much light as possible.

Most tropical trees have buttressed roots. Buttresses form a flangelike base that resembles flat, wooden triangles leaning against the tree trunk. Each tree has several buttresses that surround and seem to support it. Since tropical trees are generally quite tall and have very shallow roots, a supporting role seems logical, but the truth is that scientists aren't entirely sure what buttresses do. Some argue that the buttresses help provide water to the tree or prevent progress of clinging vines. Although human beings may be stumped, buttresses must be advantageous to tropical trees because so many species have them.

Another unusual tropical tree characteristic that isn't clearly understood is cauliflory. Some species, such as cacao, bear stemless flowers and fruit directly on their trunks, as if they were stuck on with Velcro, rather than at the end of young, leafy stems. Flowers blooming so conspicuously below the canopy are likely to attract the attention of bats and insects, but biologists aren't certain what the advantage of cauliflory might be.

Among animals, too, evolution has produced advantageous if visually peculiar adaptations. The attributes of anteaters have clearly understood purposes, however bizarre they may look. Anteaters have no teeth, nor do they need them. Their diet principally consists of termites and ants, which don't really require chewing, and the mammals are well outfitted to seek and swallow them. Anteaters have elephant-like, tubular snouts and a keen sense of smell that helps them locate their prey. They use massive, sharp, curved front claws to rip open nests and their long tongues, which are coated with sticky saliva, to lap up the insects. The three arboreal anteater species, the collared anteater (*Tamandua tetradactyla*), Northern (*Tamandua mexicana*), and silky (*Cyclopes didactylus*), have prehensile tails, which hold them fast in treetops while they use their front paws to claw open termite nests.

Deceptive Appearances

One way to hide from predators in a tropical forest is to blend in with your surroundings. Tropical insects, birds, reptiles, and even mammals have evolved masterful camouflage, and many look so much like a twig, a dead leaf, or surrounding vegetation that only the keenest eyes and noses can find them. Walking sticks (*Pseudophasmatidae*) are insects that appear just as their names imply, so when they pose still on a leafless tree branch, it is difficult to discern their pencil-thin, long, brown bodies and toothpick appendages. The green and brown patches on the curved body of leaf katydids (*Tettigoniidae*) look exactly like a green leaf with spots of fungus. Geometrid moths resemble, to an astounding degree, dead leaves on the forest floor. They are the same dull brownish color, and their wings curl at the edges, are pointed at the tips, and are lined with what appear to be veins on a leaf. To further avoid detection, the moths tuck away their legs, antennae, and eyes under their leaf-like wings.

Larger dead-leaf mimics include the horned frog (*Hemiphractus spp.*), whose brown, triangular head looks like a dried leaf with dark slits for eyes, and bushmasters (*Lachesis muta*), an extremely poisonous Neotropical snake whose skin bears a mottled pattern in various shades of brown. The bushmaster evolved this coloring not to help it hide from predators, but from prey. It lies coiled and still on the leaf litter, ready to strike out when a meal passes. Similarly, wild cats, such as ocelots and leopards, have evolved spotted coats that imitate the dappled pattern of sun and shadow in the rainforest. Since they are at the top of the food chain, the camouflage serves not to hide them from their enemies but to allow them to creep undetected upon their prey.

Appropriately named vine snakes (*Oxybelis spp.*) have quite thin and long bodies that are green or brown, perfect camouflage as they slide through tree branches and lianas in search of lizards.

Of course, for many tropical insects, frogs, lizards, and parrots that live in the forest canopy, just being green is sufficient protection.

Toxic Defenses

A reliable way to deter enemies is to be poisonous or at least have a very disagreeable taste. Tiny poison-dart frogs (*Dendrobates spp.*) appear dainty and vulnerable, but their skins contain a potent toxin, so effective that indigenous forest dwellers tip their arrows and blowgun darts with the substance. The frogs' skins are brilliantly hued, so predators learn to recognize them and avoid eating them. On the other side of the Earth, the feathers and skin of the hooded pitohui (*Pitohui dichrous*), which is found only on Papua New Guinea and a few surrounding islands,

contain the same toxin as the poison-dart frog. The hooded pitohui is the only bird that bears poisons as a defense mechanism, and like the tiny *Dendrobates*, it is brightly colored to warn off predators.

As caterpillars, *Heliconius* butterflies feed on passion flower plants and absorb the leaves' toxins, which are poisonous to other predators, but not to the *Heliconius* caterpillar. After metamorphosis, the butterflies retain the poisons, which makes them particularly distasteful to many, but not all, predators. A Neotropical vine, *Banisteriopsis caapi*, produces potent chemical compounds to discourage herbivores from munching on its leaves. When human beings drink an infusion concocted from the vine's bark, they are transported to another dimension, a world of hallucinatory visions.

Venomous snakes, like the pit viper, may use their poisons to deter predators, but seldom deliver a heavy dose. They primarily use their venom to subdue prey. Since snakes lack appendages to hold down kicking, biting, squirming, and flying victims, they have evolved specialized mouth muscles, toxins, and grooved or tubular fangs to deliver a dose that will immobilize their prey. Each poisonous snake manufactures a complex mixture of venom and other chemicals, which takes time to replenish again after the snake delivers a heavy dose. Rather than deplete their supply of venom, snakes will escape when threatened, or if that's not an option, deliver a relatively light amount of venom in defense, so they will still have enough left over for their next meal.

Faking It

To trick predators and attract pollinators, many species of animals and plants have evolved certain properties that mimic other species. For example, many nontoxic Neotropical butterflies have the same coloring as the toxic-tasting *Heliconius*. Two other types of pitohuis look very much like the striking black-and-copper-colored hooded pitohui, but aren't actually poisonous—they just look that way to cautious enemies. Species of long-horned beetles (*Cerambycidae*) that taste just fine have evolved to look like other beetle species that taste terrible to predators, while other long-horns look very much like stinging ants, bees, or wasps.

Plants can also be good dupers. Passion vines (*Passiflora spp.*) have stipules that resemble the eggs of *Heliconius* butterflies. *Heliconius* caterpillars sometimes eat the eggs laid by other *Heliconius* butterflies, so the females will avoid those plants that already seem to be occupied by eggs. Thus, the plant has avoided potential caterpillar predation. An orchid of southern Africa, *Disa draconis*, has evolved tubes that look like the nectar tubes of other flowers. They contain no nectar, but flies keep visiting the orchid, poking into the empty tubes and getting their faces covered with pollen, which they brush off on the next *D. draconis* they fruitlessly investigate. In different regions, the orchid's tubes are different lengths to match the tongue lengths of specific pollinators. Since one species of the orchid has three-inch tubes, scientists are searching for an as-yet-unknown fly with a very long tongue.

THE LOSS OF AN IRREPLACEABLE ECOSYSTEM

Although Earth's tropical forests have been disappearing for centuries, ever since human beings began to build shelters and learned to farm, the majority of the loss has occurred only during the second half of the twentieth century. The first tropical trees to fall were those in dry forests, the easiest to burn. Most dry forests disappeared a century ago, and today this is one of the most endangered ecosystems on Earth. During the 1970s, tropical countries lost a total of nearly 30 million acres of rainforest each year. During the 1970s and 1980s, 18 tropical countries lost more than 95 percent of their forests, and another 11 lost 90 percent. The once entirely forested regions of West Africa, Central America, and Southeast Asia are now more than 90 percent deforested. Today, rainforests are disappearing at a rate of more than an acre a second, which is 40 times faster than only 10 years ago.

How could so much forest be lost so quickly? In most regions of the world, clearing land for agriculture is the principal cause of deforestation. People need land to grow crops or to feed themselves and their families, and in many countries, no laws prevent people from entering a forest, cutting it down, burning the dry vegetation, planting seeds, and building a simple home. People often follow a river into a forest, but if a road is built—even a simple dirt path—it accelerates the arrival of impoverished people who are desperate for free land. The plot of land they clear becomes "theirs" simply because they live on it.

As discussed previously, the soils that support so much biodiversity are actually quite thin and poor.

These would-be farmers can grow crops in the ashes of burned rainforest for a few years, but eventually, the nutrient-poor soils give out, and colonists must move farther into the forest and start over. What happens to the land they abandon? Often, wealthy ranchers buy the land (or take it) and use it for highly inefficient pasture to graze livestock. On average, six acres of pastureland in the tropics are needed to feed just one cow.

Cattle ranching is a particularly strong tradition in Latin America. During the 1960s and 1970s, millions of acres of rainforest in Central America were converted to pasture land, as the price of beef increased. In 1950, 8.6 million acres in Central America was devoted to cattle. By 1983, the number had increased to 26 million acres. Most of the beef was exported to the United States. As health conscious North Americans began eating less beef, and environmentalists threatened to boycott fast food chains that bought "rainforest hamburger," beef prices fell, and hundreds of ranchers abandoned their pastureland. Thus, thousands of acres of former pastureland began to recover, supporting more flora and fauna, although nothing like the original level of biodiversity. If beef prices rise again, it's likely these second-growth forests will quickly be razed.

People who need wood for fuel also cause deforestation. Nearly three billion people depend on wood as their main source of energy. More than 90 percent of the wood harvested in Africa, 80 percent in Asia, and 70 percent in Latin America is for fuel. Poverty is what drives billions of people—mostly women and children—to gather wood for fuel. They have no access to electricity. Poverty is also what drives people to cut down forests to plant crops. Meanwhile, the population in the world's poorest nations continues to grow, meaning there will be many millions more people who need land on which to grow food and firewood to cook it. Today more than 1.7 billion people live in 40 nations that have dangerously small amounts of forest cover. By 2025, the number of people living in nations whose forest cover is critically low could reach 4.6 billion.

In many tropical nations, the best farmland is owned by wealthy, politically well-connected people. That forces the poor into forests, where the land will never yield enough food to allow them to improve their standards of living. The poor often have few options other than to cut trees on hillsides to plant corn and beans. During driving tropical rainstorms, tons of soil on these denuded hills washes away, polluting downslope sources of drinking water, often carrying away crops, homes, livestock, and people as well.

Commercial farmers plant crops, often for export, on previously deforested land, or they deforest forested land to plant monocultures, including rice, coffee, bananas, oil palm, and sugar. Timber companies cut down the valuable hardwoods in a forest, a process that usually destroys all surrounding vegetation, and, of course, jeopardizes the wildlife that depended on that lost flora. Commercial logging of tropical forests, which constitutes an $8 billion-a-year industry, has doubled in the past 30 years, accounting for some 14 million acres of forest lost each year. Illegal logging adds millions of acres more to that total. Logging roads allow colonists access to previously inaccessible areas, and the cycle described above begins again.

Development projects, such as dams, new settlements, highways, and large-scale mining and petroleum projects are also mighty engines of deforestation. Meanwhile, the developing tropical countries owe the developed world millions of dollars in loans, so they are forced to pay off these debts by selling their natural resources and then have little left over to invest in conservation.

The loss of tropical forests is clearly devastating to the wildlife and the approximately 50 million indigenous people who depend on forests directly for food and shelter. Another 2.5 million people live in cleared areas adjacent to tropical forests and also depend on them for fuel wood, food, and other resources. But human beings who live far from the falling trees are also affected.

Perhaps the most important environmental service that forests provide to people is as a regulator of water and a protector of watersheds. The tangle of branches in the forest canopy, and all the branches and vegetation below, break the force of driving rains. By the time the rain hits the ground, its earth-shattering forces are depleted, and the water trickles off trees and shrubs into rivulets, gathers into streams, and gradually empties into rivers. Without the trees, downpours can quickly dissolve pastures and cropland into rivers of mud. Instead of trickling, water rushes headlong into rivers, which can quickly overflow. Deforestation is the reason that the flooding caused by 1998's Hurricane Mitch was so destructive, leaving 11,000 people dead and many thousands more homeless in Central America. In 1999, thousands of people lost their lives to mudslides after days of rain pummeled denuded mountain slopes near the Caribbean coast of Venezuela.

Floods killed more than 3,000 people in China in 1998 and caused more than $20 billion in damage. The government acknowledged that logging of the Yangtze watershed, home to nearly 400 million people and with only 15 percent of its original forest cover, played a role in the tragic damage. Now logging of old-growth forest is banned in the watershed.

Billions of people depend on rivers, streams, and lakes for drinking water. Without forest cover, the rains that crash onto deforested hillsides wash away tons of topsoil and carry it along with agrochemicals, animal waste, and garbage directly into these sources of drinking water. The washed-away soil and debris dumped into reservoirs also affects hydroelectric power plants, whose useful life span is cut short by years, due to siltation. Trees also prevent wind erosion by breaking the force of gusts.

More than 70 percent of the Panama Canal watershed is deforested. When rains pummel naked hillsides, the clay soils are washed into the rivers and lakes that supply the fresh water that moves ships through the canal's system of locks. About $150 million must be spent on dredging each year, but the canal is filling up with silt faster than it can be removed, while deforestation continues. Meanwhile, 1.5 million Panamanians depend on the canal watershed for their drinking water.

The soils in tropical forests are always damp because trees block out the drying sun. During periods of low rainfall, forests continue to release collected moisture slowly, but relatively steadily. Without forests, the constant trickle stops; collecting rivers and lakes shrink; and the fresh fish populations dwindle or disappear.

It is difficult to fix a dollar amount on what deforestation costs as a result of erosion, drought, flooding, and loss of drinking water. But people do sell thousands of material goods provided directly by forests, products the profiteers often take at little or no cost to them. Rainforests provide such valuable resources as timber, wood pulp for paper, firewood, medicines, fibers, latexes, resins, and fruits. Rainforest travel is a growing industry, bringing tropical countries millions of dollars from tourists eager to pay for the privilege of seeing the natural wonders of these majestic ecosystems.

Countries like Costa Rica and Colombia have begun to sell one of their forests' most important environmental services: the ability to absorb carbon dioxide released from burning fossil and organic fuels that would otherwise enter the atmosphere, exacerbating what is called the "greenhouse effect," the gradual warming of the Earth. Since Northern consumers are responsible for so much fossil fuel burning, tropical countries are selling "carbon bonds" to Northern utility companies, promising to conserve their forests to help remedy the damage done by the release of carbon dioxide into the atmosphere.

Aside from the material goods and completely free services tropical forests provide to people, they surely have a value simply because so much other life depends upon them, because they themselves are life, splendid creations of nature. That human beings are unable to attach a dollar amount to this value, or even fully understand it, cannot possibly justify people's destruction of entire ecosystems.

Unfortunately, however, deforestation continues, with the resultant loss of life. In August 1999, the International Botanical Congress released a report that summarized the current state of mass plant and animal extinctions. *Plants in Peril: What Should We Do?* predicts that between one-third and two-thirds of all species, most of them native to the tropics, will be lost during this new century. If deforestation continues at the current rate, only 5 percent of tropical forests will remain, nearly all in protected areas, within 50 years. As a result, species will disappear at a rate of magnitude three to four times higher than the natural extinction rate, which is about one species each year.

In face of complicated obstacles presented by ignorance, poverty, the lack of alternatives to deforestation, greed, and growing populations, people have made valiant and creative, if not totally successful, efforts to save rainforests. Tropical biodiversity conservation is still in its experimental phase, as experts—fully aware that time is running out—invest energy, funds, daring, and passion into experiments aimed at saving biodiversity without jeopardizing human beings. Many of these promising initiatives are described in Part Four: Saving Tropical Forests.

The campaign to conserve tropical biodiversity is not open only to experts; all of us have a place in the struggle. According to Daniel R. Katz, founder and chairman of the board of the Rainforest Alliance, "We are all stewards of the planet. As our education about the need to conserve biodiversity grows, so should our individual responsibility to do what we can to conserve our natural resources. If individually we do what we can, collectively we will make history—we will have saved the greatest celebration of life on Earth. Not to mention saving ourselves."

Further Reading

Food and Agriculture Organization, *State of the World's Forests: 1997.*

Forsyth, *Portraits of the Rainforest.*

Forsyth and Miyata, *Tropical Nature.*

Gardner-Outlaw and Engelman, *Forest Futures: Population, Consumption and Wood Resources.*

Goulding, *Amazon: The Flooded Forest.*

Holldobler and Wilson, *The Ants.*

Janzen, *Costa Rican Natural History.*

Kricher, *A Neotropical Companion.*

Mitchell, *The Enchanted Canopy.*

Moffett, *The High Frontier.*

Moore, "What Makes Rainforests So Special?"

Myers, *The Primary Source.*

Newman, *Tropical Rainforest.*

Sunquist, "Blessed Are the Fruit-Eaters."

Terborgh, *Diversity and the Tropical Rain Forest.*

Wilson, *The Diversity of Life.*

Wolfe and Prance, *Rainforests of the World.*

World Commission on Forests and Sustainable Development, *Our Forests, Our Future.*

Tropical Forest Distribution Maps

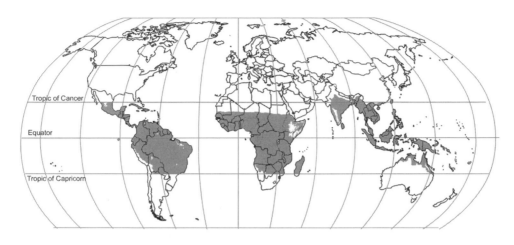

Tropical Forest Distribution 8,000 Years Ago.
Shading shows original extent of forest cover in the world's Tropics.
Source: World Conservation Monitoring Centre. Adapted by Oscar Cuevas.

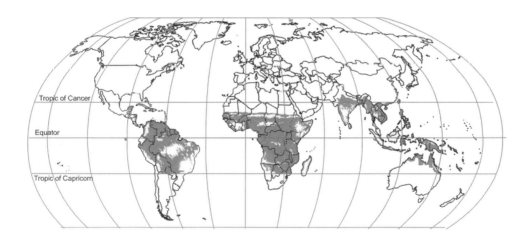

Tropical Forest Distribution in 1999.
Shading shows current forest cover in the Tropics.
Source: World Conservation Monitoring Centre. Adapted by Oscar Cuevas.

The World's Tropical Forests: Where They Are; How Much Is Left; How Fast They Are Vanishing

In visualizing how much forest has been lost, remember that once these countries, which all fall between the Tropics of Cancer and Capricorn, were virtually entirely forested. So, the number in the first column, the total land area in square miles, is also the amount of original forest cover.

Country	Land area (mi²)	Total forest cover, 1980 (mi²)	Total forest cover, 1995 (mi²)	% change 1980–1995	Average annual change, 1900–1995 (mi²/yr)
Africa					
Angola	481,226	96,199	85,692	−11	−915
Benin	42,699	21,817	17,853	−18	−232
Burundi	9,912	1,004	1,224	22	−4
Cameroon	179,644	83,272	75,648	−9	−498
Central African Republic	240,470	122,956	115,530	−6	−494
Comoros	861	n.a.	35	n.a.	−2
Congo, Democratic Republic of	875,081	465,566	421,686	−9	−2,856
Congo, Republic of	131,819	77,472	75,413	−3	−162
Côte d'Ivoire	122,748	46,814	21,110	−55	−120
Djibouti	8,948	n.a.	85	n.a.	0
Equatorial Guinea	10,827	7,330	6,875	−6	−39
Ethiopia and Eritrea	424,600	56,437	52,415	−7	−239
Gabon	99,461	74,926	68,936	−8	−351
Gambia, The	3,860	409	351	−14	−3
Ghana	87,830	42,356	34,825	−18	−452
Guinea	94,848	29,185	24,577	−16	−290
Guinea-Bissau	10,854	8,419	2,309	−73	−201
Kenya	219,688	5,242	4,987	−5	−12
Liberia	37,180	18,883	17,397	−8	−104
Madagascar	224,474	66,832	58,309	−13	−502
Malawi	36,315	15,699	12,889	−18	−212
Mauritius	784	42	46	9	0
Mozambique	302,659	72,186	65,087	−10	−448
Nigeria	351,557	65,369	53,191	−19	−467
Reunion (France)	965	n.a.	344	n.a.	0
Rwanda	9,523	822	965	17	−2
Sao Tome and Principe	387	n.a.	216	n.a.	0
Senegal	74,317	31,158	28,491	−9	−193
Seychelles	175	n.a.	15	n.a.	0
Sierra Leone	27,645	7,778	5,053	−35	−166
Somalia	242,153	3,034	2,910	−4	−4
Togo	20,995	6,083	4,806	−21	−73
Uganda	77,065	27,062	23,561	−13	−228

Country	Land area (mi²)	Total forest cover, 1980 (mi²)	Total forest cover, 1995 (mi²)	% change 1980–1995	Average annual change, 1900–1995 (mi²/yr)
United Republic of Tanzania	341,066	146,695	125,489	−14	−1,247
Zimbabwe	149,324	36,963	33,621	−9	−193
Asia/Pacific					
American Samoa (USA)	77	n.a.	0	n.a.	0
Australia	2,965,368	n.a.	157,905	n.a.	66
Bangladesh	50,246	4,856	3,899	−20	−35
Bhutan	18,142	11,484	10,638	−7	−35
Brunei	n.a.	n.a.	1,675	n.a.	−12
Cambodia	68,137	52,048	37,944	−27	−633
China	3,586,326	487,896	514,627	5	−336
Federated States of Micronesia	270*	n.a.	n.a.	n.a.	n.a.
Fiji	7,052	3,239	3,223	n.s.	−15
French Polynesia (France)	1,413	n.a.	0	n.a.	0
Guam (USA)	212	n.a.	0	n.a.	0
Hawaii (U.S. state)**	6,423	2,461	2,316	−6	−18
India	1,147,651	224,880	250,919	12	27
Indonesia	699,266	480,477	423,793	−12	−4,184
Kiribati	282	n.a.	0	n.a.	0
Laos People's Democratic Republic	89,089	55,854	47,999	n.a.	−571
Malaysia	126,820	83,237	59,718	−28	−1,544
Maldives	115*	n.a.	n.a.	n.a.	n.a.
Marshall Island	70*	n.a.	n.a.	n.a.	n.a.
Myanmar	253,814	126,998	104,803	−17	−1,494
Nauru	8*	n.a.	n.a.	n.a.	n.a.
Nepal	55,198	21,539	18,613	−14	−212
New Caledonia	7,056	n.a.	2,694	n.a.	−4
Northern Mariana Islands (USA)	1,950*	n.a.	n.a.	n.a.	n.a.
Palau	630*	n.a.	n.a.	n.a.	n.a.
Papua New Guinea	174,804	143,380	142,585	−1	−513
Philippines	115,094	43,054	26,117	−39	−1,011
Samoa	1,092	n.a.	525	n.a.	−8
Singapore	235	15	15	0	0
Sri Lanka	24,947	8,083	6,933	−14	−77
Solomon Islands	10,804	9,469	9,222	−3	−19
Thailand	197,204	69,955	44,892	−36	−1,270
Tonga	278	n.a.	0	n.a.	0
Tuvalu	10*	n.a.	n.a.	n.a.	n.a.
Vanautu	4,705	n.a.	3,474	n.a.	−31
Vietnam	125,639	41,159	35,192	−14	−521
Latin America/Caribbean					
Anguilla (United Kingdom)	60*	n.a.	n.a.	n.a.	n.a.
Antigua and Barbuda	170*	n.a.	35	n.a.	0
Argentina	1,056,362	140,994	131,016	−7	−344
Aruba (Netherlands)	75*	n.a.	n.a.	n.a.	n.a.
Barbados	166	n.a.	0	n.a.	0
Belize	8,801	7,905	7,573	−4	−27
Bolivia	418,571	214,547	186,477	−13	−2,243
Brazil	3,264,213	2,318,941	2,127,397	−8	−9,858
British Virgin Islands (United Kingdom)	58	n.a.	15	n.a.	0
Cayman Islands (United Kingdom)	100	n.a.	0	n.a.	0
Chile	289,037	31,216	30,463	−2	−112
Colombia	400,938	222,996	204,534	−8	−1,011
Costa Rica	19,709	7,431	4,817	−35	−158
Cuba	42,391	7,712	7,110	−8	−93

Country	Land area (mi²)	Total forest cover, 1980 (mi²)	Total forest cover, 1995 (mi²)	% change 1980–1995	Average annual change, 1900–1995 (mi²/yr)
Dominica	290	n.a.	178	n.a.	0
Dominican Republic	18,675	5,528	6,107	10	−100
Ecuador	106,860	55,476	42,989	−23	−730
El Salvador	7,998	602	405	−33	−15
French Guiana (France)	34,926	n.a.	30,841	n.a.	−3
Grenada	131	n.a.	15	n.a.	0
Guadeloupe (France)	652	n.a.	309	n.a.	−4
Guatemala	41,854	19,489	14,826	−24	−317
Guyana	75,984	71,784	71,707	n.s.	−35
Haiti	10,638	147	81	−45	−3
Honduras	43,190	22,079	15,884	−28	−394
Jamaica	4,180	1,992	676	−6	−62
Martinique (France)	409	n.a.	147	n.a.	−2
Mexico	736,754	213,933	213,794	n.s.	−1,961
Montserrat (United Kingdom)	39	n.a.	12	n.a.	0
Netherlands Antilles (Netherlands)	309	n.a.	0	n.a.	0
Nicaragua	46,860	28,004	21,462	−23	−583
Panama	28,730	14,529	10,808	−26	−247
Paraguay	153,358	65,180	44,494	−32	−1,262
Peru	494,080	272,956	260,789	−4	−838
Puerto Rico (USA)	3,420	n.a.	1,062	n.a.	−8
Saint Kitts and Nevis	262*	n.a.	42	n.a.	0
Saint Lucia	235	n.a.	19	n.a.	0
Saint Vincent and the Grenadines	151	n.a.	42	n.a.	0
Suriname	60,216	57,518	56,823	−1	−46
Trinidad and Tobago	1,980	787	621	−21	−12
Turks and Caicos Islands (United Kingdom)	192*	n.a.	n.a.	n.a.	n.a.
Venezuela	340,471	199,824	169,821	−15	−1,942
Virgin Islands (USA)	138	n.a.	0	n.a.	0

**The data for Hawaii were obtained from the Natural Resources Conservation Service, except for the land area, which is from the *Statesman's Yearbook*. Statistics are for 1982 and 1992, rather than for 1980 and 1995 (columns two and three). Percentage change in column four is for the period 1982–1992, and average annual change (column five) is for the period 1987–1992.

Sources: Land areas (column one) are from the Food and Agriculture Organization's *State of the World's Forests 1997,* except for those marked with an asterisk, which were found in the *Statesman's Yearbook: The Essential Political and Economic Guide to All the Countries of the World, 1998–99.* Figures for total forest cover in 1980 and 1995 (columns two and three) are from Table 11.1 of the World Resources Institute's *World Resources 1998–99.* Column four, percentage change in forest cover from 1980 to 1995, refers to the amount of forest lost only during these 15 years. Figures for this rate and for the average annual change, 1990–1995 (column five), are from *State of the World's Forests 1997* and the *Statistical Yearbook, 1995.* "n.a." indicates that data were not available. "n.s." indicates that the value is so small as to be classified as not significant. The data for forest cover includes tree plantations. In the few cases where the numbers in column four or five are positive rather than negative (such as with Mauritius and India), it reflects the amount of tree plantations in the country or the growth of secondary forests on lands that were deforested and abandoned. The zeroes in the fifth column do not indicate that the country has stopped deforestation, but rather that so much forest has been razed that there are no longer any trees left to lose.

PART I
Tropical Forest Wildlife

Tropical wildlife species were chosen for this section on the basis of superlatives: They are the best known, the most endangered, the most fascinating, the tiniest, the most recently discovered, the biggest, the most colorful. Animals are listed alphabetically by Latin names of their class, order, family, genus, and species, with common names in parentheses, when they exist. Check the index to find a particular animal quickly.

Amphibians *(Amphibia)*

Amphibians are vertebrates that live in both aquatic and terrestrial habitats. Their scientific name is derived from this dual environmental strategy—*amphibios* is Greek for "double life"—although some species are entirely terrestrial, and others spend all their lives in the water. The nearly 5,000 species lack fur and feathers, and have two lungs and a three-chambered heart. Most begin life as larvae and metamorphose into adults.

ANURA (frogs and toads)

Frogs and toads include at least 4,300 species, nearly as many as in the order of mammals. They live on all continents except Antarctica and in nearly every habitat. All "toads" can be considered frogs, but most belong to the family of *Bufonidae*. In general, frogs have smooth skin, long limbs, and live in water, while toads are stout, have bumpy or warty skin, and favor dark and damp places, but not water.

Arthroleptidae

These are small frogs that lie within the leaf litter of the forest floor in sub-Saharan Africa. The 70 species are called "squeakers" because of their high-pitched voices.

Arthroleptis stendodactylus (common squeaker)

The common squeaker is a small, terrestrial frog that is widely distributed across central Africa, primarily in dry forests, but also in the western rainforests. Averaging about one-and-a-half inches long, these frogs have plump bodies with relatively short legs and no webbing between their toes and fingers. Their abdomens are consistently white and granular, but their backs vary considerably in coloration, with a mottled combination of brown, gray, and red that usually contains an hourglass pattern. Their eyes protrude slightly from the sides of their heads, and a dark line runs along the top of the nose between the eyes. Though capable of climbing trees, common squeakers spend most of their time on the forest floor, where they feed on ants, termites, beetles, and other insects. Active mostly at night, they may also emerge from their resting nooks by day following heavy rains, when males hide in thick vegetation and emit their continuous metallic peeps. Upon mating, females hide clumps of between 30 and 80 fertilized eggs in crevices or burrows in the moist earth; metamorphosis takes place within the egg, so that completely developed froglets hatch out of them.

Further Reading:
Mattison, *Frogs and Toads of the World*.
Stewart, *Amphibians of Malawi*.

Trichobatrachus robustus (hairy frog)

The hairy frog is found in Cameroon and Equatorial Guinea. Its unique attribute is that males develop hairlike structures on their flanks and thighs during the mating season. They sit under water on egg masses laid in streams, and the hairs apparently augment respiration through the skin, so the egg-sitter can stay submerged for long periods.

Further Reading:
Cogger and Zweifel, *Encyclopedia of Reptiles and Amphibians*.

Bufonidae

These toads are found in a variety of habitats worldwide except Australia, Madagascar, and ocean islands.

The 400 species—more than half of which belong to the genus *Bufo*—are heavy-set, short-legged and have wartlike glands on their bodies and legs. They usually hide out in holes in the ground during the day, coming out to feed at night. Their size ranges from about one inch to 10 inches.

Further Reading:
Cogger and Zweifel, *Encyclopedia of Reptiles and Amphibians.*

Bufo marinus (cane toad)

The cane toad is also known as the giant or Neotropical toad. It is one of the most common and widely distributed tropical amphibians. Native to South and Central America, *B. marinus* thrives in an array of ecosystems, from rainforest to farms and towns, where choruses of its loud, staccato call are common night music. A large toad—body length averages four to six inches—it has a brown, bumpy back and pale belly. It is a voracious and opportunistic predator, feeding on an array of insects, small vertebrates, and even vegetable matter, which is unusual for a toad or frog. Because of its gluttony, it was introduced to southern Africa, Australia, and much of Asia to control insect pests on sugar plantations, but it has adversely affected native toad species in most of those countries. Active at night, cane toads can sometimes be spotted hunting beneath lights in towns and on farms. They mate year round, laying long strings of eggs in swamps, ditches, and other standing water; females can lay as many as 35,000 eggs per year. These toads have few predators because adults, eggs, and tadpoles are toxic.

Further Reading:
Badger and Netherton, *Frogs.*
Chocran, *Living Amphibians of the World.*
Mattison, *Frogs and Toads of the World.*

Atelopus zeteki (golden frog)

The golden frog is a toad, in spite of its common name, although its small size and bright coloration make it resemble a poison-dart frog (*Dendrobates*). Ranging in length from 1 to 15 inches—females are about 50 percent larger than males—*A. zeteki* has a narrow body, thin limbs, and bright yellow skin, which is usually marked with black spots or bands on its back. This bright coloration warns potential predators that these toads are toxic, and they are consequently able to hop around the forest floor by day undisturbed, hunting for small insects. Indigenous to the mountain forests of western Panama, the golden frog is only one of more than 40 species in the genus *Atelopus*. Most of them are brightly colored, and they are distributed through Central and South America. Unlike most toads and frogs, among which only males call out at night, female atelopids also sing, and both sexes look for each other along the edges of streams, where eggs are laid and fertilized. Eggs hatch within days, and the tadpoles have several adaptations that allow them to survive in the quick-flowing waters of mountain streams.

Further Reading:
Badger and Netherton, *Frogs.*
Chocran, *Living Amphibians of the World.*
Mattison, *Frogs and Toads of the World.*

Dendrobatidae

These are small, often colorful frogs found in rainforests of Central and South America. Nearly all the 170 species are diurnal.

Dendrobates, Epibpedobates, and *Phyllobates spp.* (poison-dart frogs)

Poison-dart frogs are quite small, with an average size of one-and-a-half inches, and are usually green, red, orange, yellow, blue, or black, or a combination of these hues, sometimes in solids or startling patterns. Among the decaying leaf litter, they shine like tiny jewels. *Dendrobates* have only one known predator, a snake (*Leimadophis epinephelus*), which is immune to the poisons. Deforestation is far more damaging to the frogs' population. They live near rivers and streams from Nicaragua to Bolivia, the Guianas, and southeastern Brazil. Unlike most other frogs, they are diurnal. They hunt among the leaf litter for ants, small insects, and arthropods.

The toxins in their skin glands are bitter-tasting alkaloids; some 200 different toxic alkaloids have been identified. At least three species, *Phyllobates terribilis, Phyllobates aurotaenia,* and *Phyllobates bicolor,* contain poisons so toxic that they are harmful or even deadly to humans. The toxins can cause convulsions, paralysis, and eventually death if they enter the bloodstream of an animal. As its name suggests, *Phyllobates terribilis,* a bright yellow frog that grows to one foot in length, is the largest *Dendrobates,* and the most deadly. Just touching it can cause a nasty irritation to human skin. The Chocó indigenous people of Colombia use the poisons in hunting by tipping the ends of their blowgun darts with the toxins that they secrete from the frogs. The Chocó usually hang frogs over a fire by pushing a sharp stick in their mouths. The heat forces the poison to come to the surface of the frog's skin, where it appears as a frothy, white substance. The Chocó simply dip a dart into the froth. One frog can provide enough toxin to coat 50 to 100 darts. Only 55 of the 135 species of poison-dart frogs are known to produce toxins, and hunters use only three of these to tip their arrows.

Many poison-dart frog species are extraordinarily caring parents. The females lay from 30 to 40 of their eggs on the forest floor, coated in a jellylike mass to keep them moist, and then watch over them—sometimes the fathers also tend the eggs. When the tadpoles hatch, they squirm their way onto the parent's back and are carried around, safe from predators, until they are deposited in a safe place. Some species really go the extra mile: Mothers haul the tadpoles up into the trees—not an easy journey for these ground-dwellers—in search of tank bromeliads, which hold pools of water. They deposit the tadpoles, a few at a time, in several bromeliads. If they choose a deposit spot that already contains a tadpole, it does a lively little dance that causes the bromeliad to vibrate and alerts the parent that the safe haven is already taken. By not putting all its tadpoles in one bromeliad basket, the *Dendrobates* ensures greater survival success. But the tadpole-care duties don't end there. The female returns to the forest floor, but makes the trip back up to the bromeliad incubators once every five days or so and drops several infertile eggs in the pools of water with her tadpoles. The eggs provide nutrition to the developing young, which are fully grown frogs in two to three months.

Further Reading:
Forsyth and Miyata, *Tropical Nature.*
Moffett, "Poison-dart Frogs: Lurid and Lethal."

Epibbpedobates tricolor

This is a half-inch poison-dart frog found in south-western Ecuador that has attracted renewed interest from scientists. In 1976, John Daley, a researcher with the National Institutes of Health, found that an extract from the tiny frog's skin could block pain 200 times more effectively than morphine. In 1998, scientists at Abbott Laboratories in Illinois were able to isolate a compound from the frog's poison that yielded a powerful painkiller without damaging poisons. Best of all, it has none of morphine's addicting side effects. The potential new drug is now going through a series of trials required before it can be used.

The drug may be *Epibbpedobates tricolor*'s legacy. When Daley first visited Ecuador in the mid-1970s, he was able to collect and take home more than 750 of the bright red and blue frogs. Only a few years later, development had destroyed much of the frogs' habitat and further collecting became illegal. Less than 6 percent of the frogs' original habitat remains. In altered habitat or captivity, the frog no longer produces its poison. Scientists speculate that some item in the frog's diet, such as an ant or beetle, probably provides the poison.

Further Reading:
Recer, "Poison from South American Frog Leads to Discovery of Powerful Painkiller."

Centrolenidae

Frogs are almost all arboreal and live in cloud and rainforests from southern Mexico to Bolivia. Most of the more than 120 species are no more than an inch long. Most of the 65 species are no more than an inch long and have translucent skin on their chests through which their internal organs are visible.

Centrolenella fleischmanni (glass frog)

The glass frog is a small, secretive tree frog found in the cloud forests of Central America, where its high-pitched peeps fill the misty night air. It is a delicate, light green frog less than one inch long, with thin limbs, suction disks on its fingers, and a blunt head topped with large, protruding eyes. It is one of about 65 species of glass frogs found in Central and South America, most of which have translucent skin on their undersides, through which the heart and other organs are visible. Like most glass frogs, this species spends its days hidden in the foliage, emerging at night to hunt for small insects and search for a mate. Glass frogs mate on vegetation overhanging streams, where the pale masses of eggs are stuck to the underside of a leaf or branch. Males usually guard eggs and may even urinate on them to keep them from drying out. Upon hatching, tadpoles drop into the water, where they develop into frogs. This is also one of many frog species that has mysteriously disappeared from, or grown scarce within, parts of its range.

Further Reading:
Badger and Netherton, *Frogs.*
Chocran, *Living Amphibians of the World.*
Mattison, *Frogs and Toads of the World.*

Hylidae

This family of frogs includes about 770 species, at least 500 of which live in the Americas, particularly the tropics. Most species are arboreal; their toes have sticky pads to help them climb. In the Neotropical genus *Phyllomedusa*, the first finger is actually able to touch the other three, like a thumb, so that they can grasp onto twigs.

Further Reading:
Cogger and Zweifel, *Encyclopedia of Reptiles and Amphibians.*

Agalychnis callidryas (red-eyed leaf frogs)

The red-eyed leaf frogs are vividly colored, with lime-green bodies, white throats and bellies, orange hands and feet, and sides marked with dark blue, with ivory borders, and vertical bars. The iris of the frog's eye is deep red. Females are fairly large, measuring nearly three inches long, while males are less than half that size. Found from Mexico to Panama, they have enlarged toe pads and webbed feet. Excellent climbers, they live in the forest canopy, spending days flat against green leaves and hunting for insects and other small prey at night. The red-eyed leaf frog is endangered, due to deforestation, pollution, and global climate change. Its habitat needs include a particular temperature range during the day and night and a relative humidity of 80 to 100 percent.

Further Reading:
Janzen, *Costa Rican Natural History.*
Norman, *Common Amphibians of Costa Rica.*

Gastrotheca riobambae (marsupial frog)

The marsupial frog is so called because the female carries her fertilized eggs in a pouch on her back. A stout, light green frog with flecks of cream and black, its muscular limbs end in long toes and fingers adapted for its life in the branches and on the ground. Indigenous to the highland forests of Peru, Ecuador, and Colombia, where it preys on an array of insects, the marsupial frog is one of more than 30 species of *Gastrotheca* distributed throughout northern South America and Central America. These frogs mate with their bottoms raised so that the eggs roll down the females' back as they are laid, and the males fertilize them as they spin into her pouch. A female can carry as many as 100 eggs for three or four months until they hatch; she then releases the tadpoles in a shallow pond or ditch, holding her pouch open and scooping

them out with her hind legs. One of more than 60 species of the *Hylidae* family in which the female carries eggs and/or young on her back, the marsupial frog is one of only a dozen species of *Gastrotheca* that release their tadpoles into water—in the other marsupial species, frogs develop directly within the pouch.

Further Reading:
Halliday and Adler, *Encyclopedia of Reptiles and Amphibians.*
Mattison, *Frogs and Toads of the World.*

Litoria caerulea (White's tree frog)

White's tree frog has a blunt face, bright green back and limbs, and pale, granular belly, and is common across northern Australia and New Guinea and also found in Indonesia. The frog may have been introduced in these countries and actually originated in South America, where only a small population remains. Its abundant, thick skin helps it endure dry conditions, and when a resting frog pulls its limbs together, that skin bunches up in folds over its brow and sides, giving it a truly bizarre appearance. It can survive in a variety of environments, from secondary forests to farms and gardens, where it hunts for an array of insects—it is a voracious eater. White's tree frogs are most commonly found in forests near water sources. Adults have a body length of about five inches and large hands and feet with thick toe pads that make them very capable climbers, as are most hylids. During the rainy season, males gather at swamps, ditches, and other kinds of still bodies of water, and fill the night air with deep croaks.

Further Reading:
Badger and Netherton, *Frogs.*
Mattison, *Frogs and Toads of the World.*

Hyperoliidae

Most of these small frogs live in Africa and Madagascar. Of the 300 species, some are terrestrial, others arboreal, and others can climb trees but spend most of their time in low vegetation near water.

Hyperolius puncticulatus (sedge frog)

The sedge frog is a small tree frog with a smooth back colored orange or brown and decorated with black and white spots and pale bands running laterally from the nose to the back. Its large eyes protrude from the sides of its broad head, its hands and feet have little webbing, and its thick toes and fingers end in disks, which help it climb. Sedge frogs inhabit the forests of central and eastern Africa, where they are invariably found near ponds or streams. They are most active at night, when they hunt for everything from beetles to grasshoppers. During rainy-season nights, males perch in the foliage near the water's edge and emit a loud, quacklike mating call. *Hyperolius puncticulatus* is one of more than 100 species of sedge frogs distributed through a variety of habitats across sub-Saharan Africa; all of them are small and have striking coloration.

Further Reading:
Badger and Netherton, *Frogs*.
Stewart, *Amphibians of Malawi*.

Leptodactylidae

This huge frog family has some 720 species distributed in the southern United States and most of Latin America. Characteristics among the species vary widely—they range in length from just under 1 inch to nearly 12 inches, may be completely aquatic or totally terrestrial, may lay eggs in water or on land. Some species are vividly colored; others are drab.

Assa darlington

A small, ground-dwelling frog, this species is indigenous to the montane rainforests of northeastern Australia. Its plump body measures less than an inch long, and it has short front legs and a small head. The uneven border of its back and head complements its mottled brown coloration to camouflage it well amidst the leaf litter, where it hunts for tiny insects and mates. Females bury their fertilized eggs—clumps of around 10 white balls—beneath wet leaves or moss, but the male guards the eggs. After about 10 days, the eggs hatch, and the male moves close to the clump, letting the pale tadpoles climb onto his back and wriggle their way into his lateral pouches—he may push them inside with his forelimbs. Tadpoles can take between 40 to 70 days to grow and metamorphose within these pouches, during which time the father becomes increasingly round—they emerge as completely developed froglets, ready for life in the rainforest.

Further Reading:
Mattison, *Frogs and Toads of the World*.

Eleutherodactylus bransfordii (Bransford's litter frogs)

Bransford's litter frogs are little brown frogs (only a bit longer than one inch) patterned with a variety of blotches. They are terrestrial and eat small invertebrates that they find among the leaf litter. Bransford's litter frogs live at low- and mid-elevations of Costa Rica, Nicaragua, and Panama, where they are the most common amphibian on the forest floor. Females lay eggs among the leaf litter, and the embryos develop into tadpoles within their eggs, hatching as tiny froglets. Some species of *Eleutherodactylus* skip the egg stage, producing froglets directly. With more than 400 species in the Americas, the *Eleutherodactylus* genus of frogs has more species than any other vertebrate genus.

Further Reading:
Janzen, *Costa Rican Natural History*.
Norman, *Common Amphibians of Costa Rica*.

Eleutherodactylus Iberia

One of the tiniest frogs in the world was discovered in 1997 on the island of Cuba. It measures only ⅜ of an inch, or smaller than a dime, and just a hair longer than a miniscule frog in Brazil (*Psyllophryne didactyla*). Herpetologist Alberto Estrada found the wee frog, which is black with a narrow red stripe down its back, one night on the forested slopes of eastern Cuba. He heard an unfamiliar chirping and managed to spot the frog amongst the leaf litter and ferns. Although most frogs lay about 25 eggs in the water, which hatch into tadpoles, this Cuban amphibian lays just one egg at a time, and that one on land.

Further Reading:
National Geographic, "Tiny Frog with a Sizable Name."

Eleutherodactylus diastema (tink frogs)

Tink frogs inhabit wet, lowland forests from Nicaragua to Colombia. Unlike most other frogs of this genus, they are arboreal. They are a yellowish, pale brown, with yellow throats and breasts. On damp nights in the rainforest, the male's call is loud and unmistakable, something like pinging a wine glass—a sharp "tink!" Surprisingly, this reverberating call comes from a very small frog, less than an inch long.

Further Reading:
Janzen, *Costa Rican Natural History.*
Norman, *Common Amphibians of Costa Rica.*

Physalaemus pustulosus (tungara frogs)

Tungara frogs are remarkable for their complicated mating calls. It is a whine that sounds like a sliding ricochet, often ending with a series of chucks. This tumultuous call comes from a male frog that's just a bit over an inch long. When he calls, the male's body and vocal sac inflates, alternatively. A female approaches a calling male, and he grasps her by the armpits. They usually hop a distance away, to a shallow puddle of water. The female repeatedly produces a mixture of eggs and jelly that the male picks up with his hind feet, fertilizes, and whips into a mass of foam. The white foam, with about 150 eggs inside, floats high on the water. Should the puddle dry up, the eggs or tadpoles, which hatch within 48 hours, can still survive in the foam for up to five days. The foam also protects the eggs against some small predators, such as ants. Tungara frogs are terrestrial and are found from Mexico to northern South America.

Further Reading:
Janzen, *Costa Rican Natural History.*
Norman, *Common Amphibians of Costa Rica.*

Pelobatidae

This family of 88 species of toadlike, burrowing frogs lives in North America, from Canada to southern Mexico, Europe, northern Africa, western Asia, Southeast Asia, Indonesia, and the East Indies.

Megophrys nasuta (Asiatic horned toad)

The Asiatic horned toad was named for the triangular projections over its dark eyes and similarly horn-like nose. It is a ground-dwelling inhabitant of the rain-forests of Thailand, Malaysia, Indonesia, Borneo, and the Philippines. Growing to a length of about four inches, the toad has a mottled brown back that gives it the appearance of a leaf from above—it even has ridges that resemble leaf veins—making it almost impossible to spot on the jungle floor. It spends most of its time half-buried in the leaf litter, waiting for insects and small vertebrates to wander close enough to be snatched up. Deforestation has reduced the range of this and many of the other 21 species of *Megophrys* distributed through Asia.

Further Reading:
Badger and Netherton, *Frogs.*
Halliday and Adler, *Encyclopedia of Reptiles and Amphibians.*
Mattison, *Frogs and Toads of the World.*

Pipidae

Most of the 26 species of this frog family are aquatic, with large, fully webbed feet. They inhabit tropical Africa and Central and South America.

Pipa pipa (Surinam toad)

The Surinam toad lives in varied habitats, from massive rivers like the Amazon and Orinoco to small swamps and ditches scattered across tropical South America. Its mottled brown coloration makes *P. pipa* hard to spot in those muddy waters, and its squat body makes it appear to have been stepped on. Growing to between four and eight inches in length, the Surinam toad has large webbed feet, long fingers, small, dark eyes, and a pointed nose; it is one of the few members of the *Anura* order of amphibians that lacks a tongue. It rests by day and hunts at night, using the filaments on the tips of its slender fingers to locate food—living or dead—in the muddy water bottoms. When mating, the male mounts the female, and the locked couple swims somersaults underwater as the female lays her eggs, and the male fertilizes and pushes them onto her swollen back. The female's back is covered with a honeycomb of tiny craters, where the fertilized eggs become imbedded, after which her skin closes over the tiny, egg-laden pits. Eggs develop, and the froglets—there is no tadpole stage—hatch on their mother's back, emerging from their chambers about three months after fertilization.

Further Reading:
Chocran, *Living Amphibians of the World.*
Cogger and Zweifel, *Encyclopedia of Reptiles and Amphibians.*

Pseudidae

This genus includes only five species, all found in South America; all are almost totally aquatic.

Pseudis paradoxa (paradoxical frog)

The paradoxical frog is an inhabitant of the Amazon and other South American rivers. It is one of the few frog species in which the tadpoles are larger than the adults; hence the paradox. Whereas an adult frog's body is between two and three inches long, tadpoles grow as long as 10 inches before shrinking down to less than three inches at metamorphosis. Tadpoles are dark brown with cylindrical bodies and relatively small eyes and mouths; local people catch them with fishing hooks and eat them in some areas, but since the species is common across the northern half of South America, it is not considered threatened. Adult frogs are stout, with light green backs and pale bellies. Their muscular legs are mottled brown, and large webbed feet attest to their aquatic lives. Their long, slender fingers help them search for small invertebrates in the mud and floating vegetation, whereas the eyes that protrude from the tops of their heads allow them to hunt airborne insects while watching out for anything that might be hunting for them. Their extremely slimy skin helps them slip away from predators. They lay their eggs in a foam that floats on the standing water, from which tadpoles swim free upon hatching.

Further Reading:
Chocran, *Living Amphibians of the World*.
Halliday and Adler, *Encyclopedia of Reptiles and Amphibians*.

Ranidae

Ranidae are the "true frogs" and have the widest distribution of any frog family. The more than 650 species live in North, Central, and northern South America, Europe, much of Asia, northern Australia, and most of Africa.

Conraua goliath (Goliath frog)

The Goliath frog is the largest frog in the world, reaching a body length of one foot and a weight of seven pounds. With their hind legs extended, these monsters measure more than two and a half feet long. Their dark brown and black bodies have small bumps covering the sides and back, and their large legs and webbed feet allow them to leap far and swim fast. They are found only along the large rivers in the rainforests of western Africa, where they prey on an array of insects and vertebrates. They mate during the dry months—June to August—depositing their fertilized eggs directly in the water. Naturally rare within their limited range, Goliath frogs are threatened by the destruction of their habitat through deforestation and the damming of rivers, and because they are hunted for food and the pet trade.

Further Reading:
Cavendish, *Endangered Wildlife*.
Mattison, *Frogs and Toads of the World*.

Mantella aurantiaca (golden mantella)

The golden mantella is a tiny denizen of Madagascar's endangered rainforest and one of the most brightly colored frogs in the world. Just over an inch long, it is a shimmery gold, though many individuals are more orange than gold. It has big black eyes on the sides of its head, short legs, thin arms, and large hands, with tiny suction discs on the tips of its toes and fingers. One of four species of brightly colored mantellas native to Madagascar, the golden mantella spends most of its life on the forest floor, occasionally climbing onto trunks and low branches, where it hunts for small insects. Males attract females by emitting a high-pitched peep, but unlike most frogs and toads, they don't fertilize eggs as they are being laid. Instead males deposit their sperm inside the female, much as higher life forms do. As is the case with many species native to the island nation of Madagascar, the golden mantella's range has been greatly reduced by deforestation.

Further Reading:
Badger and Netherton, *Frogs*.

Platymantis boulengeri (Boulenger's platymantis)

Boulenger's platymantis is a small, gray frog native to the rainforests of Borneo. Although a member of the predominantly aquatic *Ranidae* family, *Platymantis* is an arboreal species, with only a remnant of webbing on its feet, and slender toes and fingers that are slightly broadened at the tips, to assist in climbing. Smooth gray skin covers its compact body and muscular limbs, and its large mouth comes to a slight point. A dark brown mask extends from its mouth back over its eyes. This species' cryptic coloration makes it hard to spot on tree bark or amidst the leaf litter, where it sometimes forages. Males sing from a hidden spot on the forest floor, which allows females to find them, and females lay just a few large eggs, which they bury in the moist earth. This species needs no water to develop, since metamorphosis takes place within the egg, and froglets hatch ready for life on land.

Further Reading:
Chocran, *Living Amphibians of the World*.
Mattison, *Frogs and Toads of the World*.

Rhacophoridae

The 300 species of this frog family are found in temperate and tropical Africa, Asia, and Australia. Most are arboreal.

Chiromantis xerampelina (gray tree frog)

The gray tree frog is a large, arboreal species native to the dry forests of central and southern Africa. Although its habitat is arid for much of the year, the gray tree frog manages to preserve moisture by sleeping in a humid cranny, which it abandons at night to hunt for insects, and returns to before dawn. It also conserves water by excreting its waste as a concentrated paste. Its body is about three inches long, its mottled gray skin is covered with tiny warts and a waxy film, there is a slight hump on its back, and its eyes sit on the top of its small head. Opposable fingers that end in large disks allow this frog to grab twigs and cling to vertical surfaces, making it an expert climber. Following rainstorms, males chirp to attract females, and as they mate, the female secretes a sticky fluid that both frogs beat into a foam with their legs. The female then lays about 150 eggs, which the male fertilizes as they enter the foam nest, which is built on the branches of trees overhanging ponds or lakes. She then guards the eggs by lying on top of the nest. This species sometimes practices cooperative breeding, with dozens of frogs gathering on a single branch to create a giant foam nest. After a few days, eggs hatch, and the tiny tadpoles wriggle to the bottom of the nest, dropping into the water below.

Further Reading:
Chocran, *Living Amphibians of the World.*
Halliday and Adler, *Encyclopedia of Reptiles and Amphibians.*
Stewart, *Amphibians of Malawi.*

Rhacophorus nigropalmatus (Wallace's flying frog)

Wallace's flying frog is one of several species of flying frogs that inhabit the rainforest canopy in southern Asia. It was named after the famous naturalist Alfred Russel Wallace, who identified it in the 19th century. It is a medium-sized green frog with yellow sides and a pale belly, a three-inch body, long, muscular arms and legs, and extensive webbing on its large hands and feet. That webbing, together with flaps along its limbs, allow Wallace's flying frog to glide through the air between trees—it can sail as far as 21 feet—which is how it escapes predators. Frogs in the *Rha-cophoridae* family mate in branches overhanging streams or ponds, where the female builds a bubble nest by whipping a sticky fluid she secretes into a foam with her legs. When the eggs hatch, the tadpoles drop, or are washed by rain, into the water below. Endemic to the rainforests of Thailand, Malaysia, and Indonesia, this species has had most of the forest in its range destroyed, and much of what remains is threatened by deforestation.

Further Reading:
Badger and Netherton, *Frogs.*
Chocran, *Living Amphibians of the World.*

Rhinophrynidae

Rhinophrynidae is a family of one, *Rhinophrynus dorsalis,* commonly known as the Mexican burrowing toad, found in dry forests and pastures from southern Texas to Costa Rica. It is not a true toad, but does have some toadlike features. The blob-shaped, three-inch amphibian has a cone-shaped head, very small eyes, and short, plump legs. It can inflate its loose-skinned body to greatly increase its size. Its smooth skin is brown, scattered with pale yellow spots. Its rear legs are shaped to be digging tools, as the frog claws out a hole in the mud in which to bury itself, turning like a corkscrew as it burrows into the ground. Once buried, it turns and twists until it has a round chamber. Then the frog inflates itself so it is tightly wedged in its hole and cannot be pried loose.

When the burrowing toad is hungry for its usual diet of insects, particularly ants, it comes to the surface again at night. Herpetologists aren't sure how long it stays buried. When bothered, it excretes a sticky white goo, which may be a defense mechanism, since it has caused an allergic reaction in more than one inquisitive herpetologist. Mating takes place in water, and females deposit eggs—anywhere from 2 to 8,000—that stick together in a clump. The eggs hatch in several days, releasing larvae that have broad, flat heads and wide, slit-like mouths that extend across the entire front of the heads and are edged with barbels. The larvae are black above and shimmery, silvery gold below. Metamorphosis takes place among all the larvae at about the same time, so the newly formed froglets exit the water en masse. They immediately set to burrowing.

Further Reading:
Cogger and Zweifel, *Encyclopedia of Reptiles and Amphibians.*
Janzen, *Costa Rican Natural History.*

Sooglossidae

The three species in this family are confined to the Seychelles in the Indian Ocean. Their genetic relationship to other frogs is uncertain.

Sooglossus seychellensis (Seychelles frog)

The Seychelles frog is a pale, spotted frog with bands on its thighs. This rare species grows no longer than an inch and lives on the forest floor, where it hunts for tiny invertebrates in the leaf litter. Females hide their eggs beneath moist leaves, and both parents guard them for the two weeks it takes for them to incubate. The tadpoles are well developed when they hatch—they already have rudimentary limbs, and for a reason: The father frog backs up to newborns so they can climb onto his back. They remain on their father's back until they have completed metamorphosis, and often for a while after that, protected from dehydration by a layer of mucus that he secretes, and living off of their remaining yolk supplies.

Further Reading:
Halliday and Adler, *Encyclopedia of Reptiles and Amphibians.* Mattison, *Frogs and Toads of the World.*

URODELA (salamanders)

Salamanders is one order of animals that has greater diversity in the temperate zones of the Eastern and Western hemispheres. Nine of the ten families are found outside the tropics. Most of the 400 species live in moist, forested habitats.

Plethodontidae

This family of salamanders has a groove on each side of their heads, running from the nostrils to the upper lip, which helps them detect smells. Unlike other salamanders, the adults do not have gills or lungs, but breathe through their skin. This is the only family of salamanders that has tropical species.

Further Reading:
Norman, *Common Amphibians of Costa Rica.*

Bolitoglossa mexicana

This salamander grows to between four and six inches long, has a slender body and long tail. It is a mottled gold or reddish-brown; its belly, short legs, and small webbed feet are of a slightly lighter shade. It was first identified in rainforests of southeastern Mexico, and its range extends through the humid lowlands and mountains of eastern Guatemala, southern Belize, and northern Honduras. Rare throughout its range, it can be found in vegetation or on the ground, and it spends most of its time hiding in the leaf litter, in bromeliads, or beneath bark, rocks, or logs. It eats a variety of small insects, and mates in a hidden, moist nook, where it deposits dozens of tiny eggs.

Further Reading:
Halliday and Adler, *Encyclopedia of Reptiles and Amphibians.* Lee, *Amphibians and Reptiles of the Yucatan Peninsula.*

Bolitoglossa striatula (striated salamander)

The striated salamander originated in North America. Millions of years ago, the species began a southern migration that eventually ended in South America. *Bolitoglossa* is the only genus with species widespread in South America, and only a few other genera have penetrated the tropics from North America. The striated salamander is about five-and-a-half inches long, pale yellow, with dark stripes down its back and stomach. At night, it forages among the leaf litter for small invertebrates.

Further Reading:
Janzen, *Costa Rican Natural History.* Norman, *Common Amphibians of Costa Rica.*

Bolitoglossa subpalmata (mountain salamander)

The mountain salamander is restricted to high-elevation forests in central and eastern Costa Rica and is probably the best-studied tropical salamander. It hides and forages under leaf litter, fallen logs, and rocks and, during drier periods, crawls into crevices and root tunnels in the soil. Females deposit about 25 eggs in cavities under imbedded rocks and logs and usually the female, but sometimes the male, tends the eggs, coiling its body about them and occasionally rotating them with its forelimbs and tail. Tiny salamanders hatch after four-and-a-half months. The mountain salamander has a toxin in its skin that paralyzes the lower jaws of snakes in mid-meal, sometimes allowing the nearly ingested salamander to escape and recover.

Further Reading:
Janzen, *Costa Rican Natural History.* Lee, *Amphibians and Reptiles of the Yucatan Peninsula.* Norman, *Common Amphibians of Costa Rica.*

GYMNOPHIONA

The *Gymnophiona* is an order of limbless amphibians that are well adapted for burrowing. Some of the 155 species have scales; some lack tails.

Caeciliidae

The 150 species of this amphibian are found in South and Central America, Africa, India, and the Seychelles. They have recessed mouths and no tails.

Schistometopum thomensis (Sao Tome caecilian)

Sao Tome caecilian is found only on the island of Sao Tome, off the western coast of Africa. Its ringed, thin body is bright orange; it can grow up to a foot long; and its small head has tiny black eyes and a narrow, rounded nose. It lives in the humid soil of the forest floor, where it hunts for worms and other small invertebrates. Fertilization is internal, and females retain their eggs, which develop and hatch inside them, so that they bear live young—a trait common to about half of the world's caecilian species. Due to its small range and habitat destruction, the Sao Tome caecilian is considered an endangered species.

Further Reading:
Chocran, *Living Amphibians of the World.*
Halliday and Adler, *Encyclopedia of Reptiles and Amphibians.*

Siphonops annulatus (South American caecilian)

The South American caecilian is a bright blue, limbless amphibian with a thick, wormlike body more than a foot long. Its smooth skin is lined with wide, pronounced rings. Like most caecilians, it is practically blind and deaf, but it has an excellent sense of smell, which it uses to find the worms and other soil-dwelling invertebrates it eats. Little is known about caecilians, since they spend almost their entire lives underground, emerging only during heavy rains. One of the most widely distributed of the 150 species of caecilians that exist in the humid Tropics, this species is distributed throughout the Amazon basin and south into northern Argentina. It has been found in primary forests, farms, and even construction sites. Scientists don't know much yet about its reproductive habits, aside from the fact that fertilization is internal; when mating, the pair joins their back ends so that the male can insert his sperm, and females lay eggs.

Further Reading:
Halliday and Adler, *Encyclopedia of Reptiles and Amphibians.*
Chocran, *Living Amphibians of the World.*

Ichthyophidae

This family hold about 37 species with short tails, found in Southeast Asia, India, Sri Lanka, Sumatra, Borneo, and the Philippines.

Ichthyophis glutinosus (Ceylonese caecilian)

The Ceylonese caecilian is found only in the island nation of Sri Lanka. Its dark brown or gray, snake-like body has cream-colored lateral bands, and narrow, discrete rings. It can grow as long as one-and-a half feet, and its narrow head has small, black eyes, below which are tiny sensor tentacles, a common trait of caecilians. It inhabits the soil of forests and some farmland, where it feeds on worms, insects, and small vertebrates. They breed from March to September, when the female will lay 10 to 20 oblong eggs in an earthen nest near water. She guards her eggs until they hatch, and her tiny, gilled offspring slither into the water, where they complete metamorphosis before emerging again for an adult life in the ground.

Further Reading:
Chocran, *Living Amphibians of the World.*
Halliday and Adler, *Encyclopedia of Reptiles and Amphibians.*

The Disappearing Frogs and Toads

In 1987, the U.S. biologist Martha Crump counted 1,000 tiny golden toads (*Bufo periglenes*) in Costa Rica's Monteverde reserve, a protected cloud forest. She watched at night, as hundreds of astoundingly beautiful male toads, glowing a brilliant gold-orange, emerged in pools of water to find mates. She returned to the reserve a year later, but found only 10 toads. In 1989, she could find only one. Later that year she attended a conference and mentioned the curious disappearance to her colleagues. Many had similar stories to tell—so many, in fact, that the scientists raised a global alarm. Dozens of species of frogs and toads were missing in action worldwide.

During the 1990s, herpetologists documented increasing declines, but could not point to the culprit. They did feel sure that the cause was environmental, even though many amphibian species were missing from seemingly protected habitats, like the Monteverde reserve.

Over a decade of studies have revealed several factors, all of them caused by humans. In the northwestern United States, where the Cascade frog (*Rana cascadae*) and several other species are in decline, amphibians are suffering because of increased levels of ultraviolet-B radiation penetrating a thinner ozone layer. The radiation destroys the animal's eggs, laid in high-altitude pools of water. A scientist examined 2.4 million Cascade frog eggs in 1992; 95 percent of the embryos were dead. Meanwhile, populations of the mountain yellow-legged frog (*Rana muscosa*) in California's Sierra Nevadas have been affected by the introduction of predatory fish species, particularly trout.

Habitat destruction and alteration, pesticide poisoning, radiation, even culinary habits—the penchant for frog's legs in Europe has driven declines on the continent and India, and now in Indonesia and Bangladesh—are understandable, if disturbing, reasons for amphibian die-offs. But what can explain disappearances in relatively undisturbed areas, where intertwining forest canopies provide shade and shelter?

In Monteverde, western Panama, and near the Costa Rica-Panama border, at least 21 local species of amphibians have vanished from their mid- to high-elevation forest habitats. In 1997, the herpetologist Karen Lips discovered that hundreds of frogs at a biological research station in western Panama had died due to massive skin infections of a kind of fungus that is normally found throughout the animals' watery habitats. The fungus covered the frogs' skins, suffocating them. Although the fungus was known to attack some invertebrates, there was no record of it assaulting a vertebrate species. Why weren't the frogs able to ward off a common fungus that had previously never posed a problem? Lips and other scientists surmise that global warming and declining rainfall are so stressing the frogs that they are no longer able to resist the fungus.

The biologist Alan Pounds, who lives in Monteverde, has been puzzling over the golden toad's vanishing act for more than a decade. Once one of the reserve's most famous residents and a prime tourist attraction, the tiny, gleaming toads are now considered extinct. Pounds and collaborators Michael Fogden and John Campbell found that other amphibians and reptiles living at high elevations in the mountaintop reserve had also suffered declines in populations, while a number of bird species that normally live at lower altitudes, such as toucans (*Ramphastidae*), were beginning to migrate up. These changes of habit coincided with unusually dry and warm weather produced by the weather pattern known as El Niño, and a rise in sea-surface temperatures. Although this pattern is a naturally occurring weather phenomenon, human-induced global warming generates dramatic fluctuations in rainfall similar to those produced by El Niño. The temperature change and lack of rain have caused the bank of clouds that daily settles on the reserve to rise. As the clouds have crept up the mountain's slope, mists and condensation have decreased. So some species have migrated upslope, toward the receding clouds and the moisture they need to survive. But the golden toad already lived nearly at the mountain's peak, so heading toward the heavens was not an option. It simply had nowhere to go.

Further Reading:
Lips, "Mass Mortality and Population Declines."
Phillips, "Where Have All the Frogs and Toads Gone?"
Souder, "Frog Decline Linked to Climate Shift."
Tuxill, "The Latest News on the Missing Frogs."

Arachnids (*Arachnida*)

Arachnids fall under the phylum *Arthropoda* and include the order of spiders, mites, ticks, daddy longlegs, and scorpions, as well as other orders—some 73,400 species in all, found worldwide. Like other arthropods, they have segmented bodies, tough exoskeletons, and jointed appendages. They lack jaws, and most inject digestive fluids into their prey, then suck out the liquefied remains. For further information, see The Arachnology Homepage <http://www.ufsia.ac.be/Arachnology/Arachnology.html>.

ARANEAE (spiders)

Spiders include about 34,000 species, found worldwide. They differ from insects, another arthropod group, in that they have eight legs rather than six, and their bodies are divided into two parts, rather than three. They are predators that feed almost entirely on other arthropods, especially insects. Many weave webs from silk that they produce. Most are terrestrial.

Araneidae

These are the orb weavers; about 3,000 species are found worldwide.

Mastophora dizzydeani (bolas spider)

The bolas spider gets its name from the Spanish word "bola," or ball, after the sticky ball at the end of a silk line that these spiders use to fish for their airborne prey. Although in the same family as the orb weavers, this species has lost the ability to weave a web and must thus rely on the "bolas" to capture food. An inconspicuous gray-brown spider with an oval body less than a half-inch wide and relatively short legs, the bolas spider spends the day resting, hidden on a branch. As soon as it grows dark, it produces a small, extremely sticky ball, which it suspends below it on a thread of silk. The ball contains a substance that smells like the pheromones of a certain moth species, and when one of those moths approaches to investigate, the spider swings the ball toward it. If the ball sticks to the moth, the spider will let it flutter for a while, and then draw it up to inflict a paralyzing bite, after which it wraps the moth in silk, and feeds on it. If it doesn't catch anything, the spider will eat the ball at the end of the night. One of nearly two dozen species of bolas spiders found in Latin America, *M. dizzydeani* was named after the famous U.S. baseball pitcher Jerome "Dizzy" Dean.

Further Reading:
Hogue, *Latin American Insects and Entomology.*
Preston-Mafham, *The Natural History of Spiders.*
Preston-Mafham, *Spiders of the World.*

Nephila clavipes (golden silk spider)

The golden silk spider is one of the most conspicuous spiders of the American tropics, since it is a large orb weaver whose ample webs are a common sight in secondary forests, plantations, and rural towns. An undisturbed forest gap may contain dozens of this species' golden webs, which are usually more than three feet in diameter. At the hub of the webs sits the female, whereas its outer reaches may be dotted with the tiny forms of males and kleptoparasites—smaller spiders that subsist on food they steal from the host spider. The female has an olive-brown, elongated oval body, about one-and-a-half inches long. Her back is decorated with cream-colored or yellow spots, and she has conspicuous tufts of dark hair on her two-inch-long legs. Males are about one-tenth the size of females, weigh even less, and lack the leg hairs. They live in webs woven by females, eating the smaller insects caught in them. The golden silk spider's diet includes flies, bees, cicadas, moths, and butterflies, though females will release unpalatable butterflies from their webs. The spiders quickly attack edible insects, injecting them with digestive poison, wrapping them in silk, and then sucking out their insides. Females are messy eaters, often slobbering juice onto the carcass, thus leaving food for kleptoparasites. Females lay a round egg sac, which they attach to a leaf or branch tip—each sac contains hundreds of eggs. This is the only American species of the genus *Nephila*, which has several similar species common in Africa and Asia.

Further Reading:
Hogue, *Latin American Insects and Entomology.*
Levi, *A Guide to Spiders and Their Kin.*
Preston-Mafham, *Spiders of the World.*

Nephila maculata (giant wood spider)

The giant wood spider is a large orb weaver common throughout the Asian tropics. It is famous for the strength and size of its webs. From India to Papua New Guinea and southern China, the yellowish webs of this and other *Nephila* species can be found hanging over forest paths, along streams and rivers, and in plantations. The large females can usually be seen at the center of these webs, with smaller males and parasitic spiders on their peripheries, and "garbage lines" of silk strung with the dry carcasses of devoured prey often dangling beneath them. Those webs capture an array of large insects, including dragonflies, moths, cicadas, beetles, as well as small birds, and their silk is so strong that it is used as fishing nets in some countries. Females of this species are 10 times as large as males, with elongated, oval, two-inch-long bodies, and leg spans more than twice that. Females are dark with light bands on their legs and golden patterns on their bodies, and males are reddish-orange, with black markings. The male approaches the female cautiously to avoid being eaten and will often spend days covering her back with pheromone-laden silk before attempting copulation, though he sometimes simply mounts the female when she is engrossed in eating. Females lay more than 1,000 eggs, from which tiny, golden babies emerge. In some areas, people eat the bodies of adult females.

Further Reading:
Preston-Mafham, *The Natural History of Spiders.*
Preston-Mafham, *Spiders of the World.*

Ctenidae

Called the wandering spiders, the *Ctenidae* are mainly large, tropical spiders numbering about 400 species. Their eyes are arranged in three rows.

Phoneutria fera (banana spider)

The banana spider is a large, gray-brown spider found in the forests and farms of the Amazon basin. Also known as a wandering spider because it is a vagrant hunter, it has been discovered hidden among bunches of bananas on plantations, which resulted in its common name. Adults may have a leg span of more than four inches. The spider's legs and body are covered with short hairs, with longer reddish hairs around the mouth, and its eight eyes are arranged in three rows. An aggressive hunter, the banana spider hides in a dark nook during the day and wanders the forest floor at night in search of a variety of insects and small vertebrates. When threatened, it will rear up on its back four legs and raise its two pairs of forelegs, which have dark undersides broken by pale bands. Its venom is the most toxic of any South American spider, and bites are sometimes fatal, which together with its reputation for hiding in fruit and people's shoes, makes it an animal much feared.

Further Reading:
Hogue, *Latin American Insects and Entomology.*
Preston-Mafham, *Spiders of the World.*

Salticidae (jumping spiders)

The family of jumping spiders is the largest spider family, with about 4,000 species, most of them found in the tropics. They hunt during the daytime by stalking then pouncing on prey, and they have the best vision of all spiders.

Portia spp.

These jumping spiders are found in Australia, Africa and Asia—15 species in all. Unlike most spiders, *Portia spp.* both build webs and capture prey away from their webs. They also have slow and choppy movements, rather than the rapid scurry of most spiders. Jumping spiders use their acute vision to stalk prey. While most spiders prefer insects as food, *Portia* fancies other spiders. They also look very unspider-like, more like a small piece of detritus that might be blown into a spider web by the wind. To capture spiders that are often twice as large as they are, they invade other spiders' nests and use mimicry and deception to capture the web owner. *Portia* lands on a spider's web and plays it like a harp, plucking at the silk threads with its legs and flicking its abdomen up and down, precisely duplicating the movements of a trapped insect. When the duped web weaver approaches, *Portia* strikes with a lethal bite to the head. Clever *Portia* often approaches its prey when the wind is blowing, or when a caught insect is struggling to free itself, since these movements mask its arrival onto the web. Sometimes *Portia* will park itself above a spider's nest, then drop a silk line alongside the web, without touching it. Then, when parallel with the spider in the web, it swings in to make a kill.

True to form, *Portia* also mates in a manner different from most other spiders. Instead of the male making the overtures and first display, the females initiate interactions by displaying first. In some species, once a female has been mounted, she drops a silk thread, and mating takes place while the pair is suspended on the dragline. During or after mating, females often swing around brusquely with fangs extended, in a sometimes successful attempt to eat their suitors.

Further Reading:
Jackson, "Mistress of Deception."
Jackson and Wilcox, "Spider-Eating Spiders."

Theraphosidae

The spiders in this family are large and hairy. The family includes about 800 species most commonly found in tropical forests, particularly in South America.

Theraphosa lablondi (bird-eating spider)

A large, brown, hairy spider found in the rainforests of northeastern South America, this fierce-looking creature commonly reaches a leg span of more than 10 inches and a weight of more than 3 ounces, making it the biggest spider in the world. Commonly called tarantulas, they should not be confused with the true tarantulas: smaller spiders of the family *Lycosidae* that are found only in southern Europe. They are called bird-eating spiders because this species, and many other large *Theraphosidae* species, occasionally captures nesting or sleeping birds, though their varied diets consist mostly of an array of insects, lizards, frogs, toads, mice, and other spiders.

Resting by day in crevices beneath rocks or tree roots, amidst epiphytes, or in burrows they've dug, bird-eating spiders hunt at night, roaming the forest floor or the canopy. When they encounter prey, they run up to or pounce on it—some species can jump as far as three feet—and raise the front of their bodies up, then strike downward, jamming their external fangs into the victim and injecting a liquid into them that is either a venom or digestive juice. Although most species are harmless to people, some are known to inflict serious bites, whereas the hairs of others contain irritants.

Males are nearly as big as females but are nonetheless cautious during copulation, to avoid being eaten. Females lay hundreds of eggs in a loose cocoon, which they guard until the spiderlings hatch and leave the nest in three to five weeks.

Tarantulas have many predators in the wild, including the hawk wasp (*Pepsis spp.*), which will paralyze the spider with a sting and lay an egg on it, so that its larvae feed on the living spider. A less gruesome, but more serious, threat to some species is their collection by people to sell as pets or mounted souvenirs.

Further Reading:
Hogue, *Latin American Insects and Entomology.*

Selenocosmia crassipes (trap-door spider)

The trap-door spider is a large, hairy denizen of northern Australia's tropical forests, grouped with that country's many trap-door species, although its burrow has no door. A relative of the South American bird-eating spiders, it looks quite similar to *Theraphosa lablondi*, but unlike them, it is largely a sedentary hunter, waiting for prey to get caught in the sticky sheet of silk it spreads around the entrance to its burrow. Nevertheless, it may sometimes emerge at night to hunt in the surrounding forest. When it encounters prey, or an enemy, this spider raises up the front of its body, exposing its fangs, and pounces on its victim, grasping with its front legs as it thrusts its fangs down and injects its highly toxic venom. Adults are large—about three by five inches—dark brown, with a reddish sternum, and have long, bristly legs that end in claws set in thick tufts of hair. They rarely leave their burrows, which are sinuous tunnels as deep as two feet, at the back of which they keep the remains of their prey. As the spider grows, it expands the burrow regularly, digging side pockets in which it buries its refuse. The species takes several years to reach sexual maturity, and when a male spider does, it will leave its burrow at night to search for females, entering their burrows to mate. Males die shortly after mating, and females deposit eggs in a saclike cocoon inside their burrows. Spiderlings spend their first few months in their mother's burrow, after which they emerge on a humid night to disperse and dig burrows of their own.

Further Reading:
York, *Spiders of Australia.*

Theridiidae

These spiders are known as cobweb weavers or comb-footed spiders, as they have a comb on their fourth leg, which they use to wrap their prey in silk. There are about 2,500 species worldwide.

Theridion grallator (happy face spider)

The happy face spider is a striking looking spider that inhabits the forests of several Hawaiian islands, including Oahu, Molokai, Maui, and Hawaii. This is a tiny, yellow, silk-spinning spider, with a body less than one-fifth of an inch long. The abdomens are sometimes marked with colorful patterns that look like a cartoon face, with two spots for eyes and a larger blotch for a mouth. The markings may be red, black, and/or white and arranged so the face may appear to be smiling, frowning, angry, foolish, or blasé. Happy face spiders that have no patterns on their abdomens are the most common, so the spiders that have markings may have evolved them as a way to trick birds accustomed to searching for unadorned yellow bodies. The female happy face spider is attentive to her offspring, which is unusual for spiders.

Further Reading:
Line, "Put on a Happy Face."

RICINULEI (tickspiders)

The hooded tickspiders are heavy-bodied and tiny, measuring an average of 0.3 inches long. There are only 25 known species.

Ricinoidae

Cryptocellus foedus (ricinuleid)

The ricinuleid is a tiny, rare relative of the spider found in the moist leaf litter of the rainforest floor and caves of Brazil. Its armored body is only 0.08 inches long and 0.04 inches wide, and its oval abdomen has a checkered pattern on the back, reminiscent of a turtle's carapace. It has four pairs of thick, segmented legs, the first pair of which is relatively short, and the second pair of which is longer than the rest. It is rare even in the few places where it has been found, and consequently little is known about this, or any of the other approximately 25, species of ricinuleid that have been identified—all of them in the dense forests of South America, Central America, and West Africa. The largest ricinuleids are less than 0.4 of an inch long, though their fossil record dates back to the Carboniferous Period, when they were quite common. They belong to one of the rarest orders of animals in the world. All ricinuleid species have extremely thick cuticles and hoods that cover their mouth parts; they are blind, and use their short forelegs to feel their way around and forage. They feed on other arthropods.

Further Reading:
Denny and Arias, *Insects of an Amazon Forest.*

Birds (*Aves*)

Warm-blooded vertebrates with feathers, primarily adapted for flight, birds are vital seed dispersers and pollinators in tropical forests. About 8,600 species are found, worldwide, and they all lay eggs that are nearly always incubated by the female or both parents.

Names of orders in this class are arranged in a traditional way of listing birds, known as the "Wetmore sequence," named after the U.S. ornithologist Alexander Wetmore. Under each order, families and species are listed alphabetically by their Latin names.

CASUARIIFORMES (emus, cassowaries)

Casuariidae (cassowaries)

Cassowaries arc a family of three species of huge, flightless birds that stand about five feet tall and weigh as much as 130 pounds. All species have shiny black, coarse, and shaggy plumage and a bare face and neck whose skin is blue, red, orange, or yellow. The cassowary head is crowned with a bony casque, used as a prow to push aside obstacles as it moves through the forest understory. Cassowaries have stout and powerful legs, and the innermost of their three toes has a long, sharp claw, used in defense. They are solitary birds that feed almost exclusively on fallen fruit. Males are smaller than females, and once the female has laid the eggs on the leaf litter, she has sole responsibility for caring for the several eggs and then the chicks. Cassowaries are valued for their meat and feathers and are often raised in captivity. All three species are found in Papua New Guinea, while the southern cassowary (*Casuarius casuarius*), which has a high, narrow, and curved casque and either one or two divided and brightly colored wattles, is also found in Australia's rainforests. It can be spotted at forest edges and at riverbanks and is a good swimmer, willing to fearlessly cross large rivers. On Papua New Guinea, cassowaries are important seed dispersers because the island lacks most of the large terrestrial mammals that do this important job in other rainforest regions.

Further Reading:
Beehler, Pratt, and Zimmerman, *Birds of New Guinea*.
Pizzey, *A Field Guide to the Birds of Australia*.
Wolfe and Prance, *Rainforests of the World*.

PELECANIFORMES (pelicans, boobies, cormorants, and allies)

Anhingidae (anhingas)

Anhingas is a small family, with just four species, widely dispersed over the Eastern and Western Hemispheres. Males have mostly glossy, greenish-black plumage; females are brown, gray, or white with black. They have long, sharply pointed bills, long necks and tails, but short and stout legs. About 35 inches long, they frequent tree-lined freshwater swamps, lakes, and slow-moving rivers. They feed on small fish, amphibians, and aquatic reptiles, spearing them with a sudden dart of their lancelike bills. Anhingas swim with only their narrow heads and snakelike necks out of the water. Their feathers are not entirely waterproof, so after a swim they perch in a tree with their wings and tails outstretched to dry. They build large nests of sticks in trees, often among the nests of other water birds, and line them with green leaves or moss. Both parents feed regurgitated food to the three to five hatchlings. When the young are alarmed, they drop into the water below their nests, later scrambling back, if they survive.

Further Reading:
Howell and Webb, *A Guide to the Birds of Mexico and Northern Central America.*
Stiles and Skutch, *A Guide to the Birds of Costa Rica.*

Anhinga anhinga

This New World species is found in the southeastern United States south to Ecuador, northern Argentina, and Uruguay. The northernmost breeding birds migrate south during the winter. *A. melanogaster* is found throughout eastern Asia south to Australia. *A. rufa* occurs in suitable habitat in most of Africa.

Further Reading:
Beehler, Pratt, and Zimmerman, *Birds of New Guinea.*
Howell and Webb, *A Guide to the Birds of Mexico and Northern Central America.*
Stiles and Skutch, *A Guide to the Birds of Costa Rica.*
Williams and Arlott, *Birds of East Africa.*

CICONIIFORMES (herons, storks, ibises, and allies)

Ciconiidae (storks)

Storks are large wading birds with long legs and long, thick, pointed bills and predominantly white plumage. The 17 species are found worldwide in temperate and tropical regions, but only three species are native to the New World. Males and females share parenting duties, building the nest, incubating the eggs for 28 to 36 days and feeding the chicks, which remain in the nest for more than 100 days in the larger species.

Jabiru mycteria (jabiru)

The jabiru is the largest flying bird in the Americas, at nearly five feet long. Its plumage is entirely white, its head and neck naked and black with a broad red band at the base of the neck, and its legs and bill are black. Jabirus are found singly or in small groups in freshwater marshes and lakes, where they forage slowly for prey, particularly mud eels. They are residents in southeast Mexico to northern Argentina and Uruguay. Their nests are huge platforms of sticks, which are often used and added to year after year,

and built high in trees. The birds, in spite of their size and an ungainly takeoff, fly gracefully, often soaring.

Further Reading:
de Schauensee and Phelps, *A Guide to the Birds of Venezuela.*
Howell and Webb, *A Guide to the Birds of Mexico and Northern Central America.*
Stiles and Skutch, *A Guide to the Birds of Costa Rica.*

Threskiornithidae (ibises and spoonbills)

Ibises and spoonbills are found on all continents, in both tropical and temperate zones. Most of the 33 species of *Threskiornithidae* are residents of marshes or shorelines, but some are birds of the forest or savanna.

Eudocimus ruber (scarlet ibis)

The scarlet ibis, true to its common name, has flamed-red plumage, with black at the tips of its wings. Like all ibises, the 23-inch scarlet has a long, thin, downward-curving bill and feeds on fish, insects, and invertebrates. Scarlet ibises are found in South American mangrove swamps, from Colombia to Brazil, in the Amazon's flooded forests, and in Trinidad. Their nests are bulky platforms of sticks in mangrove treetops; they often nest in colonies with other ibises and herons. Both parents share in building the nest, incubation, and caring for the young.

Further Reading:
de Schauensee and Phelps, *A Guide to the Birds of Venezuela.*
Raffaele, *A Guide to the Birds of the West Indies.*
Sick, *Birds in Brazil: A Natural History.*
Stiles and Skutch, *A Guide to the Birds of Costa Rica.*

FALCONIFORMES (diurnal birds of prey)

Cathartidae (American vultures)

American vultures spend much of their time soaring the skies in search of carrion, their principal food. The seven species of American vultures are found only in the New World. Instead of building nests, they lay their eggs on the ground in a protected place, on the floor of a cave, or in a cavity of some sort, such as a hollow tree stump. Except for the king vulture, they are all black with bare heads and necks.

Further Reading:
Stiles and Skutch, *A Guide to the Birds of Costa Rica.*

Sarcoramphus papa (king vulture)

King vultures are standouts from other black-and-gray-feathered vultures because of their mostly creamy-white plumage. Their rump and tail feathers are black, and they have a blue-gray ruff about their necks. Their heads and necks are rainbow-hued, with splashes of orange, yellow, and blue, while their wattles, flaps of skin that drape over their beaks, are red-orange. Found from Mexico to central Argentina and on Trinidad, they perch in the uppermost branches of emergent canopy trees and soar high up in the tropical skies. Unlike other vultures, they don't have great senses of smell to guide them to carrion. Instead, they watch the movements below of other vultures, whose excited actions give away the location of food. Then, the 30-inch monarch drops from its lofty heights like a bullet, while all other vultures make way; hence, it earns the name "king."

Further Reading:
de Schauensee and Phelps, *A Guide to the Birds of Venezuela.*
Stiles and Skutch, *A Guide to the Birds of Costa Rica.*

Accipitridae

This family includes 205 species of vultures, harriers, harrier-eagles, sparrowhawks, goshawks, buzzards, eagles, and kites found worldwide. They are all diurnal raptors and strong fliers, with powerful feet and hooked claws.

Further Reading:
de Schauensee and Phelps, *A Guide to the Birds of Venezuela.*
Stiles and Skutch, *A Guide to the Birds of Costa Rica.*

Harpia harpia (harpy eagle)

The harpy eagle, with its seven-foot wingspan, legs as big around as a child's wrist, and nine-inch-long feet fitted with talons the size of grizzly bear claws, is the most powerful eagle in the world. These raptors were once common in forests from southeastern Mexico to northern Argentina and southern Brazil, but now they are extremely rare throughout their range, because of deforestation and shooting. Human beings are both afraid and suspicious of powerful animal species like the harpy eagle. *H. harpia* mainly preys on arboreal mammals, particularly monkeys and sloths, which it can snatch from tree branches in flight. Females are about one-third larger than males and can carry off heftier prey, but males are the more agile fliers. Working together, the pair hunts quite efficiently. Females lay one or two eggs, but only one chick will survive. Parents care for their offspring for 10 months or more. Fledglings don't even attempt to fly until they are three months old. Because raising chicks requires so much time and energy, it's not surprising that harpy eagles produce eggs only every two or three years.

The Peregrine Fund, a conservation group in the United States, has managed to hatch harpy eagle eggs at its research center in Idaho. The chicks were hand-raised with puppets, so they wouldn't become accustomed to people, and in 1998, four six-month-old harpies were released in the rainforest of Soberanía National Park in Panama.

Further Reading:
Jukofsky, "New Chick in Idaho."
The Peregrine Fund, USA <http://www.peregrinefund. org>.
Rettig, "Remote World of the Harpy Eagle."

Eutriorchis astur (Madagascar serpent-eagle)

The Madagascar serpent-eagle is found only in undisturbed, primary rainforest on the island of Madagascar. It is one of the rarest birds in the world, since so little rainforest remains on the island. In fact, until 1993, it was thought to be extinct. Biologists with the Peregrine Fund, a U.S. conservation group, spotted the eagle, and four years later, one of them discovered an active nest. Thanks to the discovery, in 1997 the government created a new national park to protect about 745,000 acres of rainforest habitat for the raptor.

Further Reading:
Langrand, *Guide to the Birds of Madagascar.*
The Peregrine Fund, USA "First Madagascar Serpent-Eagle Nest Discovered" <http://www.peregrinefund.org>.

Harpyopsis novaeguineae (New Guinea eagle)

The New Guinea eagle hunts small mammals—including tree kangaroos and wallabies—in beech forests that line lower mountain slopes on New Guinea. It either swoops down and plucks mammals from the ground or extracts them with its claws from tree cavities. The 32-inch eagle in turn is hunted by people for its tail and wing feathers, still used in traditional headdresses. This practice has greatly depleted its population. The New Guinea eagle has broad, banded wings and a long, banded tail. Its underparts are white and unmarked, and it bears a short, broad crest.

Further Reading:
Bald Eagle Information <http://www.baldeagleinfo.com/eagle/eagle6.html>.
Beehler, Pratt, and Zimmerman, *Birds of New Guinea*.

Pithecophaga jefferyi (Philippine eagle)

The Philippine eagle is also called by a less charismatic name, the monkey-eating eagle. Probably fewer than 200 of these powerful eagles remain on the largely deforested Philippine islands. They have a wingspan of nearly seven feet and prey on a variety of forest animals, including monkeys. A pair of Philippine eagles may have a territory as large as 23,000 acres. The eagles, the primary predator of the Philippines, prefer to build nests among epiphytic ferns or emergent forest trees on steep mountain slopes. They lay just one egg, rearing one offspring every two years. Deforestation, hunting, and trapping are the main reasons for their decline, although the raptors have always been rare. Even before forest clearing, the population probably did not exceed 6,000.

The Philippine Eagle Conservation Program, which is based in Mindanao, is trying to breed captive eagles, but so far, no young have hatched. The program has also started an "Adopt-A-Nest" campaign; the organization pays local people for every occupied nest they find.

Further Reading:
Bald Eagle Information <http://www.baldeagleinfo.com/eagle/eagle6.html>.
World Wide Fund for Nature, Switzerland <http://www.panda.org / resources / publications / species / underthreat/ philippineeagle.htm>.

Stephanoaetus coronatus (crowned eagle)

The crowned eagle is a massive, 34-inch raptor, with black plumage on its upper parts, and blotched black, orange, and white plumage underneath. It has a pronounced, rounded crest. The crowned eagle is a rare bird, most commonly found in the highland forests of Kenya and the Democratic Republic of Congo. Its presence depends on that of its principal food source, monkeys. When a crowned eagle approaches, monkeys will hurl themselves into the air and drop 60 feet to the ground in a frantic attempt to escape the feared predator.

Further Reading:
Williams and Arlott, *Birds of East Africa*.
Wolfe and Prance, *Rainforests of the World*.

GALLIFORMES (chickenlike birds)

Cracidae

These 44 species of turkeylike game birds generally inhabit warm, humid forests in the Americas. They are largely arboreal, plucking fruits and leaves from trees and vines, or they might descend to the forest floor to feast on fallen fruit; a few species mostly forage on the ground. They have long necks, broad tails, and long, looped tracheas that allow them to call loudly, if not particularly sweetly. Many species have crests, brightly colored wattles, or other head and bill adornments. Females lay only two to four whitish eggs that have rough, sometimes pitted, shells. The small clutch size and slow reproduction mean their populations are less able to withstand hunting than are temperate game birds, which have larger broods. Hunting plus deforestation are threatening the survival of several species.

Further Reading:
Howell and Webb, *A Guide to the Birds of Mexico and Northern Central America*.
Stiles and Skutch, *A Guide to the Birds of Costa Rica*.

Crax rubra (great curassow)

The great curassow is a robust bird, 36 inches long. Males and females have a prominent crest of forward-curling feathers and long tails. Males are glossy black with white bellies. The bases of their bills are yellow, adorned with a yellow, bulbous knob. Females have a variety of colorings, black or chestnut colored with bars of black or white, with their heads and crest barred with black and white. Their range is southern Mexico to western Ecuador, but they are rarely found outside parks. They forage on the ground, picking up fallen fruits or scratching for small animals. They travel in pairs or small groups and tend to run rather than fly when disturbed. Curassows roost in treetops, where they build small nests of leaves and twigs in a fork or depression.

Further Reading:
Howell and Webb, *A Guide to the Birds of Mexico and Northern Central America*.
Stiles and Skutch, *A Guide to the Birds of Costa Rica*.

Ortalis cinereiceps (gray-headed chachalaca)

The gray-headed chachalaca is about 20 inches long, with a small head and long neck and tail. Its head and upper neck are gray, while the rest of its body is olive and brown. The featherless sides of its throat are red. The birds live in forest underbrush, often along streams, and walk rather gracefully along tree branches in search of fruit and foliage. Their range is eastern Honduras to northwest Colombia, but they are severely hunted throughout. They are sociable birds, living in groups of six to a dozen or more.

Further Reading:
Skutch, *Birds of Tropical America.*
Stiles and Skutch, *A Guide to the Birds of Costa Rica.*

Penelope purpurascens (crested guan)

The crested guan is 34 inches long, with a large, bushy, erect crest and a red dewlap. Their plumage is mostly dark olive-brown, with white spots on their necks and chests, and their wings are a shiny green. Their facial skin is blue-gray, and their eyes and legs are red. They live in pairs or small family groups in the forests of southern Mexico to western Ecuador and northern Venezuela, where they forage in the canopy for fruits or new foliage. In spite of their size, they can walk nimbly along tree branches. Hunting as well as deforestation have affected their populations, so that the crested guan is rarely seen outside protected forests.

Further Reading:
Howell and Webb, *A Guide to the Birds of Mexico and Northern Central America.*
Stiles and Skutch, *A Guide to the Birds of Costa Rica.*

Phasianidae (pheasants and quail)

Most of the 164 species of pheasants and quail are found in the Old World, from Europe to Africa and Australia. They are chickenlike birds with stout bills and short, plump legs. Most family members build nests on the ground in rough bowls made of available vegetation.

Further Reading:
Stiles and Skutch, *A Guide to the Birds of Costa Rica.*

Meleagris ocellata (ocellated turkey)

The ocellated turkey is endemic to the Yucatán Peninsula of Mexico and the Petén area of Guatemala. It is one of only two species of turkey in the world. The other is the North American wild turkey (*M. gallopavo*), once common in U.S. forests and now enjoying a comeback, thanks to restoration programs sponsored by hunters. Male ocellated turkeys are about three feet long and look similar to the North American wild turkey, except that their coloring is much more vibrant. Their featherless heads and upper necks are bright blue with orange warts, and they have a deep red circle around their eyes. They have an inflatable forehead wattle that hangs down over their black bill and an inflatable horn on their crown that they puff up and display when courting. Their metallic blue-black feathers are tipped in blue, with their upper wings a metallic green, and they sport copper-colored feathers on their wings and tails. The female's plumage is a bit duller than the male's, more green than bronze, but she also has orange-red warts on her head and neck. Ocellated turkeys are significantly less plump than their North American relatives; males weigh about 11 pounds. They are found in forests and clearings, abandoned milpas, or farm plots, and brushy woodlands, usually in small groups, browsing for seeds, berries, leaves, and insects.

Further Reading:
Howell and Webb, *A Guide to the Birds of Mexico and Northern Central America.*
National Wild Turkey Federation, USA <http://www.nwtf.org/hunting/ocellated.html>.

GRUIFORMES (cranes, rails, and allies)

Eurypygidae

This is one of four small families whose ancestors came from the Southern Hemisphere and were divided from one another by continental drifts. The other three families are in South America and New Caledonia.

Eurypyga helias (sunbittern)

The sunbittern is the only member of its family. The 19-inch bird is mostly gray, black, and white overall, but when it spreads its wings it reveals a striking pattern of barred gray and black with tawny and black-edged chestnut patches. Its opened black and white speckled tail bears a similar pattern. The sunbittern has a long bill and legs and a large head held on a slender neck. Found from southern Mexico to northwestern Peru and the Amazon in Brazil, it inhabits fast-flowing and rocky forest streams or lakes. Hopping from river rock to rock or wading in the water, it plucks out small frogs, tadpoles, crayfish, small fish, and insects. It flies with deep heavy flaps and a long glide.

Further Reading:
de Schauensee and Phelps, *A Guide to the Birds of Venezuela.*
Sick, *Birds in Brazil: A Natural History.*
Stiles and Skutch, *A Guide to the Birds of Costa Rica.*

Rhynochetidae

The kagu (*Rhynochetos jubatus*) is the sole member of its family. It has no close relatives and is found only on the South Pacific island New Caledonia. There it occupies a variety of habitats, from dense scrub to wet forest, although it is most populous in undisturbed habitat at higher elevations. The kagu is a long-legged, gray, 23-inch-long bird with reddish orange bill and legs, and a crown of feathers on its large head. It is a ground bird that can run rapidly through the forest underbrush and is barely able to fly. It forages for worms, snails, and arthropods. Its call is a piercing cry, almost like a scream, most often heard at night, although it is not nocturnal. Once a fairly common bird, it is on the edge of extinction, due to deforestation and the introduction of exotic predators—cats, dogs, rats, and pigs. Like many islands, until humans moved in and introduced foreign species, New Caledonia originally had no predatory mammals. Political unrest on the island has benefited the bird, since many Europeans have left, and mining, logging, and hunting have abated. Local conservationists launched a kagu captive-breeding program in the mid-1980s, and chicks hatched in captivity were released into forests, while predators were captured and removed from kagu habitat. There are now an estimated 650 kagus in New Caledonia.

Further Reading:
World Wide Fund for Nature, Switzerland <http://www.panda.org / resources / publications / species / underthreat / kagu.htm>.

CHARADRIIFORMES (shorebirds, gulls, auks, and allies)

Jacanidae (jacanas)

Jacanas are found in the tropics in both the Eastern and Western Hemispheres. All eight species inhabit ponds, marshes, lakes, and lagoons and have unusually long toes and nails. These long appendages allow them to broadly distribute their weight so that they can dash over water surfaces and walk on top of aquatic vegetation. They eat the many insects, snails, seeds, and small fish that live in floating water plants.

Further Reading:
Goulding, *The Flooded Forest.*
Stiles and Skutch, *A Guide to the Birds of Costa Rica.*

Jacana spinosa (Northern jacana)

The Northern jacana is one of at least two jacana species that are polyandrous—the female will accept more than one male as her mate. It is the female that defends her territory during breeding season. The males construct a sloppy nest, a pile of leaves and stems that form a raised platform atop aquatic plants, such as the mammoth leaves of the Amazon's giant water lily (*Victoria amazonica*). The female lays four buff-colored eggs scribbled with black, crisscrossing lines, and while the male incubates them, tucking his wings under them to shield them from water, she drops more eggs in the nests of her other mates. Downy chicks hatch after about 23 days, and it is the male that protects them, even taking them under his wing to carry them to safety. The chicks leave their fathers after about three months. *Actophilornis africanus* (African jacana), found in the tropical wetlands of Africa, is also polyandrous.

Further Reading:
Goulding, *The Flooded Forest.*
Stiles and Skutch, *A Guide to the Birds of Costa Rica.*

Laridae

The family includes 80 or so species of gulls and terns, web-footed birds found near water. They breed in monogamous pairs and nearly always nest in colonies that often number many thousands of couples.

Further Reading:
Stiles and Skutch, *A Guide to the Birds of Costa Rica.*

Gygis alba (fairy tern)

The fairy tern is a pure white, fairly small bird, about 12 inches long, with long wings. It lives on tropical islands of the Pacific, Indian, and south Atlantic Oceans, where it has no natural enemies. Its unnatural enemies, however, are threatening this ethereal and tame bird. On the Seychelles, for example, barn owls were brought in to rid the islands of rats, which had sneaked onto the islands from French ships. The barn owls wiped out the rats but also the fairy terns. Imported cats and dogs that prey on nesting birds, plus development on the coasts where the birds feed and nest, have brought the fairy tern close to extinction on human-inhabited islands.

Unlike the feathers of gulls and other terns, the powdery plumage of the fairy tern is not waterproof, so it has evolved the ability to dive-bomb and snatch a small fish or squid from the ocean's surface without ever dampening a feather. It can also gather a crowd of fish at once in its awl-shaped bill, without dropping any of them. Rather than build a nest, the fairy tern lays her single, creamy-white-and-brown spotted egg in particularly precarious sites—the crotch of a tree, a hollow in a branch, or perched atop a palm tree stump. The male and female take turns incubating the egg. When the fluffy chick is born, it instinctively clasps firmly onto its birthplace with long-toed feet.

Further Reading:
Beehler, Pratt and Zimmerman, *Birds of New Guinea.*
Bruemmer, "Celestial Visions."

Rynchopidae (skimmers)

Skimmers are found along the coasts of North and South America and in tropical Africa and Asia. They include just three species, all in the same genus.

Further Reading:
Stiles and Skutch, *A Guide to the Birds of Costa Rica.*

Rynchops niger (black skimmer)

The black skimmer is a shorebird, but its hunting grounds include the flooded forest of the Amazon River and its tributaries, along with other rivers and estuaries from the southwestern United States throughout South America. This 19-inch, long-winged bird has a black body, with a white face and chest. The lower mandible of its orange and black bill is longer than the upper one, a unique configuration. As the bird flies low along calm waters, slowly beating its wings to avoid wetting them, it drops its longer lower bill into the water to quickly snap up fish and crustaceans. The African skimmer (*R. flavirostris*) is found along the rivers and lakes of much of Africa. The Indian skimmer (*R. albicollis*) is fairly common along the Ganges River system.

Further Reading:
de Schauensee and Phelps, *A Guide to the Birds of Venezuela.*
Goulding, *The Flooded Forest.*
Serle and Morel, *Birds of West Africa.*
Stiles and Skutch, *A Guide to the Birds of Costa Rica.*
Williams and Arlott, *Birds of East Africa.*

PSITTACIFORMES (parrots and macaws)

Psittacidae (parrots and macaws)

Parrots and macaws are monogamous, highly social birds found in almost every tropical forest in the world, usually flying in large and raucous flocks. There are 337 species in the parrot order and family, which includes macaws, cockatoos, lories, and lorikeets. The tiniest are the 3-inch pygmy parrots, and the largest, the macaws, are about 40 inches. Mid-size species of the New World are usually referred to as "parrots." Green is the predominant parrot color, making them so difficult to spot in the treetops, but many species sport rainbow hues. Parrots' bills are adapted for prying open hard nuts, chomping on fruits, and grinding seeds. Their legs are short and strong, and, with their bills, help them clamber among branches. They also use their feet to lift food to their mouths, a rare habit among birds.

Some species build nests, but most lay eggs in tree hollows. Often a parrot pair will begin preparing a nest site months before they actually nest. In most species, the female alone incubates the white eggs. The young are naked and blind when they hatch, and at first, just the female feeds them regurgitated food; within a few days the male joins her in feeding them directly. Smaller hatchlings remain in the nest for three to four weeks, while large macaw chicks need three or four months to develop flight feathers.

The 53 species of lories and lorikeets are found in Australia, Papua New Guinea, and several islands of the South Pacific, where they are most often seen in flocks, shrieking as they fly swiftly over the forest canopy. Their plumage is particularly intense, ranging from stunning scarlets and oranges to glossy purples and emerald greens. Males and females look exactly alike, so only surgery or DNA testing allows human beings to distinguish the two sexes. They have evolved brush-tipped tongues to allow them to easily feed on pollen, although they also sip nectar and eat soft, ripe fruits, insects, and insect larvae. Their beaks are narrower and have pointier tips than other parrot species, since they don't need to grind their soft foods before swallowing. Some lory species are primary pollinators of particular plant species, filling a niche that bats and bees hold in other tropical regions. They are strong flyers, able to island-hop to forage, roost, or nest. Like other parrots, they nest in tree cavities. At least 13 species are endangered, due to deforestation; introduced pest species, such as rats, mongooses, mosquitoes, and domestic dogs and cats; and their being smuggled north and sold as pets.

Cockatoos, found in New Guinea, the South Pacific, and Australia, are large parrots with crests and all-white or all-black plumage. Australia is home to 11 of the 17 species. They have massive beaks, which they use to crack open nuts, dig up roots, or pry food from wood, aided by a worm-like tongue. Like most parrot species, they nest in holes in trees.

Pygmy parrots of New Guinea and several Pacific islands are the smallest in the family, just over three inches long. They feed on lichens and fungi.

The colorful fig-parrots, or lorilets, are just a bit larger, and are found in Australia, New Guinea, and nearby islands. Lorilets are fruit eaters and are particularly fond of native figs, but also dine on berries, seeds, and nectar.

Principally because of deforestation and the illegal pet trade, one in four parrot species is in danger of extinction.

Further Reading:

Beehler, Pratt, and Zimmerman, *Birds of New Guinea.*
Juniper and Parr, *Parrots: A Guide to Parrots of the World.*
Stiles and Skutch, *A Guide to the Birds of Costa Rica.*

Amazona imperialis (imperial Amazon)

The imperial Amazon is endemic to the island of Dominica, where it is critically endangered. Perhaps 100 birds survive, but it is difficult to know because this is a particularly secretive parrot. *A. imperialis* has a dark maroon head, with green upper parts, edged with pale gray, a dark violet patch on the back of its neck, green and red wings, and a deep violet belly. The 18- to 19-inch bird, known locally as *sisserou*, does not fly in noisy flocks like most parrots, but stays in family groups of two to four birds. The small family unit flies quietly beneath the canopy, feeding on fruits. Deforestation, hunting, the pet trade and hurricanes are responsible for the parrot's decline, but the people of Dominica have made great efforts to control the first three.

Further Reading:

Juniper and Parr, *Parrots: A Guide to Parrots of the World.*
Raffaele, *A Guide to the Birds of the West Indies.*
World Wide Fund for Nature, Switzerland <http://www.
    panda.org/resources/publications/species/underthreat/
    imperialamazon.htm>.

Amazona versicolor (St. Lucia's parrot)

St. Lucia's parrot, found only on the tiny island for which it is named, is making a comeback. Although the bird is highly endangered, its chances of survival have been greatly improved by determined conservationists. Called *jacquot* by islanders, the parrot has a blue face, emerald green wings, and scarlet feathers about its throat and neck. When *jacquot* was chosen as the country's national symbol, after St. Lucia's independence from Britain in 1979, the new government set aside a 1,900-acre forest reserve, where nearly all the remaining parrots survive.

An island-wide educational campaign led by the RARE Center for Tropical Conservation, a U.S. nonprofit group, has turned the multicolored parrot into a hero. Today, its population seems to have stabilized at 500 birds, and news of the striking Amazon parrot's survival brings more bird-loving tourists to the island. This boost to the island's economy also helps guarantee *jacquot*'s future, so perhaps it will not meet the fate of 3 of the 12 species of Amazon parrots originally found in the Caribbean. These three, endemic to Martinique, Guadeloupe, and Grenada, are now extinct, due to hunting and loss of habitat.

Further Reading:

Blake, "Fighting For a Rare Bird."
Juniper and Parr, *Parrots: A Guide to Parrots of the World.*

Amazona vittata (Puerto Rican parrot)

The Puerto Rican parrot once numbered in the hundreds of thousands and lived throughout Puerto Rico, back when the island was covered in forests. Today the population is restricted to the wet forests of the Luquillo Mountains. A low of 13 parrots was recorded in 1975, and after a period of intense conservation and management, the wild population had increased to 48 birds in 1989. But Hurricane Hugo swept across the island that year and reduced the population to 22 birds. By the mid-1990s, the wild population had recovered to 44 parrots, with more in captivity. The Puerto Rican parrot is about 12 inches long, has green feathers with wing feathers in two shades of blue, a red forehead, and a white ring around its eye.

Beyond deforestation, the bird's endangered status is also due to the arboreal brown rat, the pearly-eyed thrasher (*Margarops fuscatus*), which preys on untended eggs and chicks, a warble fly (of the *Muscidae* family) that infests nestlings, and predation of nestlings by the red-tailed hawk (*Buteo jamaicensis*).

Further Reading:
Juniper and Parr, *Parrots: A Guide to Parrots of the World.*
Raffaele, *A Guide to the Birds of the West Indies.*
World Wide Fund for Nature, Switzerland <http://www.panda.org/resources/publications/species/underthreat/puerto_rican.htm>.

Anodorhynchus and *Ara spp.* (macaws)

Macaws are the largest parrots in the world, with some species as long as 40 inches. They have particularly strong beaks, long tails, and are festooned with brightly colored feathers of red, yellow, green, and/or blue. Like virtually all parrots, they are extremely social birds, gathering in noisy, sizeable groups. They eat seeds, nuts, berries, and leaves and can ingest fruits that contain chemicals toxic enough to kill other animals. Unlike most tropical birds, they prefer a fruit's seed, which they crack with their hooked beaks; their favorites include seeds of mahogany (*Swietenia*), kapok (*Ceiba pentandra*), and rubber trees (*Hevea brasiliensis*). They also eat large amounts of clay, apparently because mineral-rich clays neutralize plant poisons.

Macaws, which seem to mate for life, nest from January through April, usually in the holes of dead or decaying canopy trees. The birds need large expanses of rainforest to survive, since on average, only 1 tree per 63 acres of forest already has a hole large enough for a macaw nest or is sufficiently decayed so that a macaw can excavate a sizeable cavity in the tree trunk. Further, macaws have low reproduction rates; they lay only two eggs, and not always every year. Often, only one of the hatchlings survives. Male and female share in the feeding of fledglings, which stay with their parents for two or three years. The small family members interact constantly with each other, a trait unusual in birds. Mated adults preen each other and their offspring for hours, plucking off lice and ticks. Although they don't socialize much with other macaws, they chatter nonstop among themselves. Because they have few natural predators, macaws that survive their first year probably can live some 35 years in the wild, unusually long for a bird. In captivity, they may live up to 70 years.

Although most macaw species are protected, fledglings are frequently stolen and sold as pets. Their beauty jeopardizes their existence, since admiring human beings are willing to pay thousands of dollars to watch a macaw in captivity. Of the 18 species of macaws, 8 are endangered.

Further Reading:
Munn, "Winged Rainbows: Macaws."
Stiles and Skutch, *A Guide to the Birds of Costa Rica.*

Anodorhynchus hyacinthinus (hyacinth macaw)

The hyacinth macaw is the largest parrot, about 40 inches long. It has deep cobalt blue plumage and inhabits the fringes of the Amazon rainforest in Paraguay, Bolivia, and Brazil, where its population once numbered between 50,000 and 100,000 birds. The relentless demand for these birds as pets diminished the species' numbers to fewer than 3,000. More hyacinth macaws live in captivity than in the wild.

Further Reading:
Juniper and Parr, *Parrots: A Guide to Parrots of the World.*
Sick, *Birds in Brazil: A Natural History.*
Wolfe and Prance, *Rainforests of the World.* ·

Anodorhynchus leari (Lear's macaw)

Lear's macaws inhabit the very dry and thorny forest, called the "caatinga," of northeastern Brazil, in the state of Bahia. "Caatinga" means "white forest"; the trees in this ecosystem lose their leaves during the dry season, exposing the light gray—nearly white—tree bark and barren branches. In 1978, when Lear's macaw was first identified, the population stood at just 65 birds. A second population was located in 1995, and today the population is estimated at fewer than 150 birds, making this magnificent, three-foot-long parrot one of the world's most endangered species. Lear's macaw has deep, rich, indigo-blue plumage, so striking that the bird is coveted by human beings who like to collect rare and beautiful birds. Deforestation, which eliminates their food source, and illegal capture of the birds for the international pet trade are the two principal reasons the population is so low. Lear's macaws nest in burrows in sandstone cliffs and roost on cliff faces. Their primary food is the nine-inch nut of the licuri palm tree. The macaws use their heavy-duty beaks to crack open the nut shells. The birds fly anywhere from 12 to 60 miles to find their favorite food. They are swift and strong flyers, averaging speeds of 30 miles per hour. Unfortunately, farmers use licuri palm leaves to feed their cattle, which also graze on palm seedlings, so the trees are becoming scarce.

The Brazilian government has begun a major campaign against poaching and launched a Committee for the Preservation of Lear's Macaw. For more information, contact the committee: <aves@zoologico.com.br>. The best place to see Lear's macaws in the United States is Busch Gardens in Tampa Bay, Florida, the only zoological park in the world to have successfully bred Lear's macaws in captivity. Busch Gardens also supports macaw field research and educational activities in Bahia.

Further Reading:
National Wildlife Federation, USA <http://www.nwf.org/ wildalive/macaw/sciencefacts.html>.

Ara ambigua (great green or Buffon's macaw)

Great green macaws are yellowish-green, with scarlet foreheads, deep blue on the top of their wings, and red on the top of their tails, tipped with blue. They are found from eastern Honduras to western Colombia, with another small population in western Ecuador. They are under pressure throughout their range from deforestation and the robbing of young birds from their nests for the illegal pet trade. They mainly nest in cavities of *Dipteryx panamensis*, or almendro trees, and feed almost exclusively on this tree's fruit. Unfortunately, this leguminous tree is being increasingly logged.

The great green macaw is one of just two macaws found in Costa Rica, but even in this conservation-conscious country, the government has been unable to prevent the powerful logging industry from cutting down almendro trees. Just 30 breeding pairs of the brilliant green macaws remain in Costa Rica. Conservationists are trying to raise funds to buy land in great green macaw territory, in an attempt to save the parrot.

Further Reading:
International Aviculturists Society, USA <http://www. funnyfarmexotics.com/IAS>.
Juniper and Parr, *Parrots: A Guide to Parrots of the World*.
Stiles and Skutch, *A Guide to the Birds of Costa Rica*.
Vida Foundation, Costa Rica <http://www.vida.org/ contenidos/programas/lapa/conserv.htm>.

Ara macao (scarlet macaw)

The scarlet macaw is the best-known macaw and is found from southern Mexico to Peru, Bolivia, eastern Brazil and on the island of Trinidad. Its feathers are mainly a bright red along its 33-inch body, with sunny yellow and sky blue feathers on its lower wings. It has virtually no feathers on its creamy white face. In countries like Costa Rica and Belize, scarlet ma-

caws are valuable as tourist attractions: Most visitors hope to see these rainbow-colored birds and are willing to pay for the privilege. A wildlife biologist in Costa Rica found that one macaw in Carara Biological Reserve on the country's Pacific coast is worth nearly $500,000 over its 30-year lifetime, in terms of how much money tourists spend to see it. Young scarlet macaws in Carara are frequently stolen from their nests, and only 314 macaws remain in the reserve. Poachers can sell the birds in Costa Rica for about $200 on the black market. Baby birds smuggled to the United States can be sold for as much as $4,000. By showing that each of the reserve's macaws attract about $14,000 annually in tourist dollars, biologists hope to convince the government to do a better job at protecting the birds.

Further Reading:
Juniper and Parr, *Parrots: A Guide to Parrots of the World*.
Stiles and Skutch, *A Guide to the Birds of Costa Rica*.

Cacatua haematuropygia (Philippine cockatoo)

The Philippine cockatoo is all white, except for its crest, wings, and tail, which are shaded with yellow near the ears and wings and reddish pink under the tail. Fifty years ago it was a common parrot in its original range throughout the Philippines, but today it is found only in a few areas, mainly in lowland forests, including mangroves. Its total population is estimated to be between 1,000 and 4,000 birds and falling further, due to deforestation and trapping for the illegal pet trade. The latter is the more severe threat; young cockatoos are stolen from every accessible nest. Its voice is a raucous "awwk" and grating screams, and it can also mimic the human voice, another reason—along with its beauty—that the cockatoo is a popular pet. It reaches 12 inches in length and feeds on seeds, nuts, and berries.

Further Reading:
Gonzales and Rees, *Birds of the Philippines*.
Juniper and Parr, *Parrots: A Guide to Parrots of the World*.
World Wide Fund for Nature, Switzerland <http://www. panda.org / resources / publications / species / underthreat / ph-cockatoo.htm>.

Cacatua moluccensis (salmon-crested cockatoo)

The salmon-crested cockatoo is endemic to Indonesia, where today it is found only in Seram, particularly in Manusela National Park, which covers about 10 percent of the island. A 20-inch parrot, its plumage is a whitish salmon-pink, with deep pink in its underlying crest feathers and under its wings. Its tail feathers are orangish-yellow and pink at their base. These cockatoos are often solitary, or live with a mate, or sometimes in a small group. They are powerful flyers, with rapid, shallow wing beats interspersed with gliding. They fly slowly and unusually low for a parrot. In addition to eating seeds, nuts, and fruits, they use their strong bills to chew through the outer layers of young coconuts until they reach the milk and pulp. Deforestation and extensive capture for the illegal pet trade have greatly depleted their population; perhaps 2,000 birds remain. In the mid-1980s, and before export was made illegal in 1987, more than 5,000 birds per year entered importing countries. Though illegal now, trapping continues. Poachers use decoys to lure the cockatoos to traps; they are usually transported to Singapore by boat.

Further Reading:
World Wide Fund for Nature, Switzerland <http://www. panda.org / resources / publications / species / underthreat / ph-cockatoo.htm>.

Cacatua leadbeateri (pink cockatoo)

The pink cockatoo is probably the most beautiful of the cockatoos, with white plumage that is lightly washed with pale pink. It has a forward-sweeping crest that is white when closed but when spread reveals bands of deep scarlet and yellow. A rare resident of Australia, it is about 15 inches long and has a peculiar call that sounds like a quavering, falsetto cry.

Further Reading:
Pizzey, *A Field Guide to the Birds of Australia.*

Cyanopsitta spixii (Spix's or little blue macaw)

Spix's or the little blue macaw has become a symbol of tropical forest and wildlife conservation. Only one of these 20-inch parrots with gray-blue and cobalt blue plumage survives in the wild, in its natural habitat of gallery forest in northern Bahia, Brazil. The dry woodland forest is primarily comprised of caraiba trees (*Tabebuia caraiba*), which once provided Spix's macaws with perches and nesting holes, used year after year.

Although the species was probably always rare and confined to this region of Brazil, deforestation of its forest home sealed its doom. Settlers began clearing forest in the area 300 years ago; only a fraction of the woodlands remain. Added to this is the impact that the illegal bird trade has had on the Spix's macaw since the early 1970s—many unscrupulous people

are willing to pay thousands of dollars just to have a rare macaw in a cage. There are about 30 birds held in captivity by the Brazilian government's Permanent Committee for the Recovery of the Spix's Macaw. The surviving wild bird is a male. In 1995, a wild-born female held in captivity was released near the male's home tree. The male macaw had previously hooked up with a female Illiger's macaw, but their eggs were always infertile. The released female Spix's macaw joined the couple, and the threesome began flying together. But the female Spix's disappeared after seven weeks, the probable victim of a collision with an electrical transmission wire. In 1999, biologists replaced the odd couple's infertile eggs with wild Illiger's macaw fledglings, which the macaws raised as their own, feeding them and teaching them to fly. Three months later, the young birds left the nest, having developed voices identical to their foster father's. The next experiment will be to put Spix's macaw fledglings hatched in captivity in the nest hole. But conservationists will have to move fast. The male Spix's macaw is about 15 years old. Although the species can reach 35 years in captivity, no one is sure how long they can live in the wild.

Meanwhile, the small pueblo of Curaça, near the surviving male's forest home, has been transformed by all the attention. The project has paid for a new school, and villagers are proud of being the town that holds the world's only Spix's macaw. They report to biologists of the bird's activities and whereabouts.

Further Reading:
World Wide Fund for Nature, Switzerland <http://www. panda.org/resources/publications/species/underthreat/ spix_macaw.htm>.

Eos histrio (red and blue lory)

The red and blue lory is rapidly disappearing. Fewer than 2,000 survive, on the Indonesian islands of Sangihe and Talaud. Striking parrots with deep red and blue plumage, they suddenly were in great demand by the international pet trade, and traders wiped out about one-third of the population in the early 1990s. Since by the late 1980s virtually all of Sangihe had been converted to coconut and nutmeg plantations, the red and blue lories have nearly disappeared from that island. Conservationists have started an environmental education campaign on the islands, distributing posters and booklets to encourage people not to harm the red and blue lory.

Further Reading:
Lori Journal International, Netherlands <http://www.parrotsociety.org.au/clubs/lji/articles/ljiart01.htm>.

Probosciger aterrimus (palm cockatoo)

The palm cockatoo is the largest cockatoo, at about 25 inches long. The male's beak is the largest of all the *Psittaciformes*. Its plumage is dark gray with a long, black crest, and its facial skin is orangish pink, turning scarlet when excited. The palm cockatoo is a slow flyer that perches in emergent trees above the rainforests of New Guinea and Australia. It nests in hollow tree trunks, laying one white egg on a layer of splintered twigs. Its common name derives from its favorite foods, the nuts of palm trees.

Further Reading:
Beehler, Pratt, and Zimmerman, *Birds of New Guinea*.
Pizzey, *A Field Guide to the Birds of Australia*.

Psittacus erithacus (gray parrot)

The gray parrot is gray all over except for its bright scarlet tail. The 12-inch parrot flies in huge and noisy flocks, passing high overhead at dawn and dusk as it commutes from its roosting to its feeding grounds. Found in Sierra Leone, Gabon, and east to Kenya and Tanzania, it feeds on fruiting trees, particularly the African oil palm. Often hundreds, if not thousands, of gray parrots roost together at night.

Although they remain rather abundant, they are unfortunately a popular pet, since they have an uncanny ability to mimic human speech. In Nigeria, illegal trapping and smuggling of the bird, mainly to Europe and particularly to Scotland, has nearly wiped out a once healthy population. One parrot may fetch $500 on the black market; trappers are paid about $10. Recent clashes between out-of-town trappers and residents of the village of Ikodi, Nigeria, have been violent. Villagers collect and sell fallen gray parrot feathers, which are used in folk medicine and as adornments, so the loss of the birds threatens their livelihoods. In Ikodi, killing or poaching the birds or felling a nesting tree is forbidden.

Further Reading:
Raufu, "Smugglers Trap Nigeria's Endangered Parrots to Brink of Extinction."
Serle and Morel, *Birds of West Africa*.
Wolfe and Prance, *Rainforests of the World*.

Vini ultramarina (ultramarine lory)

The ultramarine lory has exceptionally beautiful plumage of several shades of brilliant blue. Originally, this seven-inch bird was found on a group of French Polynesian islands called the Marquesas. Today, probably fewer than 1,500 survive on just a few islands, with the largest population on Ua Huka. They are usually found in pairs or small flocks in the canopy, where they feed on nectar, pollen, fruit, and insects. In addition to deforestation, the accidentally introduced black rat (*Rattus rattus*), which eats the lory's eggs, caused the population crash. Since no rats live on Ua Huka or the island of Fatu Hiva, biologists are trying to release captive-bred ultramarine lories there.

Further Reading:
Hubers, "The Ultramarine Lory."

MUSOPHAGIFORMES

Musophagiformes has just one family, which is sometimes placed in the order of cuckoos, *Cuculiformes*.

Musophagidae (touracos, or plantain eaters)

Touracos, or plantain eaters, is a family of 18 species of large and conspicuous, often colorful, arboreal, and fruit-eating birds, found exclusively in forests and savannas of Africa. Quite agile, they frequently hop and run along tree branches. Their calls are loud and resonant.

Further Reading:
Serle and Morel, *Birds of West Africa*.

Tauraco persa (green-crested touraco)

The green-crested touraco is a 17-inch-long bird, whose head, mantle, and breast are green. It also has an entirely green head crest. Its back, wing coverts, and tail are a glossy violet, and its wing feathers are crimson. Often when one touraco gives its clamorous call of "kaw-kaw-kaw," others in the area all answer, so that shortly all the nearby touracos are calling in a chorus of kaws. The green-crested touraco is found from Senegal and Gambia, east to the Central African Republic and south to Gabon and Congo. Its nest is a flimsy platform of twigs, balanced in a forest tree.

Tauraco bannermani (Bannerman's touraco)

Bannerman's touraco is similar to the green-crested touraco, except that it has a red head and crest, inhabits only the montane forests of western Cameroon, and is a far rarer bird. The number of trees in its habitat was reduced by half from 1965 to 1985, and since deforestation continues, the future of the Bannerman's touraco is in question.

Conservationists are trying to protect the 25,000-acre Mount Oku forests in this part of Cameroon by working with local residents. One alternative for the farmers who cut the forests is beekeeping, since honey fetches a good price at local markets, and honey production requires a healthy forest. Bannerman's touraco has an important role in the culture of the Kom people, who live in this part of Cameroon. The touraco call is mimicked by melodies played on the xylophone, known as the Njang. When someone in the village dies, Njang music is played continually for three days.

Further Reading:
Serle and Morel, *Birds of West Africa*.
World Wide Fund for Nature, Switzerland <http://www.panda.org/resources/publications/species/underthreat/turaco.htm>.

CUCULIFORMES (cuckoos)

Cuculidae (cuckoos)

Cuckoos are found in tropical and temperate regions worldwide, but most species in northern climes migrate south during the cold months. There are 127 species. The birds are generally slender, with long tails. More cuckoo species exist in the Old World than the New and many Old World cuckoos are parasitic—they lay their eggs in the nests of other birds, which raise the often larger and hungrier cuckoo nestlings at the expense of their own offspring. Most nonparasitic cuckoos are monogamous, with males and females building the nests and sharing incubation and feeding duties.

Further Reading:
Stiles and Skutch, *A Guide to the Birds of Costa Rica*.

Opisthocomidae (hoatzin)

Hoatzins were once thought to belong with the parrots in the *Psittaciformes* order, but recent anatomical studies of this unusual species have revealed that it is probably more closely related to the *Cuculiformes*, or cuckoos. The only member of the family is *Opisthocomus hoazin*, an outlandishly strange-looking bird that inhabits the flooded and mangrove forests of the Amazon, the Orinoco River in Venezuela, and the Guianas. The hoatzin is the size of a chicken, with a bright blue face and a crest of spiked feathers that makes the bird look a bit like a beaked punk rocker. It is only slightly more adept at flying than a chicken and tends to make only short, heavy treks from roosting tree to feeding trees. The hoatzin eats only leaves and is the only bird to have a cowlike ruminant system, sacs in its stomach laden with bacteria that enable it to digest leaves. Hoatzins build crude nests, usually on branches that overhang the water. Hatchlings are born with a pair of claws at the bend of each wing, the only bird that has this adaptation. The young hoatzins use their claws to climb about and hang from branches. In the face of danger, the young birds simply drop from the nest into the water. Since there is ample peril there as well, such as predatory fish, snakes, and caimans, it is imperative that they are able to scramble out of the water quickly. Their wing claws help them pull themselves back into their nest. Once they can fly, they lose their wing claws.

Further Reading:
Goulding, *The Flooded Forest*.

STRIGIFORMES (owls)

Strigidae

Strigidae include 120 species of owls found on every continent except Antarctica. They are nocturnal hunters with eyes that face forward and are set in what are called facial disks, with radiating feathers surrounding them like the rays of the sun. They are usually tawny, tan, gray, and black, and so are easily camouflaged during the day. Most are arboreal and prey on a wide variety of small mammals, birds, reptiles, crustaceans, and fish. They rarely build nests, but instead lay their eggs in tree cavities, in buildings, or among rocks.

Further Reading:
Stiles and Skutch, *A Guide to the Birds of Costa Rica*.

Bubo virginianus (great horned owls)

The great horned owl is large, about 21 inches long, and powerful. It has grand yellow eyes and prominent ear tufts. It is a wide-ranging owl, found in Alaska and Northern Canada all the way south to Tierra del Fuego, Argentina. It is at home in rocky and wide-open areas as well as in forests. Usually great horned owls use an old nest of another large bird, laying two to three white eggs. They are aggressive and will attack any creature that ventures too close to the borrowed nest.

Further Reading:
Kielder Water Bird of Prey Centre, England <http://www.discoverit.co.uk/falconry/hornedow.htm>.
Stiles and Skutch, *A Guide to the Birds of Costa Rica.*

Otus ireneae (Sokoke scops owl)

The Sokoke scops owl is small, at only six inches long, and is mainly found in the Sokoke Forest on the coast of Kenya. There are three different kinds, or morphs, with feathers of mottled gray, brown, and reddish. Identified only in 1965, this owl is a rare bird, with just 1,000 pairs remaining. Saving the owl was the impetus behind a massive effort to conserve the Sokoke Forest, one of the most important forest tracts in Africa.

Further Reading:
World Wide Fund for Nature, Switzerland <http://www.panda.org / resources / publications / species/ underthreat/scops-owl.htm>.
Zimmerman, Turner, and Pearson, *Birds of Kenya and Northern Tanzania.*

Pulsatrix perspicillata (spectacled owl)

The spectacled owl is about 20 inches long and has a powerful bill and feet. It is dark brown and gray, with a band of white encircling its bill and another nearly circumscribing its yellow eyes, so it looks as if it's wearing glasses. The young, however, have reverse coloring: They have fluffy white plumage with a black facial disc. Female spectacled owls lay two eggs in a tree hole, but usually only one chick survives. The fledgling is dependent on its parents for nearly a year and may not attain full adult plumage for several years. Spectacled owls are found from southern Mexico to western Ecuador, Bolivia, and northern Argentina, but only in localized populations, since deforestation has so diminished their habitat. Although they inhabit dense forest, they like to hunt along streams and forest edges at night. Like many owl species, they perch on bare branches while hunting, leaning forward to scan the earth; when they spot prey, they strike with a rapid pounce to the ground or a nimble swoop. They feed on large insects and

mammals no larger than opossums, plus lizards, crabs, and some birds.

Further Reading:
The Hawk Conservancy, England <http://www.hawk-conservancy.org/priors/spectacl.htm>.
Howell and Webb, *A Guide to the Birds of Mexico and Northern Central America.*
Stiles and Skutch, *A Guide to the Birds of Costa Rica.*

CAPRIMULGIFORMES (potoos)

Nyctibiidae spp. (potoos)

Potoos are found from Mexico to Argentina and on the islands of Trinidad, Hispaniola, and Jamaica. They have long wings, long tails, large yellow or brown eyes, and intricately patterned plumage of brown, gray, buff, white, and black. They hunt at night for insects and sometimes small bats. The great potoo (*Nyctibius grandis*) is a large (20 inches long) bird that lives in the lowland forest canopy of Guatemala to southeastern Brazil, calling actively on moonlit nights with a loud, reverberating bark. The smaller common potoo (*N. griseus*) frequents more open woodlands and forest edges. The potoo often rests by day by perching quite upright in open areas on tree stumps with its feathers compressed, neck stretched straight up in line with its body and tail, and head inclined upward, motionless, like an extension of a tree stump.

Further Reading:
Skutch, *Birds of Tropical America.*
Stiles and Skutch, *A Guide to the Birds of Costa Rica.*

APODIFORMES (swifts and hummingbirds)

Trochilidae spp. (hummingbirds)

Hummingbirds are found only in the Americas. With 330 species, they compose the second-largest family of birds in the Western Hemisphere. Only a few species range to North America and temperate South America; all the rest live in the Neotropics. These are tiny birds, ranging in size from 2.5 to 8 inches, and are notable for their iridescent colors, glittering hues of greens, purples, blues, and reds. Like mini-helicopters, they can fly in all directions and hover in place, which allows them to sip nectar from long, tubular blossoms. To fly, they beat their wings at blinding speed, 60 to 200 times per second, and average about 30 miles per hour in flight. Thirty percent of a hummingbird's weight consists of flight muscles and its miniscule brain, which at 4.2 percent of its body weight is proportionately the largest in the bird kingdom.

These speedy birds require a good deal of energy, so they must consume, on average, half their weight in sugar each day. Most hunt for insects for protein, and feed on nectar to provide a quick sugar fix. A few species, such as the ruby-throated hummingbird, nest in North America and spend the cold winter months in Latin America. Thus, the smallest birds in the world fly thousands of miles to reach their winter and summer homes, plus make a nonstop dash 600 miles across the Gulf of Mexico.

Further Reading:
Stiles and Skutch, *A Guide to the Birds of Costa Rica.*

Lampornis calolaema (purple-throated mountain gem)

The purple-throated mountain gem, like all mountain gems, inhabits highland forests, especially cloud forests. This species is found from southern Nicaragua to western Panama. Males have glittering, pale blue crowns, and the upper parts of their bodies are bronze-green, while their throats, appropriate to their names, are purple. Their chests are a shimmery green, their bellies gray, and their notched tails blue-black. At four inches, they are a medium-size hummingbird. Males are particularly aggressive and defensive about their favorite flowers and often sing while flitting about blooming territory. The nest of the purple-throated mountain gem is a compact and deep cup of plant fibers, moss, lichens, and bits of tree ferns built in an understory shrub or small tree.

Further Reading:
Stiles and Skutch, *A Guide to the Birds of Costa Rica.*

Panterpe insignis (fiery-throated hummingbird)

The fiery-throated hummingbird is bold and beautiful, about 4.5 inches long, with a shiny green back, greenish-blue belly, dark blue tail, a patch of violet on its chest, a bright blue crown and deep black on the sides and back of its head. Its common name comes from its coppery-orange throat that sparkles when it catches a ray of sun. Found in the high-altitude forest canopy of Costa Rica and western Panama, fiery-throated hummingbirds are quite aggressive, chasing off many other hummer species from their favorite flowers, usually those of vines,

low-growing shrubs, epiphytes, and bromeliads. Fiercely protecting territory around clumps of flowers is common behavior among other aggressive male hummers, since the number of flowers he controls helps him attract females. Unlike most other hummingbird species, *Panterpe* males form a bond of a sort with females. They allow certain, selected females to sip on the flowers in their territories, although not the same flowers on which they themselves feed. The nectar-rich shrub *Macleania glabra* is a particular favorite, and in fact, *Panterpe* plans its sex life around this plant's flowering. Most birds in the cool highlands mate during the dry season. But the fiery-throated hummingbird breeds between July and November, the coldest and wettest months of the year, because that's when *M. glabra* flowers.

Further Reading:
Janzen, *Costa Rican Natural History.*
Stiles and Skutch, *A Guide to the Birds of Costa Rica.*

Topaza pella (crimson topaz)

The crimson topaz is a a seven-inch-long resident of northern South America, most often found along streams, foraging in the canopy of tall, flowering trees. The males are particularly striking hummingbirds, with fiery purple on their napes and upper backs, shading to a glittery gold on their upper tails. Their central tail feathers are bronze, turning to deep purple, and are very narrow, 2.5 inches long, curved, and cross each other at their tips. The crimson in the hummer's common name comes from the color of its breast and belly. The female's plumage isn't quite as colorful, but she, too, has feathers of gleaming green, crimson, bronze, and violet.

Further Reading:
de Schauensee and Phelps, *A Guide to the Birds of Venezuela.*
Howell and Webb, *A Guide to the Birds of Mexico and Northern Central America.*
Janzen, *Costa Rican Natural History.*
Stiles and Skutch, *A Guide to the Birds of Costa Rica.*

TROGONIFORMES (trogons)

Trogonidae

Trogonidae include 40 species, found in the warmer regions of the Old World and New, but mostly in the New. They are medium-size forest birds, with long, broad tails and soft plumage. Colorful trogons are found in nearly all wooded areas in the Americas.

Males have brilliant green, blue, or violet plumage on their upper torsos and chests, while posterior under-parts are contrasting yellow, orange, or red.

Further Reading:
Stiles and Skutch, *A Guide to the Birds of Costa Rica.*
Williams and Arlott, *Birds of East Africa.*

Trogon bairdii (Baird's trogon)

Baird's trogons are exceptionally handsome and found only in the rainforest canopy of Costa Rica and bordering southwest Panama. The males are far more colorful than the females, with chests, heads, and upper backs of deep violet, shading to a greenish-blue on their lower backs and tails, and lower breasts and bellies of vermilion. Like many trogons, their tails are barred with black and white, and they have lovely, rolling, mellow songs. Trogons eat both in-sects and fruits. They excavate nests in long-dead trees with soft, but not yet rotten, wood. Deforesta-tion threatens this exquisite trogon's survival, with its only remaining habitat being Corcovado National Park in southern Costa Rica.

Further Reading:
Skutch, *Birds of Tropical America.*
Stiles and Skutch, *A Guide to the Birds of Costa Rica.*

Pharomachrus mocinno (resplendent quetzal)

Resplendent quetzals are among the most beauti-ful birds in the world. The males' radiant, emerald green and ruby-red plumage has entranced pre-Columbian cultures and 20th-century birdwatchers alike. The word *quetzal* is taken from *quetzalli,* Aztec for "tailfeather," which came to mean "precious" or "beautiful." *Pharomachrus* is Greek for "long mantle," and *mocinno* is taken from the name of a now obscure 19th-century naturalist who helped bring the first specimen to Europe. The Aztec of Mexico revered quetzals, and only royalty and nobility could wear the long tail feathers in elaborate headdresses.

The birds are only 14 inches long, but the males have elegant, three-foot-long green tail feathers that stream behind them when they fly through the mists of the high-altitude cloud forests where they nest, from southern Mexico to western Panama. Using their bills, quetzals carve nests into the rotting wood of tall, dead tree trunks. They feed on fruits, insects, small frogs, and lizards, but their favorite food is the fruit of the *Lauraceae,* or avocado family (*Persea spp.*). Although some cloud-forest reserves have safe-guarded the quetzal throughout its range, a team of

scientists attached tiny radio transmitters to the birds and found that they migrate downslope, outside the reserves, in search of their favorite fruits. Much of this lower-altitude land is being rapidly deforested, so conservationists are working with governments and farmers to ensure that enough *Lauraceae* trees remain standing to guarantee healthy populations of this gorgeous bird, the national symbol of Guate-mala.

Further Reading:
Jukofsky, "Mystical Messenger."
Stiles and Skutch, *A Guide to the Birds of Costa Rica.*

CORACIIFORMES (kingfishers, motmots, hornbills, and allies)

Bucerotidae (hornbills)

Hornbills are found in sub-Saharan Africa and Southeast Asia and are named for their enormous down-curved bills. The upper bills are adorned with grotesque-looking horny protrusions, called casques. The 44 species of hornbills are black and dark brown, with white or cream on their bodies, wings, and tails, while their bills can be brown, black, yellow, or red, and their necks yellow with splashes of blue. Their reverberating calls can be loud cackles, toots, bellows, or barks. They are omnivorous, eating fruits, berries, insects, and even small animals.

Hornbills are important seed dispersers through-out their habitat. A study done by researchers with San Francisco State University and the University of California at Davis found that hornbills spread the seeds of nearly one-fourth of the tropical trees in a rainforest study site in Cameroon. Further, nearly 100 percent of the seeds that they defecate germinate successfully. With an average wingspan of some four feet, the birds will fly 100 miles or more through the forest, looking for ripe fruit. The big birds' role as seed spreaders is becoming increasingly important as deforestation and hunting wipe out populations of other dispersers, such as elephants and primates.

Further Reading:
Long, "The Shrinking World of Hornbills."

Buceros bicornis (great pied hornbill)

The great pied hornbill, at four feet long and six pounds, is the largest male of the nine species of hornbills found in India. Its plumage is black, with bands of white. It lives in Southeast Asia and in India's southern forests, where intensive logging threatens its survival. Great pied hornbills mostly eat forest fruits and heavily depend on fig trees. In India, fig leaves are traditionally fed to domestic elephants used in logging operations, and the treetops are frequently lopped off, depriving the hornbills of an important food source.

Like several other species of hornbills, the great pied nests in tree cavities, the female using her bill to seal herself in with a sticky mixture of droppings, chewed wood, mud, and bark. This unusual habit presumably serves as protection against predators. She leaves a slit in the plaster just large enough for her mate to pass in food. She also defecates through the opening, at a high velocity, so as not to soil the nest. When the young chicks are ready to emerge and fly, she pecks open the opening so they can test their wings. Great pied hornbills mate for life.

Further Reading:
Kannan and James, "Fruiting Phenology and the Conservation of the Great Pied Hornbill (*Buceros bicornis*) in the Western Ghats of Southern India."

Motmotidae spp. (motmots)

Motmots include nine species found only in the Neotropics, mainly in Mexico and Central America. They have short, broad, slightly down-curved bills, large heads, and beautiful plumage in shades of green, blue, and reddish. On most species, the two long central tail feathers lose their barbs toward the tips, revealing naked spines. When perched, they often twitch their pendulum-like tails from side to side. They catch insects, other invertebrates, or small reptiles and also eat fruit. Both sexes build the nest, an unlined burrow dug into a roadside or stream bank. Both sexes also incubate the round, white eggs for about three weeks, and both feed their young. Motmots are so named because several species have owl-like, resonant, double hoots.

Further Reading:
Howell and Webb, *A Guide to the Birds of Mexico and Northern Central America.*
Stiles and Skutch, *A Guide to the Birds of Costa Rica.*

Momotus momota (blue-crowned motmot)

The blue-crowned motmot is 15.5 inches long, with red eyes, a turquoise-edged crown, and black face mask; the rest of its plumage is shades of bluish-green. They are adaptable birds and can survive in second-growth forest, forest edges, shady gardens, and shaded coffee plantations, along with their native rainforest. Blue-crowned motmots are found from northeastern Mexico to northwestern Peru and northern Argentina.

Further Reading:
Howell and Webb, *A Guide to the Birds of Mexico and Northern Central America.*
Stiles and Skutch, *A Guide to the Birds of Costa Rica.*

PICIFORMES (woodpeckers, jacamars, toucans, puffbirds, and allies)

Bucconidae (puffbirds)

Puffbirds are a Neotropical species with large heads and thick necks, short tails, and lots of loose-fitting plumage, so that their short legs are barely visible. Their bills have a cleft hook at the tip. They have perfected an energy-saving method of foraging: They sit quietly on a perch, seemingly asleep, but their sharp eyes can detect insects and other small invertebrates that wander into their range of vision—perhaps 60 feet away. Then they strike with a rapid dart, and may beat their victim senseless against the perch before swallowing. They also eat small frogs and lizards. Both male and female puffbirds build a nest in arboreal termite nests or in burrows in the forest floor that they line with leaves.

Further Reading:
de Schauensee and Phelps, *A Guide to the Birds of Venezuela.*
Stiles and Skutch, *A Guide to the Birds of Costa Rica.*

Monasa morphoeus (white-fronted nunbird)

The white-fronted nunbird is slimmer and has a longer tail than most other puffbirds. The 12-inch-long bird is all-over dark to light gray, except for an array of short, stiff white feathers on its forehead and chin. Its curved and slender bill is bright vermilion. Found from eastern Honduras to northern Bolivia and southeastern Brazil, the nunbird has a louder call than most puffbirds. White-fronted nunbirds inhabit rainforest canopy and adjoining shady clearings, especially near rivers. From its lookout perch, it captures large-bodied insects, spiders, small frogs, and lizards and often follows monkey troops, flocks of caciques, or other energetically foraging birds to catch the prey flushed by their commotion.

Capitonidae (barbets)

Barbets are stocky birds with short tails and wings, short legs, and strong feet. Like woodpeckers, which belong to the same family, their first and fourth toes are directed backward. Bristly tufts sprout from the base of their thick bills, which gives them their common names; "barbe" is French for "beard." Many barbets are brightly colored with patterns of red, orange, yellow, and green. They have loud voices, and their song is usually just one repeated note. They are mainly arboreal fruit-eaters. Most of the 74 species are residents of the Old World, particularly Africa, where they mainly inhabit savannas and thornscrub, although a few are forest species. Only 13 species of barbets are found in the Neotropics.

Further Reading:
Serle and Morel, *Birds of West Africa.*
Stiles and Skutch, *A Guide to the Birds of Costa Rica.*

Buccanodon duchaillui (Duchaillu's yellow-spotted barbet)

Duchaillu's yellow-spotted barbet inhabits lowland forest from Sierra Leone to Gabon and Congo and can also be found in savanna woods and montane forests in the highlands of Cameroon. The six-inch barbet has black plumage spotted and barred with yellow, a yellow strip on the side of its head and neck, and a bright red crown. It eats fruits and the occasional insect. It nests in a hole in a decaying tree. Duchaillu's yellow-spotted barbet has a distinctive voice, a soft and musical purring call, that lasts a second or two.

Further Reading:
Serle and Morel, *Birds of West Africa.*

Semnornis frantzii (prong-billed barbet)

The prong-billed barbet is found only in Costa Rica and western Panama, where it prefers cool, wet, and mossy mountain forests and nearby second growth. Almost seven inches long, it has a very short tail, and a short and thick bill, whose upper mandible ends in a small hook; the lower mandible has two small prongs that fit around it. Most of their upper parts are brownish-olive, and their breasts are a golden tawny orange. The males have a tuft of shiny, long, black feathers on their napes. They are found in flocks of a dozen or more except at breeding time, when they become quite territorial, and males chase off other males. They mainly feed on fruits found in the canopy, which they swallow whole or squeeze to

extract the juice. Prong-billed barbets are monogamous and nest in cavities in dead trees.

Further Reading:
de Schauensee and Phelps, *A Guide to the Birds of Venezuela.*
Stiles and Skutch, *A Guide to the Birds of Costa Rica.*

Galbulidae (jacamars)

Jacamars are a family of 15 species of New World birds that live in forests, forest edges, along streams, and in clearings. Most species have iridescent plumage of green, bronze, gold, and metallic violet; all have long, very thin, and pointed bills, so that they can nab stinging insects and hold them away from their faces as they knock them senseless against a tree branch. They have short legs and, except for the three-toed jacamar (*Jacamaralcyon tridactyla*), two toes on each foot point forward, and two backward. Using their bills, males and females dig out burrows in earthen banks, wooded slopes, or arboreal termite nests. They line the nests with regurgitated insect parts before the female lays two to four eggs. They share incubation and feeding of the downy hatchlings.

Galbula rudicauda (rufous-tailed jacamar)

The rufous-tailed jacamar is one of the most wide-ranging of the jacamars. Found from southeastern Mexico to Ecuador and northern Argentina, the rufous-tailed is nine inches long. Its head, chest, and upper parts are a shimmering golden green; its belly and undertail coverts, cinnamon. The male's throat is white and the female's is cinnamon or buff-colored. Like all jacamars, they have a distinctive voice: a loud and sharp scream and a song that consists of several loud, rapid notes.

Further Reading:
de Schauensee and Phelps, *A Guide to the Birds of Venezuela.*
Geobopological Survey <http://www.geobop.com/Birds/Piciformes/Galbulidae/index.htm>.
Stiles and Skutch, *A Guide to the Birds of Costa Rica.*

Ramphastidae (toucans)

Toucans are famous for their disproportionately large and colorful bills, which make the bird in flight look as if it's pushing along a banana. The bills are actually light and hollow, crisscrossed inside with thin bones that make them quite strong. Their tongues are narrow, flat, long, and stiff, notched with deep indentations on both sides. The main part of their diet consists of fruit, although they also eat insects and other small invertebrates, small lizards, snakes, birds' eggs, and nestlings. There are 42 species, which include toucans and the generally smaller toucanets and aracaris. All inhabit principally low-altitude forests of the Neotropics. Their plumage is mainly green or black, with splashes of red, yellow, blue, and/or white. They range in size from the 24-inch-long toco toucan (*R. toco*) to the 11.5-inch emerald toucanet (*Aulacorhynchus prasinus*).

Further Reading:
Stiles and Skutch, *A Guide to the Birds of Costa Rica*.
Wolfe and Prance, *Rainforests of the World*.

Ramphastos sulfuratus (keel-billed toucans)

Keel-billed toucans are the "Fruit Loops" birds with the rainbow-hued bills, which they use to pluck fruit, insects, and small lizards off branches. They toss these morsels into the air and down their throats with a quick, upward flip of the head. Their plumage is mostly black, with a bright yellow bib edged in red, a yellow-green face, and a patch of red on their lower tail coverts. Keel-billed toucans, which are about 20 inches long, are found from eastern Mexico to northern South America. They travel in small flocks of about six and nest in deep tree hollows or cavities—another reason why standing, decaying trees are important to forest ecology.

Further Reading:
Stiles and Skutch, *A Guide to the Birds of Costa Rica*.
Wolfe and Prance, *Rainforests of the World*.

PASSERIFORMES (passerines or perching birds)

Passerines, or perching birds, comprise a huge order that includes about 60 percent of all the world's avian species.

Cotingidae (cotingas)

Cotingidae is a family with 65 species of forest dwellers found in Central and South America to northern Argentina. Many species have common names derived from their voices or their favorite foods: bellbird, chatterer, fruit-eater, berry-eater. They are widely varied in size and colors, and males sport examples of unusual ornamentation, such as wattles, crests, and beards. All members of the family have broad bills with slightly hooked tips, and short legs. They are generally shy and solitary.

Further Reading:
Fjeldsa and Krabbe, *Birds of the High Andes*.
Stiles and Skutch, *A Guide to the Birds of Costa Rica*.

Cephalopterus spp. (umbrellabirds)

Umbrellabirds are three species of large (15 to 20 inches long), chunky, black birds that live in Neotropical forest canopies. The male bare-necked umbrellabird (*C. glabricollis*), found in Costa Rica and western Panama, is entirely black except for the bare skin of its throat and chest, which is orange-red. Males court females in of varying, but generally small sizes, inflating the loose skin at their throats and chests to attract attention. Both males and females sport umbrella-like crests; the males' are particularly large. They feed on palm fruits, large insects, small lizards, and frogs.

The Amazonian umbrellabird (*C. ornatus*), a resident of rainforests from Colombia to the Amazon in Brazil, is also black, and the upstanding feathers of its crown form a tall crest with white shafts. It has a triangular wattle hanging from its lower throat, covered in front with metallic blue feathers. Females lack a wattle and have smaller crests.

Further Reading:
de Schauensee and Phelps, *A Guide to the Birds of Venezuela*.
Stiles and Skutch, *A Guide to the Birds of Costa Rica*.

Cotingas spp.

Cotingas spp. are mainly found in South America, except for the lovely cotinga (*C. amabilis*), which resides in southeastern Mexico to Costa Rica. *C. amabilis* is indeed a lovely bird; males are an electric turquoise blue, with a purple patch on their throats and lower chests, and black wings and tails, edged with greenish-blue. Like all cotingas, they are usually silent, making an occasional rattle when flying. But females will let loose with a loud and agonized shriek if distressed while defending their young. Cotingas are about seven-and-a-half inches long, live in forests and forest edges, eat fruits, and build a shallow, cup-shaped nest high in the canopy. Male pompadour cotingas (*Xipholena punicea*) have black-tipped white wings, but the rest of their bodies are a shiny raspberry purple. They are found in forests from Ecuador to the Amazon in Brazil.

Further Reading:
de Schauensee and Phelps, *A Guide to the Birds of Venezuela.*
Howell and Webb, *A Guide to the Birds of Mexico and Northern Central America.*
Stiles and Skutch, *A Guide to the Birds of Costa Rica.*

Procnias spp. (bellbirds)

Bellbirds are four species of mostly white cotingas that live mainly in South American forests and feed on berries and fruits. Only one species, the three-wattled bellbird (*P. tricarunculata*), is found in Central America, from Nicaragua to Panama. Males of this species have three stringlike black wattles that hang over their bills. They are chestnut-colored with white heads, necks, and upper chests. Male bearded bellbirds (*P. averano*) are mostly silvery-white, except for their chocolate-colored heads and wings. Many short and threadlike wattles hang from their bare throats. White bellbird (*P. alba*) males have one long, thin wattle that hangs over their bills. Both the bearded and white bellbirds are found in Venezuela, the Guianas, Brazil, and Trinidad. All male bellbirds have a loud, penetrating, clear metallic cry that sounds like the clanging of a bell. Females are mostly silent.

Further Reading:
de Schauensee and Phelps, *A Guide to the Birds of Venezuela.*
Howell and Webb, *A Guide to the Birds of Mexico and Northern Central America.*
Stiles and Skutch, *A Guide to the Birds of Costa Rica.*

Coerebidae (bananaquit)

The bananaquit (*Coereba flaveola*) is the sole species in the *Coerebidae* family. The four-inch bird has a gray throat, slate-gray upper parts, and a yellow belly. Its head is black except for a white stripe across the top of its eyes. The ubiquitous bananaquit inhabits forests, forest edges, gardens, and clearings from southern Mexico to northeastern Argentina and southern Brazil. It is common in the Caribbean islands, except for Cuba, where its absence is curious and unexplained. It uses its black, sharp, downward-curved bill to probe at small flowers or pierce larger ones for nectar, and it pokes into fruits for juice. Both male and female bananaquits build the globe-shaped nest made of plant fibers and moss and lined with seed down and feathers. Even when it's not breeding season, they sleep singly in the nests. The female incubates the eggs alone, but both parents feed the nest-

lings. The two or three young remain in the nest for 18 days, quite a long time for such a small bird.

Further Reading:
Howell and Webb, *A Guide to the Birds of Mexico and Northern Central America.*
Raffaele, *A Guide to the Birds of the West Indies.*
Stiles and Skutch, *A Guide to the Birds of Costa Rica.*

Drepanididae (Hawaiian honeycreepers)

The family of Hawaiian honeycreepers is placed in a separate family from Neotropical honeycreepers, which are with the *Thraupidae*. At least 8 of 54 original species, found only on the Hawaiian islands, are now extinct, due to habitat destruction and the introduction of exotic birds and mammals. All the original species are believed to have descended from one ancestor, the finches that colonized Hawaii three million years ago. Through evolutionary radiation, they adapted to specific habitat niches and, over millions of years, developed into separate species, like Darwin's finches on the Galapagos Islands. Species with thin bills are nectar-feeders, while those with thicker bills eat seeds, fruits, and insects. They range in size from four to eight inches and have red and black or greenish plumage. The iiwi (*Vestiaria coccinea*) is one of the more common honeycreepers. It has a red body, black wings with small, white patches, a black tail, and a sickle-shaped, red bill. About six inches long, the bird's song is squeaky and high-pitched. The palila (*Loxioides bailleui*) is 1 of 14 endangered Hawaiian honeycreepers. The palila feeds almost exclusively on the immature seeds of the mamane tree and is found only in the dry forests on the slopes of Mauna Kea, on the island of Hawaii, at altitudes between 6,000 and 9,000 feet. Of the world's 168 most critically endangered species, 16 are Hawaiian honeycreepers.

Further Reading:
Champlin, "Searching for Hope in the Family Tree."
Hawaii Biological Survey, USA <http://www.bishop.hawaii.org/bishop/HBS/palila.html>.
Line, "Is This the World's Rarest Bird?"
Royte, "On the Brink: Hawaii's Vanishing Species."

Formicariidae (antbirds)

Antbirds include about 260 species that live in the Neotropics and are particularly abundant in the thick rainforests of the Amazon and Guyana, Suriname, and French Guiana. They forage near or on the ground and live alone or in small family groups. They range in size from 3 to 14 inches. Smallest are the antwrens, next in size are antvireos, mid-range in size are the antbirds or antshrikes—these have a little hook at the tip of their bills—and the largest, the antthrushes, are ground foragers. Short-tailed and plump, antpittas also hop about on the ground. Antbirds come in various shades of black, brown, and buff, with spots of white on their wings and flanks. Some species have bright blue or red patches of featherless skin around their eyes. Antbirds get their names from the 28 species that regularly follow army ants. When the aggressive ants set off on a raid, the birds post themselves near the advancing column to capture the insects that are trying to escape the ants. Although some of the larger species of antbirds may eat small frogs, lizards, snakes, and fruits, most are nearly entirely insectivorous. Antbirds tend to mate for life.

Further Reading:
Kricher, *A Neotropical Companion.*
Stiles and Skutch, *A Guide to the Birds of Costa Rica.*

Fringillidae (siskins, canaries, finches, and allies)

Siskins, canaries, finches, and allies are found on all continents except Australia, with more diversity in the Old World than the New. The males are usually more colorful than the females, and many of the 125 species have melodious voices. They are small, seed-eating birds with strong, conical bills. The canaries of Africa mainly live in the dry savanna.

Further Reading:
Serle and Morel, *Birds of West Africa.*
Stiles and Skutch, *A Guide to the Birds of Costa Rica.*

Carduelis cucullata (red siskin)

The red siskin frequents lower montane forest as well as grasslands in Venezuela, northeastern Colombia, and Trinidad. It is, however, quite rare, with a current population only in the high hundreds or low thousands. This four-inch-long bird has an entirely black head and tail, while most of the rest of its body is a bright vermilion. Its wings are streaked black and light vermilion. A flocking bird, the red siskin eats fruit and seeds. During the first half of the 19th century, the red siskin was heavily trapped for the caged-bird trade. It is still in demand because it breeds with the domestic canary, resulting in offspring with lovely

songs and colorations. Capturing red siskins is now illegal, but the bird remains an endangered species.

Further Reading:
de Schauensee and Phelps, *A Guide to the Birds of Venezuela.*
World Wide Fund for Nature, Switzerland <http://www.panda.org/resources/publications/species/underthreat/red_siskin.htm>.

Linurgus olivaceus (oriole finch)

The oriole finch inhabits forested mountain highlands of Cameroon, Nigeria, Kenya, Uganda, and Tanzania. The males are bright yellow with olive upper parts and yellow underparts. They have a black head and neck and orange bill. Females are mostly olive green. These six-inch finches are generally quiet birds, emitting only occasional thin, high-pitched songs.

Further Reading:
Serle and Morel, *Birds of West Africa.*
Williams and Arlott, *Birds of East Africa.*

Icteridae (American orioles, blackbirds, grackles, cowbirds, and meadowlarks)

This family includes American orioles, blackbirds, grackles, cowbirds, meadowlarks, and relations, about 90 species total, and all are found only in the Western Hemisphere. Most of the family members are tropical residents; those that nest in temperate climates usually migrate south during the winter.

Further Reading:
Stiles and Skutch, *A Guide to the Birds of Costa Rica.*

Icterus bonana (Martinique oriole)

The Martinique oriole is endemic to the Caribbean island of the same name, where it inhabits all forests but montane cloud forests. This seven-inch-long bird is black with a reddish-brown hood and orange shoulder, rump, and abdomen. The Martinique oriole forages in the canopy for small invertebrates and fruits and builds a pendulum-like woven nest that suspends from leaves or palm fronds just 7 to 13 inches from the ground. The main cause of the bird's decline is the shiny cowbird (*Molothrus bonariensis*), which first colonized Martinique in the 1940s. The cowbird lays its eggs in other birds' nests, and the unknowing mothers feed its larger, more ravenous offspring, while their own nestlings frequently perish. A study revealed that cowbirds parasitize about 75 percent of the oriole nests on the island.

Further Reading:
Raffaele, *A Guide to the Birds of the West Indies.*

Icterus oberi (Montserrat oriole)

The Montserrat oriole is endemic to another Caribbean island, tiny Montserrat, and is even rarer than the Martinique oriole. It is mostly black except for a lemony yellow breast and abdomen. Unlike most orioles, it feeds almost exclusively on insects. Deforestation restricted the bird to the remaining forests in the Soufrire and Centre Hills, and parasitism by the shiny cowbird also took a toll. Then Mount Soufrière erupted in 1995, covering the island in ash and forcing most human inhabitants to abandon their homes. Wildlife had no such option. When island endemics are already vulnerable due to habitat degradation, natural catastrophes can be the final push toward extinction, which scientists believe will soon be—if it isn't already—the fate of the Montserrat oriole.

Further Reading:
Lovette, Bermingham, and Ricklefs, "Mitochondrial DNA Phylogeography and the Conservation of Endangered Lesser Antillean *Icterus* Orioles."
Raffaele, *A Guide to the Birds of the West Indies.*

Psarocolius montezuma (Montezuma oropendola)

Montezuma oropendolas are large birds that are about 15 inches long and found from southern Mexico to central Panama. The adult males are chestnut-colored with long, bright yellow tails and long, orange-tipped bills. The skin of their cheeks is pale blue. They have loud and varied calls, ranging from clucks, yelps, whining notes, and screams, interspersed with a melodious liquid gurgling, and a sound that echoes the crumpling of paper. Although they prefer to nest in isolated trees in open areas, they often forage for insects, fruit, and seeds in the canopy.

Like those of the smaller, chestnut-headed oropendola (*Psarocolius wagleri*), which ranges slightly farther south to northern Ecuador, the nests of Montezuma oropendolas are marvels. They are tightly woven sacks up to six feet long that are attached to branch ends. Two eggs lie on a bed of leaves at the bottom of the pouch. Oropendolas nest in colonies, so one large tree may be strewn with as many as 40 of the dangling nests. Females may raise up to three broods during the December to mid-June nesting season, but less than one half percent of the chicks will fledge. The mortality rate is high, due to predation by toucans, snakes, monkeys, and the larvae of an insect known as the botfly. To avoid botfly infestations, many oropendolas build their nests near those of several wasp species and a species of stingless but biting bees (*Trigona*).

Further Reading:
Janzen, *Costa Rican Natural History.*
Milwaukee Public Museum <http://www.mpm.edu/collect/brf.html>.
Stiles and Skutch, *A Guide to the Birds of Costa Rica.*

Indicatoridae (honeyguides)

Honeyguides are a family of small, forest-dwelling, rather plain brown, olive, gray, and white birds found in Asia and, principally, in Africa. They are parasitic, laying their eggs in the nests of other birds, such as bee-eaters and woodpeckers. They mainly eat beeswax and bee larvae.

Further Reading:
Geobopological Survey <http://www.geobop.com/Birds/Piciformes/Indicatoridae/index.htm>.
Serle and Morel, *Birds of West Africa.*
Williams and Arlott, *Birds of East Africa.*

Indicator indicator (greater or black-throated honeyguide)

The greater or black-throated honeyguide is grayish-brown, with white underneath, and has a bright pink bill. It's a resident of west, east, and central Africa, in rainforests, highland dry forest, and thornbush and acacia woodland. Both the black-throated honeyguide and scaly-throated honeyguide (*I. variegatus*) lead animals, including human beings, to beehives, so they can eat the honeycomb and larvae when their honey-loving partners break open the hive. Honey badgers (*Mellivora capensis*), baboons (*Papio spp.*), and mongooses are frequent honeyguide followers. The birds possess an organism in their stomachs that enables them to digest beeswax. Without assistance from an animal willing to follow the flick of its white tail throughout the forest, however, the birds would be unable to get at their principal source of food. Among some indigenous people of Africa, it is a punishable crime to kill a honeyguide.

Further Reading:
Serle and Morel, *Birds of West Africa.*
Williams and Arlott, *Birds of East Africa.*

Muscicapidae (Old World flycatchers)

The Old World flycatchers are a widespread and large family of more than 100 small- or medium-size insect-eating birds. They range from northern Europe and Asia south to Africa, the Philippines, and Indonesia.

Further Reading:
Beehler, Pratt, and Zimmerman, *Birds of New Guinea.*

Machaerirhynchus flaviventer (yellow-breasted boatbill)

The yellow-breasted boatbill is a small (four inches long), colorful, and quite active flycatcher that inhabits forests of New Guinea and northeastern Australia. The males are black and yellow with black, white, and gray upper parts, bright yellow bellies and breasts, white throats, and yellow eyebrows. Females are paler below and olive-gray above. Both have a disproportionately large, broad, and flattened bill. They usually travel in pairs, darting from tree to tree and snatching up small insects. Their nests are small saucers built of tiny twigs, strips of bark, and plant stems, bound together with spiders' webs and decorated with lichen.

Further Reading:
Beehler, Pratt, and Zimmerman, *Birds of New Guinea.*
Pizzcy, *A Field Guide to the Birds of Australia.*
The Rainforest Trust, USA <http://www.rainforesttrust.com/lib7cont.html>.

Myadestes palmeri (puaiohi, or Kauiai, thrush)

The puaiohi, or Kauiai, thrush is one of Hawaii's highly endangered bird species; only an estimated 200 to 300 remain. It's a rather drab-looking, seven-inch-long bird, with gray-brown feathers and long, pink legs. Its call is a simple trill. It is a prolific breeder, producing four clutches of two eggs each in a single nesting season. Biologists have taken wild eggs, allowed them to hatch, and raised them in captivity, and in 1999, released 15 offspring from the captive-raised birds into their habitat, the swampy forest near the top of the Kauiai volcano. Whether efforts like this will save Hawaii's dwindling native bird population remains unknown. The population has been decimated principally by habitat loss, but also by predators introduced to the islands by human beings: rats, mosquitoes, feral cats and pigs, and mongooses. But if the released birds survive, similar efforts, though expensive, could help save a few species

before it's too late. Of the state's 140 native bird species and subspecies, half are now extinct. Of those that remain, 50 percent are endangered. Nothing will save Hawaii's po'ouli; only three of the little gray-capped and black-faced birds remain. Meanwhile, at least seven of the released puaiohi have built nests and produced chicks, but rats took two of the nestlings.

Further Reading:
Conrow, "Born to Be Wild."
U.S. Geological Survey, Biological Resources Division, USA. "A First: Endangered Puaiohi Birds Fledge Four Chicks in the Wild" <http://www.its.nbs.gov:8000/pr/newsrelease/1999/7–9.html>.

Nectariniidae (sunbirds)

The sunbirds are a family of small, flower-probing, often brightly colored birds. Most of the 120 species are found in Africa and India, although they also live through southern Asia to Australia. They resemble the Neotropics' hummingbirds, but although they also sip flower nectar and their flights are rapid and erratic, they do not beat their wings nearly as fast as hummers and usually feed from a perch. Their habitat ranges from coastal to alpine forests. Most species can be seen in flower gardens, forest edges, and arid bush country, but some are strictly forest dwellers. In addition to nectar, their diet includes arthropods, particularly spiders.

Further Reading:
Beehler, Pratt, and Zimmerman, *Birds of New Guinea.*
Zimmerman, Turner, and Pearson, *Birds of Kenya and Northern Tanzania.*

Nectarinia rubescens (green-throated sunbird)

The green-throated sunbird lives in pairs or small family groups in the forests of southern Sudan, Uganda, western Kenya, and northwestern Tanzania. Males are velvety black with a crown and throat patch of shimmery green, edged with metallic violet. Females are dusky brown and olive.

Further Reading:
Williams and Arlott, *Birds of East Africa.*
Zimmerman, Turner, and Pearson, *Birds of Kenya and Northern Tanzania.*

Pachycephalidae

Pachycephalidae includes 46 species found in Australia, New Guinea, Polynesia, Micronesia, and east through Indonesia, the Philippines, and India.

Further Reading:
Beehler, Pratt, and Zimmerman, *Birds of New Guinea*.

Pitohui dichrous (hooded pitohui)

The hooded pitohui is particularly pleasing to see and hear. About nine inches long, it has a black head, wings, and tail; the rest of its body is bright russet. Like all five species of pitohuis, it is endemic to Papua New Guinea and a few surrounding islands. When the hooded pitohui flies, its wings make a musical fluttering sound, and its voice is soft and mournful. It frequents the forest understory and forest edges and eats arthropods and small fruits.

In 1992, scientists discovered that the hooded pitohui has poisonous feathers and skin. No other bird is known to be toxic in this way. The powerful neurotoxin, which is hundreds of times more poisonous than strychnine, is the same poison produced by poison-dart frogs in South America, and the pitohui is armed with it for the same reason: as a defense against would-be predators. Two other varieties of the bird mimic the flashy look of the hooded pitohui but lack the nasty-tasting toxin. By mimicking their cousin, they hope to repel predators as well.

Further Reading:
Angier, "Rare Bird Indeed Carries Poison in Bright Feathers."
Beehler, Pratt, and Zimmerman, *Birds of New Guinea*.

Paradisaeidae (birds of paradise)

Birds of paradise live in Australia, New Guinea, and west to the Moluccas. Of the 43 species, 38 are found in New Guinea. Most of the *Paradisaeidae* look a bit like crows or starlings, with powerful bills and feet and black or brown plumage, perhaps with shades of red, orange, or green. The males of seven species have long and brightly colored plumes that they erect in elaborate courtship displays. Males usually gather in small groups to court females—the communal area where they come together is called a lek. Birds of paradise mainly forage for fruit and sometimes for arthropods.

Further Reading:
Beehler, Pratt, and Zimmerman, *Birds of New Guinea*.

Manucodia chalybatus (crinkle-collared manucode)

The crinkle-collared manucode is a large (14-inch), crowlike bird that lives in hilly forests of New Guinea and nearby Misool Island. Its plumage is blue-black, with crimped, velvety feathers on its neck, breast, and back. Shy birds, crinkle-collared manucodes live in the middle to upper stories of the forest and forage for fruits, particularly figs. Like other manucodes, they have distinctive voices. The males have a looped trachea, not found in other songbirds, and their call is a series of up to eight hollow, haunting, and low-pitched notes—a hooo—that is sometimes answered by the female with descending, whistled notes. They are usually found in pairs and are monogamous.

Further Reading:
Beehler, Pratt, and Zimmerman, *Birds of New Guinea*.
The Rainforest Trust, USA <http://www.rainforesttrust. com/lib7cont.html>.

Paradisaea raggiana (Raggiana bird of paradise)

The Raggiana bird of paradise of New Guinea is about 13 inches long and has a floppy, undulated flight. The male has trailing orange plumes, a green chin, and a yellow crown and nape. Male Raggianas display in groups in canopy trees to attract females. The flagrant plumage of the Raggiana and other birds of paradise isn't very practical, but because they live on an island, they have few predators.

Further Reading:
Beehler, Pratt, and Zimmerman, *Birds of New Guinea*.

Paradisaea rudolphi (blue bird of paradise)

The blue bird of paradise has a black head, white bill, and white rings around its eyes. The males have black breasts and blue plumes with two black, long tail streamers. To court females, males hang upside down from a branch, spreading the long, lacy plumes along its flanks and curling their narrow tail feathers. Swaying back and forth, they make metallic, guttural calls. The 12-inch male sings every morning from an open perch in the forest canopy.

Further Reading:
Beehler, Pratt, and Zimmerman, *Birds of New Guinea*.
The Rainforest Trust, USA <http://www.rainforesttrust. com/lib7cont.html>.

Pteridophora alberti (King of Saxony bird of paradise)

The King of Saxony bird of paradise bears pearly-blue plumes that extend from each side of its head and are nearly twice the length of its nine-inch body. The males are black with a yellowish breast and belly. When courting, the males perch on a high tree branch, bounce up and down while expanding and retracting their back feathers, and make strange hissing sounds. When a female approaches, they sweep their head plumes in front of her before pursuit.

Further Reading:
Beehler, Pratt, and Zimmerman, *Birds of New Guinea.*
The Rainforest Trust, USA <http://www.rainforesttrust. com/lib7cont.html>.

Philepittidae (asities)

A small family of only four species, these tiny birds are found only in the rainforests of Madagascar. The common sunbird asity (*Neodrepanis coruscans*) is generally found in lower montane forest, while the yellow-bellied sunbird asity (*N. hypoxantha*) lives at higher altitudes. Both have tiny bodies with short tails and thin, curved bills. During breeding season, the males have glistening, brilliant blue plumage, with black above and buttercup yellow below, and turquoise blue wattles around their eyes. The common sunbird asity's blue plumage is less intense than that of the yellow-bellied sunbird. In both species, the plumage of females and males during the nonbreeding season is duller blue. The sunbirds and the velvet asity (*Philepitta castanea*) build globe-shaped nests that they hang from branch tips. The female apparently does the construction work, first building a hollow sphere and then poking a hole in the side for an entrance.

Further Reading:
Hawkins, Safford, Duckworth and Evans, "Field Identification and Status of the Sunbird Asities of Madagascar."
Langrand, *Guide to the Birds of Madagascar.*

Picidae (woodpeckers)

Woodpeckers, with 210 species, are found worldwide except in polar regions, Australia, New Zealand, Madagascar, and remote islands. Most have brilliant red, yellow, or green feathers or are black and white, and some have high crests. All sleep in holes that they carve in trees with their strong, pointed beaks. With their widely spread toes, they can clamp themselves to tree trunks and branches so they can forage for insects, fruit, nectar, and/or sap.

Further Reading:
Stiles and Skutch, *A Guide to the Birds of Costa Rica.*

Campephilus principalis (ivory-billed woodpecker)

The ivory-billed woodpecker, once found in the southeastern United States and Cuba, is now on the edge of extinction, with only a small population in northeast Cuba, if it survives at all. It is (or was) a large woodpecker, 20 inches long. The plumage is striking black and white, and the bill an ivory color. The female's crest is black, and the male's red. Despite its large size, it has a soft call, something like a toy trumpet. To find the wood-boring insects it feeds upon, it peels away big pieces of bark with its strong bill. Its dependence on large dead trees for nesting and general deforestation have led to its decline. Reports continue that it survives in hard-to-reach patches of forest, although the last sighting was in 1991. In Cuba, indigenous people once adorned their homes with the birds as protection against witchcraft.

Further Reading:
Raffaele, *A Guide to the Birds of the West Indies.*
World Wide Fund for Nature, Switzerland <http://www. panda.org / resources / publications / species / underthreat / ivory-woodpecker.htm>.

Pipridae (manakins)

Manakins inhabit thick, lowland forests of the Neotropics, including those of Trinidad and Tobago. Most of the 60 species are fairly small and robust, with short tails and short, broad bills. Adult males are more colorful than females, and in many species the males gather at courtship sites called leks, where they dance about and make loud, strange vocal or wing sounds to attract females. They mainly eat berries, plucking off the fruit on short flights without landing, but they also snatch insects from foliage.

Further Reading:
de Schauensee and Phelps, *A Guide to the Birds of Venezuela.*
Stiles and Skutch, *A Guide to the Birds of Costa Rica.*

Chiroxiphia pareola (blue-backed manakin)

The blue-backed manakin is a very lively and agile bird, about four inches long. Its territory is from Colombia south to northern Bolivia and Brazil. Blue-backed manakins like to forage in low bushes near forest streams. The male is mostly black with sky-blue patches on its back and the sides and back of its head, and a bright red forehead. To court females, males rip off the vegetation from a branch, perch themselves on the cleared area, then call to a subordinate male, which obligingly sings a duet with the dominant male. To add a touch of gymnastics to the courtship display, the two males jump repeatedly over each other's backs. When an impressed female approaches, the subordinate male departs, leaving the dominant male to complete the wooing display on his own.

Further Reading:
de Schauensee and Phelps, *A Guide to the Birds of Venezuela*. The Rainforest Trust, USA <http://www.rainforesttrust. com/lib7cont.html>.
Stiles and Skutch, *A Guide to the Birds of Costa Rica*.

Pittidae (pittas)

Pittas are brilliantly colored songbirds, about six inches long, that inhabit the floor of dense forests, where they forage in the leaf litter in search of insects and snails. They have a strong sense of smell, helpful in locating earthworms. To open a snail shell, they break it on rocks. The 24 species of pittas are found only in the Old World tropics, with most inhabiting southern Asia. They build their dome-shaped nests of twigs on the ground. They hop around more frequently than they fly, but will often deliver their loud double whistle from high in a tree.

Further Reading:
Beehler, Pratt, and Zimmerman, *Birds of New Guinea*. Wildash, *Birds of South Vietnam*.

Pitta gurneyi (Gurney's pitta)

Gurney's pitta lives in southern Myanmar and the only remaining area of lowland rainforest in peninsular Thailand. They are very close to extinction, with perhaps as few as 25 surviving pairs in the latter country. There have been no recorded sightings in Myanmar since 1914. Like most pittas, the males are brilliantly colored, with a bright blue cap, black face, yellow throat, and striped chest. BirdLife International is working with the Royal Thai Forest Department to reduce deforestation in the Gurney's pitta habitat and convince villagers to refrain from trapping the birds for the caged-bird trade. Villagers are also helping with a reforestation effort. Gurney's pitta is one of the most threatened birds of Asia.

Further Reading:
World Wide Fund for Nature, Switzerland <http://www. panda.org / resources / publications / species / underthreat / gurneypitta.htm>.

Pitta sordida (hooded pitta)

The hooded pitta inhabits forests from India through Indonesia to Papua New Guinea. Its forehead, crown, and nape are chestnut brown, its back and breasts, bottle green, and its belly, scarlet. It has a pale turquoise patch on its shoulders.

Further Reading:
Beehler, Pratt, and Zimmerman, *Birds of New Guinea*. Wildash, *Birds of South Vietnam*.

Ptilonorhynchidae (bowerbirds)

Bowerbirds live in Papua New Guinea and Australia, mostly in mountain forests. They are about 10 inches in length and have rather drab plumage of black, olive, brown or greenish, although the males of several of the 18 species have orange crests. They are primarily frugivores, with some species sometimes taking arthropods, especially to feed nestlings. Their calls are loud and harsh, but they often throw in quite accurate imitations of other bird species and a range of ventriloquial notes. To entice females, males build "bowers," or territorial display sites, which are constructed of twigs and decorated with bits of brightly colored flowers or berries.

Further Reading:
Beehler, Pratt, and Zimmerman, *Birds of New Guinea*.

Amblyornis inornatus (Vogelkop bowerbird)

The Vogelkop bowerbird has rather drab olive-brown plumage, the plainest of all the bowerbirds, but the males of this species build the most elaborate bowers. They construct a teepee-shaped hut of twigs on the forest floor. It is three feet high and five feet in diameter, with an entrance and a "lawn" in front, both of which are strewn with flowers and fruits.

Further Reading:
Beehler, Pratt, and Zimmerman, *Birds of New Guinea*. Wolfe and Prance, *Rainforests of the World*.

Ptilonorhynchus violaceus (satin bowerbird)

The satin bowerbird inhabits the rainforests of northern Australia. Quite an active bird, it hops about vigorously on the ground and leaps between branches. It flies with swoops and dips. The males have glossy, blue-black plumage with blue eyes and a bluish bill tipped with yellow. The females are olive green on their backs, with tawny wings and tails. Their stomachs have downy white feathers edged in brown.

Males go to great lengths with their bowers, first laying a mat of twigs about three feet in diameter, then building on that two parallel, arched walls of twigs 15 inches high and 18 inches long. On the inner walls, the male applies a plaster of dried, macerated leaves and wood pulp. He decorates the elaborate display with blue or yellow objects: feathers, flowers, insect parts, berries, broken glass, pieces of plastic. The colors of the objects he chooses apparently purposefully match his own coloring. Once he has constructed the bower, the male perches overhead calling loudly. If a female approaches, he nabs a display ornament in his bill and poses in a trance, with his feathers partly raised and head low, his eyes flushing lilac-pink. Then he leaps to the side, opens his wing and tail feathers, and performs a robotic, stiff-legged dance while making wheezing and whirring sounds. Mating takes place near the bower; males will attempt to mate with any female that approaches.

Further Reading:
Pizzey, *A Field Guide to the Birds of Australia.*

Rhinocryptidae (tapaculos)

Tapaculos are found only in the Neotropics, with only one species in Central America and the 27 other species in South America. With their short wings, they are weak flyers, so they usually hop or scamper over the ground, holding their tails erect. They scratch the forest floor to find food—mainly insects and small invertebrates. Some species build nests dug into grassy banks in dense forest, while others nest in tree cavities or burrows abandoned by small mammals. Shy and secretive, they are hard to find, and scientists know little about them. Bird-watchers have mainly tracked them by following their loud calls.

That, in fact, was how Brazilian ornithologists found a new species in 1997, the lowland tapaculo (*Scytalopus iraiensis*)—they heard a song they had never heard before and so set up delicate and lightweight mist nets to catch the mysterious bird. They found the bird inside the city of Curitiba, 420 miles southwest of Rio de Janeiro. The lowland tapaculo's marshy habitat in Curitiba is scheduled to be destroyed by dam construction.

Further Reading:
Astor, "Song Led to Finding New Bird Discovered in Brazil" <http://abcnews.go.com/sections/science/DailyNews/newbird980424.html>.
Stiles and Skutch, *A Guide to the Birds of Costa Rica.*

Rupicolidae (cocks-of-the-rock)

Cocks-of-the-rock are a family of only two species, both found only in northern South America. The 14-inch-long males of both species are incandescent orange with fan-shaped crests that spring forward before their eyes and fall forward to cover their bills. Cocks-of-the-rock forage in the treetops for fruit and insects and build nests of mud.

Further Reading:
de Schauensee and Phelps, *A Guide to the Birds of Venezuela.*

Rupicola rupicola (Guianan cock-of-the-rock)

The Guianan cock-of-the-rock is famous for the spectacular dance that males perform to attract the chestnut-colored females. Males gather in a group, at courtship sites called leks, and make strange noises, flap their wings about, bow, and hop up and down. After the females make their selections and the pairs mate, the males take off. This cock-of-the-rock is fairly widespread south of the Orinoco in the Amazon of Venezuela, and is also found in the Guianas, far eastern Colombia, and north of the Amazon Forest in Brazil. The other species in the family, the Andean cock-of-the-rock (*R. peruviana*), is found in the Andes from Colombia to Bolivia.

Further Reading:
de Schauensee and Phelps, *A Guide to the Birds of Venezuela.*
Houston Zoo <http://www.houstonzoo.org/birds/pages/cockoroc.htm>.

Thraupidae

Thraupidae includes tanagers, honeycreepers, chlorophonias, and euphonias. About half a dozen of the 230 species in this family migrate to temperate areas of North and South America to breed; all the rest are residents of the Neotropics, including the Antilles islands.

Cyanerpes caeruleus (purple honeycreeper)

The purple honeycreeper inhabits cloud forests and mangroves, as well as forest edges and clearings, from western Ecuador to French Guiana, Brazil, Bolivia, and Trinidad. Males are mostly a vivid violet, with black throats and wings and a black stripe through their eyes. Their legs are lemon yellow. The females are green and buff, with patches of blue on their breasts and sides. Purple honeycreepers are about four inches long and forage for fruit and insects in treetops. They also sip flower nectar and have a long (0.7-inch), curved black bill to make this easier.

Further Reading:
de Schauensee and Phelps, *A Guide to the Birds of Venezuela.*
The Rainforest Trust, USA <http://www.rainforesttrust.com/lib7cont.html>.

Euphonia minuta (white-vented euphonia)

The white-vented euphonia is a dainty, slender bird, just under four inches long, and a resident of southern Mexico, Central America and western Ecuador, northern Bolivia, and eastern Brazil. The males have yellow foreheads, shiny blue-black heads, and throats and upper parts of sunny yellow, with white feathers under their tails. Females also have yellow foreheads, but their upper parts are olive, their breasts pale yellow, and their throats and bellies white.

Euphonias forage in pairs or small groups in the forest canopy and mostly eat berries, with the occasional small insect. All euphonias eat mistletoe berries and then excrete the seeds, which usually fall, stick to branches, and germinate. Like other euphonias, the white-vented builds a globe-shaped nest of moss, ferns, plant fibers, and orchid roots, with a side entrance. The nest is built high in the canopy amid mosses and epiphytes. The female incubates her three to five eggs for 12 to 18 days. The hatchlings remain in the nest for 11 to 24 days, fed by both parents with regurgitated food.

Further Reading:
Howell and Webb, *A Guide to the Birds of Mexico and Northern Central America.*
The Rainforest Trust, USA <http://www.rainforesttrust.com/lib7cont.html>.
Stiles and Skutch, *A Guide to the Birds of Costa Rica.*

Tangara chilensis (paradise tanager)

The paradise tanager is a remarkably colorful bird in a genus known for its brilliantly hued plumage. The top and sides of its head are covered with short, apple-green feathers, its lower back is scarlet, its rump and upper tail are golden yellow, its throat and upper breast are violet, and its belly is a bright turquoise. About five-and-a-half inches long, it forages for berries and insects in groups with other small birds. The paradise tanager lives in forests of northern South America to the Brazilian Amazon.

Further Reading:
de Schauensee and Phelps, *A Guide to the Birds of Venezuela.*
The Rainforest Trust, USA <http://www.rainforesttrust.com/lib7cont.html>.

Turdidae (thrushes, robins, solitaires, allies)

Thrushes, robins, solitaires, and allies include some of the best-known birds, with the sweetest voices and most complex songs. The 300 or so species are found worldwide, except New Zealand and several South Pacific islands, but the greatest diversity is found in Old World temperate areas, particularly highland forests. Dozens of species are migrants, nesting in temperate areas and migrating to tropical areas during the colder months. Tropical residents also make altitudinal migrations—flying from high altitudes to lower.

Further Reading:
Howell and Webb, *A Guide to the Birds of Mexico and Northern Central America.*
Stiles and Skutch, *A Guide to the Birds of Costa Rica.*

Myadestes melanops (black-faced solitaire)

The black-faced solitaire is not a flashy-looking bird, but it has one of the sweetest songs in the cloud forest. Slate gray, with a black head, face, and chin, its legs and broad short bill are orange. Found only in Costa Rica and western Panama, it lives in the tangled understory of wet mountain forests, flying up into the canopy to feed on fruits and seeds or to perch and sing. Its song is a high-pitched, clear whistle, interspersed with rich, flutey notes. It builds a cup-shaped nest of moss and liverworts in hollows of mossy banks. Black-faced solitaires breed in the highlands and then sometimes move to lower elevations.

The birds are increasingly rare except in inaccessible areas or parks, because their ethereal songs so move human beings that they trap the birds and keep them in cages in their backyards to hear them all the time. In Costa Rica, residents will pay poachers up to $300 per bird. To the north, the slate-colored solitaire (*Myadestes unicolor*), a resident of the humid evergreen and pine forests of Mexico, Honduras, El Salvador, and Nicaragua, faces the same problem because of its haunting whistle.

Further Reading:
Howell and Webb, *A Guide to the Birds of Mexico and Northern Central America.*
Stiles and Skutch, *A Guide to the Birds of Costa Rica.*

Vangidae

Found only on the island of Madagascar, most *Vangidae* have glossy blue and black plumage above and whitish feathers below, and they range in size from 5 to 12 inches. Originally, there were 13 species of Vangidae on Madagascar, but few can survive outside their forest habitat.

Further Reading:
Langrand, *Guide to the Birds of Madagascar.*

Euryceros prevostii (helmet vanga)

The helmet vanga has rufous, or reddish, wings and upper parts, and its under parts are black with light brown stripes on its belly, flanks, and undertail coverts. Its tail is all black, except for two rufous tail feathers in the middle. Its fat, hooked, two-inch bill is light blue. The helmet vanga is about 11 inches long and is at home only in undisturbed forests, where it hunts for insects. It is a strong flyer and seizes its prey on the wing, as well as from branches or tree trunks. Because of its strict habitat needs and Madagascar's fast-disappearing forests, the helmet vanga is seldom seen.

Further Reading:
Center for Biodiversity and Conservation, USA <http://research.amnh.org/biodiversity/center/cmada.html>.
Langrand, *Guide to the Birds of Madagascar.*

Vireonidae

The *Vireonidae* is a family of 43 species of fairly small birds found only in the New World. Most species breed in North America and the West Indies. Most are migratory; even some that nest in the tropics still make long migratory flights.

Further Reading:
Stiles and Skutch, *A Guide to the Birds of Costa Rica.*

Hylophilus ochraceiceps (tawny-crowned greenlet)

The tawny-crowned greenlet inhabits the forests of southern Mexico to Ecuador and Brazil and does not migrate. It has yellow and brown plumage, with a buff-yellow breast, and a raised and bushy tawny crown. About 4.5 inches long, it inhabits low- to mid-level forests, in pairs or small groups, and often with mixed flocks of other bird species, particularly the golden-crowned warbler and red-crowned anttanager. It gleans spiders, grasshoppers, roaches, moths, ants, and other insects from foliage, holding larger prey beneath a foot while it dismembers them. Like other members of this bird family, the tawny-crowned greenlet's nest is a sturdy cup of plant material that's attached by its rim to a forked twig and hangs suspended. The greenlet covers its nest with green moss.

Further Reading:
Howell and Webb, *A Guide to the Birds of Mexico and Northern Central America.*
The Rainforest Trust, USA <http://www.rainforesttrust.com/lib7cont.html>.
Stiles and Skutch, *A Guide to the Birds of Costa Rica.*

Migratory Birds: Flying South, North, East, and West

In North America, from August to November, hundreds of bird species wing their way south to escape cold weather. Some 250 species—including songbirds, shorebirds, and hawks—fly as far south as Mexico, Central and South America, and the Caribbean. The massive migration of birds from their nesting grounds in the United States and Canada is one of nature's most impressive phenomena and an amazing feat of wing power.

Migrant birds must pack in the food before such a trip, sometimes doubling their body weight. The tiny blackpoll warbler (*Dendroica striata*) is just five inches long and weighs less than an ounce, even after its preflight feeding frenzy. Yet in 72 hours it can fly thousands of miles from Maine to northern Venezuela, without stopping for food, water, or even a brief rest. The three-inch, ruby-throated hummingbird (*Archilochus colubris*) nests in eastern North America then flies to wintering grounds from Mexico to Panama by passing through Florida and Cuba, through Mexico, or amazingly, by dashing across the Gulf of Mexico, a nonstop flight of 600 miles. An ornithologist working in a national park in the Dominican Republic captured a Bicknell's thrush (*Catharus bicknelli*) in a mist net and found a familiar-looking band attached to its leg. He had tagged the exact same bird the previous summer on Mount Mansfield in Vermont, 2,000 miles north.

Raptors, or birds of prey, must stay over land to catch the wind and soar, so they fly over the ever-narrowing isthmus of Mexico and Central America to reach their South American wintering grounds. In Veracruz, Mexico, so many raptors fill the skies during the northern autumn and spring months—as many as four million birds—that the phenomenon is called the "River of Raptors."

In the 1980s, ornithologists realized that the populations of some songbird species seemed to be declining. They concluded that the decline was likely caused by reduced habitat not only in their northern homes but also in their southern habitats. Migrants tend to return to the same forests every year, but if that forest has been razed, they are in trouble. Birds that stop to rest or reach their final southern destination point need food, water, and shelter fast. A cattle pasture or banana plantation won't do. Biologists are still studying the relationship between the climbing deforestation rate in the Neotropical forests and declining migratory bird populations.

New research shows that the introduction of non-native plants to North America may also be threatening songbird populations. When American robins and wood thrushes build nests in honeysuckle and buckthorn shrubs—neither of which are native to the United States—their offspring almost never escape predation. The non-native shrubs don't provide adequate protection. Honeysuckle has strong branches, so birds tend to build their nests lower to the ground than usual, making it easier for raccoons and other predators to steal eggs and nestlings. Buckthorn doesn't have protective thorns like the native hawthorn shrub does. Both honeysuckle and buckthorn are aggressive plants that have largely replaced native shrubs that provide better nest sites.

Habitat destruction in Africa and Asia also affects the populations of the migratory birds of Europe. At least 200 bird species abandon Europe and northern Asia when the cold weather moves in, flying south to sub-Saharan Africa and Southeast Asia. The woodpecker-like European wryneck (*Jynx torquilla*) breeds as far north as Scandinavia and flies south to central Africa and southern Asia for the winter. The wee willow warbler (*Phylloscopus trochilus*), which weighs no more than one-third of an ounce, breeds in Siberia and manages to fly 7,000 miles to its winter home in east African woodlands.

The European honey buzzard (*Pernis apivorus*) puts on about 10 ounces before attempting the nonstop flight of more than 4,000 miles from its northern European nesting grounds to tropical Africa. Since it has a particular diet—bees and wasps, which it robs from nests—it gains a clear advantage from its ability to make the trip without needing to stop to search for food. Not all migratory birds fly south. Cool weather south of the Tropic of Capricorn drives many bird species north. In South America, many birds of Chile and Argentina, particularly those that nest in the Andes mountain range, begin flying north in January. The blue and white swallow (*Notiochelidon cyanoleuca*) of central Chile and Argentina may fly as far north as Mexico, while Argentina's population of fork-tailed flycatchers (*Tyrannus savana*) usually migrates no farther than northern South America.

Beyond the millions of migrant birds from Europe and Asia, tropical Africa also supports many migrant birds that fly north from the southern regions of the continent during the cool and dry months of April through August. Migrant species include the rainbow-hued pygmy kingfisher (*Ceyx picta*), which winters as far north as Mozambique, and the paradise flycatcher (*Terpsiphone virdis*), a chestnut, black, and gray bird with 10-inch tail plumes that winters in Zambia and Angola.

Fewer than 15 percent of Australia's and New Zealand's land birds are migrants. Those that are include the bright turquoise sacred kingfisher (*Halcyon sancta*), which is found throughout Australia, with southern populations joining northern residents or flying on to Sumatra, Borneo, New Guinea, and surrounding islands during the winter months of April through September.

Further Reading:
Elphick, *The Atlas of Bird Migration.*
Line, "Eating on the Run."
Line, "Silence of the Songbirds."
Terborgh, *Where Have All the Birds Gone?*
Wille, "Mystery of the Missing Migrants."

Fish (*Pisces*)

The fish that live in the thousands of rivers and streams that course through the world's tropical forests play a vital role in the ecosystem. They are an essential source of nutrition for rainforest people and wildlife, and in the Amazon's flooded forest, they help disperse tree seeds. The term "fish" describes a life-form, not an order of animal. The 20,000 species of fish are all vertebrates belonging to the phylum *Chordata,* and they are divided into five to seven classes that differ from one another anatomically. The species below are listed alphabetically by the Latin names of their order, family, genus and species.

ATHERINIFORMES

This order contains about 16 families of bony fish, both marine and freshwater species.

Anablepidae

The family *Anablepidae* includes two species of four-eyed fish found in North, Central, and South American rivers.

Anableps spp. (four-eyed fish)

Four-eyed fish don't wear glasses, as their common name might suggest. Rather, their eyes are split in two, horizontally, with the lower sections adapted for seeing in the water, and the upper halves containing different lenses for seeing in air. The three species of *Anableps* inhabit the rivers and estuaries of Central America's Pacific slope and the northeastern coast of South America. Four-eyed fish average six to eight inches long and have elongated bodies with rounded tails, large pectoral fins, and wide, froglike mouths. Their big, round eyes sit atop their flattened heads. They spend most of their time floating just below the water's surface, with the top halves of their eyes protruding, hunting for a variety of small aquatic animals, while watching out for above-water predators, such as herons and other birds. Their reproduction is another unusual aspect of four-eyed fish: Fertilization takes place inside the female, which gives birth to several live young one to two inches long. Genitalia of both genders are oriented either to the left or right, in such a way that "right-handed" males can mate only with "left-handed" females, and vice versa.

Further Reading:
Berra, *An Atlas of Distribution of the Freshwater Fish Families of the World.*
Herald, *Living Fishes of the World.*
Wheeler, *The World Encyclopedia of Fishes.*

CHANNIFORMES

"Snakeheads," as this order is also called, have elongated bodies and are found in tropical freshwaters in the Old World. They are, like most fish, carnivorous, but unlike most fish, they can survive for relatively long periods out of water.

Channidae

The snakeheads are found in Africa and Asia.

Channa micropeltes (snakehead)

The snakehead is an odd-looking, air-breathing fish that inhabits the rivers of India and much of Southeast Asia. It is the largest of the 14 species of *Channa spp.*, a group distributed through tropical Africa and Asia, with some species as far-flung as Japan and Oceania. It has a cylindrical body more than three feet long and weighing as much as 44 pounds, with soft-rayed dorsal and anal fins running the length of its body and the large reptilian head that characterizes all species of *Channa*. Its body, which is covered with small scales, is a mottled brown with lateral bands of red flanked by black running down both sides. Like most snakeheads, *C. micropeltes* is an insatiable predator that eats fish, insects, crustaceans, frogs, snakes, small mammals, and water birds. Snakeheads spend much of their time hidden in the reeds or algae, where both parents will prepare a nest by biting off vegetation. Males guard the floating eggs and colorful fry until the young can swim away. The ability of snakeheads to breathe air allows them to live in oxygen-poor swamps—some species will drown if prevented from surfacing to gulp air—and to survive days on end out of water. They are an important food fish, and live snakeheads are often displayed in the stalls of Asian markets.

Further Reading:
Berra, *An Atlas of Distribution of the Freshwater Fish Families of the World.*
Graham, *Air-Breathing Fishes: Evolution, Diversity, and Adaptation.*
Herald, *Living Fishes of the World.*

CYPRINIFORMES

Found worldwide in fresh water, except in Antarctica and Australia, the *Cypriniformes* include about 3,500 species.

Characidae

The *Characidae* live in fresh or brackish waters in Africa and South and Central America. This large and diverse family includes many species that are popular as food and for aquariums. Many are brightly colored, such as tetras.

Colossoma macropomum (tambaqui)

Tambaqui have black scales on their bellies but are greenish-yellow on their backs, excellent camouflage for murky river bottoms and light surfaces. They are more than three feet long and can weigh 65 pounds. The unusual structure of their noses gives them an

excellent sense of smell: They have nasal flaps that they can raise to increase the flow of water past their olfactory cells, and so can sniff out the fruits and seeds on which they depend. They are particularly fond of rubber tree fruits (*Hevea brasiliensis*), so Amazon fishermen often park their boats by these trees, harpoons ready to stab at their own dinners. The tambaqui is the most important food fish of the central Amazon. Like many tasty fish of this region, its populations are threatened by overfishing.

Further Reading:
Goulding, *The Flooded Forest.*
Goulding, *Floods of Fortune.*

Hydrocynus vittatus (tigerfish)

Tigerfish are large, aggressive predators that live in many African lakes and rivers. A large, silvery fish with a bluish back, it has a shark-like dorsal fin and a splash of pink on the lower lobe of its forked tail. It feeds on smaller fish and has fanglike teeth that are visible even when its mouth is shut, which makes it dangerous for fishermen to handle. Nevertheless, tigerfish are popular with sport fishermen, since they are legendary fighters. While *H. vittatus* can grow as long as three feet and weigh as much as 40 pounds, its cousin the giant tigerfish (*H. goliath*) is one of Africa's largest freshwater species, reaching a length of six feet and weighing as much as 150 pounds. The giant tigerfish is found only in the Congo River system and Lake Tanganyika.

Further Reading:
Herald, *Living Fishes of the World.*
Leveque, *The Freshwater Fish of Tropical Africa.*

Pygopristis, Pygocentris, Pristobrycon, and Serrasalmus (piranhas)

Piranhas are abundant in the rivers of eastern and central South America, with the greatest number of species in the Amazon River system. Despite their vicious reputation as flesh-eaters, the diet of most piranha species consists mainly of fruit, although some species are carnivorous. There are a total of about 30 species. They vary in color, with some a silvery hue with orange bellies and throats and others virtually all black. All species have saw-edged bellies and large heads with strong jaws, with triangular, razor-sharp teeth that snap shut like steel traps. The most ominous piranha is the red-bellied (*Pygocentrus nattereri*), which has the strongest jaws and sharpest teeth, designed to rip out chunks of flesh. Red-bellied piranhas, which can reach two feet in length, also hunt in groups from a dozen up to 100. When they attack prey, one fish will latch on to it, while others from the group join in on the kill, rapidly darting in for a quick bite, then darting out of the feeding frenzy.

Further Reading:
Goulding, *The Flooded Forest.*

Semaprochilodus spp. (jaraqui)

Jaraqui is silver with a bright yellow and black striped tail. It looks and tastes a bit like carp and is a popular fish in central Amazon markets. With the aid of specially designed lips, it feeds on the dead leaves and other detritus of the river bottom. It can turn its thick and fleshy lips inside out to form a suction pad that is lined with bristles, used to scrape and filter the detritus from submerged trees in the flooded forest.

Further Reading:
Goulding, *The Flooded Forest.*

Electrophoridae

Electrophoridae are found in South America and have long eel-like bodies, although they aren't eels, and no scales.

Electrophorus electricus (electric eels)

Electric eels are large, cylindrical fish that inhabit the languid waters and swamps of South America's major river systems. One of several fish species capable of producing significant electrical shocks, the electric eel is by far the most powerful, delivering discharges of between 350 and 650 volts, which is enough to stun a horse. Those discharges are generated by electric organs flanking the spinal column, which account for about three-eighths of the eel's weight. They are an excellent defense mechanism, but they also help the eel hunt because they can stun nearby fish, which they then catch and eat. Electric eels spend the day resting in reeds or other quiet areas and hunt at night, when they may also use weak electric signals to detect prey and navigate. They can grow as long as eight and a half feet and have marble-brown, scaleless bodies with just one fin, a long, undulating anal fin that runs along most of the underside of their bodies. The fin helps the eel move both forwards and backwards.

Further Reading:
De Carli, *The World of Fish.*
Herald, *Living Fishes of the World.*

Gasteropelecidae

Gasteropelecidae live in Central and South America. They have strongly compressed bodies, with well-developed pectoral fins, and are capable of flight. There are about nine species; many are popular aquarium fishes.

Gasteropelecus sternicla (hatchetfish)

Gasteropelecus sternicla is a small, silvery fish that measures no longer than two to three inches, and, like other species in this family, is considered the only fish capable of flying. Species in other fish families may use their oversized pectoral fins to glide over the water, but hatchetfish are the only ones that actually flap theirs like wings. Hatchetfish spend most of their time floating near the surface, hunting for small insects, but when frightened, they race away, rising out of the water and flying as far as nine feet. They have translucent tails and fins, upward-oriented mouths, and narrow bodies with deep, round bellies that hold the bones and muscles for their winglike pectoral fins. When in flight, they make a sound similar to that of a hummingbird.

Further Reading:
Berra, *An Atlas of Distribution of the Freshwater Fish Families of the World.*
Herald, *Living Fishes of the World.*

LEPIDOSIRENIFORMES (lungfish)

Lepidosirenidae

Lepidosirenidae are lungfish native to western Africa, Brazil, and Paraguay, found in slow-moving freshwater and swamps.

Protopterus spp. (African lungfish)

African lungfish are gray, eel-like creatures with pale bellies, stringy pectoral and pelvic fins, and long tails that end in a point. Together with the South American and Australian lungfish, the four African species are considered "living fossils" because they differ little from fish that were common more than 400 million years ago. Although lungfish are among hundreds of fish capable of breathing air, they are some of the few species that will drown if prevented from surfacing to gulp air, and their lungs are very similar to those of land animals. They also have the rare capability to aestivate—they survive droughts by burrowing into the mud and creating a mucus cocoon that keeps them moist inside dry earth for months on end. Fish have survived more than four years within such cocoons, though they were extremely emaciated when they were finally returned to water. They average three feet in length, although fish twice that length have been caught. Lungfish are common in the oxygen-poor swamps and marginal waters of Africa's major rivers and lakes, where they prey on small fish, crabs, snails, and other animals. Lungfish, in turn, are a popular catch for African fishermen, though some tribes won't eat them. The fish spawn during the rainy months, when the females lay thousands of eggs in burrows dug by males, who guard them until the eggs hatch. Larvae have external gills, which they lose after several months. The species *P. dolloi*, which lives in the Congo River, is very similar to the South American lungfish (*Lepidosiren paradoxa*), a fact cited as proof that those two continents were once attached.

Further Reading:
FishBase <http://nypa.uel.ac.uk/ibs/sp2000/fishbase/>.
Graham, *Air-Breathing Fishes: Evolution, Diversity, and Adaptation.*
Lowe-McConnell, *Ecological Studies in Tropical Fish Communities.*

MORMYRIFORMES

Mormyriformes include two families of soft-rayed, bony fish with electricity-producing organs that live in freshwater in Africa.

Mormyridae

The family *Mormyridae* contains 18 genera and nearly 200 species ranging in size from six inches to five feet long, with greatly varying coloration and markings.

Gnathonemus spp. (elephant fish)

Elephant fish, also known as *mormyrids,* are small- to medium-size fish that live only in Africa, and are named for their distinctive snouts, which resemble a pachyderm's trunk. They use those elongated mouths, or lower lip protuberances, to dig up insects, mollusks, worms, and other small animals from the bottoms of the rivers, lakes, and swamps they inhabit. Mormyrids are thick-skinned, slimy fish, most of which have slender bodies, deeply forked tails, and large, nearly identical dorsal and anal fins at the back of their bodies, which help them swim both backwards and forwards.

One trait shared by all elephant fish is their ability to generate a weak electrical field that acts like radar, helping them navigate, forage, and identify other fish in murky waters and at night, when most species are active. Mormyrids are among the most common fish in the vast Congo River system, which drains the rainforests of western Africa, and the larger species are popular food fish all over Africa. Consequently, some species are threatened by intensive fishing and deforestation of watersheds.

Further Reading:
Berra, *An Atlas of Distribution of the Freshwater Fish Families of the World.*
Lowe-McConnell, *Ecological Studies in Tropical Fish Communities.*

OSTEOGLOSSIFORMES

This is an ancient order of prehistoric freshwater fish. There are four families found in freshwater worldwide, except for very cold regions.

Osteoglossidae

Osteoglossidae is a small family, with one species found in Africa, three in South America, and three in Southeast Asia to Australia. Most are omnivores or carnivores that live in still waters.

Arapaima gigas (pirarucú)

Pirarucú is the most magnificent predatory fish of the Amazon. It reaches more than 10 feet in length and can weigh 450 pounds, making it one of the two or three largest freshwater fish in the world. Unlike most other predatory fish, like some piranha species, pirarucú have few small bones. Firm and rich-tasting steaks one foot in diameter can be cut from a fish that weighs just 50 pounds. Only very young fish have gills, which they lose a few days after birth. Pirarucú must surface to fill their lung organs, on average every 10 to 15 minutes, although they can stay submerged for nearly half an hour if necessary. They bite with their jaws, like most predatory fish, but also give a tongue-lashing. Their huge tongues have teeth, and, when pressed against the roof of their mouths, are like a second set of jaws. Dried pirarucú tongues were traditionally used in the Amazon as seed graters.

Further Reading:
FishBase <http://nypa.uel.ac.uk/ibs/sp2000/fishbase/>.
Goulding, *The Flooded Forest.*

Osteoglossum bicirrhosum (arowhana)

The arowhana is a smooth swimmer and, at three feet long, probably the largest fish in the world that feeds on insects and spiders. It swims with wavy motions along the river edge, hunting for insects that have fallen into the water. It may also snatch small birds and bats perched on low-hanging limbs. Arowhanas have chin whiskers and eyes that are divided by a horizontal partition. The optical divider seems to help them see well both in and above the water. In addition to their superior swimming skills, arowhanas are excellent high jumpers. If they spot food on a branch above the water, they circle below the target, then compress their bodies, and hurl out of the water at precisely the right spot, so their large mouths can capture the targeted prey. Arowhanas are mouth-breeders: once the female lays the eggs and the male fertilizes them, the male scoops them up into his mouth. The young are born inside this safe cavern, living off their yolk sacs for a few weeks. Dad then releases them, but if danger threatens, the young flee back into their father's mouth, perhaps guided by the chin whiskers.

Further Reading:
Goulding, *The Flooded Forest.*

Pantodontidae

Pantodontidae are found in freshwaters of tropical western Africa. They are closely related to the *Osteoglossidae.*

Pantodon buchholzi (freshwater butterflyfish)

The freshwater butterflyfish is a tiny cousin of the massive South American pirarucú; this species is not related to marine butterflyfish. It inhabits the still, marginal waters of Africa's Congo and Niger River systems, and is small; adults average four to five inches long, with gray-brown, mottled bodies, semi-transparent fins, and large mouths that are oriented upward. They have long rays that hang down from their undersides and large pectoral fins that resemble wings, which enable them to glide more than six feet through the air. The species was actually discovered by a butterfly collector, who caught one in his net when it leaped out of a swamp. Butterflyfish spend

most of their time floating just below the surface, waiting for small insects to fall into the water, which they quickly gobble up. Eggs float on the surface and hatch within a matter of days, after which the fry feed on tiny aquatic insects.

Further Reading:
Berra, *An Atlas of Distribution of the Freshwater Fish Families of the World.*
FishBase <http://nypa.uel.ac.uk/ibs/sp2000/fishbase/>.
Herald, *Living Fishes of the World.*

PERCIFORMES

Perciformes is an order with 140 families, which have many physical variations among the families. Most are marine species, found worldwide. The smallest in the family is the gobie, at under a half-inch long, and the largest is the swordfish, which can be 25 feet long.

Anabantidae

Commonly called labyrinth fishes, *Anabantidae* inhabit tropical freshwaters of Asia and Africa. They breathe with their gills, but also possess a supplemental breathing structure, a chamber above the gills, called the labyrinth. There are 70 species of labyrinth fishes.

Anabas testudineus (climbing perch)

The climbing perch lives in ponds and rivers throughout the Asian tropics, from India to the Philippines. Its wide distribution is at least partly because it can survive for extended periods out of water. A perch-shaped, gray-green fish averaging nine inches in length, *Anabas* is a cousin of several African species of the genus *Ctenopoma* and is also related to gouramis and fighting fish. It is one of many fish species capable of breathing air, but one of few capable of crawling across the land to find a new pond or river home. Climbing perch pull themselves forward using strong pelvic fins and spikes beneath their gill covers. Such overland migrations often take place en masse, at night, or during a rainstorm, when there is less danger of the fish drying out; they can travel nearly 200 yards in one night. Their name comes from a legend that they climb palm trees to drink their sap, and indeed, fish have been found high in trees. But biologists now believe that birds of prey actually left these treetop-stranded fish in branches.

Climbing perch are an important food fish

throughout Asia because they have few bones and can survive out of water. In Malaysia and India, people carry them in moistened clay pots, in which the fish can stay alive for several days. The species' ability to breathe air enables it to inhabit oxygen-poor waters, and during drought, it may burrow into the mud and aestivate. Although it has gills, it will drown if prevented from gulping air. Climbing perch eat an array of insects, crustaceans, and plant matter.

Further Reading:
Berra, *An Atlas of Distribution of the Freshwater Fish Families of the World.*
Graham, *Air-Breathing Fishes: Evolution, Diversity, and Adaptation.*
Wheeler, *The World Encyclopedia of Fishes.*

Belontiidae

This is a family of about 50 species of fighting fishes, found in freshwater in tropical Africa, India, Myanmar, Sri Lanka, and Southeast Asia.

Betta splendens (Siamese fighting fish)

The Siamese fighting fish is one of more than a dozen species of fighting fish native to Southeast Asia. Many of this genus are bred for the aquarium trade. In the wild, this two-inch fish with oversized fins is usually a dull red or green, but years of careful breeding have resulted in the colorful specimens sold in pet stores. Fighting fish are so named for the aggressiveness of males—females are quite docile—which will fight to the death if placed together in an aquarium. In India and Indonesia, two males are customarily pitted against each other for sport, with large sums of money often being bet on the matches.

In spite of their pugnaciousness, male fighting fish are caring fathers. They build bubble nests by blowing mucus-coated bubbles, and following mating, they pick the fertilized eggs up from the bottom and carry them to their floating nests, which they then guard until their young hatch and swim away. Some *Betta* species, such as the Java and Sumatra fighting fish, are mouthbreeders—males carry the eggs and young in their mouths until they can swim freely.

Further Reading:
Berra, *An Atlas of Distribution of the Freshwater Fish Families of the World.*
Herald, *Living Fishes of the World.*

Centropomidae

Most of the *Centropomidae* live in brackish or marine waters, although there are a few freshwater species.

Lates niloticus (Nile perch)

The ancient Egyptians worshipped the Nile perch and featured the fish in tomb paintings and other art. Sometimes Nile perch were even mummified and buried. These massive fish are just as common in the jungle-lined branches of Africa's Congo and Niger River systems as in the Nile River. One of eight species of the genus *Lates*, the Nile perch is Africa's largest freshwater fish, with records of individuals longer than six feet and weighing 250 pounds, although the average weight is 60 pounds for females and 25 pounds for males. It is a voracious predator, eating a variety of medium-size fish, and is in turn sought after by African fishermen, who have overfished it in some areas. Since they know the fish is such a strong fighter, African fishermen will often let a hooked Nile perch wear itself out by towing their canoe around for two hours or more before trying to reel in the fish.

Lates spp. have stout, mottled bodies covered with scales, large dorsal and pectoral fins, small heads, and large mouths with protruding jaws, filled with rows of tiny teeth. They spend their days resting in deep water and head to the shallows to hunt at night. Spawning takes place in quiet, marginal waters. The eggs float just below the surface for a day before hatching, and the tiny fry spend their first days devouring plankton, then moving on to prawns and small fish as they grow.

Further Reading:
Leveque, *The Freshwater Fish of Tropical Africa.*
Lowe-McConnell, *Ecological Studies in Tropical Fish Communities.*

Gobiidae

Simple and often colorful fishes, *Gobiidae* have small and slender bodies and comparatively large fins. There are more than 700 species that inhabit shallow coastal waters and estuaries worldwide. The Philippine goby, the world's smallest fish at one-half inch long, is a member of this family.

Periophthalmus chrysospilos (mudskipper)

An air-breathing fish that climbs over land and up tree roots, the mudskipper is practically amphibious, spending as much of its life on land as in the water.

These atavistic creatures provide a glimpse of evolutionary events 350 million years ago, when the first fish abandoned the water and began to adapt to life on land. It is a six- to ten-inch, elongated fish with bumpy skin—mottled brown with gold flecks—and a bulbous head containing the enlarged gill chamber that allows it to breathe air. *P. chrysospilos* is common in mangrove forests, the coastal forests of tropical tidal zones of the Malay Peninsula, Sumatra, and Java. It is one of 12 species of *Periophthalmus*, one of several mudskipper genera distributed from West Africa to Australia and Japan. Quite active on land, mudskippers crawl on dexterous pectoral fins, below which lie modified pelvic fins that function as suction cups, which allow the fish to climb vertical mangrove roots. They feed on an array of crustaceans, insects, worms, small fish, and plant matter, and will burrow into the mud to escape birds, large fish, and other predators. Their big, protruding eyes give them a froglike appearance. Since they don't have tear ducts, they must regularly roll their round eyes down into moist sockets.

Mudskippers often gather in groups of 20 to 30, but when mating, males can be fiercely territorial. Males dig large, funnel-like nests in the mud, then raise their brightly colored dorsal fins, stand on their tails, and leap into the air to attract females. The couple enters the nest for fertilization, after which the female lays her eggs at the bottom and guards eggs and fry until they can swim.

Further Reading:
Banister, *Encyclopedia of Aquatic Life.*
Graham, *Air-Breathing Fishes: Evolution, Diversity, and Adaptation.*

Osphronemidae

Osphronemidae includes just one species, the giant gourami.

Osphronemus goramy (giant gourami)

The only member of its family, the giant gourami is native to Borneo, Java, Sumatra, and the Malay Peninsula, but is now distributed from India to the Philippines. Its range has been expanded by human beings, who have introduced this delicious, meaty fish to an array of nations and aquatic habitats. It is a pale, spade-shaped fish with a disproportionately small head, long, slender pelvic fins, and large dorsal and anal fins along the back half of its narrow body. The giant gourami can grow to two feet long and weigh several pounds. It eats only aquatic plants and can breathe air, which enables it to live in oxygen-poor waters. Like most of the smaller gourami species, which are members of the families *Belontiidae* and *Helostomatidae, Osphronemus* places its eggs in bubble nests and guards them till they hatch.

Further Reading:
Berra, *An Atlas of Distribution of the Freshwater Fish Families of the World.*
Graham, *Air-Breathing Fishes: Evolution, Diversity, and Adaptation.*
Herald, *Living Fishes of the World.*

Toxotidae

Only one genus, *Toxotes*, found from eastern India to Papua New Guinea and Australia, is a member of this family.

Toxotes jaculator (archer fish)

The archer fish, an unusual fish found in the coastal estuaries and lowland rivers of the Asian tropics, is famous for its ability to shoot down insects perched on overhanging foliage by spitting drops of water at them. *Toxotes* does this by snapping its gill covers shut and pushing a stream of water up through a tube formed by its tongue pressed against the top of its mouth, which has a groove in it similar to the barrel of a gun. An adult archer fish can hit a target as far as 10 feet away. Just as remarkable as its range is the fish's ability to aim accurately from under water—no small task considering that it must compensate for the distortion caused by the refraction of light on the water. Adult archer fish average seven to nine inches long and have narrow bodies that are rounded at the back and come to a point at the head. They have large, protruding eyes, an upwardly oriented mouth, and silvery scales with several dark, vertical bars along their upper flanks. Juveniles have luminous green or gold spots on their sides, which may help them see each other in turbid waters, in order to school.

Although juvenile archer fish primarily eat small aquatic animals, adults prey both on crustaceans and aquatic insects captured in the water and insects they shoot down, such as beetles, spiders, caterpillars, and flies. If their prey is near the surface, they may leap out of the water and snatch it. They are occasionally caught by fishermen and are considered good eating fish, whereas juveniles are captured for the aquarium trade.

Further Reading:
Banister, *Encyclopedia of Aquatic Life.*
Berra, *An Atlas of Distribution of the Freshwater Fish Families of the World.*
Wheeler, *The World Encyclopedia of Fishes.*

POLYPTERIFORMES

Found in freshwater of Africa; 9 of the 10 species of *Polypteriformes* inhabit the Zaire River basin.

Polypteridae

Polypteridae have long bodies covered with thick, bonelike scales.

Polypterus spp. (bichirs)

The bichirs are related to lungfish and average one to two feet in length; some species reach four feet. They are found in the swampy areas of equatorial west Africa's lakes and rivers, including the Congo and Senegal River systems. They are also called "living fossils" because they have hardly evolved since the Cretaceous period, 135–65 million years ago. Their ability to breathe air allows them to live in oxygen-poor waters and to survive long periods out of the water. They are capable of crawling over land for short distances, using their large pectoral fins. Those species that live in drier regions can survive droughts by aestivating underground. Bichirs have snake-like heads, elongated bodies covered with hard, diamond-shaped scales, and five to eight dorsal finlets for which the genus was named: *Polypterus* means "many fins." They rest in vegetation by day and hunt at night for smaller fish, amphibians, and invertebrates, which they locate using a highly developed olfactory system; their eyesight is rather poor. There are 10 species of bichir ranging in coloration from yellow-brown to gray-green.

Further Reading:
Berra, *An Atlas of Distribution of the Freshwater Fish Families of the World.*
FishBase <http://nypa.uel.ac.uk/ibs/sp2000/fishbase/>.
Graham, *Air-Breathing Fishes: Evolution, Diversity, and Adaptation.*

RAJIFORMES

This order of eight families, *Rajiformes*, includes rays, banjofishes, and sawfishes.

Potamotrygonidae

Potamotrygonidae is a family of freshwater rays.

Potamotrygon laticeps (freshwater stingray)

The freshwater stingray lives in several rivers of Colombia and Venezuela and is one of 18 species of rays distributed through the major river systems of the American tropics. All belong to this uniquely freshwater ray family. This stingray's flat, circular body has a diameter of about 12 inches, with a slight protuberance at the snout and a narrow, pointed tail just over a foot long, on the back of which is a barbed, poison-filled spine. While the ray's mouth and reproductive organs are on its pale, smooth underside, its eyes flank a smooth head that rises slightly out of its flat back, which is covered with tiny spines and colored a mottled dark brown and black. These bottom feeders eat a range of crustaceans and small invertebrates and give birth to live young. When resting, they partially bury themselves in the sand or mud of the river bottom, which poses a danger for fishermen and swimmers—if stepped on, a ray will lift its tail up and sting the person with its spine. These injuries are extremely painful and may be fatal.

Further Reading:
Berra, *An Atlas of Distribution of the Freshwater Fish Families of the World.*
Wheeler, *The World Encyclopedia of Fishes.*

SILURIFORMES

Siluriformes, the order of catfish, has about 30 families, most of which inhabit freshwater bodies. The majority of the 2,000 species live in Africa and South America.

Doradidae

Doradidae are thorny catfishes that live in South America. They have overlapping plates that cover the sides of their bodies. Most are under 40 inches long, and they are common aquarium fish.

Lithodoras dorsalis (rock-baçu)

Rock-baçu is a large catfish of the lower Amazon River basin. It is endowed with heavy armor: bony plates along its flanks, each of which has one thick, backward-curving spine that can cause serious injury to most animals. It also has long, serrated spines on its dorsal and pectoral fins. The rock-baçu eats fruit, mollusks, and the leaves of arum plants.

Further Reading:
Goulding, *The Flooded Forest.*

Malapteruridae

Malapteruridae are the electric catfish of Africa.

Malapterurus electricus (electric catfish)

Electric catfish are so named for their ability to generate electrical shocks as strong as 350 volts. They are common in the rivers and lakes of western Africa and the Nile. The ancient Egyptians knew about this species—it appears in tomb paintings dating from 2750 B.C.—and the Arabs used its shocks for medical treatments as early as the 11th century. An extremely plump, pale gray fish with small eyes, the electric catfish has no dorsal fin, a large, round tail, and a small mouth full of tiny teeth. Three pairs of barbs drape around its mouth. Although young electric catfish can be docile, adults are known for their pugnacity. They can reach four feet long and weigh as much as 55 pounds. They eat smaller fish, which they hunt at night by approaching a school and emitting an electrical discharge. This stuns or kills several fish, which the catfish quickly gobble up. The fish has an electric organ, consisting of modified muscle tissue, that forms a gelatinous membrane just below the skin and runs the length of its body. Primarily used for hunting, the electrical jolts also serve as a protective mechanism, and the species is consequently unpopular with fishermen, though it is edible. Water pollution and other environmental degradation threaten the fish in some areas.

Further Reading:
Berra, *An Atlas of Distribution of the Freshwater Fish Families of the World.*
Herald, *Living Fishes of the World.*
Leveque, *The Freshwater Fish of Tropical Africa.*

Pangasiidae

Pangasiidae inhabit freshwater from Pakistan to Borneo. They have compressed bodies and usually two pairs of chin barbels.

Pangasianodon gigas (Mekong catfish)

The Mekong catfish is one of several species of giant Asian catfish, which are some of the largest freshwater fish in the world. Growing to a length of 8 feet and a weight of 246 pounds, the Mekong catfish has smooth, pale skin, with a dark gray back, a sharklike dorsal fin preceded by a dangerous spine, and a large anal fin just below its slightly forked tail. Its small eyes are set low in its massive head, near its only pair of short barbels, and its gaping mouth is devoid of teeth; it eats only aquatic plants. An inhabitant of the Mekong River system, this massive animal spends most of the year in that river's lower reaches, migrating upriver at the end of the year to spawn in Lake Tali, in southern China. *P. gigas* is a valuable food fish, and years of intensive fishing have greatly reduced its numbers, leading to its current status as an endangered species. Its larger cousin, *P. sanitwongsei,* can reach a length of nearly 10 feet, but since it is a bottom feeder, its meat is valued less than that of the Mekong catfish. Nevertheless, it is eaten, and overfishing of *P. sanitwongsei,* which is found only in Thailand's Menam Chao Phya basin, has vastly reduced its numbers.

Further Reading:
Berra, *An Atlas of Distribution of the Freshwater Fish Families of the World.*
FishBase <http://nypa.uel.ac.uk/ibs/sp2000/fishbase/>.
Herald, *Living Fishes of the World.*
Kottelat, *Freshwater Biodiversity in Asia.*

Siluridae

Found in Asia, Europe, and Africa, *Siluridae* have compressed bodies, long anal fins, and short or missing dorsal fins.

Wallagonia attu (mulley)

The mulley is a large catfish found in the major rivers of the Asian tropics, from India to Indonesia. It has a slightly flattened head and a rounded, scaleless body tapered toward the tail, with a pair of long barbels, slender pectoral and dorsal fins, and a gray anal fin that runs more than half the length of its body. Its back and head are light green, and its sides are cream-colored with a patch of pink on the gill covers. Its average length is about 3 feet, although individuals as long as 6.5 feet, and weighing around 120 pounds, are not uncommon. The mulley feeds almost exclusively on fish and is known as a voracious predator that hunts by day, when it can be seen ripping through schools of small fish near the surface, snatching prey with its sharp teeth, and leaping out of the water. It is a popular food fish in most countries and as a result has been overfished in some areas.

Further Reading:

Banister, *Encyclopedia of Aquatic Life.*
Kottelat, *Freshwater Biodiversity in Asia.*

Fish of the Flooded Forest

The Amazon River basin supports about 2,500 species of freshwater fish, 8 times that of the Mississippi and 10 times that of all of Europe. During the rainy season, the Amazon's many tributaries flood their banks, covering at least 40,000 square miles for four to seven months each year. Many fish species leave the rivers they inhabit during nonflood months and swim throughout the flooded forest. Unlike fish anywhere else in the world, many of the fish species of the flooded forest depend on fruits and seeds to which the floods give them access as a primary source of food.

Further Reading:
Goulding, *The Flooded Forest.*
Wolfe and Prance, *Rainforests of the World.*

Insects (*Insecta*)

> *Insecta* is the largest class of animals in the phylum *Arthropoda*, which is the largest animal phylum. About 800,000 insect species have been identified, which is about 75 percent of known animal life. They are found in every land and freshwater habitat on Earth where food is available. Insects have segmented bodies, jointed legs, and external skeletons. Their bodies have three units: the head, which is where the mouth, eyes and a pair of antennae are located; the thorax, which generally has three pairs of legs and one or two pairs of wings; and the reproductive organs.

COLEOPTERA

The order *Coleoptera* (beetles) is the insect order with the largest number of species, with approximately 330,000. It is, moreover, the largest order in the entire animal kingdom. One of every four identified animal species in the world is a beetle. They originated some 240 million years ago and have since evolved into a hugely varied order of insects. They come in all sizes, many colors and forms, live in virtually every habitat, and may be predatory, herbivorous, omnivorous, or waste feeding. Most, but not all, have wings, and the forewings are strong and hard, acting as covers to protect the delicate, membranous hind wings and abdomen.

Further Reading:
Wootton, *Insects of the World.*

Cerambycidae (long-horned beetles)

This family, long-horned beetles, is mostly small, mundane beetles, although some are spectacularly large and fierce-looking. Their common name comes from their elongated antennae, which in many males are at least half the length of their bodies, and in some species about two or three times longer. The bigger species have large, powerful, and sometimes serrated mandibles, perfect for gnawing on wood. The sound of long-horned beetles gnawing on tree branches can be heard from several feet away. Some longhorn species have evolved intriguing protective devices, such as barbs on their long antennae. Others bear markings that make them resemble lichen on tree barks, while some species can emit foul-smelling or -tasting secretions. Rather than taste unpleasant themselves, some species look like other insects that do, or they resemble stinging ants, bees, or wasps.

Further Reading:
Janzen, *Costa Rican Natural History.*
Wootton, *Insects of the World.*

Acrocinus longimanus (harlequin beetle)

The harlequin beetle gets its name for the elaborate patterns of black, yellowish, and red markings on its back. They average 2.5 inches long and range from southern Mexico to South America. The female deposits 15 to 20 eggs after making an incision in a tree, often one infested with bracket fungus, which affords good camouflage. Once they hatch, the larvae bore into the wood. They mature within eight months and tunnel downward in the tree trunk to excavate a pupal cell. After four months, the adult beetle emerges through a fresh hole it gnaws above the entrance hole.

Further Reading:
Janzen, *Costa Rican Natural History.*

Titanus giganteus (titanic longhorn)

The titanic longhorn, a native of the northern Amazon, is the largest longhorn and one of the largest beetles on Earth. Their bodies measure up to eight inches and the antennae another four inches. Dark brown to black, they have massive jaws that can snap a pencil in half. They have been located in northern Brazil and French Guiana, but little is known of their biology.

Further Reading:
Hogue, *Latin American Insects and Entomology.*

Xixuthrus heyrovski

Xixuthrus heyrovski is just slightly smaller than the titanic longhorn and is found on Fiji. Island residents eat the beetles' huge larvae as a delicacy.

Further Reading:
Wootton, *Insects of the World.*

Elateridae (click beetles)

Click beetles include about 7,000 widely distributed species. They make a clicking noise when seized and lie on their backs to play dead. To right themselves, they bend forward at the thorax to hook their spines into notches on their abdomens. When they release their spines, there's a clicking sound, and the beetles leap into the air.

Further Reading:
Wootton, *Insects of the World.*

Pyrophorus spp. (headlight beetles)

Headlight beetles have two round, luminescent organs on their undersides that provide light, like fireflies. The organs consist of a light-producing layer and a reflector layer, which contains crystals that scatter the light outward. The glow is the result of a reaction in the outer layer between the compound lucifein and an enzyme, luciferase.

Both male and female beetles shine a blue-green light in flight. *Pyrophorus* are found from southern Mexico to southeastern Brazil and in the West Indies. About an inch long, they feed on rotting fruit and plant parts.

Further Reading:
Hogue, *Latin American Insects and Entomology.*

Scarabaeidae (scarabs)

The family of scarabs includes about 20,000 species found worldwide, but the Neotropics are particularly rich in these beetles. Most are small to medium-size, although several are among the very largest insects. The club of the straight antennae of scarabs has three to nine flat plates that may be spread apart. Scarab larvae are pale-colored grubs that usually feed on plant roots; some feed on dung, fungi, or rotting organic matter. The larvae of larger species that develop in rotting palm logs are popular local snacks in several tropical countries. Adult scarabs feed on fruit, nectar, and flowers; some species eat decaying organic matter.

The subfamily *Scarabaeinae* are beetles that generally feed on or lay eggs on feces of other larger animals. Dung rollers (*Canthon spp.*) of the Neotropics form balls of dung and bury them in burrows. The adults may eat the dung ball or the female may lay a single egg on it, which the hatched larvae later consume. The beetles have a highly developed technique that they use to roll the dung balls; sometimes two scarabs may cooperate in the rolling process, one pushing, and the other pulling. Female dung diggers (*Phanaeus spp.*), found only in the New World, first dislodge a piece of dung, then transport it to a chamber underground, sometimes with the help of a male. She shapes it into a sphere, coats it with soil, and lays a single egg inside. When hatched, the larva feeds on the dung until the sphere becomes quite hollow. It metamorphoses into an adult within its protective sphere.

Dung diggers are mostly metallic green, blue, or violet. The male has a single, upward-projecting horn on his head, while the female has a wide, shovel-shaped flange on the front of her head, which she uses to scoop dung or excavate a chamber for larvae. In Africa, most dung beetles are of the genus *Aphodius.*

Further Reading:
Hogue, *Latin American Insects and Entomology.*
University of Florida, Dept. of Entomology & Nematology, USA <http://www.ifas.ufl.edu/eny3005/lab1/Coleoptera/Scarabaeid.htm>.

Dynastes spp. (Hercules beetles)

The Hercules beetles of the Neotropics include some of the largest beetles in the world, the size of the males being augmented by quite long, down-curving front horns that extend beyond and curve down on another horn. The body length is about three inches, but the horn can be of equal length. They can weigh up to 1.5 ounces. The horns are used to defend territory and mating rights and then to carry off the prized female. Nocturnal, these beetles feed on sweet fruits and the sap from wounded trees; the larvae feed on soft, decaying wood of dead trees. Four species inhabit the Neotropics.

Further Reading:
Hogue, *Latin American Insects and Entomology.*

Goliathus spp. (Goliath beetles)

Goliathus spp. includes five species, all confined to the Tropics of Africa. They are about four inches long and, when in flight, sound a bit like tiny whirring helicopters. Each of the beetle's six legs has a pair of claws that allow it to attach itself securely to a tree trunk. *Goliathus* beetles have large eyes and sensitive, clubbed antennae. Males have a forked horn on their heads that they use in battles over females. Most are black and white, with contrasting black, vertical stripes on their thoracic shield; some species are yellow and brown. Female Goliaths lay their eggs in rotting wood or other decaying plant material. When the larvae hatch, they feed for months on the wood, and often reach six inches in length and weigh up to three ounces. The larvae then build cocoons and, after several months, a beetle breaks out of each cocoon. Like other scarab beetles, the Goliath breaks down dead plant material in the forest, returning nutrients to the soil.

Further Reading:
Living Planet <http://www.geocities.com/CapeCanaveral/Launchpad/9191/Scarabaeidae.htm>.

Megasoma elephas (rhinoceros beetle)

The rhinoceros beetle lives in lowland forests from southern Mexico, extending south to northern Colombia, and east to northern Venezuela. They weigh about one ounce, and measure up to five inches long. They get their common name from the males' long, slender, upward-turning head horn. People sometimes remove the horns of rhinoceros and Hercules beetles, grind them up, and take the powder as an aphrodisiac. They are only actually useful, however, to the beetles themselves, as fighting equipment. Male beetles battle over feeding sites and females. The larvae, which can also reach three inches, feed on the pulp of dead tree trunks and may take three to four years to mature, all within the log. That

means the host log must be quite large, since the rate of decay in tropical forests is too fast to allow *Megasoma* development in small logs or branches. As a result, *M. elephas* is a rare insect, since the beetle cannot survive in forests that have been logged.

Further Reading:
Hogue, *Latin American Insects and Entomology.*
Janzen, *Costa Rican Natural History.*

Pelidnota and Plusiotis spp. (precious metal scarabs)

The precious metal scarabs are so called because they are the color of polished metals. *Plusiotis batesi* is the color of solid gold, while *Plusiotis chrysargyrea* is silver. *Pelidnota virescens* is the color of burnished copper, and *Pelidnota sumptuosa* shines like stainless steel. About 200 species of these jewel-colored scarabs inhabit the Americas from the southern United States to Peru. Most live in high-altitude cloud forests or northern dry forests. They seem to be nocturnal, but little is known of their habits.

Further Reading:
Hogue, *Latin American Insects and Entomology.*

DICTYOPTERA

This order is commonly divided into two suborders: the cockroaches (*Blattodea*) and the mantids (*Mantidae).*

Blattodea (cockroaches)

The suborder of cockroaches contains four families and 4,000 species, of which only 25 to 30, or less than 1 percent, could be considered any sort of pest to human beings. Cockroaches have inhabited the Earth for at least 250 million years. They are the ancestors of termites and are more similar to their prehistoric ancestors than is any other living insect. The vast majority of species live in the world's tropics, and while they may seem invincible in your kitchen, many species face extinction as more tropical forests are lost.

The largest known cockroach in the world is *Megaloblatta longipennis* of South America, with a wingspan of seven inches, and the heaviest is the Australian species *Macropanesthia rhinocerus*, which weighs up to 1.5 pounds. Another large species is the death's head cockroach (*Blaberus craniifer* and *Bla-*

berus discoidalis), native to South America and about three inches long. The dark markings on their backs look a bit like a skull. The smallest cockroach, at 0.16 of an inch, is *Attaphilla fungicola*, a New World species that lives in the nest of leaf cutter ants and feeds on the fungus they farm. *Gromphradorhina portentosa*, the hissing cockroach, is native to Madagascar and is the roach most commonly kept as a pet around the world. It is a handsome reddish-brown, and when picked up or otherwise annoyed makes a hissing sound by expelling air rapidly from its trachea. Cockroaches are the food of many small mammals, birds, reptiles, and amphibians, plus mites and parasites.

Although some cockroaches are woodborers, most inhabit the dark rainforest floor, skittering about the leaf litter. Males, which are smaller than females, insert a hard case loaded with sperm inside a female's body. A few days to a few weeks later, depending on the species, the females produce an egg case called the ootheca. Initially, it protrudes from her abdomen, then withdraws into a chamber called the brood pouch, where it remains until the eggs are ready to hatch. In the giant cockroach (*Blaberus giganteus*), a woodborer and the largest species in the Neotropics, this takes about 60 days, and about 40 offspring, called nymphs, then emerge. Most female cockroaches produce two or three broods annually. The female of the Neotropical species *Xestoblatta hamata*, which inhabits the forest floor, is particularly fecund—she produces 25 eggs every 8 to 10 days and, unlike most cockroaches, mates repeatedly during her adult life. Nymphs are whitish when they hatch, darkening in a few hours as their cuticles harden. They shed their exoskeletons as they develop until they reach full size and develop wings in about six months.

Further Reading:
Earth Life Web <http://www.earthlife.net/insects/six.html>.
Janzen, *Costa Rican Natural History*.
North Carolina State College of Agriculture and Life Sciences <http:// www.cals.ncsu.edu / ncsu / cals / course / ent425 / compendium/roach.html>.

Mantidae (praying mantids)

Praying mantids are predatory insects that live throughout the tropics and in the sunnier areas of temperate zones, including southern Europe and North America. Most species are tropical, and although there are other mantid families, most belong to *Mantidae*. The majority of mantids are medium-size, about five inches long. The smallest are less than one-half inch long, and the largest is a species in Sri Lanka that is almost 10 inches long.

The mantid's common name comes from the bug's at-rest stance, with its forelegs folded up before it. These piously positioned forelegs are actually powerful weapons, often lined with rows of sharp spikes. A large South American species feeds on small birds and reptiles, as well as other insects. Mantids are able to move only the top part of their bodies, which means they can follow the approach of prey without moving the bottom part, which might startle the unsuspecting potential victim. They also have excellent eyesight. They catch prey—mostly insects—with their powerful forelegs, hold them in place, and eat them, using their strong jaws. They can feed on one small insect, pin it down with one leg, and reach out and nab another bug with the other leg.

Mantids can fly, but do so mostly at night. Bats are a common predator, and some mantid species have developed the ability to hear the high-pitched sonar that bats use to navigate. They do so via a hollow chamber in their bodies; mantids have only one ear. If they detect a bat while flying, they will drastically change directions, sometimes spiraling to the ground. Many mantid species have evolved to blend in amazingly well with their surroundings, to resemble a flower or a twig. The dead-leaf mantis (*Deroplatys dessicata*) looks, not surprisingly, just like a withered and brown dead leaf. When young, the orchid mantis (*Hymenopus coronatus*) of Malaysia bears the same hues as the orchids on which it perches to wait for prey, although it loses this camouflage as it matures.

Like cockroaches, mantids go through incomplete metamorphosis. They do not have a maggot or caterpillar, but nymphs that look like miniature, wingless adults hatch directly from the eggs. Like cockroaches, they produce egg cases, from which may hatch 30 to 300 nymphs, depending on the species. Female mantids of some species sometimes eat males during copulation. In fact, in a number of species the females cannot receive the case containing the male's sperm until she removes his head.

Further Reading:
Earth Life Web <http://www.earthlife.net/insects/six.html>.

DIPTERA (true flies)

The true flies have only one pair of wings (except for a few wingless species) and are agile aviators, able to rapidly fly backward, forward, or tack sideways. Their wings are transparent and they have sucking, piercing, and lapping mouthparts. There are about 120,000 species of *Diptera*.

Further Reading:
Raven and Johnson, *Biology*.

Culicidae (mosquitoes)

Mosquitoes are found throughout the tropics; Latin America is home to almost one-third of the world's more than 3,000 species. Because some are carriers of diseases harmful to human beings, a good deal is known about these blood-sucking insects. They breed in swamps, coastal marshes, and mangroves, but are most often found in any freshwater accumulation, especially where there is a good deal of vegetation—along ponds, marshes, lakes, and rivers. Their larvae and pupae can survive in miniscule amounts of water, including the bucket-shaped centers of bromeliads and other cupped plants or even in the tiny puddles that accumulate on fallen leaves. The annoying whine that mosquitoes make while flying is the sound of wings beating nearly 600 times per second. Mosquitoes are important sources of protein for many birds, reptiles, amphibians, and other arthropods.

The adult females of most species will bite wherever they can find exposed skin on mammals and amphibians, or through the feet and base of bird beaks, or in between reptile scales. Not all species need blood meals, but the females of most species must feed on blood before laying eggs. She removes just a drop of blood, and injects her saliva, which contains anticoagulants and anesthetic to avoid detection. The saliva causes an allergic reaction; the reaction among human beings is usually a raised, red, and itchy bump, except in cases when pathogens also exist in the saliva.

One of the greatest killers of humankind, malaria, is carried by *Anopheles* mosquitoes, found in the Old and New World. When biting, the mosquito rotates its head more forward than do other species, as though it's doing a headstand. Because there is no record of malaria in the Neotropics before the Spanish Conquest, the vectors were probably carried over by early explorers. In 1585, Sir Francis Drake came to the West Indies with 29 ships, 1,500 sailors, 800 soldiers, and at least several *Anopheles* mosquitoes, which he carried to the Caribbean and Central America. By the time they reached the New World, hun-

dreds of sailors had succumbed to malaria and Drake had to abandon his mission.

Further Reading:
American Mosquito Control Association <http://www.mosquito.org/mosquito.html>.
Newman, *Tropical Rainforest*.
Wootton, *Insects of the World*.

Aedes aegypti (yellow fever mosquito)

The yellow fever mosquito is originally from West Africa, but has spread throughout the world. A small, black insect marked with silvery-white bands on its legs and abdomen, it is probably the most domesticated of all the mosquitoes, more likely to be found around human dwellings than forests, and with a clear predilection for human blood. In the late 1950s, an outbreak of yellow fever killed 15,000 people in Ethiopia, nearly 10 percent of the population. *A. aegypti* also carries dengue fever, and encephalitis is transmitted by both *Aedes* and *Culex* species.

Further Reading:
Hogue, *Latin American Insects and Entomology*.
Newman, *Tropical Rainforest*.
Skaife, *African Insect Life*.

HEMIPTERA

Hemiptera include leafhoppers, planthoppers, treehoppers, cicadas, aphids, psyllids, whiteflies, and scale insects. There are some 60,000 species.

Further Reading:
Raven and Johnson, *Biology*.

Coccidae

Coccidae are worldwide common pests of cultivated plants. Most species are sedentary during most of their life cycles and secrete a thick, waxy coating that shields their bodies from enemies.

Further Reading:
North Carolina State College of Agriculture and Life Sciences <http://www.cals.ncsu.edu/course/ent425/compendium>.

Laccifer lacca

Found in India and Southeast Asia, *Laccifer lacca* are the natural source of shellac. The sticky coating is the only commercial resin of animal origin. The bugs coat themselves with the secretion and leave a residue on tree branches, where it is collected and used as shellac and red dye. The word "lac" is derived from the Hindi and Persian words for "one hundred thousand," which refers to the number of these tiny insects needed to produce lac, known as sticklac in its raw form, seedlac when it is semiprocessed, and shellac in its final finished form. About 300,000 insects are needed to produce 2.2 pounds of sticklac. Because it bonds well and is nontoxic, hard and strong, the processed shellac is widely used as a paint primer or sealer, in inks, in hair lacquers, as a protective varnish for papers (e.g., playing cards), and in other coatings and waxes.

Further Reading:
North Carolina State College of Agriculture and Life Sciences <http://www.cals.ncsu.edu/course/ent425/compendium/homopt~1.html>.

Reduviidae (assassin bugs)

Most species of assassin bugs live in the tropics, although some also are found in northern climes. Their name comes from their manner of quickly dealing death to their insect prey.

Further Reading:
Janzen, *Costa Rican Natural History.*

Apiomerus pictipes

This common species of assassin bug has been identified in Colorado, New Mexico, and Mexico, south to Bogotá, Colombia. The adults are a half inch long with a narrow head and a beak that folds underneath the head. The bugs are black with various shades of rust, orange, gold, or red. *A. pictipes* parks itself on a shrub and waits for insect prey. The bug advances slowly toward its prey, which can be nearly any kind of insect, slowly raises its forelegs, and then swiftly seizes the victim and inserts its beak, from which it injects a paralyzing saliva. The bug then sucks on the captured insect for an hour or more. The bugs often wait at the edges of termite trails to nab workers that pass by. Females seem to be the better hunters, perhaps because they need protein to produce eggs.

The bugs give off a distinct scent when alarmed, probably to warn off would-be predators. Their bite is sharp enough to ensure that few predators will attempt a capture more than once. Mating is a tricky matter, since both males and females may end up feeding on each other. Once they separate, they back off warily.

Further Reading:
Forsyth, *Portraits of the Rainforest.*
Janzen, *Costa Rican Natural History.*

HYMENOPTERA (ants, bees, wasps)

Ants, bees, and wasps incorporate approximately 100,000 species, many of which are social insects. Many species have stingers or chewing and sucking mouthparts and well-developed eyes.

Further Reading:
Raven and Johnson, *Biology.*

Apidae spp. (honeybees)

Like all bees, honeybees have hairs adapted for collecting pollen. There are eight species of true honeybees (some entomologists say only six), found worldwide. Honeybees are on average under an inch long, which makes them relatively large insects, so they are fairly easy to spot. Social insects, they have distinct castes, with different members of the population specializing in certain functions, and all communicating with one another. The only sexually mature female is the queen, which can produce up to 1,500 eggs per day. The larvae are fed with "bee milk," a secretion from young adult worker bees. The tens of thousands of worker bees are sexually inactive females, which have stingers to defend the hive. The colony also includes a few hundred drones, or sexually mature males, which die after mating.

Worker bees that find food sources fly back to the nest and communicate the exact location to other workers by doing a special bee dance. The worker bees' stomachs hold enzymes that convert nectar into a honeylike substance. This is regurgitated into storage cells in the hive, and fanned by other workers in order to evaporate excess water. The resulting substance is honey. To produce one pound of honey, bees must visit two million flowers. The honeybee hives are made of wax and are vertical and double-sided. Some species produce a single, exposed comb the size of an open hand; others, a comb the size of a door. Some honeybee species produce nests with multiple combs, usually in an enclosed place. When a hive's population grows too large, a second queen is produced, and the colony divides. One group stays with the original hive; the second leaves with a new queen to build a new hive.

No honeybee species is native to the Americas. *A.*

laboriosa hails from the Himalayas; *A. dorsata*, *A. florea*, and *A. andreniformis* are native to Southeast Asia and the Philippines; *A. cerana* to Japan, Southeast Asia, and India; *A. vechti* and *A. cerana* are from Borneo, and *A. mellifera*, the common honeybee, comes from northern Europe and south throughout Africa. Adaptable *A. mellifera* has been introduced virtually everywhere on Earth.

An average honeybee colony has around 20,000 members. They store vast amounts of honey in the nest, which helps insulate them against cold temperatures and provides a storehouse of food during times of year when flowers are scarce. Human beings are just one of many animal species that enjoy eating honey. Since several bee species build nests in confined places, beekeepers provide hives to encourage the insects to provide readily accessible supplies of honey.

Further Reading:
Michener, *The Social Behavior of the Bees*.
University of Michigan Museum of Zoology, Animal Diversity Web <http://animaldiversity.ummz.umich.edu/accounts/apis/a._mellifera$narrative.html#TOC>.

Formicidae (ants)

Out of 8,804 known species of ants, 2,162 are found in the Neotropics, 2,500 in sub-Saharan Africa, and another 2,080 in temperate and tropical Asia. The biologist Edward Wilson, who with Burt Hölldobler won a Pulitzer Prize for the book *The Ants*, believes that there may be as many as 20,000 different species in the world.

Further Reading:
Wilson and Hölldobler, *The Ants*.

Eciton spp. (army ants)

Army ants include 150 species, most of which live in the Neotropics. Like most ant colonies, Army ants have a single queen who lays all the eggs, while female workers tend the young and gather food for the entire colony. *E. burchelli* is perhaps the most studied species of army ant, found in lowland forests from Mexico to Brazil and Peru. Rather than a nest, the colony lives in an above-ground swarm, called a bivouac. Common sites for bivouacs, which are formed by the ants linking their legs and bodies together with their strong, clawed front legs, are under fallen logs or in between tree buttresses. An average colony includes about 450,000 workers in a bivouac, with thousands of immature ants and the mother queen in the center. During a brief interval during the dry season, about 1,000 males and several virgin queens are also sheltered.

As morning light filters down on the bivouac, it evolves into a churning mass and breaks into a raiding column. The workers that find themselves in front lay down a chemical scent, then tumble back into the column, replaced by new workers. Unlike other army ant species, *E. burchelli* are "swarm raiders," meaning that foraging workers disperse in front into a fan-shape swarm. Most other species are "column raiders," marching forward along narrow trails. The purposeful sorties of *E. burchelli* are devastating for any creature unable to escape. They overtake and sting to death spiders, scorpions, beetles, grasshoppers, and sometimes injured or weakened snakes, lizards, and nestling birds. Prey is carved into smaller pieces and passed back by the columns to the bivouac. Swarming worker ants may capture more than 100,000 arthropods a day to keep the colony well fed. Marching alongside the charging worker ants are larger soldier ants, whose massive mandibles help defend against enemies.

In Africa, army ants are called "driver ants" and belong to the genus *Dorylus*. The queen of this genus is the largest of all ants, ranging in size from 1.5 to two inches long, and is capable of delivering up to two million eggs each month. *Dorylus* bivouacs are more stable than those of most other army ants. After a march, the colony, which includes millions of workers, may settle down deep in the earth, digging tunnels and chambers. Sorties may take several days, but the sedentary period can last from six days to one or two months. The advancing, tumbling column is dozens or even hundreds of workers wide, with ants rushing forward and then falling back, replaced by other advancers. The swarm moves at a rate of about 65 feet per hour. Bites from their huge, shearing mandibles can kill monkeys or domestic livestock like pigs and chickens. Since they can saw apart prey, pieces of larger animals can be passed back along the column of workers.

Further Reading:
University of Exeter <http://www.ex.ac.uk/bugclub/raiders.html>.
Wilson and Hölldobler, *The Ants*.

Azteca spp.

Azteca spp. can easily be identified by their strong odor. Found only in the Neotropics, they nest in trees, mainly species that have hollow stems or trunks, in which the ants can tend mealybugs (*Pseudococcidae*). The mealybugs, which the ants raise like livestock, provide them with sugar and vitamins. Other species have mutually beneficial relationships with tree species. The trees provide the *Aztecas* that live in the hollow chambers between the nodes of their trunks with minute capsules filled with high-energy carbohydrates. In exchange for the housing and free food, the ants will attack any pest or vine that touches the host tree. Some species build external nests constructed of plant fibers and secretions. *Aztecas* are aggressive and will destroy the nests of other ants that try to colonize nearby; some species use noxious chemical sprays to deter the competition. They can probably afford to behave in such an unneighborly and energy-expending manner because they have a ready supply of mealybugs to provide instant nutrition.

Further Reading:
Janzen, *Costa Rican Natural History.*

Acromyrmex spp. and *Atta spp.* (leafcutter ants)

Leafcutter ants live only in the Neotropics. The genera *Acromyrmex* includes 24 species, and *Atta* 15 species. Some 50 million years before human ancestors thought to poke seeds into the ground, leafcutter ants were practicing sophisticated farming. The crop they cultivate is fungi. Worker ants saw off pieces of leaf and carry them back to the nest, where they are cut into smaller pieces, licked and chewed, spit out, and sowed in a garden of sorts. Tropical plant leaves often contain highly toxic chemicals evolved to discourage nibbling by insects. But the ants inject the chewed leaves with a species of fungus that breaks down these poisons and produces nutrients, sugars, and proteins, as food for the ants. The ants are selective about the leaves they choose for growing fungus, never taking much vegetation from any one tree. Scientists suggest that their pickiness is related to the toxins found in many flora species. Leaf predation stimulates many plants to increase their unpalatability, so the leafcutters take relatively little and move on to a new plant. Leafcutters are the most voracious herbivores in the Neotropics, collectively consuming far more vegetation than any other animal group.

One *Atta* colony may contain more than a million workers, which makes them one of the largest and most complex societies on Earth. Different workers within one colony have mandibles of different lengths and sizes, as each is adapted to do a particular job.

The smallest take care of the eggs and larvae; larger ants are in charge of foraging; and the largest, the soldier ants, defend the nest from attack. Small worker ants, called "minima," ride on the leaf fragments carried by larger worker ants, ready to snap their pincers at a small fly, *Apocephalus spp.* These flies hover above the larger worker ants, trying to lay their eggs on the back of the ants' necks. When the larvae hatch, they eat the ants' brains, so protection from the tiny, hitchhiking bodyguard ants is a lifesaver. Both workers and soldiers are sterile servants of the queen ant. According to E.O. Wilson, leafcutter ants "are among the most advanced of all the social insects." In Neotropical forests, leafcutters are considered to be "keystone species," whose loss would result in the loss of many other, dependent species. The main predator of leafcutter ants is the armadillo (*Dasypodidae*).

Further Reading:
Forsyth and Miyata, *Tropical Nature.*
Newman, *Tropical Rainforest.*
Wade, "For Leafcutter Ants, Farm Life Isn't So Simple."
Wilson and Hölldobler, *The Ants.*
Wolfe and Prance, *Rainforests of the World.*

Oecophylla spp. (weaver ants)

Weaver ants are among the most abundant ants of the Old World tropics. An ancient genus, its ancestors date back 30 million years. *O. longinoda* occurs across most of forested, tropical Africa, while *O. smaragdina* is found from India to northern Australia and the Solomon Islands. They live in huge colonies, sometimes one half million strong, with just one queen. In a masterpiece of cooperation, the weaver ants construct their nests in canopy treetops, sometimes among the branches and leaves of three to four wide-crowned trees. To undertake such a colossal construction job, workers first choose a pliable leaf, form a living chain, and together, as a single unit, pull on the leaf. Once they have pulled back a corner of the leaf, other workers assist, sometimes with several living chains formed in rows, an amazing effort of intricate coordination. Once the leaves are folded over like tents, workers form rows to hold the leaves together. Then another group of workers carry larvae in from other nests and, holding them between their mandibles, use them as sources of silk to bind the leaves together. In effect, they are tailors, weaving a silken tent by manipulating, via light pressure, mature ant larvae as if they were shuttles on a loom.

Further Reading:
Myrmecology <http://www.myrmecology.org/>.
Wilson and Hölldobler, *The Ants.*

Meliponidae (stingless bees)

Stingless bees is something of a misnomer because they do have internal stings, but no muscles that can extrude them to use as weapons. They are located throughout the tropics; for centuries, they have been provided with hive-boxes to encourage them to produce honey, which is used more often as a traditional medicine than as food. The honey of stingless bees was an important part of the Mayan culture in Central America. The Maya dedicated ceremonies to the bee gods Noyumcab and Ah-Mucencab to ensure a copious honey harvest. Stingless-bee nests are surrounded by a durable outer layer made up of mud, dung, plant resins, chewed-up wood fibers, saliva, and a more papery inner layer. Most stingless bees build nests in hollow tree cavities or abandoned sections of termite mounds. The nests have entrance tubes protected by guard bees. While they cannot sting, they do have strong bites and caustic acids in their saliva. The tubes, which may be sealed off at night, lead into chambers where the honey is stored.

Further Reading:
Island Crop Management <http://nanaimo.ark.com/~cberube/beevolv2.htm>.

Vespidae (social wasps)

Social wasps are venomous stinging insects whose members include hornets and yellowjackets. They construct nests of chewed-up wood or foliage, which they use for only one year. They feed on nectar and sap. Researchers study social insects because they are ecologically so successful, being the most populous social organisms on Earth.

Mischocyttarus mastigorphorus

Mischocyttarus mastigorphorus is the only known social insect species ruled by males. The wasp is about one-half an inch long and lives in small colonies of two to three dozen individuals in the misty cloud forests of Central America. Whereas in other social insect species the females rule, and the males are at the bottom of the ladder, *M. mastigorphorus* males rule the nest, biting and chasing off females and eating a lion's share of the food gathered by workers. And they do very little work around the house.

Further Reading:
University of California at Riverside Dept. of Entomology <http://insects.ucr.edu/ent133/ebeling/ebel9–2.html>.
University of Washington <http://www.washington.edu/newsroom/news/>.

ISOPTERA (termites)

Termites, like *Hymenoptera*, are social insects, but with important differences: *Hymenoptera* larvae are helpless, while termite nymphs are fully active from the very start. Also, while ant, bee, and wasp workers are always sterile females, termite workers may be either male and female and, in their early stages, capable of mating. In termite communities, the adult males, the kings, mate with the queens at regular intervals and stick around afterward. In contrast, male *Hymenoptera* usually die right after mating. Termites feed, groom, and protect each other, and—a characteristic of a truly social animal—the offspring of one generation assists the parents in raising the next.

That colonies of ants and termites are so similar is remarkable. Termites evolved from cockroaches, independently from ants, and much earlier, about 220 million years ago. Termite colonies comprise three castes: workers, soldiers, and reproductive, with workers making up the bulk of the population. The workers locate and provide food and water to the other caste members, dig tunnels, and repair nests. They exist solely to serve other members of the colony.

Winged termites are called alates, and are the only termites to have eyes. Each species of termite produces a group of alates at a particular time of year. Their wings break off after several months (depending on the species), and the alates, now called delates, find mates. Each couple starts a new colony, of which they are the king and queen. Alates are preyed upon by more species of amphibians, reptiles, birds, and ground-foraging mammals than any other group of insect.

Entomologists have identified 2,753 termite species worldwide. Termites are soft bodied, colorless, or pale white. They have strong mouthparts designed for chewing leaves, seeds, and wood. Termites of sexual castes have two pairs of wings, but they are delicate and incapable of carrying the insect on long flights. The soldiers of some species have large heads and massive mandibles for defense, or an extension on the head through which they can spray a gummy secretion to ward off intruders.

Simple termite nests may hold just a few dozen insects, while the mound-building species, such as Africa's *Macrotermes*, live in communities of hundreds of thousands. These termites cultivate a type of fungus on fecal matter, which they deposit in underground gardens, similar to those tended by leafcutter ants. Termites that eat dry wood can cause considerable damage as they bore holes in wooden homes and furniture. In the rainforest, their capacity to eat wood is extremely important; they are probably the most important animals that contribute to decomposition. They are the only insect that can easily digest cellulose, the main component of plants. In tropical forests, termites may consume up to one-third of the total annual production of dead wood, leaves, and grasses. Some 2,000 to 4,000 termites (sometimes as many as 10,000) inhabit every 200 to 400 square feet of forest.

Ants are the principal termite enemy, but there are also specialized mammal predators, such as the aardvark of Africa, pangolins of Africa and Asia, and anteaters of the Neotropics.

Further Reading:
Natural History Museum, England <http://www.nhm.ac.uk/science/entom/project3/index.html>.
O'Toole, *Encyclopedia of Insects.*
University of Toronto, Urban Entomology Program, Canada; <http://www.toronto.edu/forest/termite/iso1.htm>.
Wolfe and Prance, *Rainforests of the World.*
Wootton, *Insects of the World.*

Termitidae

The *Termitidae* include more than 1,600 species, or more than 70 percent of all termite species identified in tropical and subtropical regions worldwide. Most species build nests underground or inside dead wood, but a few construct elaborate mounds and tree nests.

Further Reading:
O'Toole, *Encyclopedia of Insects.*

Cubitermes spp.

This soil-feeding termite species of the African rainforest builds mushroom-shaped nests with a series of caps that allow the rain to run off them.

Further Reading:
O'Toole, *Encyclopedia of Insects.*

Hospitalitermes spp.

Inhabiting Malaysia, these snouted termites forage on lichens and other vegetation on the forest floor and tree trunks. As with other snouted termites, the columns of foraging insects are flanked by soldiers that, in spite of being sightless, can accurately fire a sticky and poisonous substance through their snouts at menacing ants. Most snouted termite soldiers have jaws much smaller in size than soldiers of other species.

Further Reading:
O'Toole, *Encyclopedia of Insects.*

LEPIDOPTERA (butterflies and moths)

The order *Lepidoptera* includes about 160,000 species of moths and butterflies worldwide (except Antarctica); about 90 percent are moths. What distinguishes butterflies from moths is the former's clubbed antennae and lack of connection between their fore- and hindwings. Unlike moths, butterflies hold their wings vertically over the back of their bodies while at rest, are often brightly colored, and are active during the day.

The wings, bodies, and legs of adult butterflies and moths are covered with dusty scales. Light refracted by these scales gives the insects their marvelous array of colors. No other animal goes through such a profound metamorphosis from juvenile to adult than *Lepidoptera*, from minute egg to caterpillar (or larva) to chrysalis (or pupa) to a full-grown, graceful adult. The larvae are feeding machines, equipped with chewing mandibles and bodies that protect a large gut. Nearly all are herbivores. Most adult butterflies sip flower nectar through a proboscis, but some species skip the pretty flowers and feed on rotting fruit, dead animals, dung, and fungus. Most butterflies live only a few weeks.

Unlike other members of *Insecta*, *Lepidoptera* do not carry plant or animal diseases, nor are any of them parasites or predators of human beings. However, many hundreds of species, mostly moths, and particularly the larvae, do cause damage to plants. They eat huge quantities of plant material, and they in turn are consumed by fungus, invertebrates such as spiders, beetles, and ants and vertebrates such as frogs, birds, and monkeys.

Further Reading:
Janzen, *Costa Rican Natural History.*
Pyle, *Handbook for Butterfly Watchers.*
Ross, "Butterfly Carnivale."

Butterflies

Nymphalidae (brushfoots)

Brushfoots are named for their two front legs, which are smaller than the other four and resemble miniature hairbrushes. This is the largest family of butterflies, and its members live worldwide. Some species feed on nectar and others on a variety of fluids, including urine, blood, sweat, feces, and fermented fruit juices. The undersurfaces of their wings are drab, but designed to appear like dead leaves, lichens, or the staring eye of a dangerous vertebrate. The top surfaces are often shimmery and brightly colored. Some species have evolved to be quite distasteful to potential predators; other species mimic this defense.

Further Reading:
DeVries, *The Butterflies of Costa Rica and Their Natural History.*
Ross, "Butterfly Carnivale."

Actinote leucomelas

The males of *Actinote leucomelas* are black with iridescent blue on the upper surface of their wings, and a yellow patch on the underside of their forewings. Found from Mexico to Panama, the subfamily to which it belongs, *Acraeinae*, is more diverse in the Old World tropics, with about 50 species in the New World. Mating of New World species has never been recorded, but in the Old World species, it happens like this: The male seizes a female while she's in flight, grasping her with his legs. The pair tumbles in a struggle to the ground. During mating, the male deposits a plug on the tip of the female's abdomen, presumably to prevent any further mating.

Further Reading:
Janzen, *Costa Rican Natural History.*

Caligo memnon (owl butterflies)

Owl butterflies range from Mexico to Colombia and have a wingspan of about four inches, making this one of the largest butterflies in the world. The underside of their wings is gray and black with two large spots, encircled by white, on their hind wings. One theory is that the "owl eyes" scare off predators. Lepidopterist Philip DeVries suspects that the spots are actually targets—birds snatch at the spots on the rear wings, which allows the owl butterfly to slip out of the beak's grasp, losing just a small wing piece.

Further Reading:
DeVries, *The Butterflies of Costa Rica and Their Natural History.*
Janzen, *Costa Rican Natural History.*

Danaus plexippus (monarch)

The monarch is famed for its impressive feat of endurance: Each year, during the northern winter months, it migrates south to California or Mexico, sometimes covering 80 miles a day at speeds of up to 11 miles per hour. The longest recorded trek was more than 3,000 miles. Because they stop to breed during their southerly migrations, the monarchs that fly north in the temperate spring months may be five generations removed from the original migrants. When they reach their breeding grounds, each female lays about 400 eggs on milkweed plants and then dies. The monarch's chrysalis is jade green with a golden crown.

The monarchs that winter in Mexico's fir forests in the central highlands are those that were hatched east of the Rockies in the United States, or about 90 percent of all monarchs—more than 100 million butterflies. When they reach their destination, they descend from the sky and alight upon the trunks of the fir trees, layer upon layer of black, orange, and white. There are nine known monarch wintering sites; Mexico has designated five of them as reserves. But this status may not be enough, since one of the sites, the Chivati-Huacal sanctuary in Michoacán, is being illegally logged. Studies have shown that butterfly mortality from bird predation is highest at wintering sites that have low tree density. Sites that are deforested also have higher temperatures and lower humidity, which causes the monarchs to become active and use up precious fat reserves needed for the long flight back to their breeding grounds.

Monarchs' conspicuous coloring of black and white stripes with yellow spots warns predators that they contain nauseating chemicals, gained when the larvae feed on milkweed, which also contains poisons. The monarchs' milkweed habit has involved them in the controversy over the use of genetically modified foods. A hybrid corn called Bt-corn contains bacterium genes that make a natural pesticide that is quite effective against the European corn borer, a moth that burrows into corn stalks and can destroy entire crops. Bt-corn pollen does not harm humans, honeybees, spiders, ladybugs, and many other beneficial insects, but it does seem to affect monarchs and perhaps other butterfly species. In a Cornell University laboratory, half the monarch caterpillars that fed on Bt-pollen-dusted milkweed died within four days.

Further Reading:

DeVries, *The Butterflies of Costa Rica and Their Natural History.*

Ebersole, "Trouble for Monarchs?"

Heliconius spp. (passionflower butterflies)

Passionflower butterflies have brightly colored, elongated wings. The bright red, orange, blue or green patterns of the 70 species warn predators that their bodies are infused with toxins acquired from the larvae's host plant, the passionflower. Heliconians are the only butterflies that supplement their nectar diet with pollen. Though their brains are miniscule, they are nevertheless able to memorize the location of food sources and potential sites for laying eggs and mating. Males engage in the unusual sex practice of pupal mating—they sniff out and perch on female chrysalides, awaiting the emergence of their future mate, or they may puncture the pupa and mate before the female has a chance to exit. Except for a few species found in the Old World, nearly all *Heliconius* are found in the Neotropics, from the southern United States and all of Latin America, plus the West Indies. The greatest diversity is in the Amazon Basin of Peru and Brazil.

Further Reading:

DeVries, *The Butterflies of Costa Rica and Their Natural History.*

Pyle, *Handbook for Butterfly Watchers.*

Ithomiinae (clearwings)

Clearwings are a *Nymphalidae* subfamily, found only in the Neotropics, most species of which have elongated wings that lack the pigment of most other species, making them difficult to see when they are perched. They fly slowly and for short distances, and with their transparent or semiopaque wings, appear as lazy ghosts. They choose among a small number of plant species whose flowers produce not only sweet nectar, but also alkaloids that taste noxious to predators. Male clearwings use the alkaloids to send sexy messages during courtship.

Some species' wings are marked with yellow, orange, and black. These species dine on bird droppings. Although most butterflies pursue a mate on their own, male clearwings gather in a traditional group in a mating area known as a lek. They release sex pheromones that lure females and other males that join the fun. Eventually, scores may gather, partying for a week. Clearwings live relatively long lives, for butterflies, slowly flying about their forest habitats for up to a year.

Further Reading:

DeVries, *The Butterflies of Costa Rica and Their Natural History.*

Janzen, *Costa Rican Natural History.*

Ross, "Butterfly Carnivale."

Morpho peleides

Morpho peleides is a spectacular butterfly found from Mexico to Colombia. The underside of their wings is drab, but the topside is an iridescent blue. With their slow, loping flights, they look like large, flashing, azure sequins zigzagging against the green fabric of the forest. Their radiant display may be a defense mechanism—so much flash and dazzle makes it difficult for predators to get a lock on their target.

Further Reading:

DeVries, *The Butterflies of Costa Rica and Their Natural History.*

Janzen, *Costa Rican Natural History.*

Papilionidae (swallowtails and birdwings)

The 534 species of swallowtails and birdwings have been located in nearly every habitat worldwide. The numbers of some species, however, have become dangerously low due to habitat destruction and overcollection.

Homerus spp.

These swallowtails are so spectacular and rare that collectors have hunted them nearly to extinction in Jamaica's two remaining and isolated rainforests. The largest butterfly in the Western Hemisphere, the *Homerus* has wings with bands of black and yellow and patches of shimmery blue that are six inches across from tip to tip. The butterfly's common name derives from the long tails. Illegal dealers pay $50 or more to villagers whose monthly income is half that; in Japan a *Homerus* swallowtail can fetch nearly $2,000.

Further Reading:

Sustainable Development Networking Programme, Jamaica <http://www.jsdnp.org.jm/papilioho.htm>.

Papilio antimachus

A swallowtail, and the largest butterfly in central Africa, *Papilio antimachus* has a wingspan of nearly nine inches. *P. ophidicephalus*, the emperor swallowtail, with a five-inch wingspan, is the largest butterfly in eastern Africa. When bothered, the caterpillar, like that of most swallowtail species, rears back its head and shoots out a forked, yellow organ from behind its head. It emits an unpleasant smell, something like spoiled pineapple, to deter enemies.

Further Reading:
Skaife, *African Insect Life*.

Papilio dardanus (male mocker swallowtail)

The male mocker swallowtail, prevalent throughout Africa south of the Sahara, has the usual yellow and black markings of swallowtails, with the same long tails at the end of both wings that give this species its common name. But the females look quite different from the males—so different that only laboratory studies proved that they belonged to the same species. In fact, many females look different from one another. Some are deep purple, with large yellow patches at the base of each hind wing, closely resembling *Amauris albimaculata*, which belongs to a completely different family of butterfly. Others have the appearance of several other species that may be found in the same range. The "model" species has an unpleasant taste to predators, so the female mocker butterfly has evolved to resemble these other distasteful species, even flying in the same manner as their models. Because the males don't need to pause to lay eggs, they don't need the protection that this mimicry affords.

Further Reading:
Skaife, *African Insect Life*.

Ornithoptera alexandrae (Queen Alexandra's birdwing)

Queen Alexandra's birdwing is one of about 30 birdwing butterfly species that are found in India, Southeast Asia to New Guinea, the Solomon Islands, and the rainforests of eastern Australia. All birdwing larvae depend on one genus of vine (*Aristolochia*), the only food they eat. As deforestation has diminished the numbers of *Aristolochia*, so has the population of birdwings fallen. The Queen Alexandra's birdwing is the largest butterfly in the world; the wingspans of the females can reach nearly 11 inches. The males are

an iridescent green and black; the females dark brown with white spots. They sip nectar from a variety of forest flowers. Queen Alexandra's birdwings now inhabit only one area of Papua New Guinea, which is the only area where the vine *Aristolochia dielsiana* remains. Conservationists' campaigns have encouraged residents to plant the vine in their gardens.

Further Reading:
Orsak, "Killing Butterflies to Save Butterflies: A Tool for Tropical Forest Conservation in Papua New Guinea."
People of the Tropical Forest <http://www.ulb.ac.be/soco/apft/GENERAL/PUBLICAT/ARTICLES/DON.HTM>.

Moths

Noctuidae (owlet moths)

The owlet moths are the largest family of *Lepidoptera*, with about 20,000 species worldwide. Most are dull in color, although some tropical species are brilliantly hued. Night flyers, owlet moths are attracted to light. They feed on fruit and sap.

Further Reading:
Hogue, *Latin American Insects and Entomology*.

Ascalapha odorata

The batlike appearance and large size of *Ascalapha odorata* have engendered various nicknames throughout its range in the lowland to mid-elevation Neotropics. In Mexico, it's called "la mariposa de la muerte" ("the butterfly of death"), and elsewhere "la bruja negra" ("the black witch"). With six-inch wingspans, the adults' dark brown wings are marked with wavy lines and eyespots. They feed at night on ripe and rotting fruits high in the canopy. Although the moth doesn't breed in the United States, it does seem to migrate north, as every year, usually from August to October, reports of spottings occur in California and even as far north as New York and Minneapolis. Not much is known about *A. odorata* migrations; scientists assume these wayward moths simply strayed too far from their usual flight routes.

Further Reading:
Janzen, *Costa Rican Natural History*.

Saturniidae (emperor moths)

With about 800 species, emperor moths are the largest moth family in the world. Every continent has some well-known species, such as North America's cecropia (*Hyalophora cecropia*) and luna (*Actias luna*). Most species weave thick, silken cocoons, sometimes used as a source of commercial silk. The largest are the atlas moths (*Attacus spp.*) of Asia and the great-tailed Hercules moth (*Coscinoscera spp.*) of Papua New Guinea's and Australia's rainforests. The wing-span of female atlas moths can reach nearly 10 inches. *Saturniidae* have colored rings on their broad wings, and most species have a small proboscis, or tongue; some have none at all. Adult emperor moths find mates and avoid predators, and females lay up to 200 eggs. They don't eat. The caterpillars must eat enough food to provide adequate nutrition for the adult moths during their entire lifetimes.

Most emperor moths are various shades of yellow, orange, red, or brownish. Their bodies are quite hairy, and the males have feathery antennae. The second segments of the legs of some species have spines or claws. Most are nocturnal. Parasitic flies and wasps feed on the larvae, and owls, other insect-eating birds, and bats prey on adults. Camouflage protects some species, and colored eyespots on their wings can also ward off enemies. When the moths are at rest, the brightly colored spots on the rear wings are covered. But when the moths are scared, they jerk forward their forewings and the suddenly exposed, blazing spots deter, or at least momentarily confuse, potential predators, long enough for the moths to take flight.

Further Reading:
Pinhey, *Emperor Moths of South and South Central Africa.*

Sphingidae (hawk moths)

Hawk moths have narrow wings that allow them to hover over flowers; they are the only moths with this ability. Found worldwide, they have long, coiled tongues that are perfect for sipping nectar from deep blooms, such as orchids. Some species are diurnal, but most are active at twilight. Hawk moth caterpillars have a horn on the next to last segment of their abdomens, the purpose of which is unknown.

Further Reading:
Skaife, *African Insect Life.*

Acherontia atropos (death's head hawk moth)

The death's head hawk moth, common in Europe and Africa, gets its name from a yellow, skull-shaped mark on its thorax. No insects have voices, but this moth can squeak when bothered. Lepidopterists aren't sure, but think it possible that the moth makes the high-pitched sound through its proboscis by a double-sucking action, then by blowing through a basal membrane, something like a trumpeting elephant. The larvae make cracking noises. In addition to sipping nectar, the moth likes to dip its proboscis into honey. It mimics the chemical scent of honey-bees to gain access to the hive, then punches holes in the cells' wax covering to feed on the honey.

Further Reading:
Natural History Museum, England <http://www.nhm.ac.uk/science/entom/project5/index.html>.
Skaife, *African Insect Life.*

NEUROPTERA (lacewings, dobsonflies)

Lacewings and dobsonflies have biting mouthparts and two pairs of similar wings that are held high over their abdomens when the insects are at rest. The wings are etched with veins. This is the most primitive order of insects that goes through complete metamorphosis.

Myrmeleontidae (ant lions)

Ant lions are thin insects about two inches long with broad heads and prominent compound eyes. Their transparent wings, etched with veins, are pointed at the ends. They resemble dragonflies or damselflies, with longer, more prominent, and clubbed antennae, although the antennae are shorter than in most other lacewing species. They are not strong flyers and are active only in the evenings. The 2,000 species are found worldwide, more commonly in tropical regions. The largest ant lion is of the African genus *Palpares*, which has a wingspan of more than six inches.

Female ant lions lay their eggs in dry soil. Once the larvae hatch, they begin their search for an appropriate spot to dig a pit. They scoot backward just below the surface of the soil, leaving faint lines in the dirt called "doodles," which is why ant lions are sometimes called "doodle bugs." Once it finds the right spot, the larva begins to dig a funnel-shaped pit, using its tapered abdomen and flat head to push and excavate sand. When a passing ant or other small arthropod slips into the pit, the larva, which is hiding at the bottom beneath the soil, uses its long, piercing mandibles to seize the trapped bug. The ant lion sucks the prey's body fluids, then tosses the remains from the pit with a flick of its head.

Further Reading:
The Antlion Pit <http://www.antlionpit.com/welcome.
html>.
Janzen, *Costa Rican Natural History.*

ODONATA (dragonflies and damselflies)

Dragonflies and damselflies are ancient animals, dating back some 300 million years. With 5,000 species, they are widely dispersed on Earth and are often brightly colored scarlets and blues. They are particularly beneficial insects to human beings because they eat gnats, flies, and mosquitoes. Dragonfly and damselfly hatchlings feed on mosquito larvae in the water. Now-extinct *Odonata* were much larger than the species found today; the largest insect known to have inhabited the Earth, *Meganeura monyi*, had a wingspan of nearly 30 inches.

Dragonfly and damselfly nymphs live in water. Damselfly nymphs may spend only a few months underwater, while some species of dragonfly nymphs spend as much as three or four years submerged. They then crawl up a plant stem or onto a rock, and their outer skin breaks open. A dragonfly or damselfly emerges. Some species live only a few weeks, during which they keep themselves busy staking out territory, hunting for food, and courting and winning a mate. Unlike most male insects, dragonfly and damselfly males have their copulatory organs at the front part of their abdomens rather than at the ends. These insects commonly mate while flying in tandem, often just over the surface of the water, so the female can dip her abdomen into the water and deposit her eggs. Some species migrate great distances, as much as 900 miles. They hunt for small insects, particularly mosquitoes, gnats, and flies, which they can scoop right out of the air by folding their long, bristle-covered legs into a basket-shaped trap beneath their bodies as they fly.

Anisoptera (dragonflies)

Dragonflies are powerfully built, with large, bulging compound eyes that take up most of their heads. Their hindwings are a bit smaller than their forewings (*anis* means "unequal" in Greek), and at rest they hold these four large, multiveined and membranous wings at right angles to their long bodies. Their wingspans can be up to six inches. They fly fast and maneuver elegantly— they can zoom straight up, hover, dart to the side, stop on a dime, fly backwards, and change directions in midflight. They are also speedy eaters, able to consume their own weight in food in only 30 minutes. Dragonfly nymphs are short and stubby with thick abdomens that hold a special gill chamber. The nymphs breathe by pumping water through the chamber and can propel themselves by expelling water through the gills. The world's smallest dragonfly is *Nannophya pygmaea*, a species found in Malaysia whose body length is only 0.6 of an inch.

Further Reading:
Wootton, *Insects of the World.*
Worldwide Dragonfly Association <http://powell.colgate.
edu/wda/dragonfly.htm>.

Zygoptera (damselflies)

Damselflies have wings that are of equal size and usually folded together when the insects are at rest. They are more delicately built than dragonflies and have a more fluttering flight. Like dragonflies, they have large eyes, but they are more widely set apart on their heads. They have excellent vision and can spot prey at 25 feet and zoom in for the kill at 60 miles per hour. The largest member of the order *Odonata* is a damselfly, *Megaloprepus coerulatus*, with a wingspan of about 7.5 inches.

Damselfly nymphs are able to breathe underwater via three feathery structures called gills, located at the ends of their abdomens. The gills also serve as fishtails, allowing them to wiggle through the water. In 1999, scientists discovered a new species of damselfly in the diminishing rainforests of Cebu, an island in the Philippines, termed *Cebu recionemism*. Damselflies and dragonflies in the Philippines are increasingly rare, due to the misuse of agrochemicals and the draining of the ponds and swamps where the insects live. Spraying with chemicals to rid areas of *Aedes aegypti*, the mosquito that carries dengue fever, has also destroyed dragonfly and damselfly nymphs. Ironically, these predatory insects are considered to be the best defense against mosquito build-up. Not only do the adults eat mosquitoes, but their nymphs eat mosquito larvae. Scientists are considering reintroducing dragonflies and damselflies to freshwater areas.

Further Reading:
Bengwayan, "New Damselfly Found in Threatened Rainforest Could Fight Disease."
Wootton, *Insects of the World.*

ORTHOPTERA

The order *Orthoptera* includes grasshoppers, crickets, katydids, and locusts—some 20,000 species in all. They are known for their jumping ability. Many have two pairs of wings; others are wingless. Among the largest of the insect orders, the adults have biting and chewing mouthparts.

Coolooidae

Coolooidae is a family of insects discovered in 1976 in Cooloola National Park in Queensland, Australia. Scientists have identified several different species, which are commonly called Cooloola monsters. They are a little over one inch long, with broad bodies and a flat, shovel-shaped head with very short antennae. They are clearly related to crickets and grasshoppers, but scientists couldn't decide to which family they belonged, so they gave the strange new bug its own family. The Cooloola monster burrows in the sandy and moist soils of coastal rainforests and eucalypt forest in Queensland. Females, which have swollen abdomens and tiny feet and claws, spend their entire lives underground, while the males appear above ground at night. Not much is known about this insect, but it is likely it is predatory, probably feeding on beetle larvae and other insects.

Further Reading:
O'Toole, *Encyclopedia of Insects.*

Tettigoniidae (katydids)

Katydids are closely related to crickets and grasshoppers, but have longer antennae, often two or three times the length of their bodies, that are covered with sensory receptors that allow them to navigate after dark. There are 4,000 katydid species worldwide; 2,000 of them are in the Amazon rainforest, where they have evolved numerous ways to avoid being eaten. The Crayola katydid's body carries repulsive-tasting chemicals, probably obtained from the leaves it chooses to eat, and like many toxic animals, its is brightly colored to warn potential predators. Its wings are bright green, its body yellow with aqua stripes; its legs are blue and yellow with red tips. In contrast, katydids of the *Cycloptera* genus are almost invisible against foliage, having evolved to look exactly like dead brown or green leaves.

Another mimic is the wasp katydid, which looks just like a stinging wasp. As do wasps, they have the much shorter antennae, nearly transparent forewings, and narrow abdomens. They even bob like wasps, but they have no sting. The peacock katydid takes no chances: It mimics a leaf, but when attacked, opens its wings to reveal two round blotches that look like large eyes. The startled predator may hesitate long enough to allow the insect to escape. *Markia hystrix,* the lichen katydid, feeds on masses of pale, long-fibered lichens in the cloud forests of Central America and has evolved to look just like its food. It has spines on its legs that are the same color, shape, and size as lichen fiber, and its buff and brown body and wings bear a pattern that blends in perfectly with its surroundings. The champion at frightening predators has to be the three-inch-long spiny devil katydid, whose legs and face are lined with yellow thorns, and whose head bears a multipronged crown of thorns. If that doesn't scare away an enemy, the spiny devil has mighty jaws that can cut through flesh.

Further Reading:
Castner, "Please Don't Eat the Katydids."

Mammals (*Mammalia*)

Mammals are a group of vertebrates in which the young are nourished with milk from secreting glands of the mother. Other distinguishing mammal features include the presence of fur or hair; a lower jaw that is directly attached to the skull; three tiny bones that transmit sounds across the middle ear; and a diaphragm that separates the heart and the lungs from the abdomen. Red blood cells in mammals lack a nucleus, which all other vertebrates have. With a couple of exceptions, all mammals give birth to live young; that is, they are viviparous. Mammals are highly successful animals that have invaded virtually every ecological niche, although the Tropics hold the greatest diversity of species. There are about 4,000 species of mammals, arranged in 20 orders and about 120 families.

ARTIODACTYLA

The mammals in the order *Artiodactyla* are known as even-toed ungulates because their weight falls evenly on their third and fourth toes. Most have antlers or horns. Their natural habitats are found in nearly every region, except for Australia, the Pacific islands, and Antarctica. They include familiar beasts of burden and animals that are sources of meat, hair, and leather: sheep, goats, pigs, cows, and camels. Although many species inhabit grasslands or savannas, others are forest animals. They browse on fallen fruit and nuts or on grasses and leaves.

Further Reading:
Redford, *Mammals of the Neotropics*.
Ultimate Ungulate Page <http://www.ultimateungulate.com/>.

Bovidae (bovines)

Bovines range widely in size, from small and delicate to huge and heavy. Of the 99 species, many have been domesticated.

Cephalophus monticola (blue duiker)

The blue duiker is a shy and small antelope, only 13 inches from hoof to shoulder, and weighing an average of 10 pounds. Its horns point straight backward. It is found in Africa's western coastal rainforests from Nigeria to Gabon, and east to Kenya and south to South Africa. The name is derived from the Afrikaans word for "diver," due to the small antelope's ability to plunge swiftly into dense overgrowth. An omnivore, it eats leaves, fruits, seeds, insects, and sometimes stalks and captures small birds and rodents. Duikers are hunted by eagles, lions, leopards, cheetahs, hyenas, jackals, and other carnivores. People hunt duikers for their meat, skins, and horns, which are used in some areas as charms against evil spirits. When they are alarmed, duikers call in a high-pitched whistle and can dart deftly and rapidly away from predators. The blue is the smallest of the 17 duiker species, all of which are forest dwellers. The largest duiker is the yellow-backed (*C. silvicultor*), which lives throughout tropical Africa. It reaches 35 inches in height and has long, yellowish hair that stands erect on its back.

Further Reading:
African Wildlife Foundation, USA <http://www.awf.org/animals/duiker.html>.
Kingdon, *Field Guide to African Mammals*.

Pseudoryx nghetinhensis (forest goat)

The forest goat was identified in 1993 by a British biologist and five Vietnamese scientists who found the animal in a nature reserve southwest of Hanoi, along Vietnam's mountainous border with Laos. Known by locals as the "forest goat," or *sao la*, the mammal looks something like a cross between a cow and a goat, but with long, smooth, and erect horns, like those of an antelope. It looks most like an oryx, with a dark brown dorsal stripe and white markings on its face and throat. The forests just south of the Vu Quang Nature Reserve were heavily bombed during the Vietnam War, but the forest goat's habitat, montane and evergreen forests, was left miraculously unscathed. Nevertheless, the animals have been heavily hunted and probably only a few hundred remain. The Vietnamese government has extended the Vu Quang Reserve from 40,000 acres to nearly 150,000 acres to protect the forest goat. This was the first new large mammal to be identified since 1937, when scientists described a species of wild cattle from Northern Cambodia.

Further Reading:
Basler, "Vietnam Forest Yields Evidence of New Animal."
World Conservation Monitoring Centre, England <http:// www.wcmc.org.uk / infoserv / countryp / vietnam / app4. html#VUOX>.
World Wide Fund for Nature, Switzerland <http://www. panda.org / resources / publications / species / underthreat / vuquangox.htm>.

Cervidae (deer)

Deer are graceful animals with long, slender legs, long necks, and short tails. Males have antlers that are shed and regrown each year. *Cervidae* are ruminants: They have complex stomachs in which the plant material they eat is fermented. After eating, they must regurgitate and more thoroughly chew the matter that was stored in one of their stomach chambers before it passes to the other three chambers for digestion. Worldwide, there are about 42 species of deer. Deer usually give birth to one or two young, which are able to walk soon after birth. Mothers leave their fawns in secluded areas while they feed. The fawns' white-spotted coats serve as camouflage in dappled light.

Further Reading:
Emmons, *Neotropical Rainforest Mammals.*
Reid, *A Field Guide to the Mammals of Central America and Southeast Mexico.*

Mazama americana (red brocket deer)

Red brocket deer are small at 2.5 feet tall, with a rounded body, arched back, slim legs, and triangular head. The shiny and short fur of the upperpart of its body is a bright reddish brown, while its throat and chest are whitish and its belly orangish. The underside of its five-inch tail is white. Found from southern Mexico to northern Argentina and on the islands of Trinidad and Tobago, it is fairly common in forested areas, except where heavily hunted. With their small size, rounded backs, and the male's short, straight, and unbranched antlers, they can move easily through dense forest, but will also forage in clearings. They eat fruit, flowers, fungi, and vegetation. Red brocket deer usually slip quietly away when alarmed, but may also run off with a whistling snort of alarm. Good swimmers, they may enter water to escape predators.

Further Reading:
Emmons, *Neotropical Rainforest Mammals.*
Reid, *A Field Guide to the Mammals of Central America and Southeast Mexico.*

Odocoileus virginianus (white-tailed deer)

The white-tailed deer inhabits regions from southern Canada south to northern Brazil. It is widespread and fairly common where not overhunted. Some populations have been wiped out by hunters. This medium-sized deer stands about three feet tall and has long legs, a long flat back, and a long, narrow head. Its upper parts are grayish-brown or reddish, while its belly, inner thighs, chest, and throat are white, and it has white facial markings. The tail is brown above, and white on the underside. Males have curved, branched antlers. They are active day or night, but are most often seen at night, dusk, or dawn, when they venture into clearings to feed. When disturbed, it snorts as it raises and fans out its tail and bounds away, head held high. In addition to leaves and twigs, white-tailed deer feed on fallen fruits and nuts. The white-tailed deer of Central America are much smaller than those of the United States, and the male's antlers are about half the size of those in the North. They are more likely found in dry forests and forest edges than in lowland rainforests.

Further Reading:
Janzen, *Costa Rican Natural History.*
Reid, *A Field Guide to the Mammals of Central America and Southeast Mexico.*

Giraffidae

Giraffidae include only two species, one of which is a forest-dweller. In prehistoric times, the long-necked giraffe (*Giraffa camelopardalis*) may have been one of the first artiodactyls to combine a move out of the forest with enlarged body size.

Okapia johnstoni (okapi)

The okapi looks a bit like a zebra, but is actually a type of giraffe, one that inhabits forests. The okapi is about five to six feet high, with short, sleek, and deep chestnut or purplish hair, white stripes on its legs and flanks, and a white face. Males have two small, hair-covered horns. Solitary, shy, and seldom-seen animals, they live in the forests of Congo, Democratic Republic of Congo, and Uganda, where they feed on fruits and leaves. Like the giraffe, they have long and muscular tongues. Scientists estimate that about 30,000 okapis survive; some 5,000 live in the Okapi Wildlife Reserve in the Congo. Deforestation and hunting for the bush-meat trade is threatening its survival.

Further Reading:
Kingdon, *Field Guide to African Mammals*.
Geobopological Survey <http://www.geobop.com/Mammals/ Artiodactyla/Giraffidae/Okapia_johnstoni/index. htm>.

Hippopotamidae

Hippopotamidae include two species, *Hippopotamus amphibius* and *Hexaprotodon liberiensis*, both found in Africa. In prehistoric times, hippos also lived in India.

Hexaprotodon liberiensis (pygmy hippopotamus)

The pygmy hippopotamus lives near forest rivers, lakes, or swamps in West Africa. At 475 pounds, it is dainty compared to the common hippo (*Hippopotamus amphibius*), which can weigh as much as five tons. It is also less aquatic; only its front toes are webbed, and it spends much of its time on land, feeding on vegetation and fallen fruit and digging up roots and tubers. During the day it snoozes and wallows in muddy water with other hippos, while at night it grazes alone on land. Its five-foot-long, rotund body is carried on stumpy legs, and its sun-sensitive skin is light brown. Due to overhunting by human beings, its only predator, it is threatened with extinction.

Further Reading:
African Wildlife Foundation, USA <http://www.awf.org/ animals/hippo.html>.
Haltenorth and Diller, *Collins Field Guide: Mammals of Africa*.
Kingdon, *Field Guide to African Mammals*.

Suidae (wild pigs)

Wild pigs originated in the Old World, although they have been introduced, either as wild or domestic species, worldwide. Eight species are located in Africa and Asia.

Babyrousa babyrussa (babirusa)

Babirusa inhabits forested riverbanks on only a few Indonesian islands: Sulawesi, Togian, Sulu, and Buru. It is one of three pig species in tropical Asia; the other two are *Sus barbatus* (bearded pig) of Malaysia, and the Javan pig, also existing on only a few Pacific islands. The babirusa is about three feet long, with a foot-long tail, and it can weigh up to 200 pounds. Its hairless hide hangs in loose folds and ranges from gray to brown. In males, the upper tusks, which are longer than those of any other pig, grow through the roof of its snout, then curve backward, reaching a maximum length of 12 inches. It also has lower tusks that protrude from the sides of its mouth. The babirusa is a plant eater and has a digestive system especially adapted to easy digestion of vegetation. Like most pigs, it is fond of wallowing in the mud, but it is an excellent runner and swimmer. The population is estimated at about 4,000 animals; it's considered a vulnerable species.

Further Reading:
Geobopological Survey <http://www.geobop.com/Mammals/ Artiodactyla/Suidae/2.htm>.
Ultimate Ungulate Page <http://www.ultimateungulate. com/babirusa.html>.

Hylochoerus meinertzhageni (giant forest hog)

Giant forest hogs are chunky pigs covered with long black hair. About five feet long, it can weigh up to 400 pounds. Males have large, naked cheeks, flat muzzles, thick tusks, and broad snout discs. Populations of giant forest hogs are scattered across most of central tropical Africa, mainly in mosaics of forest and grasslands. They mainly graze on grasses, but also eat fallen fruit, leaves, buds, and roots. Females usually travel with up to three generations of offspring, giving birth to as many as 11 playful and competitive piglets once a year. Giant forest hogs are easy targets for hunters, especially hunters accompanied by dogs. They are endangered species in much of their range, and rare in the rest.

Further Reading:

Kingdon, *Field Guide to African Mammals.*

Haltenorth and Diller, *Collins Field Guide: Mammals of Africa.*

Potamochoerus porcus (red river hog)

The red river hog has bright russet hair, with white hair above its eyes, on its cheeks, and at its jawline. Its tapering, leaf-shaped ears have long white tassels at their tips. This four-foot-long hog, which weighs a good 175 pounds, lives in rainforests from Gambia to the eastern Democratic Republic of Congo. They are also found on the islands of Zanzibar, Madagascar, Mayotte, and Mafia. The males have short, sharp tusks. The hogs use their snouts to dig up roots and tubers, their main food, but they also eat fallen fruits and invertebrates. Red river hogs often congregate in groups of up to 15 animals and sometimes in temporary crowds of up to 60. When males confront each other, they grunt, paw the earth, switch their slender, 14-inch tails back and forth, and chomp their jaws together, which releases pheromone-loaded secretions from the corners of their eyes, genitals, and necks. Red river hogs are a principal prey of bushmeat hunters, and that, with deforestation, has drained their populations. They are seldom found now outside protected areas.

Further Reading:

Kingdon, *Field Guide to African Mammals.*

Haltenorth and Diller, *Collins Field Guide: Mammals of Africa.*

Tayassuidae

Three species of pig-like ungulates make up *Tayassuidae,* all found only in the Neotropics.

Tayassu spp. (peccaries)

Peccaries are fairly large and robust, weighing in at about 90 pounds, with small legs, large heads, and no tails. They travel in groups, trotting along in a single file, and will release a strong, musky odor when alarmed. They are shy animals but sometimes stampede in panic when they see people, which can be disconcerting to both species. They mainly eat fruits, nuts, and palm seeds. Collared peccaries (*T. tajacu*), which have gray bristles, are found from the southern United States south to northern Argentina, and white-lipped peccaries (*T. pecari*) live from southern Mexico to northern Argentina. Both species are widely hunted. White-lipped peccaries, whose meat is important in the ceremonies of the Yanomami indigenous people of Brazil and Venezuela, are in particular trouble in Brazil, due to overhunting and the introduction by gold miners of domestic pigs. The domestic pigs are dominant—they fight off peccaries and take over territory. The domestic pigs were allowed to run wild, so they are now feral.

Further Reading:

Emmons, *Neotropical Rainforest Mammals.*

Redford, *Mammals of the Neotropics.*

Tragulidae

The remains of a family that was widespread in the Old World 40 to 25 million years ago, *Tragulidae* includes only four species.

Hyemoschus spp. (chevrotains)

Chevrotains are the smallest hoofed mammals and live in Asia and Africa. Three species are found in south central and Southeast Asia, while the largest species, the water chevrotain (*Hyemoschus aquaticus*), of west and central Africa, is the largest, at about 28 pounds. Chevrotains look like small, hunchbacked deer, with long, thin legs, but they are more closely related to camels. They are brown, and some have rows of white spots or stripes. The Asian chevrotains, which weigh from 6 to 10 pounds, eat mostly grass, leaves, berries, and fallen fruit, while the water chevrotain of Africa feeds almost exclusively on fruit. Forest dwellers, they are solitary and shy and stay close to sources of water. To escape predators, they might splash into the water or rapidly hightail it like rabbits into dense forest underbrush.

Further Reading:

Haltenorth and Diller, *Collins Field Guide: Mammals of Africa.*

Kingdon, *Field Guide to African Mammals.*

Minnesota Zoo <http://www.wcco.com/partners/mnzoo/chevrotain.html>.

CARNIVORA (carnivores)

Carnivores are adapted to find, chase, catch, kill, and eat other vertebrates. Although some eat only meat, others feed primarily on insects and fruit, with some meat included. The jaws and teeth of most carnivores are perfect for delivering a powerful bite or puncture. They are indigenous to all continents except Australia (although wild dogs were brought there by aboriginal people centuries ago) and Antarctica. Of the eight families, five are found in South America.

Further Reading:
Emmons, *Neotropical Rainforest Mammals*.
Redford, *Mammals of the Neotropics*.

Canidae (dogs)

The family of dogs includes 35 species that are found on all continents except Antarctica. Excellent runners, wild dogs are long legged with distinctive paws, and have straight backs, bushy tails, long snouts, and wet noses. Their teeth are well designed for grasping and biting, with four canine teeth at the front corners of their mouths. Though large, they aren't as sharp as felines' teeth. Rather than delivering a "killing bite," they slay their prey by disabling it with bites or shaking it to break its back. They have acute senses of hearing and smell and can track prey by following a scent trail. Several species hunt in packs, which requires sophisticated strategizing, organization, and communication.

Canids are the most vocal of the carnivores, with whines, howls, and barks. Most canids inhabit savannas, grasslands, or woods with open canopies. Only two species are deep rainforest dwellers; both are found in the Neotropics, and both are rare.

Further Reading:
Emmons, *Neotropical Rainforest Mammals*.
Redford, *Mammals of the Neotropics*.

Atelocynus microtis (short-eared dog)

The short-eared dog is extremely rare throughout its range: east of the Andes in Colombia, Ecuador, Peru, southern Brazil, and northern Paraguay. It has short, stiff, gray hair, a large head, and, obviously, small ears. It probably hunts alone, but everything known about this wild dog comes from observing captive species.

Further Reading:
Emmons, *Neotropical Rainforest Mammals*.

Speothos venaticus (bush dog)

Bush dogs probably hunt in small groups, and there is likely a strong bond between the male and female. They communicate with barks and whines. They have short legs, webbed feet, and short, bushy tails. Their long hair is reddish-tan or tawny, becoming nearly black toward their tails. They are just over two feet long and weigh about eight pounds. Bush dogs den in burrows or hollow tree trunks at night; they are mainly diurnal. Their prey includes larger rodents, including agoutis. By hunting in packs—they tend to live in family groups of up to 10 individuals—they can take mammals considerably larger than themselves, such as capybaras. Males carry food to nursing females. Only captive animals have been observed closely; they are so rare that even indigenous people have seldom seen a bush dog. Their habitat is in western Panama to northern Argentina.

Further Reading:
Emmons, *Neotropical Rainforest Mammals*.
Redford, *Mammals of the Neotropics*.
World Conservation Union, Canid Specialist Group <http://users.ox.ac.uk/~wcruinfo/csgweb/sppaccts/speothos.htm>.

Felidae (cats)

Cats of many sizes are major predators in all rainforests, with their sharp teeth shaped for killing and eating meat and their retractable claws and massive shoulder muscles that allow them to seize and drag down prey. With their excellent vision and hearing, they can hunt at night or dawn, capturing animals by lying in ambush or stealthily approaching, and then suddenly pouncing.

There are 35 wild cat species on all continents but Australia and Antarctica, and all are either endangered or threatened with extinction. Deforestation has greatly reduced their habitat, while international trade in patterned cat skins was responsible for the slaughter of hundreds of thousands of wild cats. Trade is now prohibited, and some cat populations are beginning to rebound. Unfortunately, the illegal trade in the bones and body parts of wild cats, used in folk medicines in Asia, is actually increasing.

Sheep and cattle ranchers around the globe have a particular aversion to wild cats. Because their ranges impinge on former cat habitat, the felines naturally want to help themselves to calves. Translocation doesn't work; the cats either make their way back to their home territory or are killed by residents of the territory into which they were moved. In Africa, some ranchers are using guard dogs, while others are experimenting with "conditioned taste aversion." Dead cows are injected with a nonlethal substance that will make a cat ill for a short time, and theoretically, the felines will develop an aversion to beef. In Venezuela, some ranchers have erected solar-powered electric fences around their livestock.

Further Reading:
Emmons, *Neotropical Rainforest Mammals.*
Sunquist, "New Look at Cats!"

Catopuma badia (Bornean bay cat)

The Bornean bay cat is mostly a mystery. Information was based on skulls and skins collected in the 1880s until 1992, when trappers in Sarawak, Borneo, captured a female and brought it to the Sarawak museum. That's been the only evidence that the cat still survives in the deep forest of Borneo. The bay cat has a bright chestnut-red coat, speckled with black markings, and, based on limited specimens, it has a head-to-body length of two feet, with a foot-long tail.

Further Reading:
Sunquist, "New Look at Cats!"
World Conservation Union, Species Survival Commission, Switzerland <http://lynx.uio.no/lynx/catfolk/badia01.htm>.

Felis aurata (golden cat)

Golden cats are powerful cats of the forested regions of tropical Africa, including mangrove and riverine forests. Nearly three feet long, with a 12-inch, ringed tail, the cats' soft fur may be red, yellow, or smoky gray. All golden cats have spots on their bellies and inner limbs; upper parts may be spotted as well. In some regions, the spots are the size of freckles; in others, large, bold rosettes. Their prey includes monkeys, rodents, and birds. Because they have mighty jaws, they are also able to take down prey as large and strong as duikers (*Cephalophus monticola*). Golden cats are solitary and mainly arboreal, sleeping in trees during the day, and are most active at dusk and at night. Not much is know about their population status, but due to deforestation and the frequent presence of their skins in markets, they are probably in trouble.

Further Reading:
Haltenorth and Diller, *Collins Field Guide: Mammals of Africa.*

Kingdon, Jonathan, *The Kingdon Field Guide to African Mammals.*
World Conservation Union, Species Survival Commission, Switzerland <http://lynx.uio.no/catfolk/aurata01.htm>.

Felis concolor (puma, mountain lion, cougar)

Known variously as puma, mountain lion, or cougar, this cat ranges from Canada all the way to southern Argentina and Chile. It is a tawny, yellow-brown to reddish color, with white fur around its mouth. Pumas are about four feet long, and their two-foot-long tails darken to a black tip. They are active day and night and are terrestrial (though good climbers), feeding mainly on deer, agoutis, pacas, and smaller prey like snakes and rats. They often cover uneaten portions of their prey with sticks and mark the spot with urine. In the tropics, these shy, solitary, and wary cats are found throughout rainforests and above the tree line in the Andes. Pumas will attack livestock where their habitat has been encroached upon and their natural prey is reduced. For that reason, ranchers hunt them and have eliminated many populations throughout their range.

Further Reading:
Emmons, *Neotropical Rainforest Mammals.*
Redford, *Mammals of the Neotropics.*
Reid, *A Field Guide to the Mammals of Central America and Southeast Mexico.*

Felis pardalis (ocelot)

The ocelot has a sandy brown or buff coat with black rosettes or long ovals that have tawny brown centers; its underbody is white with black spots. Found from southern Texas south to northern Argentina, ocelots are mostly nocturnal and terrestrial. They may climb trees to rest during the day, but are more likely to sleep on the ground, hidden in thick vegetation. Their favored prey includes small rodents, rabbits, opossums, iguanas, and birds. Ocelots are about 2.5 feet long, with 14-inch, banded tails spotted with black. Their populations are low due to habitat loss and because they were once heavily hunted for their skins.

Further Reading:
Emmons, *Neotropical Rainforest Mammals.*
Reid, *A Field Guide to the Mammals of Central America and Southeast Mexico.*

Felis wiedii (margay)

The margay ranges from northern Mexico south to northern Argentina. Medium-size at two feet long, the cat has a rather bushy 17-inch tail. The long and thick fur on its upper parts are gray-brown to tawny, spotted with black rosettes and long ovals that have brown centers. Its under parts are white with black spots and stripes. They are mainly nocturnal, usually spending the day hidden in trees. The most arboreal of Neotropical cats, it hunts in trees and can partially rotate its hind feet and so can run down trees head-first, like a squirrel. The margay's principal prey are mice, opossums, and squirrels. Although their coats aren't as valuable as that of the ocelot, margays are still hunted and, as a result, endangered.

Further Reading:

Emmons, *Neotropical Rainforest Mammals.*
Reid, *A Field Guide to the Mammals of Central America and Southeast Mexico.*

Herpailurus yaguarondi (jaguarundi)

The jaguarundi is spotless and the most variable in color of the wild cats; it may be dark gray, black, chocolate brown, reddish, tawny yellow, or chestnut, with gray being the most common shade. It is about the size of an otter, at 2.5 feet long, with a 19-inch, slender tail. In addition to being found in forests, the lithe jaguarundi is also seen in agricultural areas. It seems to adapt fairly well to human disturbance, and its coats are less valuable than those of the spotted cats, but it is still an endangered animal. It is mainly diurnal and terrestrial, although it's a good climber. Jaguarundis eat birds, reptiles, and small mammals and range from southern Texas through much of Argentina.

Further Reading:

Emmons, *Neotropical Rainforest Mammals.*
Reid, *A Field Guide to the Mammals of Central America and Southeast Mexico.*

Neofelis nebulosa (clouded leopard)

The clouded leopard is named for the blurry spots on its coat. The marks on its tawny to silvery-gray coat are partially edged in black, and the insides are a darker color than the background pelt color. Not much is known about this elusive and rare wild cat, but it appears to be at home in the trees as well as on the ground. It hides silently in tree branches at night, dropping down on prey underneath. Scientists com-

pare clouded leopards to the extinct saber-toothed tigers because their canine teeth are quite long, well adapted for shredding meat. Prey includes birds, monkeys, small deer, and wild boars. They are also fair swimmers. Adult males are about six feet long and weigh about 45 pounds. They inhabit dense tropical forests of Southeast Asia and are found in India, Bangladesh, Nepal, and China.

Conservationists are unsure how many clouded leopards remain in the wild, but because their pelts and bones are popular items on the black market, the animals are certainly endangered. Clouded leopards were even found on the menu of restaurants in Thailand and China.

Further Reading:

Breining, "The Ghosts of Way Kambas."
World Conservation Union, Species Survival Commission, Switzerland <http://lynx.uio.no/catfolk/nebul01.htm>.

Panthera onca (jaguar)

The jaguar is the largest of the American cats. Its body is about five feet long, with a 1.5-foot tail. It has a powerful build and a glossy, tawny coat with black spots. Jaguars may hunt night and day, but are most active just after dusk and before dawn. Their diets are wide-ranging: porcupines, armadillos, peccaries, brocket deer, anteaters, monkeys, coatis, and tapir, and, near water: turtles, caiman, and fish. In their search for food, they don't confine themselves to the forest floor, but also climb trees and swim rivers. They either stalk their prey or ambush it, sometimes by stretching out on the branch of a tree and suddenly dropping down on a passing animal. Shy and solitary, they are seldom seen and require large expanses of forest cover. Once found from the southwestern United States to northern Argentina, the endangered jaguar is rare in most of its range and has been completely wiped out from the United States and Uruguay. Only a few hundred still roam the forests of Mexico and Central America; the population is higher in the larger, intact forests of the jaguar's South American range. In addition to loss of habitat, threats to the jaguar's survival include over-hunting for the fur trade, elimination by ranchers, and probably loss of available prey.

Further Reading:

Emmons, *Neotropical Rainforest Mammals.*
Redford, *Mammals of the Neotropics.*
Wolfe and Prance, *Rainforests of the World.*

Panthera pardus (leopard)

The leopard lives in woodlands, lowland forests, and heavily vegetated areas of Africa, south of the Sahara, tropical Asia, and Indonesia. It has a lithe and powerful body, about five feet long, with a three-foot tail. Males can weigh up to 140 pounds. Leopards are solitary and nocturnal, sleeping by day in secluded spots or in trees. They are quite agile climbers and are the most successful hunters of the cat family. With their yellowish coats pattered with black spots and rings, they are well camouflaged and masters at creeping silently through vegetation and pouncing on prey, delivering a bite on the back of the neck and dragging the victim into a tree, even if it is three times the cat's weight. Their prey includes impalas, gazelles, duikers, warthogs, rodents, ground birds, monkeys, baboons, frogs, and fish. Females give birth to one to three cubs; they are very dedicated mothers, caring for their young until they are about two years old, leaving them only to hunt.

Leopards are highly endangered. Although it is illegal to sell or buy leopard skins, they are still in great demand on the black market. Recently, officials in India were shocked when they confiscated 18,000 leopard claws from poachers. Since each leopard has 18 claws, five on its hind feet and four on each front foot, 18,000 claws represent the slaughter of 1,000 wild cats. Only about 10,000 leopards remain in the wild in India. Many people wear leopard claws around their necks in the belief that they protect against evil and bring good luck.

Further Reading:
Israel and Sinclair, *Indian Wildlife.*
Kingdon, *Field Guide to African Mammals.*

Panthera tigris (tiger)

Tigers once staked out territory from eastern Turkey to Southeast Asia, as well as the islands of Sumatra, Java, and Bali. Due to hunting and habitat destruction—tigers need large expanses of forest cover—today they are found no farther west than India and have vanished from Java and Bali. Of the original eight subspecies, people have wiped out three, and a fourth, the Chinese tiger, is near the brink of extinction. At the turn of the last century, there were about 100,000 tigers, but by the year 2000, fewer than 6,000 remained in the wild.

In the Tropics, about 500 Sumatran tigers are found in Indonesia; fewer than 1,500 Indo-Chinese tigers in the rainforests of Malaysia, Myanmar, Thailand, Cambodia, Laos, and Vietnam; and about 5,000 Royal Bengal, or Indian, tigers in India, Bangladesh, Nepal, and Bhutan.

The U.S. government and international conservation groups are working with the government of India to stop the serious poaching problem—the animals are no longer hunted for sport or their skins, but for the parts hunters once left behind. Nearly every part of the tiger is still used in traditional Asian medicine. Tiger hunting is illegal everywhere, and trade in tigers and tiger products is banned under a global trade treaty known as the Convention on International Trade of Endangered Species of Wild Fauna and Flora, or CITES, which also covers all the wild cats, dozens of other animals, including fish, frogs, birds, reptiles, etc., as well as dozens of plant species. But habitat destruction continues, and illegal trade is booming. The skins of the Indian tiger (*Panthera tigris tigris*) are mostly sold to Arab countries, where they may fetch $15,000 each. The bones and other parts are ground to powder and used as folk cures. One tiger may produce about 24 pounds of powdered bone, which, at $500 per gram, can bring more than $5 million. The local poachers are usually paid about $100. Indian tiger genitalia are sold to wealthy Asians; one tiger penis can cost $1,700 in Taiwan, where a bowl of tiger-penis soup sells for $300. Japan is the only Asian country that still permits the legal sale of tiger parts. Japan did not sign the CITES treaty. It is illegal to bring tiger parts into Japan, but it is not illegal to sell those that are there.

Tigers hunt at night, and may cover 18 miles or more in their silent search for prey, which has also dwindled dramatically due to habitat loss. When their natural prey is scarce, tigers may take livestock from villages. They will eat human beings only when extraordinarily hungry or ill, and unable to capture faster animals, including mammals, birds, frogs, and fish. They need massive amounts of meat to stay alive—an adult Bengal tigress with two cubs must have 6,800 pounds of meat a year. A tiger makes a large kill every week or so. After stealthily stalking its prey, it bounds forward to deliver a fatal bite with its saber-sharp teeth.

The Sumatran tiger (*P. tigris sumatrea*) is the smallest of all the tigers, weighing an average of 260 pounds and growing up to eight feet in length. By contrast, a male Siberian tiger (*P. tigris altaica*) can weigh up to 675 pounds and reach 11 feet in length. Recent DNA testing on Sumatran tigers has convinced some scientists that it may be a separate species. The Royal Bengal tiger is the most populous of the tiger subspecies; up to 4,000 survive in India, 350 in Bangladesh, 300 in Nepal, and 50 in Bhutan. The rarest is probably the Javan tiger (*Panthera tigris sondaica*), if it still exists at all. None have been spotted on Java, a densely populated island, since the 1970s, when five were thought to survive. The last Bali tiger (*Panthera tigris balica*) was shot in 1937.

Further Reading:
Breining, "The Ghosts of Way Kambas."
Burns, "Medicinal Potions May Doom Tiger to Extinction."
Environment News Service, "Urgent Action Afoot to Save Vanishing Tigers" <http://ens.lycos.com/ens/jan2000/2000L-01-17-02.html>.
Matthiessen, "The Last Wild Tigers."
World Wide Fund for Nature, Switzerland <http://www.panda.org / resources / factsheets / species / frame.htm?fct_tigers.htm>.

Prionailurus viverrinus (fishing cat)

The fishing cat is found in forests near rivers and streams and in mangrove forests from India to the Thai peninsula and south to Sumatra and Java. A pair of dark lines run from the top of its eyes up over its back, where the solid lines break into rows of parallel black spots. Its coat is olive gray, and it has a stocky and powerful build. At 2.5 feet long, it has a short and muscular tail. A proficient angler, the cat is also a strong swimmer. It catches fish and other aquatic animals by snatching them out of slow-moving water with its paws or by diving right in to seize them with its jaws. It also preys on land-based animals, such as rodents, young deer, and wild pigs. The fishing cat is threatened by destruction of its wetland habitat, by water contamination, and by hunting and trapping for its fur.

Further Reading:
Big Cats Online <http://dialspace.dial.pipex.com/agarman/intro.htm>.
World Conservation Union, Species Survival Commission, Switzerland <http://lynx.uio.no/catfolk/viver01.htm>.

Mustelidae

Mustelidae are found on all continents except Antarctica and Australia. The about 67 species worldwide include weasels, martens, otters, badgers, and skunks. Fossil records show that they first appeared in North America and Europe. They usually have long, slinky bodies, long tails, broad heads, small ears and eyes, and short legs. Their soft and glossy fur has

made them popular in the fur trade. They have powerful senses of smell and are swift and skillful hunters. With sharp and crushing sets of teeth, some species can kill prey much larger than themselves with a devastating bite to an animal's head or neck. Most species sleep and raise young in burrows or hollow logs.

Further Reading:
Emmons, *Neotropical Rainforest Mammals.*
Redford, *Mammals of the Neotropics.*

Eira barbara (tayra)

The tayra looks like a large mink, with dark brown fur and an 18-inch, black bushy tail. It is active during the day, feeding on rodents and other small mammals, nestling birds, lizards, eggs, fruits, and honey. Found from southern Mexico to northern Argentina, it is an excellent climber, and while it prefers forests, it is also found in coastal areas and savannas.

Further Reading:
Emmons, *Neotropical Rainforest Mammals.*
Kricher, *A Neotropical Companion.*
Redford, *Mammals of the Neotropics.*

Pteronura brasilensis (giant river otter)

At an average of five feet long, the giant river otter is indeed remarkably large, for an otter. Its tail is nearly two feet long, its eyes large, ears small, feet large and well webbed. Its short and glossy fur is dark brown when dry and a shiny black when wet. Each otter has a different set of cream and tan spots on its throat. They are always found near or in lowland forest rivers, streams, or lakes of South America, east of the Andes, and south to northern Argentina. They travel in small groups of five to nine, often squabbling over food with growls and squeaks. They mainly eat large fish, but will also dine on snakes and small caiman. At night, the group sleeps in a large burrow dug into a lakeshore or riverbank. Hunting for their pelts and habitat loss has decimated populations of giant river otters, once common in nearly all freshwater streams of their range. In the Amazon, there are perhaps no more than 2,000 to 5,000 remaining.

Further Reading:
Emmons, *Neotropical Rainforest Mammals.*
Redford, *Mammals of the Neotropics.*

Procyonidae

The family *Procyonidae* includes about 16 species, all found only in the Neotropics. They are all medium-size omnivores and excellent tree climbers. All raise their young in arboreal nests.

Bassaricyon gabbii (olingo)

The olingo is a nocturnal and solitary mammal, with gray-brown fur on its back and cream-colored fur on its chest and stomach. It has a 15-inch body and a 17-inch, bushy, faintly ringed tail. It has a short, pointed muzzle, short, rounded ears, short legs, and broad feet with short, curved claws. Although long, its tail is not prehensile like that of the kinkajou (*Potos flavus*), an animal it resembles. The olingo holds its tail straight out from its body or arched upward. It can dash through the trees with agility and speed. It ranges from central Nicaragua to western Colombia and western Ecuador and mainly eats fruit, especially figs, but sometimes invertebrates. During the dry season, it sips flower nectar.

Further Reading:
Emmons, *Neotropical Rainforest Mammals.*
Reid, *A Field Guide to the Mammals of Central America and Southeast Mexico.*

Nasua nasua (coati)

The coati has an elongated snout and a long, faintly ringed tail that it usually carries vertically upright, like cats do. Its fur is a thick, dark brown. Found from southern Venezuela to northern Argentina, it forages in trees and along the forest floor, poking its sensitive nose among the leaf litter. It eats fruit, invertebrates, and other small animals. The white-nosed coati (*N. narica*) is a similar species found in Arizona, Texas, all of Central America, and the west coasts of Colombia, Ecuador, and Peru. This coati is named for the white fur on its muzzle and chin that extends in a white stripe above its eyes. While male coatis are solitary (except at mating time), females travel in bands of 4 to 24.

Further Reading:
Emmons, *Neotropical Rainforest Mammals.*
Redford, *Mammals of the Neotropics.*

Potos flavus (kinkajou)

The kinkajou prowls the rainforest canopy at night, alone or in pairs. It has a round face, rounded ears, and wide, round eyes set wide apart. It is 20 inches long and its 20-inch tail is prehensile, allowing it to easily travel through the treetops in search of fruit, primarily figs. Kinkajous are omnivores, but seldom eat insects. Their dense and woolly fur is usually

golden brown, with creamy yellow underparts. They live in low- and midelevation forests from southern Mexico to southern Brazil and are important seed dispersers and pollinators.

Further Reading:
Emmons, *Neotropical Rainforest Mammals.*
Reid, *A Field Guide to the Mammals of Central America and Southeast Mexico.*

Procyon cancrivorous (crab-eating racoon)

The crab-eating racoon inhabits eastern Costa Rica, south to Uruguay and northeastern Argentina. It looks similar to the northern racoon (*P. lotor*), found throughout North and all of Central America, but its mask of black fur, encircling the eyes, is not quite as large. Its bushy tail is about half as long as the rest of its body and has wide black and pale rings. The crab-eating racoon's fur has a yellow-reddish tinge. It is an average of 22 inches long and weighs about 14 pounds. It spends days sleeping curled up in hollow trees, hunting at night for crabs, as well as mollusks, fish and some amphibians, insects, and fruits. Because of its food preferences, it is mainly found around swamps, rivers, and beaches. It is an excellent swimmer and tree climber.

Further Reading:
Emmons, *Neotropical Rainforest Mammals.*
Redford, *Mammals of the Neotropics.*

Ursidae (bears)

Bears include eight existing species, four of which are found in the Tropics: the Asiatic black bear, the spectacled bear, the sloth bear, and the sun bear. None of them hibernates routinely, as do bears in temperate climes. The Asiatic black bear (*Ursus thibetanus*) hibernates in its more northern range, in Pakistan, Afghanistan, northern India, and China, but during cooler weather in Southeast Asia, it will sleep for only a short period or simply leave high-altitude forests for lower, warmer valleys.

Further Reading:
The Bear Den <http://www.nature-net.com/bears/index.html>.
Cyber Zoomobile <http://www.primenet.com/~brendel/ursidae.html>.

Helarctos malayanus (sun bear)

The smallest bear species, the sun bear, stands at about two feet tall from feet to shoulders, with an average length between 3.5 and 4.5 feet. It has short, dense, black fur and a large yellow or orange U-shaped blaze on its chest. Sun bears originally lived in lowland forests of Southeast Asia from Malaysia and Indonesia, and as far west as India; they are probably extinct in India and Bangladesh. Deforestation has endangered populations in Myanmar, Thailand, Laos, Cambodia, and Vietnam. Farmers often kill bears that raid coconut-tree farms, where the bears fell young trees to eat the tender palm heart. Another threat is the illegal trade in bear gallbladders, a folk remedy used in Asia. Primarily nocturnal animals, sun bears eat fruit, small rodents, lizards, birds, ants, termites, earthworms, insects, and honey. Female sun bears give birth to two cubs.

Further Reading:
The Bear Den <http://www.nature-net.com/bears/index. html>.
World Wide Fund for Nature, Switzerland <http://www. panda.org/news/press/news_153b.htm>.

Tremarctos ornatus (Andean bear)

Latin America's only bear species, the Andean bear lives in the Andean deserts, up through the misty cloud forests, to the frosty, mountain-top *paramo* grasslands. Other than pumas and human beings, no other flightless animal traverses so many different elevations. The Andean bear is also known as the "spectacled bear" because of the cream-colored fur that circles around the animal's eyes, often extending down to its throat and chest. Each animal has a distinct pattern of these "spectacle" markings. The rest of the bear's thick coat of fur is black or deep brown.

The Andean bear is a migrant from North America: It made its way south about two million years ago. Although numerous legends chronicle the bear's mystical powers, in recent decades, these mystical powers have not kept the endangered species from being slaughtered to prevent it from raiding the cornfields and killing the livestock that have relentlessly invaded its habitat. The bear's favorite foods are fruits and bromeliads, but it will also eat berries, grasses, insects, and small animals such as rodents, rabbits, and birds. It plays a key role in the health of its habitat, as it disperses tree seeds that are too large for other animals to ingest. Scientists believe the bear is primarily responsible for the propagation of at least three important hardwoods in its range.

Not much is known about its numbers, but the U.S. biologist Bernie Peyton guesses that between 18,000 and 65,000 bears survive throughout its entire range, from Panama south to Argentina. Now 58 protected areas in Latin America provide the bears with refuge.

Further Reading:
Stolzenburg, "Andean Ambassador."
Youth, "The Bear Facts."

Viverridae

Viverridae is one of the most diverse families in the order of carnivores, with 66 species distributed throughout a wide range. This family of mammals, which includes civets, linsangs, genets, and mongooses, is found only in the Old World, from southern Europe, sub-Saharan Africa, and on through the Middle East, India, and most of Southeast Asia. *Viverridae* are the only carnivores on the island of Madagascar. They inhabit forests, savannas, and deserts. They range in size from the dwarf mongoose (*Helogale spp.*), at 16 inches long, to the African civet (*Civettictis civetta*), which can be up to 57 inches long. Most civets and genets look like spotted, long-nosed cats. They have slender bodies, pointed ears, and short legs. Except for mongooses, they all have bursae, a pocket along the edge of their ears, and all have scent glands, which produce a pungent oil called civet, which has been used as a base for perfumes. Most viverrids are nocturnal hunters that feed on insects, small vertebrates, birds, eggs, crustaceans, and fruit. They usually travel alone and have excellent senses of sight, smell, and hearing.

Further Reading:
Kingdon, *Field Guide to African Mammals.*
University of Michigan, Animal Diversity Web <http:// animaldiversity.ummz.umich.edu/chordata/mammalia/ carnivora/viverridae.html>.

Arctictis binturong (bearcats, or binturongs)

Bearcats, or binturongs, are arboreal relatives of the mongoose. They reach about six feet in length, counting a three-foot tail, which is prehensile—binturongs are the only carnivores in the Old World with prehensile tails. They have long, shaggy coats, small ears, long ear tufts, and eight-inch whiskers. Although they eat meat, principally rodents, carrion, insects, lizards, eggs, and nestlings, most of their diet consists of fruit, and they are important seed dispersers in their range: the forests of Southeast Asia, Malaysia, Indonesia, and the westernmost island of the Philippines, Palawan.

Binturongs are nocturnal, sleeping curled up in the

treetops during the day. They move slowly and deliberately through the trees. They don't jump from branch to branch—at about 30 pounds, they aren't very light on their feet—but can walk upside down, hanging from boughs, and they descend trees headfirst, like squirrels. They have short, strong legs, with nonretractable claws on their feet, which, with their leathery soles, give them good traction.

Further Reading:
Online Animal Catalog <http://www.blarg.net/~critter/articles/sm_furry/mongoose.html>.
Wildlife Waystation <http://www.waystation.org/html/binturong.html>.

Crossarchus obscurus (cusimanse)

The cusimanse is one of 23 species of mongooses in Africa; 8 more are in Asia. About 13 inches long, with dense, shaggy, dark fur and an eight-inch bushy tail, the cusimanse has a long, narrow face and snoutlike nose. Like most mongooses, it is gregarious, living in family groups of up to 12 animals in the dense undergrowth of forests from Gambia and Sierra Leone east to Cameroon and Gabon. It digs in the soil for insects, earthworms, snails, crabs, and other invertebrates, and hunts mice, lizards, frogs, and other vertebrates and their eggs. To break eggshells or hard-shelled prey, it dashes them against a stone or tree. A terrestrial carnivore, it spends the night in a burrow it digs itself. Although their populations are still fairly healthy, cusimanses are vigorously hunted, so their status may change.

Further Reading:
Haltenorth and Diller, *Collins Field Guide: Mammals of Africa*.
Kingdon, *Field Guide to African Mammals*.
The Rainforest Trust, USA <http://www.rainforesttrust.com/lib7cont.html>.

Genetta cristata (crested genet)

The crested genet is an endangered species that lives in the rainforests of southeastern Nigeria and southern Cameroon. It is extremely shy and elusive, so little is known about its habits. Crested genets are small, about 23 inches long, with 18-inch tails, and have small crests on the back of their necks. *G. servalina* (forest genet) is another forest dweller, found in east Africa. Genets look a bit like small, pointy-nosed cats. They are pale to brownish-gray, with dark brown or black spots arranged in rows on their soft fur. Their long tails are ringed in black and white.

Like cats, they have retractable claws, helpful in climbing trees and catching prey. They feed on small insects, rodents, birds, reptiles, amphibians, fruits, and berries. They are solitary or travel in pairs. They mainly hunt on the ground, at night. Through scent glands, they produce secretions to mark territory or send sexual messages. They can also emit a foul-smelling secretion that discourages enemies. Human beings are their principal predators, but hawks frequently snatch young genets.

Further Reading:
Kingdon, *Field Guide to African Mammals*.
World Conservation Monitoring Centre, England <http://ftp.wcmc.org.uk/species/data/species_sheets/crestgen.htm>.

Nandinia binotata (palm civet)

Palm civets are fairly common in east African forests. Their mottled coat blends well with the shadowy treetops they inhabit. Its body is catlike, with powerful limbs and a lengthy tail. It has omnivorous tastes, eating fruits, insects, eggs, rodents, birds, and bats. It kills living prey with many fast and deadly bites, but actually seems to prefer fruit and is an important seed disperser. Palm civets are solitary animals that forage at night. They always defecate at dung heaps, called middens, near their customary routes of travel. The middens may also serve to mark territorial boundaries. Female African palm civets are sexually mature at one year and produce one or two litters of one to three young annually.

Further Reading:
Kingdon, *Field Guide to African Mammals*.
University of Michigan, Animal Diversity Web <http://animaldiversity.ummz.umich.edu/accounts/nandinia/n._binotata>.

Osbornictis piscivora (aquatic genet)

The aquatic genet lives near forest streams and rivers of northeastern Democratic Republic of Congo. This small and slender genet, about 18 inches long with a 15-inch bushy tail, has a reddish-brown coat with dark patches on its ears, back, and tail. It feeds mainly on fish and other aquatic animals, which it captures by first slowly approaching a quiet pool in its habitat and gently patting the water surface with a paw. Prey is detected by sight or by vibrations, and the aquatic genet then swiftly nabs its next meal in its jaws.

Further Reading:
Kingdon, *Field Guide to African Mammals.*
The Rainforest Trust, USA <http://www.rainforesttrust.com/lib7cont.html>.

Paradoxurus zeylonensis (golden palm civet)

The golden palm civet is found in southern and Southeast Asia. They may eat small mammals, birds, snakes, lizards, frogs, and insects, but they prefer fruit, particularly mango, melon, pineapple, and bananas. They are arboreal, but less so than the African palm civet, and have adapted fairly well to deforestation, taking to sleeping on bungalow rooftops.

Further Reading:
University of Michigan, Animal Diversity Web <http://animaldiversity.ummz.umich.edu/accounts/paradoxurus/p._zeylonensis.html>.

Poiana richardsoni (Central African linsang)

The Central African linsang is a secretive and rare member of the *Viverridae* family. Its slender body, covered with short, thick, pale yellow fur with rows of dark, oval spots, is about 14 inches long and its black-ringed tail, 15 inches. Its habitat is in Sierra Leone, Liberia, Ivory Coast, Cameroon, Gabon, Congo, and northern Democratic Republic of Congo. Nocturnal, the central African linsang is a good climber, spending most of its time in trees and sleeping in nests that it builds of leaves. It eats arboreal vertebrates, invertebrates, and fruits. Two species of linsang live in Asia.

Further Reading:
Haltenorth and Diller, *Collins Field Guide: Mammals of Africa.*
Kingdon, *Field Guide to African Mammals.*

Salanoia concolor (plain mongoose)

Found only in northeastern Madagascar, the plain mongoose is one of the smallest mongoose species, weighing only two pounds. It is about 15 inches long, with a bushy, seven-inch tail, and is covered with reddish brown fur speckled with black or tan. It has small eyes and ears, a pointed snout, short legs, and long claws on its feet. The plain mongoose is diurnal, and although it can climb trees, it spends most of its time on the ground, hunting mainly for insects, although it also eats lizards, frogs, small mammals, birds, eggs, and fruit. It takes shelter in burrows in the ground or in hollow trees. It mates for life and lives in family groups in territories it marks with secretions from its anal glands.

Further Reading:
Haltenorth and Diller, *Collins Field Guide: Mammals of Africa.*
Missouri Botanical Garden, USA <http://mbgnet.mobot.org/sets/rforest/animals/salano.htm>.

Viverra civettina (Malabar large spotted civet)

Unlike other civets, Malabar large spotted civets resemble dogs more than cats, with long legs and canine-shaped heads and muzzles. One of the world's most endangered animals, their total population is no more than 250. They weigh about 18 pounds. Originally they inhabited the rainforests of the Western Ghats in southwest India, but most of their habitat has been deforested. By the late 1960s, biologists assumed they were extinct, until one was sighted in 1987. A 1990 survey revealed that a small population still survives in South Malabar. Since most of their forest habitat is gone, the majority of the surviving civets probably take refuge in cashew farms, where there is a dense understory of shrubs and grasses, but these are now being cleared for rubber plantations. Little is known about this animal, but scientists guess that they eat small animals, eggs, and some plants.

Further Reading:
Animal Info <http://www.animalinfo.org/species/carnivor/vivemega.htm>.

CHIROPTERA (bats)

Bats are the only mammals that can fly. A wing membrane extends from each side of the body and hind leg to the forearm, where it is supported by elongated fingers. Some species have an additional membrane between the hind legs, which may enclose the tail. The membranes are soft, strong, flexible, and will heal if punctured. They are also remarkably elastic; when a bat closes its wings, the membranes do not fold up, but rather collapse, then easily stretch out again when the wings are opened.

The nearly 1,000 species are found on all continents but Antarctica. In any forest, there are as many species of bats as there are of all other mammal species combined. Bats are divided into two suborders, the *Megachiroptera* (meaning large-hand wing) and the *Microchiroptera* (small-hand wing). The first group is found only in the Old World. They are called fruit bats because they have evolved to feed on fruits, pollen, and nectar. Fruit bats are a vital part of the forest ecosystem, as they disperse the seeds of a wide variety of trees. They are also considered a delicacy in much of their range, served curried, marinated in coconut milk, or smoked and dished up with rice. They are even showing up in butcher shops in London. This is still legal, since fruit bats are not endangered species—yet. A fruit bat native to Guam became extinct in the 1960s because it was exported in huge numbers to feed bat-craving Samoans.

Both groups of bats are also important pollinators, and many bat-pollinated trees have evolved ways to attract the mammals to their blossoms. Flowers pollinated by bats (called *chiropterophilous,* or "bat loving") are mainly found in the canopy, held upright on stems, and are usually white with a funky odor.

The *Microchiroptera* bats, found both in the Old World and New, have developed a kind of sonar system, called echolocation, to navigate at night and in lightless caves and to zero in on flying insects. The mammals make high-frequency calls, usually above the human hearing range, out of their mouths or noses (or both), then listen for echoes to bounce from the objects in front of them. They are able to interpret the reflected sounds and form a picture of objects in front of them.

For their size, bats are the world's longest-lived mammals, some reaching 30 years old (which would be the equivalent of a 100-year-old human being). Unlike other mammals of their size, they have low reproductive rates; females of most species produce just one pup per year.

Further Reading:
Bat Conservation International, USA <http://www.batcon. org/discover/trivia.html>.
Emmons, *Neotropical Rainforest Mammals.*
Horton, "Bat du Jour."
Redford, *Mammals of the Neotropics.*
Wolfe and Prance, *Rainforests of the World.*

Megachiroptera (large-hand wing bats)
Pteropidae (flying foxes and old world bats)

Hypsignathus monstrosus (hammer bat)

The hammer bat is the largest continental African fruit bat, with five-inch forearms (a measurement taken from the elbow to the wrist of the folded wing) and an average weight of nine ounces—males are almost twice the weight of females and can actually weigh up to 15 ounces. They have brown fur and membranes, tubular noses, and white tufts at the base

of their ears. Males have inflatable sacs over the raised ridge of the nose and on each side of the neck. Male hammer bats are built for making noise, as their larynx has displaced their hearts, lungs, and diaphragm to fill their chests with a structure like a tuba. At night, their blaring honks echo through the lowland forests of Senegal to western Uganda. The males gather in a "choir," calling to attract females, who make frequent inspections before choosing a mate. The loudest, most competitive singers are preferred.

Further Reading:
Kingdon, *Field Guide to African Mammals.*

Nyctimene rabori (Philippines tube-nosed fruit bat)

With a wingspan of some 22 inches, the Philippines tube-nosed fruit bat inhabits the remaining lowland rainforests of the Philippines. It roosts in vegetation and large hollow trees and eats wild figs. Because most of its habitat has been destroyed, the largest population of *N. rabori* is on Negros Island in very narrow strips of forest. Less than 1 percent of the original lowland forest remains on Negros. The bats are named for their laterally directed, tubelike nostrils, although scientists are unsure what purpose this unusual design has. The endangered Philippines tube-nosed fruit bat has a broad dark stripe down the center of its back; it is one of the few striped bats in the world.

Further Reading:
Animal Info <http://www.animalinfo.org/species/bat/ nyctrabo.htm>.

Pteropus spp. (flying fox)

Flying foxes are fruit bats that are found in the Malay-Indonesia islands. Instead of using echolocation to hunt down insects, the larger fruit bats have a keen sense of sight and smell to lead them to blooming flowers or ripe fruit. They can attain wingspans of four feet. The bats are active at night, and roost in trees during the day, hanging from branches by one or both feet, their wings wrapped around them as they slumber. At roost sites, thousands can be seen hanging from tree branches—or heard, since they can be quite noisy when awake. They take off at dusk, and may cover some 155 miles during one night of foraging. Unlike most animals, although resembling human beings and guinea pigs, fruit bats do not generate vitamin C, so it is supplied to them in the fruits they eat.

Flying foxes are considered to be delicacies in their

native habitat, which hasn't helped their population. Nine species of Pacific Island flying foxes are endangered or threatened. In Indonesia, some 30,000 are killed for consumption each year, and it's a growing industry.

Further Reading:
Bat Conservation International, USA <http://www.batcon. org/home/cover.html>.
Wolfe and Prance, *Rainforests of the World.*

Pteropus conspicillatus (spectacled flying fox)

The spectacled flying fox lives in New Guinea and other nearby islands and in the coastal rainforests of Australia. With four-foot wingspans, they are black with a circle of straw-colored hair around their eyes and along their muzzles.

Further Reading:
Bat Conservation International, USA <http://www.batcon. org/home/cover.html>.

Pteropus niger (black flying fox)

Found only on the island of Mauritius, the black flying fox, like many island animals, is extremely sensitive to any alterations in its habitat. Only a few hundred animals remain. Because so few wild fruit trees are available, the bats fly out of their roosting caves at night to feed on cultivated fruit. Farmers protect their crops with slingshots and rifles. Perhaps the rarest bat on Earth is the Rodrigues flying fox (*P. rodricensis*), found only on the neighboring island of Rodrigues. A 1976 census counted only 150 of the mango-feeding bats.

Further Reading:
Animal Info <http://www.animalinfo.org/species/bat/ pterrodr.htm>.
Hill and Smith, *Bats: A Natural History.*

Microchiroptera (small-hand wing bats)

Craseonycteris thonglongyai (bumblebee bat)

There is just one species in the famly of *Craseonycteridae:* the bumblebee bat, *Craseonycteris thonglongyai*, the smallest known mammal, weighing just two grams when fully grown, or less than a penny. It was first found and described in 1974 and is known to live only near the Kwai River in Thailand. It has a pig-like snout, no tail, relatively large eyes, and tiny ears nearly hidden by its fur. The little bat roosts in limestone caves in small groups and eats insects and

spiders on the wing. It can hover like a hummingbird. Because of uncontrolled deforestation in its foraging habitat, it is extremely endangered.

Further Reading:
Bat Conservation International, USA <http://www.batcon. org/discover/trivia.html>.

Molossidae (free-tailed bats)

The free-tailed bats are a family so named because they have long, thick, naked tails that extend beyond the tail membrane. The family includes about 80 species, found worldwide, but mainly in the Tropics. Their long, narrow wings, aerodynamically similar to those of falcons, carry free-tailed bats to great heights, as high as 10,000 feet, in search of large insects, such as moths and beetles. And they fly fast; some species catch tailwinds that allow them to speed at 60 miles per hour. They normally roost in narrow crevices of caves between rocks or in tree holes, but are quite adaptable and will roost by the hundreds in tiny spaces in building eaves. The bats seen at dusk flying swiftly over rooftops in tropical countries are usually free-tailed bats.

Further Reading:
Bat Conservation International, USA <http://www.batcon. org/discover/trivia.html>.
Emmons, *Neotropical Rainforest Mammals.*
Redford, *Mammals of the Neotropics.*

Noctilio leporinus (greater bulldog bat)

The greater bulldog bat is one of the largest Neotropical bats, with forearms that are three inches long and quite large feet that have sharp, curving claws. Their common name comes from their bulldoglike faces, which have drooping lips and jowls that expose large, canine teeth. Found from the Pacific coast of Mexico south to northern Argentina and Uruguay, they feed on insects and fish. They use their echolocation to detect fish disturbing the surface of the water, then swoop down and snatch them with their claws. Their echolocation is so exact that they can detect a tiny fish fin as thin as a human hair, protruding only two millimeters above a pond's surface.

Further Reading:
Bat Conservation International, USA <http://www.batcon. org/discover/trivia.html>.
Emmons, *Neotropical Rainforest Mammals.*

Otomops martiensseni (giant mastiff bat)

Found throughout Africa, the giant mastiff bat is four inches long, with three-inch forearms. Its long, wide ears extend along the whole length of its head, and its upper lip is flanged. Giant mastiff bats catch beetles and large insects high above the ground or water. They congregate by the tens of thousands in breeding caves, which they will periodically vacate en masse.

Further Reading:
Kingdon, *Field Guide to African Mammals.*

Phyllostomidae (leaf-nosed bats)

Leaf-nosed bats are a family of bats that have fleshy folds around their nostrils that make them look as if they are balancing small, pointy leaves on their noses. This strange-looking structure acts as a megaphone, allowing them to target their echolocation sounds, which are emitted through their nostrils. Their broad wings allow them to fly at slow speeds and navigate through the forest's dense understory. The 140 species of leaf-nosed bats are found only in the New World, from southern California and Arizona, the Gulf Coast of Texas, south to northern Argentina, as well as the Caribbean islands. Like most bats, the females give birth to a single offspring. In some species, males and females roost separately during breeding season, with females and their young forming large nursery colonies.

Further Reading:
Emmons, *Neotropical Rainforest Mammals.*
Redford, *Mammals of the Neotropics.*

Desmodontinae (vampire bats)

Vampire bats are a subfamily of *Phyllostomidae.* They don't exist in Transylvania, but are found from southern Sonora, Mexico to northern Argentina, Uruguay, and northern Chile. The three species in this subfamily feed on the blood of warmblooded vertebrates by pricking a sleeping animal with their two large razor-sharp teeth. An enzyme in the vampire bat's saliva causes blood to flow without clotting, and the bat laps up this nutritious, protein-filled fluid. They lick up only about two tablespoons of blood; the host animal usually never even wakes up. Vampire bats can walk on their feet and thickened thumbs and are the only kind of bat that can launch itself into the air from the ground. Scientists turned high-speed cameras on the bats and found that their takeoffs are powered by pectoral muscles, with stabilizing help from their long thumbs.

Vampire bats that come home to roost having been unsuccessful in finding a meal are often generously fed by colony mates, which regurgitate a bit of blood. The anticoagulant found in the bat's saliva has been synthesized and is used in medication for human heart patients.

Further Reading:
Bat Conservation International, USA <http://www.batcon.org/discover/trivia.html>.
Emmons, *Neotropical Rainforest Mammals.*
Redford, *Mammals of the Neotropics.*

Stenodermatinae (Neotropical fruit bats)

Another subfamily of bats, Neotropical fruit bats, feed on fruit, supplemented by flower nectar when fruit is scarce. They are extremely important in dispersing seeds, and many plant species depend on *Stenodermatinae* bats for propagation. Neotropical fruit bats help restore, or "replant," natural or man-made forest clearings. Many species bear white stripes down their backs and on their faces. Some roost inside tents they pitch by biting the sides and ribs of large leaves, usually of palm trees, so that the leaf drops down on either side of the rib to form a protective shelter.

Further Reading:
Emmons, *Neotropical Rainforest Mammals.*

Ectophylla alba (Honduran white bat)

The Honduran white bat is tiny and delicate, at just under two inches. Its head and upper body are white—it's the only white leaf-nosed bat—with gray-white on its back, and black wings, except for its yellowish forearms and thumbs. Its ears and leaf nose are bright yellow. Found from eastern Honduras to Panama, it roosts in groups of four to eight in tents it constructs from medium-sized *Heliconia* or other understory plants. The bats nibble on either side of the leaf's midrib, which causes the sides to collapse and hang vertically. Though its status is unknown, the Honduran white bat is probably vulnerable, due to deforestation.

Further Reading:
Emmons, *Neotropical Rainforest Mammals.*
Reid, *A Field Guide to the Mammals of Central America and Southeast Mexico.*

Thyropteridae (disk-winged bats)

The three-species family of disk-winged bats are all found in Latin America only. These tiny bats have disks at the base of their thumbs and under their heels that act like suction cups, the only bat in the New World with this special adaptation. Bats of an Old World family, *Myzopodidae*, have similar disks, but they aren't as well developed. Disk-winged bats use their suction cups to cling to smooth surfaces—they can slither across a pane of glass or a smooth leaf.

Further Reading:
Reid, *A Field Guide to the Mammals of Central America and Southeast Mexico.*

Thyroptera tricolor (Spix's disk-winged bat)

Spix's disk-winged bat is a delicate mammal, just an inch-and-a-half long, found in southeastern Mexico and most of Central America, south to southeastern Brazil. Its long and fluffy fur is brown on top and white or yellowish underneath. Spix's disk-winged bats roost in young, rolled-up leaves of rainforest plants like heliconias and bananas. Unlike most bat species, they roost upright and can use their suction cups to scamper up slippery leaf surfaces. At night, they launch themselves from the top of the leaf cone to hunt for insects. They stay in groups of under 10, with equal numbers of males and females, and roost on different furled leaves each day.

Further Reading:
Reid, *A Field Guide to the Mammals of Central America and Southeast Mexico.*

Vespertilionidae (common bats)

Swift and agile, common bats are great flyers. This is the most geographically widespread of all bat families, found on all continents but Antarctica, with more than 300 species. In northern zones, most species hibernate in caves during cold winter months, or migrate south. All walk well, and most dine on insects, although a few feed on fish.

Kerivoula spp. (woolly bats)

Woolly bats include eight species found throughout most of sub-Saharan Africa. They have frizzled, frosty-gray, woolly fur and eat insects they catch during their slow, weaving flights. A small bat, with forearms measuring just 1.5 inches, it roosts among dead leaves, in old birds' nests, hollow branches, or even large spider webs.

Further Reading:
Kingdon, *Field Guide to African Mammals.*

CETACEA (dolphins)

Cetacea are an ancient order of primarily marine mammals found worldwide.

Platanistidae (river dolphins)

River dolphins include five species, four that live in freshwater rivers and one in saltwater. Two freshwater species are found south of the Tropic of Cancer: the Amazon River dolphin (*Inia geoffrensis*) and a smaller species, *Sotalia fluviatilis*, commonly known as the "tucuxi," or estuarine dolphin, of the Amazon and Orinoco River basins in South America. A fifth member of the river dolphin family is found in the salty estuaries and coastal waters of Brazil, Uruguay, and Argentina. All are in trouble from overfishing and, particularly, pollution, as well as from dams and irrigation projects that restrict them to isolated stretches of water, separated from potential mates.

Further Reading:
Morell, "Looking for Big Pink."

Inia geoffrensis (Amazon River dolphin)

The Amazon River dolphin is located throughout the Amazon and Orinoco River systems in South America, which course through seven countries: Bolivia, Brazil, Colombia, Ecuador, Guyana, Peru, and Venezuela. It averages about 6.5 feet in length, and is pink, from grayish pink to rose-colored to flamingo-bright. River dolphins found in darker water, where sunlight can't penetrate well, are pinker than those in clearer water, where they lose pigmentation from exposure to the sun's rays. The dolphins also flush to a bright pink color when excited. Unlike most other dolphin species, they have unfused neck vertebrae, which gives them the ability to turn their necks. They can also paddle forward with one flipper and backward with the other, so they can easily navigate the twists and turns of the flooded Amazon forests. They have two other characteristics that other dolphin species lack: molarlike teeth so they can chew their prey, and small hairs on their snouts, which may assist them in feeling for crustaceans buried in muddy river bottoms. They eat about 50 different species of fish, plus crustaceans and the occasional turtle.

River dolphins are less sociable animals than are the saltwater species. They usually swim alone or in pairs, most likely a mother and calf. Dolphin calves are born between July and September, after a gestation period of about 11 months, and when the Amazon and Orinoco tributaries are receding. As the waters fall back into their banks, fish become more concentrated, aiding the mother in finding food fast while she nurses her newborn. Few river dolphins are hunted by human residents, since local cultural beliefs hold that the mammals possess magical powers. But these beliefs are fading, and people who are moving in on dolphin habitat regard the animals as competitors for fish. Additional threats to the mammal's survival are pollution, deforestation, and construction of large hydroelectric dams on the rivers.

Further Reading:
Goulding, *The Flooded Forest.*
McGuire and Winemiller, "Occurrence Patterns, Habitat Associations, and Potential Prey of the River Dolphin, *Inia geoffrensis*, in the Cinaruco River, Venezuela."
Morell, "Looking for Big Pink."

DERMOPTERA

Dermoptera is the order of *Cynocephalidae*, the colugos, lemurlike, gliding mammals.

Cynocephalidae

Cynocephalidae include two species of colugos, nocturnal species sometimes called flying lemurs, although they are neither related to lemurs nor able to fly. *Cynocephalus volans* is found in the Philippines, and *C. variegatus*, which is a smaller species with longer ears and a larger skull, inhabits areas of Southeast Asia, particularly Indonesia. Colugos are slender, 13 to 17 inches long, and covered with short and soft fur. They have long limbs and tails and large claws. A furred membrane called the patagium is attached to their necks and sides of their bodies, extending along their limbs to the tips of their toes and tail. The patagium expands like a parachute when the colugo extends its legs, allowing the mammal to glide from tree to tree. The glide of one colugo was measured at 443 feet. After a glide, the colugo lands flat against a trunk. The soles of the animal's paws form sucking disks so they can easily adhere to tree trunks. Colugos are skillful climbers and almost entirely arboreal. At twilight and dawn they glide through the forest, feeding on leaves, flowers, buds, and fruit. During the day, they hang from their long claws from tree branches or hide in hollow trees. In the Philippines, they are the principal prey of the endangered Phil-

ippine eagle (*Pithecophaga jeffreyi*), which plucks them from the air mid-glide.

Further Reading:
University of Michigan, Animal Diversity Web <http://animaldiversity.ummz.umich.edu/accounts/cynocephalus/c._variegatus.html>.
Wolfe and Prance, *Rainforests of the World.*

DIPROTODONTIA

Diprotodontia is one of four orders of Australian marsupials.

Macropodidae

The second-largest order of marsupials (*Didelphidae* is the largest), *Macropodidae* includes about 54 species. Family members are found in Australia and New Guinea, plus a few surrounding islands in the Pacific. Their hind limbs are quite powerful, and their hind feet are long and narrow, with the fourth toe the longest and strongest. Macropodids graze on vegetation and have complex stomachs to help them convert grass into glucose. Most species are nocturnal and move fast by hopping.

Further Reading:
Biodiversity Group of Environment Australia <http://www.anca.gov.au/>.
University of Michigan, Animal Diversity Web <http://animaldiversity.ummz.umich.edu/chordata/mammalia/diprotodontia/macropodidae.html>.

Dendrolagus lumholtzi (tree kangaroo)

Tree kangaroos are related to the land kangaroo, but have shorter, wider hind feet and longer, narrower tails. It is arboreal, with thick, curved claws to help it climb. Tree kangaroos feed on leaves, flowers, grass, and fruit. They are the only kangaroos that walk, rather than hop. They are solitary and nocturnal, spending the day asleep, curled in the treetops of New Guinea and northeastern Australia. The joey stays in its mother's pouch for about 33 weeks.

Further Reading:
Slater and Parish, *A First Field Guide to Australian Mammals.*

Thylogale stigmatica (red-legged pademelon)

The red-legged pademelon lives in dense eucalyptus forests and rainforests along the eastern coast of Australia and north to New Guinea. It is 18 inches long, with thick, soft, reddish and cream-colored fur and a short, thick tail. A solitary animal, it eats fruits and vegetation, grazing from late afternoon through the night. In the early morning, it settles down in the same resting spot by swinging its 15-inch tail between its extended hindlegs, sitting on the tail and leaning back against a tree sapling or rock. As it dozes off, its head droops forward until it rests either on its tail or on the ground beside it. Pademelons communicate with one another by thumping their heels on the ground.

Further Reading:

Marsupial Society of Victoria <http://home.vicnet.net.au/ ~marsup/fact_sheets/redlegged_pademelon.htm>.

Slater and Parish, *A First Field Guide to Australian Mammals.*

Hyracoidae (hyrax)

Small and furry, hyraxes look a bit like rabbits with rounded ears and no tails. Scientists believe that 55 million years ago, they shared a common ancestor with elephants. Three of the six species are tree hyraxes (*Dendrohyrax spp.*), forest dwellers in Africa. They are about 16 inches long and weigh about five pounds. Nocturnal, they feed on leaves and fruit, and are often found in pairs. Their nighttime call is memorable; starting off as a whistle, it rises to a high-pitched squeal, and then peaks in a scream that sounds very much like that of a terrified human being. Their predators include leopards, pythons, and civets. Tree hyraxes have soft, thick, and long hair. Like all hyraxes, they have stumpy toes with hooflike nails. The bottoms of their feet are rubbery, giving them the traction that they need to climb trees. Hyraxes have long gestation periods for an animal of their small size, some seven or eight months. Fossils that indicate hyraxes the size of oxen once roamed the Earth may explain their prolonged pregnancies.

Further Reading:

Kingdon, *Field Guide to African Mammals.*

African Wildlife Foundation, USA <http://www.awf.org/ animals/hyrax.html>.

MACROSCELIDIDAE

Macroscelididae is the order of elephant shrews, once placed in the same order as other shrews (*Insectivora*). Actually, scientists aren't sure if they are more closely related to insectivores, rodents, rabbits, or primates. There are 15 elephant shrew species in Africa. Their name comes from their long, thin, and trunk-like noses. Their heads are long and pointy, and they have long slim legs and hunched backs. They are about 10 inches long, not including their hairless tails, and weigh less than two pounds. Their hair is short and coarse, and varies with species from olive, to brown or reddish brown. Coat colors often match soil types quite closely. Some elephant shrews bear contrasting patches of color; others have bold white spots or black stripes. They can live in open plains, but prefer dense forests, where they can find abundant food and water.

Rhynchocyoninae (giant elephant shrews)

Giant elephant shrews are more exclusively tropical than are the soft-furred elephant shrews (*Macroscelidinae*) and more commonly found in forests. A gland on the underside of giant elephant shrews' tails emits a strong, musky odor they use to mark territory, but the scent also seems to make the giant elephant shrew an unsavory meal for most predators. Their principal enemies are snakes and raptors. They are swift runners and have strong senses of smell, sight, and hearing. They eat ants, termites, beetles, spiders, worms, and other invertebrates that they flick up with their extremely long tongues. Although giant elephant shrews form pairs, they don't spend much time together, although they do keep track of each other via their scent markings.

Shakespeare was right: It is unlikely that shrews can be tamed. They are quite aggressive, and if aggravated, both males and females will scream, snap, and kick in a wild blur of activity. Sometimes they dash about in a wide circle or leap two feet into the air. This is apparently not aggressive behavior, but scientists cannot explain why they expend themselves this way. They are diurnal, sleeping at night in small nests they make amongst the leaves on the forest floor. Elephant shrew fossils have been identified to be as old as 30 to 35 million years.

Further Reading:
African Wildlife Foundation, USA <http://www.awf.org/animals/shrew.html>.
Kingdon, *Field Guide to African Mammals.*

PERISSODACTYLA (odd-toed ungulates)

Odd-toed ungulates are herbivores that spend much of the day eating. They are unable to digest the cellulose in plants, so their stomachs have chambers that house microorganisms that digest this material for them. Because this system is not very efficient, they must eat prodigious amounts of plant material to obtain sufficient energy. Perissodactyls emerged from a primitive group of ungulates 65 million years ago. These early forms had five toes on the fore- and hind feet, but over time, the toes were reduced to three on rhinoceroses and just one on horses.

Further Reading:
Emmons, *Neotropical Rainforest Mammals.*
Kingdon, *Field Guide to African Mammals.*

Rhinocerotidae (rhinoceros)

The rhinoceros includes three species found in south central Asia and two in Africa, south of the Sahara Desert; the largest numbers are in Tanzania and Zimbabwe. Most species favor savannas and shrubby regions, but others live in dense forests. The African species usually frequent open savannas more often than do the Asiatic species. All rhinos tend to stay within easy walking distance of a water source, and they are exceedingly fond of wallowing in muddy pools. Herbivores, the African species graze on grasses, while the Asian species rip leaves off trees and bushes. They also dig up and eat earth for the minerals they need, particularly salt. The African species are generally more aggressive, fighting adversaries with their one or two horns, which protrude near the nasal cavities of their elongated heads. A baby can weigh up to 110 pounds at birth; adults can weigh in at 1,000 pounds.

All five species are endangered due to habitat destruction and overhunting. Human beings hunt rhinos in particular because nearly all parts of the massive animal's bodies are used in folk medicine. The most prized part of the rhino is its horn, which is used as a dagger holder or is ground up and taken as an aphrodisiac or fever reducer.

Further Reading:
World Wildlife Fund, "Kenya's Rhinos Benefit from WWF Donation."

Dicerorhinus sumatrensis (Sumatran rhinoceros)

The Sumatran is the smallest rhino, but even a small rhino is large, at five feet tall and 1,600 to 2,000 pounds. Covered with stiff, reddish hair, it has two short horns and ear tufts. Two deep skin folds encircle its body between its legs and trunk. Its leathery skin is an average of one-half inch thick. Sumatran rhinos are thought to live in rainforests of northern Myanmar, Thailand, Malaysia, Sumatra, and Borneo, but they are extremely rare, shy, solitary, and seldom seen. They feed on young tree saplings, leaves, twigs, fruit, and bark. Favored foods are mangoes, bamboos, and figs. They consume more than 100 pounds of food daily. To prepare a sapling for consumption, the rhino bites it off, tramples it, and then dines. To mark their territory, Sumatran rhinos spread urine and feces along trails they have trampled in the forest, and twist and break saplings. They are tireless walkers, moving mostly by night. Male and female stay together for just a short time during mating periods. Births occur, after a 12- to 16-month gestation period, from October to May, the time of heaviest rainfall. Conservationists believe fewer than 500 Sumatran rhinos survive, as deforestation and hunting for their horns has depleted their populations. The animals are very sensitive to disturbances. The governments of Malaysia and Indonesia have established a few forest reserves to protect them.

Further Reading:
American Zoo and Aquarium Association, Species Survival Plan, USA <http://aza.org/Programs/SSP/ssp.cfm?ssp=71>.
Kingdon, *Field Guide to African Mammals.*
University of Michigan, Animal Diversity Web <http://animaldiversity.ummz.umich.edu/chordata/mammalia/perissodactyla/rhinocerotidae.html>.

Rhinoceros sondaicus (Java rhinoceros)

The Java rhinoceros once ranged widely through India, Bhutan, Bangladesh, China, Myanmar, Thailand, Laos, Cambodia, Vietnam, Malaysia, and Indonesia. During the war in Vietnam, the defoliant Agent Orange destroyed much of the animal's forest habitat. That loss, as well as poaching for its horn, has nearly wiped out the rhino. Scientists believed that by the 1960s, the only population of surviving Java rhinoceroses were in Ujung Kulon National Park on Java, Indonesia, where about 50 to 60 currently reside. But in 1999 a tiny population was rediscovered in Vietnam, where the World Wildlife Fund is working with Vietnamese authorities to set aside protected areas for the Java rhino, perhaps the most endangered animal on Earth.

Further Reading:
O'Connor, "Thought Extinct, a Few Javan Rhinos Are Seen in Vietnam."

Tapirdae (tapirs)

Tapirs include four species, three in the Neotropics and one in Southeast Asia. They have large, robust bodies, thick necks, stumpy tails, and elongated noses, or trunks, which they use to bring leaves to their mouths. Able swimmers, they also use their trunks, a fusion of upper lip and nose, to squirt water in defense. Tapirs are the only members of the *Perissodactyla* order in the Neotropics and the largest terrestrial mammals there. They have large teeth that are well designed for grinding up plants.

Further Reading:
Emmons, *Neotropical Rainforest Mammals.*
Wolfe and Prance, *Rainforests of the World.*

Tapirus bairdii (Baird's tapir)

Baird's tapir ranges from southern Mexico to the west coasts of Colombia and Ecuador. They stand about three feet tall, with a body length of about six feet, and weigh an average of 500 pounds. They have white or gray cheeks and chests, with reddish hair. When hungry, they wave their long snouts in the air to pick up plant scents, then stroke plant leaves with their trunks until they find one that smells just right. They have a widely varied plant diet, munching leaves of dozens of species. In this way, they can avoid stomachaches as a result of feasting too much on any single plant, which may be laced with toxic chemicals. Tapirs communicate with one another with a whistling call, which is easily imitated by hunters, who can then track them down. Tapir meat is prized as a delicacy. They are endangered animals throughout their range due to overhunting and deforestation.

Further Reading:
Emmons, *Neotropical Rainforest Mammals.*
Forsyth and Miyata, *Tropical Nature.*

Tapirus indicus (Malayan tapir)

The largest tapir species, the Malayan tapir weighs up to 800 pounds. From toe to shoulder, they measure about three feet, with an eight-foot body length. They are black and white and found in swampy and lowland forests of southern Myanmar, Thailand, Malaysia, and Sumatra. Malayan tapir populations are near extinction because of deforestation. They are hunted in some areas, although not in Sumatra, where they are considered a form of pig and are therefore avoided by the predominantly Muslim population.

Further Reading:
Emmons, *Neotropical Rainforest Mammals.*
Williams, "Super Snoots."
World Wide Fund for Nature, Switzerland <http://www.panda.org/resources/publications/species/underthreat/tapir.htm>.

Tapirus pinchaque (mountain tapir)

Mountain tapirs are found only in the Andes of Colombia and Ecuador, with a small remnant population in Peru. They are probably extinct in their former range of Venezuela. They stand about three feet high and are six feet long. Their habitat is the montane cloud forest, where trees and shrubs are stunted from the continual wind and cold. Hunting, habitat loss, and capture for the zoo trade all contribute to the continuing decline of their populations.

Further Reading:
World Wide Fund for Nature, Switzerland <http://www.panda.org / resources / publications / species / underthreat / mountaintapir.htm>.

PHOLIDOTA

Pholidota are the scaly anteaters, or pangolins, of the Old World tropics, composed of just one order.

Manidae (pangolins)

Pangolins include seven species, four in Africa and three in Southeast Asia. Their name is a Malay word that means "one who rolls up." Like the Neotropical anteaters, to which they are unrelated in spite of a resemblance, they feed on ants and termites. Some species are arboreal and have prehensile tails; others live in burrows. They have no teeth, but do have long snouts, attenuated tongues (which can extend almost 10 inches), and strong, clawed limbs—all for digging into nests and slurping up ants and termites. Their sticky tongues are kept moist by enormous salivary glands, so that they must drink frequently. Like armadillos, they are covered with hard, sharp scales, and when threatened, will curl up tightly.

Smutsia gigantea (giant pangolins)

Giant pangolins are about four and a half feet long, counting their two-foot tails, and inhabit lowland forests of western equatorial Africa, east to Uganda. Nocturnal animals, they live in burrows up to 130 feet long and 16 feet deep. Deforestation, hunting, and the use of their scales in folk medicine (as love charms) all make their future uncertain.

Further Reading:
Kingdon, *Field Guide to African Mammals*.
University of Michigan, Animal Diversity Web <http:// animaldiversity.ummz.umich.edu/accounts/manis/ m._gigantea>.

PRIMATES

The order of primates includes 232 species worldwide; nearly all live in the Tropics. One exception is *Homo sapiens*, which thrives (at least in number) nearly everywhere on Earth. Jamaica held the only nonhuman primate species in the Caribbean, but they disappeared, thanks to human pressures, by the early 18th century. The island's monkeys were the last primate species to become extinct, until September 2000, when scientists declared that Miss Waldron's red colobus (*Procolobus badius waldroni*), which once lived in West Africa's forests, was extinct. Many other primate species are in serious danger of the same fate. In fact, a January 2000 report from Conservation International and the World Conservation Union warns that 25 species of primates are on the edge of extinction, and perhaps 25 more face very uncertain futures. Deforestation and hunting are the principal threats to the primates, but illegal capture for the pet trade and export for biomedical research also imperil some species.

The areas with the most endangered primates are Vietnam, Madagascar, Brazil's Atlantic coast forest, and Guinea in West Africa. The most endangered primates include langurs, lemurs, red colobus monkeys, sifakas, tamarins, capuchins, muriquis, gorillas, and orangutans.

Further Reading:
Tuxill, John. "Death in the Family Tree."

Callithricidae (marmosets and tamarins)

Marmosets and tamarins are smaller than other Neotropical primates, with the largest of the species weighing about as much as a large squirrel. Because they are too small to be of interest to human hunters and their prey is mainly insects, they are often the species most common near human settlements.

Callithrix spp. and Cebuella spp. (marmosets)

Marmosets are found from Panama south to Brazil. About the size of a rat, they weigh less than a pound. Unlike monkeys, marmosets and tamarins have non-prehensile tails, heads adorned with ear tufts, tassels, ruffs, manes, or mantels of hair, and hands and feet that have claws instead of nails. Marmosets eat insects and fruit and are also fond of tree sap, gums, and resins. They use their scent glands to communicate with one another and usually travel in small groups composed of a female, one or more males, a few young adults, and juveniles, which are usually carried on the backs of males.

Further Reading:
Emmons, *Neotropical Rainforest Mammals*.

Callithrix saterei (Satere marmoset)

Scientists identified the Satere marmoset in 1996 in the Amazon rainforest of Brazil, making it one of the most recently discovered mammals in the world. They are the size of squirrels, with golden orange-colored fur and no pigment in their skin. Satere is the name of a group of indigenous people living near where the marmoset was found. In 1998, scientists with Conservation International identified the dwarf marmoset (*Callithrix humilis*), just six inches long with a nine-inch tail, in the central Amazon forest, only a 50-minute plane ride from the bustling city of Manaus, Brazil. The dwarf marmoset's habitat seems to be exclusively along rivers and creeks. Numbering no more than a few thousand, it has the smallest range of any primate in the Amazon.

In April 2000, a Dutch scientist identified two more marmosets in forests 190 miles from Manaus, Brazil. *C. manicorensis*, the Manicore marmoset, has a silvery white upper body, a light gray cap on its head, yellow to orange underparts, and a black tail. The other was named *C. acariensis*, or Acari marmoset. It has a white upper body and under parts, a gray back with a stripe running to its knee, and a black tail with a bright orange tip. The Manicore and

Acari are rivers, both tributaries of the Amazon, which run through the marmosets' habitat. Since 1990, 13 new primates have been identified in Brazil, bringing its total number of primate species to 81, more than any other country in the world. Of the 81, 39 are endemic to Brazil.

Further Reading:

BBC News, "New Monkey Species Discovered" <http://news.bbc.co.uk/hi/english/world/americas/newsid_724000/724041.stm.

Domingues-Brandao and Ferreira-Develey, "Distribution and Conservation of the Buffy Tufted-Ear Marmoset, *Callithrix aurita*, in Lowland Coastal Atlantic Forest, Southeast Brazil."

Line, "New Branch of Primate in the Family Tree."

Leontopithecus spp. and *Saguinus spp.* (tamarins)

Tamarins are among the smallest Neotropical primates and are mainly found in the South American Amazon. There are about 18 known species; 12 are of the *Saguinus* genus. They eat insects and fruits and live in small groups, raising their young cooperatively. Like marmosets, their facial hair distinguishes the species from each other—different species have ear tufts, ruffs, manes, or mustaches.

Further Reading:

Emmons, *Neotropical Rainforest Mammals.*

Leontopithecus rosali (golden lion tamarins)

Golden lion tamarins live in the rapidly diminishing Atlantic coast forests of Brazil. They are entirely golden, with a full mane encircling their heads. They weigh just over a pound and are about 14 inches long. Feeding mainly on fruit and nectar, they also use their long digits to pry insects out from under bark and leaves. At night they sleep in tree cavities. A 28,000-acre forest reserve protects the golden lion tamarin's habitat; about 830 survive. Predators include hawks and other raptors, wild cats, and large snakes. They were once popular as pets, but due to conservationists' education efforts, few tamarins are sold for the pet trade today. The golden lion tamarin is one of the few primates that have been successfully introduced from captivity into the wild. Reintroduction has contributed 200 animals to the wild population. Scientists believe that for the species to survive, the population should number at least 2,000 and that 55,000 acres of habitat should be set aside. Each golden lion tamarin group of two to nine primates requires about 100 acres of territory.

One hope for the golden lion tamarin is ecotourism. Villagers who live along the coastal Atlantic forest are finding that showing travelers who are eager to glimpse a flash of gold in the trees can be more profitable than farming. They charge a fee and take the binocular-toting tourists into the forests they know so well, in the hopes of spotting one of the rare and beautiful tamarins.

There are three other species of lion tamarins that some scientists argue are subspecies: black-faced (*L. caicaras*), which scientists identified in 1990; golden-headed (*L. chrysomelas*); and golden-rump (*L. chrysopygus*). All live in forest fragments in southeast Brazil and are nearly extinct because 95 percent of their habitat has been destroyed.

Further Reading:

Emmons, *Neotropical Rainforest Mammals.*

Line, "New Branch of Primate in the Family Tree."

Saguinus oedipus (cotton-top tamarin)

The cotton-top tamarin is endemic to Colombia. It is highly endangered from a combination of deforestation, hunting, and capture for the local and international pet trade and the biomedical industry. The long fur on its back is brown, draping over its sides in a fringe. Its limbs and under parts are white and its face black. The common name for this small (nine inches long, with a 14-inch tail) primate comes from the shock of white hair that grows in a narrow peak on its crown and drapes back. During the late 1960s and early 1970s, some 30,000 cotton-top tamarins were taken from the wild and imported to the United States for use in biomedical research. Finally, in 1973, exportation was prohibited. Captive species continue to be used for colitis and colon cancer research.

Further Reading:

American Zoo and Aquarium Association, Species Survival Plan, USA <http://www.aza.org/programs/ssp/ssp.cfm?ssp=28>.

Emmons, *Neotropical Rainforest Mammals.*

Cebidae (Neotropical monkeys)

The Neotropical monkeys include at least 55 species, and unlike Old World monkeys, they are all arboreal. All eat fruit; some species are also partial to leaves and others to bugs.

Further Reading:

Emmons, *Neotropical Rainforest Mammals.*

Aloutta spp. (howler monkeys)

As their name suggests, howler monkeys have particularly earsplitting calls, which echo through the rainforest at dawn and dusk. Males have a sac at their throats that acts as a voice resonator. The howling marks their territories, warning off approaching troops. An average group of howlers includes three adult males, eight females, and a varying number of juveniles.

The red howler (*A. seniculus*), red-handed howler (*A. belzebul*), black howler (*A. caraya*), and brown howler (*A. fusca*) are confined to northern South America, while the mantled howler (*A. palliata*) ranges from northern Peru north to Central America and eastern Mexico. The Mexican black howler (*A. pigra*) is found in Mexico's Yucatán Peninsula, Guatemala, and Belize. Howlers range in color from black to brown to purplish red, are about 20 inches long, with 22-inch tails, and weigh an average of 13 pounds. They are the only Neotropical monkey that eats large quantities of leaves, preferably young greenery. They also feed on fruit. Fig trees are an important source of both fruit and vegetation.

Kenneth Glander of the Duke University Primate Center found that howler females involuntarily preselect the sex of their offspring. They give birth to one baby at a time, and the female newborns of mothers low on the social caste of the howler troop usually die of malnutrition or are killed by dominant males. By eating a particular diet, Glander says, the females can affect the ion concentration of their reproductive systems, thus attracting the Y-chromosomes that result in males. Howlers are endangered due to habitat loss and hunting.

Further Reading:
Kricher, *A Neotropical Companion.*
Reid, *A Field Guide to the Mammals of Central America and Southeast Mexico.*

Aotus spp. (night or owl monkeys)

The night or owl monkey is the only nocturnal primate in the Neotropics. These rare animals live in forests of Panama to northern Argentina. Fairly small, at 14 inches, with 14-inch, nonprehensile tails, they weigh only two pounds. Their legs are longer than their arms, which helps them leap through the canopy. Their soft, thick, woolly fur is gray or brownish, while their underparts, including the bases of their tails, are a pale orange. To allow them to gather as much light as possible, their round eyes are quite large. Their favorite food is fruit; by foraging at night they don't have to compete with larger diurnal fruit eaters. They also eat foliage, flower nectar, and some insects. During the day, small groups of two to four owl monkeys sleep in tree hollows or hidden in vine-covered trees. At night they call loudly to keep the group together. They are monogamous and adaptable; they are frequently seen around human settlements that aren't far from forests. While they are seldom hunted, they were exploited in the past for biomedical research.

Further Reading:
Emmons, *Neotropical Rainforest Mammals.*
Redford, *Mammals of the Neotropics.*
Reid, *A Field Guide to the Mammals of Central America and Southeast Mexico.*

Brachyteles arachnoides (muriqui or woolly spider monkey)

The muriqui, or woolly, spider monkey is a golden beige monkey with a black face fringed with a pale ring. The largest Neotropical monkey, its body is about 21 inches long, and its prehensile tail adds another 28 inches. It weighs about 28 pounds. One of the most endangered primates in the world, the muriqui is endemic to Brazil, where it is extinct over most of its former range. A small and endangered population of about 700 to 2,000 animals remains in the Atlantic coastal forest. The muriqui mainly eats leaves, plus some flowers and fruit. They are usually in small groups of two to four animals, and males stay with their natal families for their entire lives. Other threats to their survival, along with deforestation, are hunting for food and the use of infants as pets. Because they are so isolated, inbreeding and genetic loss is also a long-term threat. Muriqui is the name given to the monkey by the Tupi indigenous people.

Further Reading:
Emmons, *Neotropical Rainforest Mammals.*
World Wide Fund for Nature, Switzerland <http://www.panda.org/resources/publications/species/underthreat/muriqui.htm>.

Cacajao rubicundus (bald-headed red uakari)

The bald-headed red uakari have striking, bright red, hairless faces and very shaggy coats. They live in just one area of the Amazon flooded forest and are great leapers—able to safely clear an 80-foot gap between tree branches. They travel in feeding groups of about eight. Their incisor teeth allow them to open up large and tough-husked fruits. Related species that also live in the flooded forests are the black uakari (*C. melanocephalus*), found in parts of Colombia, Venezuela, and Brazil; the red uakari (*C. calvus*), whose range is Brazil and Peru; and black-headed uakari (*C. melanocephalus*), which lives in parts of Brazil, Colombia, and Venezuela. Uakaris probably eat more kinds of fruit and seeds than any other Neotropical monkey.

Further Reading:
Animal Info <http://www.animalinfo.org/species/primate/cacacalv.htm>.
Goulding, *The Flooded Forest.*
Wolfe and Prance, *Rainforests of the World.*

Cercopithecidae

Cercopithecidae include 81 species of Old World monkeys and baboons. They are found from Gibraltar, northwest Africa, throughout sub-Saharan Africa, and through central and Southeast Asia, including China and Japan. Nearly all *Cercopithecidae* are diurnal and arboreal, although baboons are primarily terrestrial.

Further Reading:
Anderson and Jones, *Orders and Families of Recent Mammals of the World.*

Cercocebus spp. (drill-mangabeys)

The drill-mangabeys are swamp-forest specialists, found along rivers and wetlands of western Africa. There are six species, all with relatively long muzzles. They are sometimes referred to as "eyelid monkeys" because of their white upper eyelids. The red-capped or collared mangabey (*C. torquatus*) of Guinea, east to Cameroon, is a 20-inch-long primate with a 24-inch tail. It is slate gray, with a white underside, large black ears, and a white collar that surrounds a bright red cap. Its face is long and dark, so its white eyelids are particularly conspicuous.

Red-capped mangabeys hold their tails high, so the white tail tips hover over their heads. They eat fruits and nuts, supplemented with some plants. Active and inquisitive monkeys, they travel both on the ground and in lower tree branches. Enemies of young mangabeys include leopards, eagles, and large owls, but the biggest threat to the species is the destruction of their habitat, as people replace palm and riverine forests with farmland. The species is also intensively hunted.

Further Reading:
Haltenorth and Diller, *Collins Field Guide: Mammals of Africa.*
Kingdon, *Field Guide to African Mammals.*

Lophocebus spp.

This genus of mangabeys is called the baboon-mangabeys. They differ in appearance and habitat from the drill-mangabeys as they are mostly all black and brown and have shorter, more pointed muzzles, with deep-set eyes; they are nearly entirely arboreal and found mainly in central Africa. The baboon-mangabeys also prefer trees near water, but are less dependent on riverine fruit trees. The 20-inch-long black mangabey (*L. aterrimus*) has a ragged, slightly prehensile, 32-inch tail. Not much is known about its population status, but it is hunted for meat throughout its range.

A recent study of mangabeys in Uganda revealed that the monkeys living in secondary forests that had been logged as much as 25 years ago weighed an average of 15 percent less than those that inhabit relatively pristine and intact forests. According to the Purdue University biologists who conducted the study, the secondary-forest mangabeys traveled in smaller groups that tended to be less stable. Since mangabeys normally weigh just 15 to 20 pounds, the weight difference could have serious implications for the overall population.

Further Reading:
Environmental News Network, "Logged Forests Supply Second-rate Digs, Study Shows" <http://www.enn.com/news/enn-stories/2000/03/03072000/monklog_10719.asp>.
Haltenorth and Diller, *Collins Field Guide: Mammals of Africa.*
Kingdon, Jonathan, *The Kingdon Field Guide to African Mammals.*

Cercopithecus aethiops (vervet monkey)

The vervet monkey is mainly arboreal, traveling in groups of up to 30 individuals. It is found in most of Africa, where its principal habitats are riverine forests, although it is adaptable and can live in a variety of other ecosystems. The vervet is 20 inches long, and the dark-tipped tail is 26 inches in length. They are gray to olive-black, with darker hands and feet. The male's scrotum is a bright blue, while the penis is reddish. They have sharp canine teeth and a varied diet, preferring grasses, various parts of the acacia tree, fruits, seeds, lizards, and fledgling birds. Their predators include leopards, pythons, and large eagles.

Further Reading:
Kingdon, *Field Guide to African Mammals.*

Colobus, Pilocolobus, and *Procolobus spp.* (colobus monkeys)

Colobus monkeys are long-tailed, tree-dwelling primates. They are found in central Africa, in the forests of Nigeria to Ethiopia and Tanzania. There are three groups: the black and white colobus (*Colobus guereza*); the red colobus (*C. badius*), which ranges in color from white and red to gray and orange; and the rare olive colobus (*Procolobus verus*), which lives only in the forests of coastal western Africa. They average 20 inches in length, with 24-inch tails. "Colobe" in Greek means "cripple"; colobus monkeys have no thumbs, although this trait enhances, rather than hinders, their ability to swing through treetops. They are the most arboreal of all African monkeys, rarely descending to the ground. They bounce up and down on tree branches as if on a trampoline, to get liftoff for an airborne leap of as much as 50 feet. Predators include leopards, large eagles, and human beings.

In September 2000, biologists proclaimed that a subspecies of red colobus, *Procolobus badius waldroni*, was extinct. The monkeys once lived in rainforests of Ghana and the Ivory Coast.

Further Reading:
Kingdon, *Field Guide to African Mammals.*
Tuxill, "Death in the Family Tree."
University of Michigan, Animal Diversity Web <http://animaldiversity.ummz.umich.edu/accounts/colobus/c._guereza>.

Colobus guereza (black and white colobus)

The black and white colobus is a 28-inch-long monkey with a glossy black coat. Its face and rump are girdled by white fur, and its sides and rear also have a U-shaped mantle of white. Its elegant, 30-inch tail has a brushy white tip, and, although not prehensile, it serves as a rudder as a colobus leaps through trees. The young are completely white, turning black after infancy. Found in west central and eastern Africa, black and white colobus monkeys inhabit the forest canopy, secondary forests, and wooded grasslands. They have large, compartmentalized stomachs adapted for easy digestion of leaves, but they will also dine on flowers, twigs, buds, and seeds. They travel

in troops of six to nine, usually with only one male. With their long, strong hind legs, they can easily leap from branch to branch. Their rumps are well callused, so they can sit comfortably on slender branches. Habitat destruction and hunting threaten their survival. Their skins are commonly used as trimming for coats and dresses, or made into rugs or wall hangings.

Further Reading:
Kingdon, *Field Guide to African Mammals.*

Procolobus kirkii (Zanzibar red colobus)

Zanzibar red colobus are tree dwellers, and the island of Zanzibar, off the coast of East Africa, was once completely covered with trees. Now the island is mostly deforested, so the monkeys are confined to the forests and mangrove swamps of Jozani Forest Reserve. They live in groups of about 30. Zanzibar red colobus females give birth on average every three and a half years, which is nearly twice as long between births as mainland species. They eat fruit and leaves, and, surprisingly, charcoal, almost an ounce a day. They eat the leaves of mango and almond trees, exotics that have been planted in and around the forest reserve. The leaves of these trees are rich in protein but also contain toxins that most animals avoid. But charcoal can neutralize the toxins. The monkeys snatch the charcoal from kilns or chew on charred wood and tree stumps. They also have unusually long feet and fingers, but only a nub of a thumb—all this helps them swing from branch to branch.

Further Reading:
Struhsaker, "Zanzibar's Endangered Red Colobus Monkeys."

Macaca spp. (macaques)

Macaques include 19 species of diurnal monkeys found mainly in Asia. They are the second most prolific and widespread primate in the world, after human beings. Seven species are endemic to the Indonesian island of Sulawesi. They all probably evolved from the same common ancestor, as did their closest living relative, the pig-tailed macaque (*M. nemestrina*), which resides in Borneo. The ancestral monkey probably swam from Borneo to Sulawesi millions of years ago when sea levels were lower. Descendants from the ancestor moved across the island, gradually becoming more and more isolated from each other on the island's fingers of land and eventually evolving into entirely different species. While macaques are more arboreal than baboons, they are as at home on the forest floor as in trees. They are omnivorous and have large cheek pouches where they store extra supplies of food. In Sulawesi, the macaques are endangered by deforestation and the fact that people in the northern part of the island consider the primates a delicacy. Served with chili peppers, macaques are a favorite Christmas dinner for the wealthy. Other than human beings, its only other enemy on the island is the reticulated python.

Further Reading:
Israel, and Sinclair, *Indian Wildlife.*
Kinnaird, "Macaque Island."

Macaca silenus (lion-tailed or macaque, wanderoo)

The lion-tailed macaque, or wanderoo, is found in the wet forests of southern India. It has a black-gray coat with a ruff about its face. It weighs an average of 20 pounds and is about a foot long, and its tail, which has a slight tuft at the tip, is another 8 inches long. Lion-tailed macaques live in troops of 10 to 20 individuals, including 1 to 3 dominant males. Diurnal and omnivorous monkeys, they are primarily arboreal—more so than other macaques—although they also are comfortable on the ground, as well as in rivers. The lion-tailed macaque was never a common primate, and its numbers have steadily declined as teak, coffee, and tea plantations replaced its habitat in the southern third of India. It was also once captured for the pet trade, zoos, and research. It doesn't adapt well to human disturbance and won't travel through plantations. Probably fewer than 4,000 survive.

Further Reading:
Animal Info <http://www.animalinfo.org/species/primate/macasile.htm#profile>.
Israel and Sinclair, *Indian Wildlife.*

Mandrillus spp. (mandrills)

Elusive animals, mandrills are the largest monkey, with the male weighing in at about 60 pounds. They are about 2.5 feet long, with very short tails, really mere tufts of hair three inches long. The males are also among the most colorful mammals on Earth, with slashes of brilliant blue and red on their black faces and scarlet-colored penises and circles around their anuses. The male with the brightest facial colorings is the most dominant within a troop and sires most of the offspring. Found only in the forests of Gabon, southern Cameroon, and Congo, they are endangered, due to habitat loss and illegal hunting; mandrills are considered a delicacy. The monkeys spend much of their time on the forest floor, scavenging for insects, mushrooms, tubers, fallen fruit, small animals, and reptiles. They may walk six miles on an average day. They sleep in trees at night.

The Wildlife Conservation Society has attached radio collars to three female mandrills in the Lop Reserve in Gabon to try to learn more about the mystery monkeys' habits and devise a strategy to save them from extinction.

Further Reading:
Kingdon, *Field Guide to African Mammals.*
Lahm, "Devils of the Forest."

Miopithecus ogouensis (Northern talapoin)

Africa's smallest monkey, at just three pounds and squirrel-sized, the Northern talapoin has bright yellow-olive fur on its crown, back, and long, thin tail. The outer surfaces of its limbs, hands, and feet are golden. Its head is large for the size of its body, and it has large, round eyes. Talapoins' forest homes in central and western Africa are usually near swamps or streams, which accounts for their strong swimming abilities. They've even been spotted diving off low-hanging branches (presumably) for fun, to hunt for food in the water, or to escape predators. Although shy, they seem to adapt well in secondary forests, not far from human-made clearings, where food is plentiful and predators few. Gregarious monkeys within the species, they live in troops of 12 to 20, but may come together even more at night to roost until the treetops are filled with more than 100 talapoins. They feed on leaves, seeds, fruit, water plants, insects, eggs, and small vertebrates. Their predators include leopards, genets, large birds of prey, and large snakes.

Further Reading:
Haltenorth and Diller, *Collins Field Guide: Mammals of Africa.*
Kingdon, *Field Guide to African Mammals.*

Nasalis larvatus (proboscis monkey)

The proboscis monkey has cinnamon-colored fur and a long white tail. Endemic to the island of Borneo, male proboscis monkeys have pendulous noses as long as three inches that give the species its common name. Locally they are called "orang belanda," or "Dutchman," since the indigenous peoples think that they resemble hairy, pot-bellied, and big-nosed colonists. Scientists are unsure what purpose such a large nose might serve—females might prefer big noses or it may be that the proboscis helps radiate excess body heat. They spend most of their time in treetops, particularly in mangrove forests, mainly eating young leaves, flowers, and seeds; they are strict vegetarians. These are hefty monkeys—males can weigh up to 50 pounds—yet despite their size, they are also agile tree leapers and fast swimmers. Like colobus monkeys, they have multi-chambered stomachs that help them digest the plant matter on which they gorge much of the day. Bacteria in their sizeable guts help them to break down all this vegetation. They live in groups of 12 to 30. The proboscis monkey is an endangered species. Deforestation has destroyed most of their habitat; perhaps 3,500 of the primates remain.

Further Reading:
International Primate Protection League, USA <http://www.ippl.org>.
University of Michigan, Animal Diversity Web <http://animaldiversity.ummz.umich.edu/accounts/nasalis/n._larvatus>.

Papio spp. (baboons)

Baboons are black when they are born, then turn brown. Some species have bright red or blue skin on their faces, chests, or rumps. They travel in small and large well-organized troops, from 10 to more than 100. Large males dominate the family troops and are responsible for settling quarrels and protecting the group from predators, leopards in particular. If an enemy approaches, the dominant males bark in alarm, and the troop rushes up into the nearest trees. Troops are centered around the females and their tight-knit families. In 1900, there were likely more baboons living in Africa than people. While their populations have plummeted, they are still the most populous monkey on the continent. They live in forests, but also in savannas and even deserts. Unlike most other primates, which are arboreal, baboons live on the ground. They usually bed down for the night in trees. They eat nearly anything—plants, insects, and meat.

Further Reading:
Canadian Museum of Nature, Natural History Notebooks <http://www.nature.ca/notebooks/english/baboon.htm>.
Haltenorth and Diller, *Collins Field Guide: Mammals of Africa.*
Kingdon, *Field Guide to African Mammals.*

Pygathrix avunculus (Tonkin snub-nosed monkeys)

One of the rarest primates in the world, Tonkin snub-nosed monkeys weigh about 18 pounds, are arboreal, and are found only in rainforests on steep limestone hills in northern Vietnam. They eat leaves and fruit. Males form bachelor groups, as well as groups with one dominant male and several females. Probably fewer than 200 survive. Their decline is attributable to hunting for meat, use in folk medicine, and deforestation. By 1986, more than 70 percent of their original habitat had vanished.

Further Reading:
Animal Info <http://www.animalinfo.org/species/primate/pygaavun.htm>.
Tuxill, "Death in the Family Tree."

Daubentoniidae (aye-aye)

Daubentonia madagascariensis (aye-aye)

D. madagascariensis is the only member of this family, and many biologists consider it to be the most bizarre of all the primates. Aye-ayes are real loners, spending most of their solitary lives in treetops of the eastern coast and northwestern forests of Madagascar. The only time they interact with one another is during courtship, mating (obviously), and the year or so a baby is dependent on its mother. Aye-ayes are the size of a large racoon, with short, white hairs plus long, white-tipped black hairs. They have large black ears and pale faces. They have evolved a rather unusual middle finger, thin and knobby and able to bend in any direction. The aye-aye uses this serpentine digit to make life easier—to splash water into its mouth, for example, or to tap logs for hollow spots, so it can chew an opening with its razor-sharp teeth and probe about for tasty grubs.

Aye-ayes are killed as pests on farms or simply because they are so strange-looking; some people in Madagascar think they are evil omens. There are several healthy populations in a few protected areas on the island.

Further Reading:
Duke University Priamte Center, USA <http://www.duke.edu/web/primate/aye.html>.
Haltenorth and Diller, *Collins Field Guide: Mammals of Africa including Madagascar.*
Stewart, "Prosimians Find a Home Far from Home."

Hominidae

Hominidae is the family to which *Homo sapiens*, chimpanzees, orangutans, and gorillas belong. Until recently, *Hominidae* was restricted to human beings, while other apes were included in the family *Pongidae*. In the past few decades, however, studies of fossils, genes, and physiology have revealed the recent and common ancestry of humans and apes. Humans and apes share compact bodies without tails; large, rounded heads with large brains; and long forearms with hands and strong fingers. *Hominidae* includes five species; four of them are restricted to equatorial Africa, Sumatra, and Borneo. The fifth, *Homo sapiens*, not only lives all over the planet, but the species has also found a way to launch itself into the heavens and deep into the seas.

Further Reading:
University of Michigan, Animal Diversity Web <http://animaldiversity.ummz.umich.edu/chordata/mammalia/primates/hominidae.html>.
Kingdon, *Field Guide to African Mammals.*

Gorilla gorilla

Gorilla gorilla is hunted now more than ever, as eating "bush meat" is a fad for the wealthy and elite in many African countries. Although hunting is illegal, the $60 or so paid for each gorilla makes it difficult for many hunters to resist, and enforcement is rare. *G. gorilla beringei* (mountain gorilla) is a long-haired subspecies that lives only in the high altitude forests along the borders of Congo, Uganda, and Rwanda. Probably no more than 200 remain. They live in family groups under the leadership of an older male, which may weigh more than 300 pounds. Although they look fierce, they are gentle and curious vegetarians. Due to violent civil war in Rwanda and Congo, it has been nearly impossible for scientists to continue to monitor the gorillas. Rebels have killed at least six since 1997. Unarmed guards, paid in part by a German aid agency, patrol Kahuzi-Biega National Park in Congo, where some 70 gorillas survive, but they do not dare patrol more than 5 percent of the reserve, since much of the park is occupied by militiamen. Researchers from the United States and Rwanda are now using aircrafts and remote-sensing technology to map gorilla habitat and track their movements.

Further Reading:
Glanz, "Tracking Gorillas and Rebuilding a Country."
McNeil, "The Great Ape Massacre."
Schmidt, "Soldiers in the Gorilla War."

Pan spp. (chimpanzees)

Chimpanzees are the animal most closely related to human beings, both physically and genetically. Their DNA is 98.5 percent the same as that of humans, although their brains are half the size of the human brain. There are two species, the common chimpanzee (*P. troglodytes*) and bonobos, or the pygmy chimpanzees (*P. paniscus*); both inhabit forests of tropical Africa.

Chimpanzees are endangered due to loss of habitat and hunting, which has surged in recent years. At least 4,000 of the primates are shot and sold each year to meet the demand for chimpanzee meat. One hundred years ago, millions of chimpanzees could be found in Africa; today, fewer than 200,000 remain.

Further Reading:
Angier, "Chimps Exhibit, er, Humanness, Study Finds."
Kingdon, *Field Guide to African Mammals.*
McNeil, "The Great Ape Massacre."

Pan troglodytes (common chimpanzee)

Found from southern Senegal to western Uganda and Tanzania, the common chimpanzee is an expressive communicator, using facial expressions, vocalizations, posture, and touch to deliver its messages. A young chimp can make 30 different sounds, each with a different meaning. Chimpanzees average 2.5 feet in height, weigh about 65 pounds, and have long, black hair, a bare face, and protruding ears, lips, and brow ridges. Their ears, palms, and soles are also hairless. They have long arms and no tails.

Chimpanzees gather in bands of 15 to 120 individuals on large home territories, where they remain for years. They are diurnal and omnivorous, eating a wide selection of foods—mostly fruit, but also leaves, bark, stems, insects, birds, honey, eggs, and sometimes, small mammals. They use tools to assist them in eating and drinking, as weapons and in play. In 1997 the Dutch scientist Rosalind Alp found a group of chimps in Sierra Leone using tools in a new way—for comfort. Chimpanzees in east African forests like to feed on the fruit and flowers of the kapok tree (*Ceiba pentandra*), which is covered with sharp thorns. To avoid stepping on the spines, the chimps break off sticks and grip them with their toes, making a kind of shoe, or they use the sticks as seat cushions, sitting on them as they eat.

Scientists recently concluded that chimps display cultural variation—that is, chimpanzees living in Sierra Leone have different courting, grooming, and tool-using habits than do chimp populations in the Côte d'Ivoire. The customs are passed on to young chimpanzees by their parents. The conclusion was reached by a group of nine primatologists, including Jane Goodall of the Gombé Stream Research Center, Richard Wrangham of Harvard University, and Toshisada Nishida of Kyoto University, who pooled the observations they had made over decades. They found, for example, that in the Tai forest of the Ivory Coast, it was customary for one chimp to groom another by picking off a parasite and squashing it on its own forearm. But chimpanzees in Tanzania remove the offending bug and crush it with a leaf.

Female chimps become fertile when they are about 12, although they may be sexually active when they are about 8. Likewise, male chimps are sexually mature at 15, but sexually active from the age of 5. After an eight-month gestation period, females give birth to an infant born with open eyes, but completely dependent on its mother for at least six months. Baby chimps are born with a clinging reflex and travel on their mothers' backs, clutching on to their hair. Even in the wild, chimps may live for 40 years.

A group of scientists at the University of Alabama have proven a link between the disease known as the Acquired Immunodeficiency Syndrome (AIDS) and chimpanzees. The HIV-1 (human immunodeficiency virus) that began to spread in humans in the early 1940s came from an immunodeficiency virus in *Pan troglodytes*. The virus likely spread when hunters cut themselves butchering chimp meat or ate raw meat. Chimpanzees don't seem to get sick from the virus, so they may hold a link to a cure for AIDS. This link and the need for more research make saving chimpanzees from extinction quite important to human health.

Further Reading:
Kingdon, *Field Guide to African Mammals*.
McNeil, "The Great Ape Massacre."
Weintraub, "Chimps Find a Way to Eat in Comfort."

Pan paniscus (bonobo, or pygmy chimpanzee)

The pygmy chimpanzee, or bonobo, is the rarest great ape species. Bonobos are found within the forests south of the Congo River, in Congo, a country that is suffering grave civil unrest. Probably fewer than 20,000 remain in the wild, and Congo's turbulence, with habitat destruction and hunting, threatens the animals' survival.

Bonobos are about two feet tall and weigh an average of 60 pounds. They have long, slender limbs.

Their heads are rounder, and they have less pronounced browridges and muzzles than do *P. troglodytes*. They travel in groups that number 100 or larger, so they need vast stretches of forest to supply adequate food. They mainly eat fruit, but also munch on leaves, flowers, seeds, invertebrates, and the occasional small mammal.

The stance, stride, and gestures of bonobos are more similar to human beings' than to other apes'. Scientists have observed the different groups of the animals coming together in the forest for what seem to be meetings or discussions, with lots of vocal exchanges. Catching up on the latest news? Making plans? Scientists are unsure; because bonobos live in dense forests in hard-to-reach and insecure areas, studies are relatively rare.

Until recently, most research centered on bonobo sex life, and that research has been conducted by watching the animals behave in captivity—a far from natural state. Bonobos form strong social ties to each other through sexual interactions, and to human beings, their sexual behavior seems remarkably intense, verging on the kinky. Research shows that the bonobo's overt sexuality is an integral part of their social structure, and grooming, play, and lots of sex—between opposite sexes, same sexes and themselves—is how they bond to one another.

Further Reading:
Bonobo Protection Fund <http://www.gsu.edu/~wwwbpf/bpf/>.
Kingdon, *Field Guide to African Mammals*.

Pongo pygmaeus (orangutan)

Orangutans are less sociable than other primates. The males travel alone and the females are accompanied by an infant or one juvenile. One reason for their solitary behavior is that their preferred fruits are widely dispersed—there's just not enough to go around for a large troop. They may travel for miles throughout the forest in search of fruit. They also eat bark and insects. They once ranged throughout Southeast Asia, but now can be found only in Borneo and Sumatra. Adult males, often weighing as much as 200 pounds, are the largest canopy animals; "orangutan" is Malay for "man of the forest." Orangutans depend more on trees than do other primates, and in Borneo and Sumatra they have lost more than 80 percent of their forest habitat in just the past two decades. Illegal logging in two of Indonesia's protected areas—Tanjung Putting and Gunung Leuser

National Parks—further threatens the primates. Fewer than 20,000 orangutans survive—some 30 to 50 percent fewer than just a decade ago. Agricultural development has destroyed much of their habitat, and as a result, the primates are coming into conflict with human beings. If they find themselves isolated in a small patch of forest surrounded by farmland, they help themselves to crops. They are also popular—although illegal—pets, particularly in Taiwan, the biggest importer of baby orangutans.

Further Reading:

Orangutan Foundation International, USA <http://www.orangutan.org/>.

Public Broadcasting Service, Nature on the Web <http://www.pbs.org/wnet/nature/orangutans/html/body_save.html>.

Tuxill, John. "Death in the Family Tree."

Hylobatidae

Hylobatidae is a family of only 1 genus and 11 species, all found in tropical forests of Southeast Asia, including the islands of Sumatra, Borneo, and Java.

Hylobates spp. (gibbons)

Gibbons live in the rainforests of China, the Malaysian Peninsula, Myanmar, and Sumatra. They are slender, long-limbed, and extremely agile, able to swiftly swing from branch to branch. They can also leap across gaps in the forest canopy, as far as 30 feet. Few predators can catch a gibbon. They have small, round heads, and like all apes, no tails. They have five toes, including an opposable big toe, and four long fingers and a smaller, opposable thumb. These primates live in small family units consisting of a male and female (gibbons mate for life), and young (under seven years) offspring. They sleep huddled together and sitting up in treetops. Each morning, they announce their presence by shrieking loudly, then go off to search for fruits, leaves, and flowers, and the occasional insect, small bird, spider, or bird egg. Each gibbon family requires territory of about 40 acres. There are seven species of gibbons, which weigh an average of 15 pounds. The males, which are slightly larger than the females, are about three feet long.

Further Reading:

International Center for Gibbon Studies <http://izoo.org/icgs/animals.html>.

University of Michigan, Animal Diversity Web <http://animaldiversity.ummz.umich.edu/chordata/mammalia/primates/hylobatidae.html>.

H. hoolock (Hoolock gibbon)

Hoolock gibbons live in the forests of Assam, India, Bangladesh, south China, and Myanmar. Perhaps 170,000 of the animals remain. They are about three feet tall when standing erect, with small legs and extremely long arms. Agriculture and logging have fragmented the gibbons' habitat so they are forced to descend from trees to cross clearings. This makes them vulnerable to hunting. Gibbon meat and bones are used in traditional Asian cures.

Further Reading:

Israel and Sinclair, *Indian Wildlife*.

World Wide Fund for Nature, Switzerland <http://www.panda.org/resources/publications/species/underthreat/hoolock.htm>.

H. lar (white-handed gibbon)

The white-handed gibbon lives from the mid- to upper canopy of forests in southern China, through eastern Myanmar, Thailand, Malaysia, and Indonesia. They are an average of 1.5 feet long and weigh about 12 pounds. Their fur is dark brown or black, sometimes reddish, and they have white face rings and white hands and feet. They are rather picky eaters, preferring fruit, but only ripe fruit, and will sometimes eat plants, but only new leaves and shoots. They swing from branch to branch by their strong arms as they search out their preferred foods. They seldom descend to the forest floor. Deforestation and, in some areas, being hunted for meat, have made them an endangered species. They are also captured for the illegal pet trade.

Further Reading:

American Zoo and Aquarium Association, Species Survival Plan, USA <http://aza.org/Programs/SSP/ssp.cfm?ssp=35>.

International Center for Gibbon Studies <http://izoo.org/icgs/animals.html>.

H. moloch (silvery gibbon)

Its very dense and long, silver-gray fur gives the silvery gibbon its name. They are endemic to Java. Their diet consists of fruits, flowers, and insects. The silvery gibbon has lost 98 percent of its forest habitat on the island of Java; only 400 to 2,000 of the animals remain. The largest populations, but all fewer than 100 gibbons, are protected in three national parks.

Further Reading:

International Center for Gibbon Studies <http://izoo.org/icgs/animals.html>.

H. syndactylus (siamang)

Siamangs are the largest, darkest, and loudest of the gibbons. Native to Sumatra and the Malay Peninsula, they are covered with dense and shaggy black hair. Males are about three feet long and weigh up to 40 pounds. Unlike other gibbons, siamangs have webbing between their second and third toes. Also unique among the gibbons is their throat sac, which they can inflate to the size of their head. The sac is a resonating chamber for their vocal chords, allowing them to bellow with force. The siamang's hoots can be heard some two miles away in the forest. A siamang troops' territory consists of about 40 acres, and they assert ownership for about 30 minutes each morning by whooping loudly.

Further Reading:

International Center for Gibbon Studies <http://izoo.org/icgs/animals.html>.

Indridae

Indridae are slender-limbed, apelike primates, with long hind limbs that enable them to leap from tree to tree, where they feed on leaves. They have fairly large ears, round eyes, and thick, woolly fur. All five species of *Indridae* are strict vegetarians and are found only in Madagascar.

Further Reading:

Haltenorth and Diller, *Collins Field Guide: Mammals of Africa.*

University of Michigan, Animal Diversity Web <http://animaldiversity.ummz.umich.edu/chordata/mammalia/primates/indridae.html>.

Avahi laniger (avahi)

The avahi is the only nocturnal member of the *Indridae* family and also the smallest, weighing about two pounds. Its habitat is mountain rainforest of eastern and northwestern Madagascar; deforestation has made its status endangered. Its fur is brown with beige markings, and it has a round, owlish face with no fur. During the day, small family groups of avahis—only the male, female, and their young offspring—sleep in the dense foliage of trees and tall bushes. Avahis mate for life, and the female gives birth to a single young. In addition to leaves, they eat buds, bark, and fruit.

Further Reading:

Haltenorth and Diller, *Collins Field Guide: Mammals of Africa.*

Lemuridae (lemurs)

Lemurs are descendents of primates that once lived in tropical forests of Africa and Asia and subtropical forests of Northern America and Europe. But then monkeys came along, some 35 million years ago. Because they were more adaptable, they caused the extinction of lemurs from every forest except on Madagascar, an island off the east coast of Africa. Much of what we know about lemurs has come from watching captive animals in a large, forested enclosure at the Duke University Primate Center in North Carolina.

Lemurs range in size from the lesser mouse lemur (*Microcebus murinus*), which belongs to the family *Cheirogaleidae* and weighs about two ounces, to the gray gentle lemur (*Hapalemur griseus*), which is the size of a healthy house cat. The animals spend their entire lives in trees, except for the ring-tailed lemur (*Lemur catta*), which also walks about the forest floor. Lemurs eat insects, leaves, fruit, birds' eggs, birds, and reptiles.

Because Madagascar has only 10 percent of its forests left, lemurs are extremely endangered. At least a dozen species of lemurs have disappeared from the island since the first human beings arrived 1,500 years ago. Those that were wiped out were the larger species, including one the size of a gorilla. If lemurs have disappeared from a forest in Madagascar, the forest has little chance of recovering, since the peculiar primates are the primary seed dispersers. A powerful cyclone that hit Madagascar in March 2000 will have a lasting impact on lemurs as well as on human beings. The animals feed high in the forest canopy, but the storm smashed through trees, eliminating their food source. Lemurs that venture out of reserves looking for food risk being killed by hungry human beings.

Further Reading:

Cohn, "Duke Primate Center Fosters Research."

Duke University Primate Center, USA <http://www.duke.edu/web/primate/home.html>.

Environmental News Service, "Lucky Lemurs in Madagascar's Newest Park; Others at Risk" <http://ens.lycos.com/ens/oct99/1999L-10–08-01.html>.

Oaks, "Lemurs among Madagascar Victims."

Hapalemur aureus (golden bamboo lemur)

Identified in 1987, the golden bamboo lemur is one of Madagascar's most endangered lemurs. It is found in rainforests that have abundant stands of bamboo, particularly of the endemic species, *Cephalostachium viguieri*, a favorite food. The young bamboo shoots that this lemur eats are high in toxins, which would be lethal to most other mammals. The 2.5-foot golden bamboo lemur has short legs, a long, bushy tail, a round face with a black mask, small, round ears, and thick, soft, golden fur. Although the total population is unknown, an estimated 1,000 live in Ranomafana National Park.

Further Reading:
World Conservation Monitoring Centre, England <http://www.wcmc.org.uk/species/data/species_sheets/golden.htm>.

Loridae

Loridae is the family of lorises, pottos, and angwantibos. The six species of lorises and pottos are small, arboreal primates that live in Africa and Asia. They are nocturnal, spending the dark hours silently and deliberately searching for fruit or insects in the treetops. They have strong hands and feet, but very stubby tails. To some eyes, they must at least once have looked rather silly; "loris" means "clown" in Dutch.

Further Reading:
University of Michigan, Animal Diversity Web <http://animaldiversity.ummz.umich.edu/chordata/mammalia/primates/loridae.html>.

Loris tardigradus (slender loris)

The slender loris is a resident of southern Indian and Sri Lankan forests and is named for its long and slim legs. Like all lorises, it moves slowly and can remain motionless for hours. Its single young is born with its eyes still shut and just a thin layer of fuzz covering its body. The slender loris is endangered because of deforestation and capture by humans for local sale. A persistent myth holds that the eyes of the slender loris are magical love charms, and they are also used as cures for eye diseases.

Further Reading:
University of Michigan, Animal Diversity Web <http://animaldiversity.ummz.umich.edu/accounts/loris/l._tardigradus.html>.

Nycticebus coucang (slow loris)

A rainforest primate of India, Indonesia, and the Malay Peninsula, the slow loris has short, brown, and dense fur, a white nose, and white stripes between its round eyes, which are circled by dark rings. Although an excellent climber, this foot-long loris cannot leap from tree to tree.

Further Reading:
Minnesota Zoo <http://www.wcco.com/partners/mnzoo/loris.html>.
University of Michigan, Animal Diversity Web <http://animaldiversity.ummz.umich.edu/accounts/nycticebus/n._coucang.html>.

Perodicticus potto (potto)

The potto is found in tropical forests of Africa, from Sierra Leone to western Kenya. It spends nearly all its life in the forest canopy, eating mainly fruit, along with some insects, fruits, and leaves. It's a slow-moving primate covered with thick brown hair. Golden pottos (*Arctocebus aureus*), which inhabit dense lowland forests and savannas of Cameroon, Congo, Equatorial Guinea, Gabon, and Nigeria, are tiny, weighing only seven ounces. They have golden or russet-colored fur on their upper sides and are creamy white below. Their principal food is insects, but they occasionally munch on fruit as well. They are nocturnal and usually solitary. They are sensitive to heat and will die if exposed to the sun or lack water.

Further Reading:
Kingdon, *Field Guide to African Mammals.*
University of Michigan, Animal Diversity Web <http://animaldiversity.ummz.umich.edu/accounts/perodicticus/p._potto.html>.

Tarsiidae

Tarsiidae is a family with one genus, *Tarsius*, and five species found in Southeast Asia, including Borneo, Sumatra, and the Philippines.

Tarsius spp. (tarsiers)

Tarsiers are named for their long tarsal bones in the ankle area, which allow these tiny primates to leap as far as nine feet through the trees. They are about six inches long and weigh only four ounces. Their silky fur is buff, gray, or brown on their backs and gray or buff on their underparts. They have huge goggle-eyes, sensitive ears, and the exceptional ability to rotate their round little heads nearly 180 degrees. They are the only primate that eats nothing but live prey. They spend most of the night vaulting through the trees in search of insects and small reptiles. Their small family groups are composed of a monogamous adult pair and one or two immature offspring. Three of the five species, including the most recently discovered, *T. dianae*, live in 570,500-acre Lore Lindu National Park on the Indonesian island of Sulawesi. Another, the spectral tarsier (*T. spectrum*), was named for its eerie nocturnal calls.

Further Reading:

The Nature Conservancy, "A Little Night Musician," <http://www.tnc.org/infield/intprograms/tarsier.htm>.

The Nature Conservancy, "Tarsiers—Singing in Sulawesi," <http://www.tnc.org/infield/intprograms/tarsier.htm>.

University of Michigan, Animal Diversity Web <http://animaldiversity.ummz.umich.edu/chordata/mammalia/primates/tarsiidae.html>.

PROBOSCIDEA

The order of elephants, *Proboscidea*, has a relatively easy to trace evolutionary history because elephant ancestors fossilize so well. Before 65 million years ago, they shared ancestry with rhinoceroses, horses, and hyraxes. The African and Asian elephant ancestors separated when the continents split away from each other. Fossils from 35 million years ago show an early elephant that looks a bit like the modern tapir. By 25 million years ago, the animal's weight increased, so the legs became thicker. Tusks and trunks gradually lengthened. Elephants' direct ancestors split into three lines (African, Asian, and the mammoths) about 5 million years ago, and reached their greatest diversity, with 10 different species, about 1 million years ago. As they were reaching their evolutionary peak, a new predator appeared: human beings. Today, the elephant's survival is in question.

Further Reading:
Kingdon, *Field Guide to African Mammals*.

Elephantidae (elephants)

Elephants include two species, the African elephant, which lives south of the Sahara, and the Asian elephant, found in India, Nepal, and Southeast Asia. Both species have long, muscular trunks that are almost like a fifth limb. Males of both species and females of the African species have huge tusks derived from their upper incisors. People slaughter elephants for these teeth, which are the source of ivory.

As might be expected from the largest land animal on Earth, elephants require huge amounts of food, some 800 pounds daily. Elephant teeth are peculiar—the giant animals don't have many teeth to fill their large mouths, and as they age, their worn-out front teeth fall out and are replaced by back teeth that slowly shift forward. If an elephant lives so long that it uses up all its teeth, it will starve to death. They prefer to eat grasses, which they gather and rip from the ground with their trunks, which convey the snack to their mouths. They also consume bark, which they scrape off trees with their teeth, as well as roots, leaves, and branches and stems of trees and bushes.

Because elephants spend about 16 hours a day eating a large variety of vegetation, they need access to large areas to find adequate food supplies. Whenever human beings move into elephant habitat, the result is almost always damage to crops, pastures, and property. Elephants are particularly susceptible to plant toxins, so they prefer to dine in clearings or secondary forests, which always contain fewer poisonous plants than do primary forests.

In spite of their huge, floppy ears, they have poor hearing and also poor eyesight. Their sense of smell, however, is quite sharp. Elephants communicate with one another vocally or by stamping on the ground. They can live as long as 70 years in the wild.

Further Reading:
Dublin, McShane, and Newby, "Conserving Africa's Elephants."

Sayer, Harcourt, and Collins, *The Conservation Atlas of Tropical Forests: Africa*.

University of Michigan, Animal Diversity Web <http://animaldiversity.ummz.umich.edu/chordata/mammalia/proboscidea.html>.

Elephas maximus (Asian elephant)

The Asian elephant is more closely related to the extinct mammoth than to the African elephant. Found in India and Southeast Asia, they have huge, domed heads with relatively small ears, an arched back and a single, fingerlike protuberance that is located at the tip of the trunk—the African elephant has two. A large male may weigh as much as six tons.

Unlike female African elephants, the Asian females do not have tusks. They can be found in the forest, but prefer to graze in more open, grassy areas. Ivory poaching has decimated populations. Only about 45,000 Asian elephants survive. Centuries ago, the species was wiped out in southwestern Asia and most of China. In Thailand, only 1,500 remain in the wild, but more than twice that number live in captivity.

Elephants were traditionally used as beasts of burden and as religious symbols. Since logging was banned in Thailand in 1989, many captive elephants no longer have jobs, and their owners can no longer afford to feed them. The World Wildlife Fund and the Thai government have launched a project to see if these domesticated animals might be released into protected areas. They have attached radio collars to several elephants to see if they can integrate themselves into wild herds and find food without raiding crops.

In Vietnam, only about 100 wild elephants remain, most in fragmented forests in the central and southern regions of the country. That's down from about 1,800 elephants counted in 1990. A boom in the ivory trade had an impact, but deforestation is the main reason for the precipitous decline. Since 1992, more than half of the 100,000 acres of the elephant's native forests in the southern provinces were converted to farmlands. The result has been more interactions between wild elephants and people, with unfortunate results: 13 villagers have been trampled.

A century ago, more than 10,000 Asian elephants lived throughout Sri Lanka; only 2,500 to 3,000 remain today. A herd of 100 live in Handapanagala Forest and migrate 60 miles to Yala National Park each November, when the forest reserve is lashed by downpours. In May, they walk back to Handapanagala to spend the summer near water sources. Wildlife officials warn that a proposed golf course in Sri Lanka would block this migration route and be disastrous for the elephants. Sri Lanka's wildlife department was the first in the world to release into the wild elephants that were raised in captivity. In 1997, three elephants were set free in Uduwalwe National Park. A wild herd accepted the newcomers within minutes.

Further Reading:

Associated Press, "Vietnam's Elephants in Danger" <http://www.latimes.com/HOME/NEWS/SCIENCE/ENVIRON/tCB00V0033.html.

Elephant Information Repository <http://elephant.elehost.com/index.html>.

University of Michigan, Animal Diversity Web <http://animaldiversity.ummz.umich.edu/accounts/elephas/e._maximus>.

Loxodonta africana (African elephant)

The African elephant ranges over an estimated 3.7 million square miles in 37 African countries, from Ethiopia and Sudan south to South Africa, and from Senegal on the Atlantic coast east to Kenya. There are two subspecies of elephants in Africa, the forest elephant (*L. africana cyclotis*) and the savanna elephant (*L. africana africana*). The savanna elephant is also found in forests. The forest elephant is smaller, with more petite and rounded ears, and its slender ivory tusks are generally straighter than those of the savanna species.

About half of the elephants remaining in Africa live in the forests of the Congo River basin. An estimated 550,000 African elephants remain on the continent, and only 20 percent of the elephants' range falls within officially protected areas. During the 1980s, at least 120,000—and perhaps as many as a quarter of a million—elephants were slaughtered for their ivory tusks. A 1990 Convention on International Trade of Endangered Species of Wild Fauna and Flora (CITES) ban on ivory trade led to increases in many elephant populations, but they are still classified as endangered. In 1997, amid much controversy, the ban was loosened to allow strictly controlled ivory trade in Botswana, Zimbabwe, and Namibia.

In 1999, Zimbabwe sold 30 tons of ivory worth $3.5 million, much of it to Japan, where tusks are carved into much-sought-after ornaments. Japanese traders bought another 30 tons from Botswana and Namibia. The same year, more than 350 elephants were illegally slaughtered for their ivory in Zimbabwe. Elephants were shot with AK-47 automatic rifles, and chainsaws and axes were used to hack off the tusks. There are about 70,000 elephants in Zimbabwe, where they are considered a nuisance, as they tend to wander out of national parks or safari areas and trample crops and damage homes and sheds of impoverished farmers. In Kenya in 1999, 29 elephants were poached for their ivory in Kenya's Tsavo National Park; that's five times the average total during each of the previous six years. At the April 2000 CITES conference, delegates applauded the last-minute withdrawal by Botswana, Namibia, South Africa, and Zimbabwe of controversial proposals that would have permitted new ivory sales.

Researchers have discovered a new threat to African elephants in Zimbabwe: "floppy trunk disease." The disease starts at the tip of the animal's trunk and moves up, eventually causing the entire trunk to be paralyzed, a disaster for elephants, since they use this appendage for eating, drinking, and communicating. Some elephants have died from the disease, although others seem to have recovered. Scientists believe a plant may be the cause. Because elephants are losing habitat so fast, they are being forced to eat plant species they usually would eschew, and these species probably contain chemicals that cause the trunk paralysis.

Elephants are considered to be "keystone species," meaning they play vital roles in keeping ecosystems healthy. When they crash around the forest, they bring down smaller trees, creating light gaps that allow grasses and seedlings to grow and thus providing food for a variety of other animals. There are certain fruits that are simply too large for other animals to ingest. Where there are no elephants to eat the seeds of these fruits, the trees that bore them will likely disappear as well.

Further Reading:
Dublin, McShane, and Newby, "Conserving Africa's Elephants."
Prasad, "Sri Lanka Caters to Tourist Golfers at Elephants' Expense."
Reuters, "Toxic plant may be behind elephant trunk disease" <http://ens.lycos.com/ens/jan2000/2000L-01-17-01.html>.
Sayer, Harcourt, and Collins, *The Conservation Atlas of Tropical Forests: Africa*.
Sugal, "Elephants of Southern Africa Must Now 'Pay Their Way.'"
Tucker, "Poachers Kill 350 Elephants."

RODENTIA (rodents)

Rodents are found throughout the world's rainforests, and most are small and ratlike. Rodents are characterized by their teeth—they have one large pair of sharp incisors in the front of each jaw. There are about 1,750 rodent species worldwide, almost twice as many as the next largest order, bats.

Further Reading:
Emmons, *Neotropical Rainforest Mammals*.

Agoutidae

Agoutidae live in lowland rainforests and mangrove forests from southern Mexico to northern Argentina.

Agouti paca

Agouti paca likes to stay near water sources and is an excellent swimmer—when pursued, it is likely to jump into the nearest stream. It has a medium-size, piglike body, short legs, and a stump of a tail. Its coarse hair is reddish to chocolate brown, and it has several lines of irregular white spots on its flanks. Pacas are terrestrial and nocturnal; they spend the day

in small burrows and forage for fruits at night. Although they form monogamous pairs, they den and forage on their own. Unfortunately for the paca, its meat is delicious to human beings, who have hunted the animal nearly to extinction in several Central American countries. Since pacas can be bred and raised in captivity, conservationists are working with some farmers to establish small paca farms, so the wild animals can be left alone.

Further Reading:
Emmons, *Neotropical Rainforest Mammals*.
Redford, *Mammals of the Neotropics*.

Dasyproctidae (agoutis)

Agoutis, found from southern Mexico to northern Argentina, are similar to the *Agoutidae* except they lack the spotted patterns and are more slender.

Dasyprocta spp. (agoutis)

The agoutis include seven species of large rodents with long slender legs, large rumps, and humped backs. They average two feet long and are covered with straight reddish orange, black, tawny, brown, or black hairs, depending on the species. Diurnal and terrestrial, they run with bounding, bouncing gaits. When alarmed, they let loose with a series of grunts, squeals, or barks and may stamp their hind feet on the ground. They feed on fruits, nuts, and seeds. Agoutis are intensely hunted for their meat throughout their range. Most agouti species are found only in South America: the red-rumped (*D. agouti*), the black (*D. fuliginosa*), the brown (*D. variegata*), Azara's agouti (*D. azarae*), and black-rumped (*D. prymnolopha*). One of the agouti's favorite foods is the seeds of the Brazil nut tree (*Bertholletia excelsa*). This rainforest giant is completely dependent on the agouti, because it's the only animal with teeth sharp enough and jaws powerful enough to break open the fruit's extremely hard shells. Most nuts are lost to agouti stomachs, but those that the mammal buries and forgets about have a chance to germinate. More than the Brazil nut tree depends on the agouti's chiseling dentures. Mammals such as spiny rats (*Echimyidae*), porcupines (*Erethizontidae*), and peccaries (*Tayassu*) steal the agouti's buried Brazil-nut hoards during lean seasons. In Brazil, the poison-dart frog (*Dendrobates*) relies on empty Brazil nut pods that have filled with rainwater as a home for eggs and tadpoles.

Further Reading:
Emmons, *Neotropical Rainforest Mammals.*
Taylor, "Strange Bedfellows: The Agouti's Nutty Friend."

Echimyidae

A diverse family of large rats, *Echimyidae* comprises about 70 species. Most are restricted to South America, but three species can be found in Central America as well.

Proechimys spp. (spiny rats)

Spiny rats are the most numerous terrestrial mammals in Neotropical forests. Their brown fur is stiff, bristly, and snowy white on their underparts. They are about nine inches long, with six-inch tails. Found from Honduras south to Paraguay, they are shy, solitary, terrestrial, and nocturnal. They primarily eat seeds, fruits, and fungi, and sometimes leaves and insects, foraging slowly along the forest floor. Prolific breeders, they are the major prey of ocelots, bushmasters, and other rainforest hunters.

Further Reading:
Emmons, *Neotropical Rainforest Mammals.*
Wolfe and Prance, *Rainforests of the World.*

Erethizontidae (New World porcupines)

New World porcupines include about 15 species that range from the Arctic coast to northern Argentina. All the Neotropical species have stiff, sharp, barbed spines on their upper parts, which detach easily when touched and become painfully embedded in the skin of the offending animal. Their tails are spineless, muscular, and easily wrapped around tree branches in a spiral twist. The animals' muzzles are bulbous, hairless, and pink. They are nocturnal and arboreal, feeding mainly on fruits and tender leaf parts. Like Old World porcupines, many species are highly adaptable and can live in forests and more open areas. The spines of some species are covered with long, soft, dark fur. The only species found in Mexico and Central America are Rothschild's porcupine (*Coendou rothschildi*) and the Mexican hairy porcupine (*C. mexicanus*). Perhaps the rarest of *Erethizontidae* is the brown hairy dwarf porcupine (*C. vestitus*), found only in a small and now mostly deforested area of Colombia, east of Bogotá.

Further Reading:
Emmons, *Neotropical Rainforest Mammals.*
Redford, *Mammals of the Neotropics.*

University of Michigan, Animal Diversity Web <http:// animaldiversity.ummz.umich.edu/chordata/mammalia/ rodentia/erethizontidae.html>.

Hydrochaeridae

Only one genus and species, *Hydrochaeris hydrochaeris*, the capybara, belongs to *Hydrochaeridae*.

Hydrochaeris hydrochaeris (capybara)

Capybaras' Amerindian name means "master of the grasses." It is the world's largest rodent. It looks a bit like a pig but with no tail, and its front legs are shorter than its hind limbs. Found from Panama south to northern Argentina, many of its local populations have been wiped out by hunting. It grazes on grasses along rivers and flooded meadows, and when disturbed, plunges right into the water. Its feet are partially webbed, in order to propel the animal along in water or on swampy ground. Capybaras live in small groups of about a dozen animals and feed in the late afternoon and off and on during the night, dozing in the morning. Like cats, they catch quick naps throughout the day. Several of the group generally stand watch, and if they spot a threat, they bark an alarm to warn others. The group will dash into the water and sink below the surface; capybaras can hold their breath for about five minutes. Young capybaras are preyed upon by wild cats, caimans, snakes, and raptors, and people extensively hunt the animals for their meat and for their skins, which are made into leather. They are farmed in some countries, which takes some pressure off wild populations.

Further Reading:
Emmons, *Neotropical Rainforest Mammals.*
Goulding, *The Flooded Forest.*
Redford, *Mammals of the Neotropics.*

Hystricidae (Old World porcupines)

Old World porcupines are large, heavy, and slow-moving animals like their New World counterparts. Since they can't dash from danger, they rely on their menacing quills for defense, but their spines, although effective, lack the barbs of the New World species. The 11 species live in a variety of habitats in sub-Saharan Africa, India, Southeast Asia and some Mediterranean countries.

Further Reading:
University of Michigan, Animal Diversity Web <http:// animaldiversity.ummz.umich.edu/chordata/mammalia/ rodentia/hystricidae.html>.

Artherurus africanus (brush-tailed porcupine)

An inhabitant of rainforests of central Africa, from Sierra Leone to Kenya, the brush-tailed porcupine feeds on fallen fruits, roots, tubers, and stems, and lives in groups of up to 20 animals. It sleeps in burrows during the day and is active most of the night. The bristly fur on its limbs and face merges with very sharp, thick quills on its back, with shorter quills on its rump and tail. The brush-tailed porcupine is 18 inches long, with a seven-inch tail. The tail has hollow white quills at the tip, which the porcupine rattles when disturbed.

Further Reading:
Kingdon, *Field Guide to African Mammals.*

Muridae (murid rodents)

The largest family of mammals, with about 1,140 species worldwide, murid rodents include most of the familiar rats and mice. The arboreal species usually have broad, pink-soled feet with sharp, curved claws, long, hairy tails, small ears, large eyes, and long whiskers. Ground dwellers have long, narrow, black-soled hind feet, long legs, large ears, long, naked tails, and pointed snouts.

Crateromys heaneyi (panay cloudrunner)

The panay cloudrunner was identified in 1996 in a mountaintop forest of Panay in the Philippines. This 26-inch-long, two-pound rodent looks like a large squirrel, with charcoal gray fur. Although it is an agile climber, it moves rather slowly. It eats only fruit and lives in cavities in tree trunks. No one is sure how many panay cloudrunners remain, but logging is quickly destroying their known habitat.

Further Reading:
Line, "A Newfound Mammal of Philippine Treetops Gets High-Flown Name."

Oryzomys spp. (rice rats)

The rice rats include about 20 species found from coastal New Jersey south to southern Argentina. They are fairly common and easily seen in forests, scurrying about the leaf litter. If they are startled, they will bound away with a series of high jumps. They range in size from four to seven inches long, with three- to eight-inch tails. Their soft and fine fur is brown, tawny, or gray. Nocturnal, solitary, and terrestrial, they eat insects, fruits, and seeds.

Further Reading:
Emmons, *Neotropical Rainforest Mammals.*
Redford, *Mammals of the Neotropics.*

SIRENIA

Sirenia are mammals that slowly cruise tropical and subtropical waters and includes five species of manatees and dugongs. They are solitary animals, often called "sea cows" because they graze on aquatic vegetation.

Trichechidae (manatees)

These entirely aquatic mammals include two families and four species worldwide. Manatees are huge—they can be 15 feet long and weigh more than a ton. They look like gigantic potatoes with tiny eyes and ears and a paddlelike tail that they use to propel themselves.

Trichechus inunguis (Amazonian manatee)

The Amazonian manatee is the largest mammal in the Amazon, weighing about 800 pounds. Related to marine manatees but adapted to freshwater life, it eats aquatic vegetation, but because these greens are low in nutrients, it must consume huge quantities, some 110 pounds daily. Amazonian manatees particularly chow down during the rainiest months in the Amazon, when high waters allow them to easily reach aquatic grasses and other plants. They build up a fat reserve that tides them over during the four- to six-month period when rains and food are sparse. All this consumption means they also defecate and urinate in copious quantities. When they were abundant, they were important fertilizers of the Amazon.

Manatees are gentle and slow-moving mammals that live entirely under the water. When they rise to the water's surface to breathe, only their nostrils emerge.

Manatees were once common in Amazonian waters. Their only predators are human beings, who have managed to nearly wipe them out. When the Portuguese arrived in the 16th century, they developed a taste for greasy manatee meat and slaughtered them by the thousands. Illegal hunting continues today, although the mammals are so scarce, even hunters find them difficult to locate. Another species of manatee, *T. senegalensis*, exists in western Africa, in coastal waters, lagoons, river mouths, and inland lakes. It is also endangered because humans have slaughtered so many.

Further Reading:
Goulding, *The Flooded Forest.*
Haltenorth and Diller, *Collins Field Guide: Mammals of Africa.*
Wolfe and Prance, *Rainforests of the World.*

XENARTHRA

Anteaters, armadillos, and sloths make up *Xenarthra*. What they have in common are teeth—or rather, the lack thereof. They are the last of a group of animals that once inhabited South America when it was an island continent, unconnected to the rest of the Americas. Once the Americas were joined by the Central American isthmus, dominant animals from North America slowly migrated south, causing the disappearance of many South American species.

Bradypodidae spp. and *Megalonychidae spp.* (sloths)

Sloths are among the slowest-moving animals on Earth. They spend nearly all their time in treetops of Neotropical forests, slowly chewing leaves, particularly those of *Cecropia* trees. Instead of toes, their feet have either two (the *Megalonychidae* family) or three (*Bradypodidae*) curved claws, so they can easily hook themselves onto tree branches and hang upside down. They can rotate their heads nearly 90 degrees, and their mouths are shaped so they seem to be smiling slightly. They have long, coarse fur that is light brown but often appears green due to the algae that thrive there. Female sloths give birth to just one baby, which spends its early months clinging to the hair on its mother's stomach. Both eagles and people hunt sloths for their meat.

Further Reading:
Emmons, *Neotropical Rainforest Mammals.*

Bradypus torquatus (maned sloth)

The maned sloth is found only in the Atlantic coastal forest of Brazil. It probably once inhabited this entire region of Brazil, but only 3 percent of the Atlantic coastal forest remains. In addition to deforestation, hunting has made this a highly endangered species. Occasionally the maned sloth is caught and kept as a pet.

Further Reading:
Emmons, *Neotropical Rainforest Mammals.*
World Wide Fund for Nature, Switzerland <http://www.panda.org/resources/publications/species/underthreat/manedsloth.htm>.

Bradypus variegatus (brown-throated three-toed sloth)

The most common of the sloths, the brown-throated three-toed sloths are found from southern Honduras to northern Argentina, east of the Andes. Unlike other arboreal animals, it comes all the way to the ground to defecate. Once a week or so it languidly descends, digs a hole with its stumpy tail while clinging to the tree trunk, defecates in it, and covers it over. The whole enterprise takes about a half-hour. This sloth is virtually unable to walk but can swim and sometimes is spotted crossing rivers.

Further Reading:
Kricher, *A Neotropical Companion.*
Wolfe and Prance, *Rainforests of the World.*

Dasypodidae (armadillos)

Armadillos began to migrate north thousands of years ago from South America. Today there are 20 species in South and Central America. One species, the nine-banded, long-nosed armadillo (*Dasypus novemcinctus*), as is found all the way north to Oklahoma, as well as in the islands of Grenada, Margarita, Trinidad, and Tobago. Their heads, backs, sides, and, with some species, their legs and tails are covered with bony plates that act as protective armor. Around the center of their bodies, the plates are separated by skin, so they can bend at will. They feed mainly on ants and termites and snuffle about on their stubby legs, their long snouts to the ground, in search of food. With their triangular-shaped heads and rather large, perky ears, they are downright homely, but in an endearing sort of way.

Further Reading:
Emmons, *Neotropical Rainforest Mammals.*
Reid, *A Field Guide to the Mammals of Central America and Southeast Mexico.*

Myrmecophagidae (anteaters)

Anteaters have long, tubular snouts, far-reaching tongues coated with sticky saliva, and huge and powerful curved front claws, especially the third claw. These attributes allow them to sniff down, dig out, and lick up their principal prey: ants and termites. Found only in the New World, from Mexico to Argentina, anteaters give birth to a single young, which rides along on its mother's back, clinging to her fur.

Myrmecophaga tridactyla (giant anteater)

A distinctively large and shaggy forest animal, the giant anteater measures about four feet, with another two feet for its full and furry tail. Because its front claws, especially the third, are so long, it must walk on its knuckles. Its fur is dark brown or black, white on its forelegs, black bands at the wrists, and a wide black band bordered by two thin white stripes from its neck down to its chest and up to its shoulder.

Tamandua tetradactyla (collared anteater), *T. mexicana* (Northern tamandua), and *Cyclopes didactylus* (silky or pygmy anteater)

These other three species of anteaters are arboreal, although the first two also spend time on the forest floor. They use their long, thick, prehensile tails like a fifth paw to hold them fast while they use their front paws to claw open a termite nest or beehive. The silky anteater is just seven inches long, with a tail of about the same length. Completely arboreal, it has hair that is long, golden, shimmery, and soft.

Further Reading:
Emmons, *Neotropical Rainforest Mammals.*
Redford, *Mammals of the Neotropics.*

Reptiles (*Reptilia*)

Reptiles are air-breathing vertebrates that have scales rather than fur or feathers. The class includes some 6,000 species of crocodiles, alligators, snakes, lizards, and turtles. Many inhabit temperate areas, but the majority of reptiles are found in the Tropics. They have internal fertilization and occupy a middle evolutionary position in between amphibians and birds and mammals, the warm-blooded vertebrates. They may be viviparous, giving birth to live young; oviparous, or egg-laying; or ovoviviparous, producing eggs that develop within the maternal body and hatch either within or immediately after being expelled.

CROCODYLIA

The order *Crocodylia* includes 23 species of alligators, caimans, crocodiles, and gharials, found throughout tropical and subtropical countries. Humans have slaughtered untold numbers of crocodilians, remnants of an age when reptiles ruled the Earth. People hunt many species for their skins, which are fashioned into shoes, handbags, belts, and wallets, supporting an international trade worth more than $500 million annually. Often people kill crocs simply because the animals are large, fierce-looking, and sometimes aggressive predators. As a result, many species are threatened or endangered. Since they are at the top of the food chain, adult crocodiles are considered "keystone species," meaning their disappearance would throw entire ecosystems out of balance. Smaller crocodilians eat aquatic insects, small fish, and crustaceans; as they grow larger, they eat turtles, birds, larger fish, and mammals. In the wild, they may live up to 60 years.

Further Reading:
Crocodile Specialist Group <http://www.flmnh.ufl.edu/natsci/herpetology/crocs.htm>.

Crocodylidae

The 13 species of *Crocodylidae* live in most of sub-Saharan Africa, Madagascar, India, Sri Lanka, Southeast Asia, Northern Australia, some East Indies islands, and tropical South America.

Crocodylus acutus (American crocodile)

The American crocodile is the most widely distributed of the four Neotropical crocodile species, found from Mexico, Central America and the Caribbean islands, south to Colombia, Venezuela, Peru, and Ecuador. A small population also remains in southern Florida. They can reach lengths of 21 feet, but it's rare to find animals more than 13 feet long.

Further Reading:
Beletsky, Les. *Belize and Northern Guatemala: Ecotravellers' Wildlife Guide.*
Crocodile Specialist Group <http://www.flmnh.ufl.edu/natsci/herpetology/crocs.htm>.

Crocodylus intermedius (Orinoco crocodile)

Among the largest of the crocodiles, Orinoco crocodiles reach about 16 feet in length. They are also among the most endangered. They are found in the Orinoco River basin in Venezuela and Colombia, where hunting for their skins nearly wiped out the reptiles by the 1960s. Although hunting is now banned, poaching is common, and the population has been slow to recover. Over the past decade, captive-breeding centers in Venezuela have raised and released more than 1,300 of the crocodiles into the Orinoco.

Further Reading:
Nova Online <http://www.pbs.org/wgbh/nova/crocs/whos/ nojs.html>.

Crocodylus porosus (saltwater crocodile)

The saltwater crocodile has a high tolerance for salinity, as the common name suggests. It can be found in brackish water around coastal areas and in rivers, as well as in freshwater rivers and wetlands. Its ability to survive in saltwater probably explains its wide distribution, as this croc lives in Australia, Bangladesh, Brunei, Myanmar, Cambodia, China, India, Indonesia, Malaysia, Palau, Papua New Guinea, Philippines, Singapore, Sri Lanka, Solomon Islands, Thailand, Vanuatu, and Vietnam. These are the largest living crocodilian species, and in fact, the largest reptile in the world. Adult males can reach 23 feet in length.

Further Reading:
Crocodile Specialist Group <http://www.flmnh.ufl.edu/ natsci/herpetology/crocs.htm>.

Alligatoridae

Alligatoridae is a family of seven species. Unlike crocodiles, the teeth of their lower jaws fit inside those of their upper jaws.

Melanosuchus niger (black caiman)

The black caiman can grow up to 19 feet in length and is the second-longest animal in the rivers and wetlands of the Amazon, after the giant anaconda. Little is known about these powerful but wary reptiles, although they have hardly changed from the way their ancestors looked 80 million years ago. Only 50 years ago, they were common in the Amazon. But by the 1940s, caiman skins became the rage; at least 7.5 million black caiman skins were exported by the 1960s, when hunting caused the population to crash and was declared illegal. Poaching of black caiman and the smaller spectacled caiman (*Caiman crocodilus*) continues. Both species eat just about anything they can get their powerful jaws on—fish, birds, amphibians, insects, and small mammals. Caimans are one of the few animal species that prey on large toads, even those whose skins contain toxins poisonous to other animals. Thanks to the ban on hunting and the establishment in 1990 of the Mamairau Reserve in Brazil's flooded forest, the black caiman population is rebounding.

Further Reading:
Goulding, *The Flooded Forest.*
Thorbjarnarson, Tom. "The Hunt for Black Caiman."

SQUAMATA

Squamata is the order of snakes and lizards, which are most obviously different from each other in that lizards have limbs and snakes do not. Snakes also have only one lung or a very reduced second lung, no ears, no eyelids, no voices—yet in spite of what seem like deficiencies, they are highly successful in a variety of habitats and are found nearly everywhere on Earth except the North and South Poles. The approximately 2,700 species are solitary and independent, usually fearless, and devote most of their lives to hunting down prey, except for the seasonal—and brief—distraction of mating. Females must also take a moment to find a secure place to deposit eggs or give birth to young. Snakes do not survive well once human beings encroach on their habitat, particularly because people find these reptiles fearsome or revolting. Of the 15 families of snakes, all but two are found in tropical forests.

Lizards are scaly-skinned reptiles, and although they are closely allied with snakes, most have limbs, eyelids, and ears. They inhabit almost all areas except Antarctica and some Arctic regions, but the majority of the 3,000 species live in the Tropics.

Lizards, which belong to the suborder *Sauria*, are listed first below, followed by snakes, which fall under the suborder *Serpentes*.

Sauria (lizards)

Agamidae

The 350 species within *Agamidae* live in warmer areas of the Old World—Africa, central, south and Southeast Asia, Australia, and one species in Europe. They are the Old World counterparts of *Iguanidae* and resemble these lizards in appearance and biology. Like them, many species bear frills, flaps, and crests, and males may be brightly colored.

Draco volan (flying lizard)

The flying lizard does have wings of a sort along the sides of its body, supported by elongated ribs. In order to "fly," the lizard spreads these wings, drops from a tree, with head pointed downward, and glides to the ground. It can glide about 25 feet. In some *Draco* species, these "wings" are brightly colored and may have striking marks. Found in southern India and Southeast Asia, including the Philippines, it mainly eats small ants, termites, and other insects and is entirely arboreal. The male has a flap of skin under its head, called a dewlap, which is long, pointed, and bright yellow. The male will court a female by spreading his wings and dewlap, bobbing his head, and circling three times. The female comes to the ground only to lay eggs in a hole in the earth she makes with her head. For 24 hours, she guards the eggs, which hatch in about 30 days. There are about 40 species of *Draco* flying lizards.

Further Reading:
Mattson, *Lizards of the World.*

Amphisbaenidae (worm lizards)

Worm lizards are a large lizard family, with 130 species, but its members look the least lizardlike of all the lizards. Their long bodies are virtually the same diameter over their entire length. All but one genus (*Bipes*) lack external limbs, lack ear openings, and have their eyes hidden under skin. Rings of narrow scales encircle their bodies. It would be easy to mistake members of this lizard family for earthworms. The *Amphisbaenidae* are found in South America, Africa, and southern Europe, with more species in Africa than anywhere else. All are burrowers, and under the earth they can progress backward and forward. Some species lay eggs; others give birth to live young.

Further Reading:
Mattson, *Lizards of the World.*
Smith, *Handbook of Lizards.*

Amphisbaena alba

Amphisbaena alba is 30 inches long, stocky, and white. Found in northern South America and the island of Trinidad, it often lays its eggs in leaf-cutter ants' nests. With powerful jaws, it can eat just about any small animal it can conquer, along with carrion.

Further Reading:
Mattson, *Lizards of the World.*
Smith, *Handbook of Lizards.*

Chamaeleontidae (chameleons)

Chameleons are tree-dwelling lizards found only in the Old World—in fact, nearly half of the 135 species live on the island of Madagascar. Further, 59 of Madagascar's chameleon species are found nowhere else on Earth. Chameleons range in size from under 1 inch to 2.5 inches long. All have prehensile tails, large and independently mobile eyes, and partially fused toes. Most live in humid forest areas.

These lizards are famous for their ability to change colors and may sport hues of green, blue, yellow, red, brown, black, or white. They do not change colors to camouflage themselves from their enemies, but rather to communicate with fellow chameleons, signaling such moods and attitudes as aggression and a willingness to mate. Color changes are also responses to temperatures and light. The lizard's outer skin is actually transparent. Beneath it are cell layers that contain red and yellow pigments, called chromatophores, and beneath the pigments lie another layer of cells that reflect blue and white light. Under this is a layer of brown melanin. External light or heat levels or internal chemical reactions can cause the various cell layers to expand or contract. If the yellow cells contract, for example, blue-reflected light shines through, so the chameleon appears to be green. Most chameleons are 7 to 10 inches long. Some species have curled tails; others have heads that are helmet-shaped.

A few chameleon species have several horns on their heads, used in territorial defense. A threatened male will expand his body, puff out his throat, wave about flaps of skin on his head, and then charge while snapping his jaws. Unlike other lizards, chameleon feet have toes that are fused into opposite bundles of two and three, like pincers. Called zygodactylous toes, they are excellent aids in climbing trees. Chameleon eyes can move independently of each other, giving the reptiles a 360-degree view of their world. Large chameleons eat birds, but most dine mainly on insects. They have extremely large tongues—sometimes twice as long as their bodies—to help them scoop up bugs in hard-to-reach places. In addition to Madagascar, chameleons are found in southern Europe, Africa, the Seychelles, India, Sri Lanka, and Pakistan. Their forest habitat in Madagascar is nearly gone, so many species are probably highly endangered. Some species, such as *Chamaeleo belalandaensis,* have been collected for the pet trade, another threat to their survival.

Further Reading:
Cogger and Zweifel, *Encyclopedia of Reptiles and Amphibians.*
Holmes, Hannah. "The Lizard Wizard."
Mattson, *Lizards of the World.*

Iguanidae

Iguanidae is the largest family of lizards of the New World, to which they are restricted except for two genera in Madagascar and one in Fiji and Tonga. Nearly all the commonly seen lizards in the New World belong to this family, which contains 650 species found in a range of habitats.

Anolis spp.

The largest genus in the iguana family, the anole has more than 110 different species in the West Indies alone, another 40 in Mexico, and dozens more in Central and South America. One species is native to North America. Anoles have long tails, long legs and digits, narrow heads, and are an average of eight inches in length. Males have brightly colored throat fans that they can open via a flexible rod of cartilage just inside the skin that swings up or down from where it is attached at the middle of the throat. Females of some species also have these fans, which are usually yellow, orange, or red. Although their bodies are usually brown or green, some species can change their color quickly. Most species are arboreal, and they have evolved pads on their feet to help them climb. All anoles are egg layers. Most lay a single egg at regular intervals during the breeding season or throughout the year.

Further Reading:
Mattson, *Lizards of the World.*
Smith, *Handbook of Lizards.*

Basiliscus basiliscus (brown basilisk)

The brown basilisk is found from Mexico to Ecuador. A mostly insectivorous tree-dweller, it is famous for its otherworldly ability to scramble across water while keeping its body above the surface. Scientists recently discovered how they manage the trick: First, the lizard slaps the water with a hind foot, then strokes down rapidly, creating a cavity of air in the water. Finally, the basilisk pulls the foot up through the air pocket before the space can fill with water. This is accomplished with lightning speed; a startled basilisk can dash across six feet of water in just a second or two. Five species are water walkers, but among them, the brown basilisk is the most adept.

These lean lizards grow to about 28 inches long, which includes their long, slender tails. Adult male basilisks have one crest along the backs of their heads and necks and another in the middle of their backs and tails. They are brown with darker bands across their bodies. Diurnal and omnivorous, they dine from a huge menu that includes insects and other invertebrates, such as shrimp and scorpions, lizards, snakes, fish, small mammals, and birds, and a few flowers and fruits. They generally eschew twigs and leaves. A favorite food is the hatchlings of *Iguana* and the eggs of *Ctenosaura* lizards. They have few predators—only raptors by day and alert opossums by night. To avoid nocturnal disasters, basilisks sleep in branches hanging over the water, so if disturbed, they can drop into the drink.

Further Reading:
Janzen, *Costa Rican Natural History.*

Ctenosaura similis (spiny-tailed iguana)

This spiny-tailed iguana inhabits a range that extends from southern Mexico to Panama. They average 13 inches in length, with an equally long tail that is covered with large spiny scales that are used as defense weapons. Adults are brown, marked with a series of darker-colored bands. They are most often found in dry and lowland forests, but are adaptable, so they may also be spotted in vacant city lots and gardens. They are both arboreal and terrestrial, although they like to spend the night in tree hollows and are assisted with their climbing by sharp and strong claws. They are primarily vegetarian, but may also be predatory, feeding on bats, rodents, other lizards and lizard eggs, small birds, and arthropods. Females lay from 12 to 90 eggs, depending on their body size and age. People consider *C. similis* to be tasty, and its flesh is also thought to have medicinal properties, such as a cure for impotency.

Further Reading:
Janzen, *Costa Rican Natural History.*
Smith, *Handbook of Lizards.*

Iguana iguana (green iguana)

The green iguana lives in forests of South and Central America and on the island of Trinidad. The largest can be more than six feet long, from tip of nose to tail, which is edged with a row of large, sharp scales. They are green or bluish-green and have a dewlap. Males are territorial and will alert intruders that they've crossed a boundary by raising up their bodies and nodding their heads vigorously. They perch amidst the forest canopy, where they feed, although they also graze on ground-level vegetation.

Further Reading:
Mattson, *Lizards of the World.*

Scincidae (skinks)

The largest lizard family, *Scincidae*, has about 1,300 species, known as skinks. They are found worldwide, but mainly in the Old World, particularly Australia, Southeast Asia, and Africa. They live in a wide variety of habitats. A typical skink is small and brown, and has shiny, relatively smooth, overlapping scales. Though diurnal, they are secretive and tend to spend most of their time concealed beneath the leaf litter or under logs and rocks. Many species are limbless or have reduced limbs. Most can lose their tails as an escape mechanism, so they have evolved brightly hued tails to draw predators' attention away from less dispensable parts of their bodies.

Further Reading:
Cogger, *Reptiles and Amphibians of Australia.*

Carlia spp.

These small and terrestrial skinks most often live in leaf litter in wooded areas of northern Australia. *C. rubrigularis* is a rainforest species that favors streams. A dark gray-brown skink, it often has a pale, narrow stripe running down its back.

Further Reading:
Cogger, *Reptiles and Amphibians of Australia.*
Cogger and Zweifel, *Encyclopedia of Reptiles and Amphibians.*

Corucia zebrata

Corucia zebrata is an arboreal skink with un-skinklike habits. It reaches 28 inches in length, is slow-moving, and is mostly nocturnal. It feeds exclusively on vegetation and seldom descends to the ground. It has a thick, muscular, prehensile tail to assist with climbing, the only skink able to use its tail this way. It gives birth to one rather large youngster.

Further Reading:
Cogger and Zweifel, *Encyclopedia of Reptiles and Amphibians.*
Mattson, *Lizards of the World.*

Prasinohaema spp.

Prasinohaema spp. are the only reptiles that have green blood pigment. Found in New Guinea and the Sol-

omon Islands, the five species are arboreal, with sticky toe pads that help them scamper up slippery surfaces. In addition to their blood, their skin, tongues, and insides of their mouths are green, for reasons unknown. They lay two eggs, and these are green as well.

Further Reading:
Mattson, *Lizards of the World.*

Tribolonotus spp.

Atypical skinks of the Pacific Islands, *Tribolonotus spp.* have quite spiny scales, unlike other skinks. They inhabit rainforests, particularly along streams. Most lay just one egg, but *T. schmidti*, of Guadalcanal, gives live birth—to just one youngster.

Further Reading:
Mattson, *Lizards of the World.*

Xenosauridae (xenosaurus)

Xenosaurus is a family of only five species. One is found in China, but the other four live on the other side of the world, where the range of *Xenosaurus spp.* is Mexico south to Guatemala in a variety of habitats, but most often in rainforests or high-elevation cloud forests. Nocturnal and terrestrial lizards, they mainly feed on insects. All five species give birth to live young.

Further Reading:
Cogger and Zweifel, *Encyclopedia of Reptiles and Amphibians.*
Mattson, *Lizards of the World.*

Serpentes (snakes)

Acrochordidae (wart, or file, snakes)

Wart, or file, snakes have adapted superbly to aquatic life, and barely leave their watery homes in rivers, lakes, or estuaries of Southeast Asia to the northern tip of Australia. All three species have a flap in the roofs of their mouths, which closes off their nostrils when they submerge, while a pad on their lower jaws closes off the notch in their upper jaws through which their tongues protrude. The unique scales of the wart snakes do not overlap and are small and granular. Their skin hangs in folds and is covered with tiny bristles, whose function is unknown. The largest species, *Acrochordus javanicus*, reaches eight feet in length, is banded in black and white, and inhabits freshwater in southern Asia and Indonesia. Wart snakes feed on fish and give birth to live young.

Further Reading:
Mattison, *The Encyclopedia of Snakes.*

Boidae (boas and pythons)

The family of boas and pythons has about 60 species, found throughout the Americas and most of Africa and Asia. They range in size from 18 inches to 30 feet. All give birth to live young.

Boa constrictor

Boa constrictors can be as long as 13 feet but average around six feet. They have a huge range—at home in wet and dry forests, fields, and savannas, from western Mexico to Argentina, several Caribbean islands, Madagascar, and a few Pacific islands. The snakes' coloring—dark, diamond-shaped blotches on gray—allows them to blend well among the leaf litter, where they wait for birds, lizards, and small mammals. Clear scales cover their eyes so they always appear to be awake—that they do not blink improves their camouflage further. Sometimes they climb trees and wait for prey there. They strike by opening their mouths wide and impaling their victim on their needle-sharp, curved teeth, while at the same time constricting the animal in their coiled bodies. The snakes then swallow their prey whole. Like most other boas, *B. constrictor* gives birth to live young.

Further Reading:
Janzen, *Costa Rican Natural History.*
Kricher, *A Neotropical Companion.*
Mattison, *The Encyclopedia of Snakes.*

Chondropython viridis (green tree python)

Green tree pythons inhabit forested regions of Papua New Guinea and the Cape York Peninsula, in far northeastern Australia. A vivid green snake with a white or yellow broken stripe running the length of its four-foot-long body, it is completely arboreal, descending to the ground only to lay eggs. It hunts rodents, birds, and bats and moves about tree branches with the help of its prehensile tail. Hatchlings are usually lemon-yellow, but can also be red or blue; all soon change to green. Deforestation has made the green tree python endangered throughout its range. The green tree python looks and behaves very much like the emerald tree boa, found on the other side of the globe.

Further Reading:
Coborn, *The Atlas of Snakes of the World.*
Mattison, *The Encyclopedia of Snakes.*

Corallus caninus (emerald tree boa)

The emerald tree boa may be the most beautiful snake on Earth, with its deep green skin above and pale yellow belly. It bears a white line down its back, with scattered white dots, and has burning yellow, cat-like, narrow eyes. Found in the Amazon basin, from Peru and Bolivia through Brazil to the Guyanas, the six-foot boa lives in trees, coiling itself tightly among the green foliage. Largely nocturnal, it has a prehensile tail to help it move about tree branches, hunting for lizards, birds, squirrels, bats, and opossums.

Further Reading:
Ditmars, *Snakes of the World.*
Kricher, *A Neotropical Companion.*
Mattison, *The Encyclopedia of Snakes.*

Epicrates spp.

This, the largest boa genus, has 10 species distributed throughout the West Indies and South America. They range in size from three to six feet long, with the smaller species tending to be more arboreal and slimmer. Several species are endemic to specific islands, where they face a dim future. *E. inornatus*, the Puerto Rican boa, and the Bimini boa, *E. striatus*, are endangered, while *E. subflavus*, found on Jamaica, is in danger of extinction.

Further Reading:
Mattison, *The Encyclopedia of Snakes.*

Eunectes murinus (anaconda)

Anaconda is unveiling its mysteries to intrepid scientists who have begun to study the world's largest snake, found in most of tropical South America and on the island of Trinidad. *E. murinus* is the largest of the four species in the *Eunectes* genus; it can exceed 20 feet in length and weigh more than 200 pounds. Recently biologists uncovered the secret of anaconda sex: During the mating season, a single female copulates with as many as 12 much smaller males, in an undulating mass poetically called a "breeding ball." The group sex can last several weeks, but it's unknown whether the resulting clutch of 70 or so baby snakes, which are born live, has more than one father. Predators such as ocelots and caimans gobble about half of the newborn.

To kill their prey, anacondas first subdue animals

such as turtles, caimans, large rodents called capybaras, and even deer, by squeezing them tightly, choking the air out of them, and sometimes drowning them at the same time. Usually they swallow their food headfirst. In spite of suggestions from Hollywood, no evidence suggests that anacondas eat human beings. People have, however, slaughtered countless anacondas, often illegally, to make boots, belts and wallets from their skins. *E. murinus* is semi-aquatic, often found in swamps and forests near slow-moving rivers.

Further Reading:
Mattison, *The Encyclopedia of Snakes.*
Rivas, "Tracking the Anaconda."

Python spp.

Pythons include 20 to 25 species of snakes found only in the Old World, in temperate and tropical regions of western Africa to China (but not on Madagascar), Australia, and the Pacific Islands. They are usually found near sources of water and are excellent climbers and swimmers. Pythons are not poisonous, but they kill by constriction and then swallow their victims whole. They don't crush prey, but rather squeeze them so tightly that they can no longer breathe. The largest pythons can squeeze and swallow pigs or small deer, but mostly take rats and similarly sized mammals. Many species have sensory depressions, or pits, on their heads that act as heat receptors, allowing them to detect and close in on prey.

People hunt pythons for their meat and skins. Overhunting has had serious consequences in some areas, as larger species prey on rodents. Where the pythons have been obliterated, rats are causing serious damage to crops. Like other snakes, pythons are preyed upon by raptors. They are generally docile snakes, although females will aggressively defend their young. All species lay eggs—from 5 to 100, depending on the species and body size.

P. curtus (blood python)

The blood python is almost always found near water, particularly wetlands, of the Malay Peninsula, Sumatra, and Borneo. Its coloring is variable, with the most striking species bearing shades of orange and pink to deep red, with bands and patches of black and yellowish gray. It can reach nine feet in length, but is usually shorter. *P. curtus* is killed in large numbers for its skin and meat; some 60,000 blood python skins are harvested each year.

P. molurus (Burmese, or Indian, python)

The Burmese, or Indian, python is a forest snake of India, Indonesia, southern China, and the Malay Peninsula. It can be as long as 16 feet and weigh some 200 pounds; the male is considerably shorter and slimmer than the female. Its back is light to dark brown, with a network of dark-bordered, cream-colored bands. Near its anus, it bears two clawlike spurs, vestiges of appendages, which are sharp enough to inflict serious injury. Its head is distinct from its body, and it has an arrowhead mark on the top of its head, pointed toward its snout. It hunts, mainly at night, for mammals, including small deer, birds, and other reptiles. In spite of its size, the Burmese python is an agile climber and may hang suspended from a tree with its tail twisted about a branch, waiting for prey. A strong swimmer, it can stay underwater for 15 minutes. Because it is among the gentlest of the pythons, it is the species most often seen in movies.

P. reticulatus (reticulated python)

The reticulated python can reach 26 feet in length, challenging the anaconda for the title of world's largest snake. Anacondas, though, are much heftier; reticulated pythons are rather slender. They range from southern Myanmar to Indonesia and the Philippines and are usually found in tropical forests, near lakes and rivers, but they have also adapted to urban environments. So, in addition to rodents, small wild cats, rabbits, and small deer, they take pigs, domestic cats, and dogs. When encountering human beings, however, they always attempt to escape rather than attack. Their backs are light to dark brown, with an intricate series of reticulating black lines down the sides. Black lines also stretch from the snakes' yellow eyes to the base of their jaws. Their bellies are dirty white or yellow.

P. regius (royal, or ball, python)

The royal, or ball, python lives in a variety of ecosystems, including tropical forests, in western and central Africa. A nocturnal snake, it often hunts for small mammals in their burrows. When threatened, it coils up into a ball, protecting its head. It averages 48 inches long, but can become as long as 75 inches. It is light brown, patterned with darker brown bands and splotches. All pythons have good senses of smell, but the ball python's is particularly sharp. It has a specialized olfactory organ in the roof of its mouth and has eight pits on its face that act as heat receptors, so it can easily close in on prey at night.

Further Reading:
Cox, *The Snakes of Thailand.*
Deoras, *Snakes of India.*
Gharpurey, *The Snakes of India and Pakistan.*
Tweedie, *The Snakes of Malaya.*

Sanzinia madagascariensis (Madagascan tree boa)

The Madagascan tree boa is an arboreal snake found only on Madagascar. Although tree boas are the most common snake on the island nation, deforestation has made this green or yellow boa an endangered species.

Further Reading:
Mattison, *The Encyclopedia of Snakes.*

Bolyeriidae (Round Island boas)

Round Island boas is a family with just two species, *Bolyeria multicarinata*, the Round Island burrowing boa, and *Casarea dussumieri*, the Round Island keel-scaled boa. Both are found only on 373-acre Round Island, which is off the northern coast of Mauritius, east of Madagascar, in the Indian Ocean. When people brought rabbits and goats to the island in the nineteenth century, the resulting habitat destruction brought nearly all the islands' native wildlife and plant species to the edge of extinction. In the 1970s and 1980s, the rabbits and goats were removed from the island, but recovery has been slow for the boas. In fact, it may be too late for the burrowing boa. Once the island lost most of its tree cover, the snake had no leaf litter and loose soil to burrow through. The keel-scaled boa seems to have recovered as vegetation has regenerated, along with an increase in the snake's prey: geckos and skinks.

Further Reading:
Mattison, *The Encyclopedia of Snakes.*

Colubridae

Among the largest, and most widespread, of all the snake families, *Colubridae* has nearly 2,000 species. In most parts of the world, colubrid snakes are the most familiar to people. They are harmless to human beings and have filled nearly every ecological niche, from rainforest treetops to mangroves to deserts.

Ahaetulla spp. (Asian vine snakes)

Asian vine snakes include eight species found in tropical forests of Southeast Asia. They range from two to six feet in length and have very slender and elongated bodies, with long heads that are distinct from their bodies. Their snouts are wedge-shaped. The snakes' large eyes have horizontal pupils shaped like keyholes lying on their sides. A groove that runs from the front of the eyes to the snout gives them binocular vision, an unusual attribute in snakes. These arboreal, lively snakes prey mostly on lizards. They have fangs at the rear of their mouths, but like all members of this family, their poison is not toxic to human beings.

A. mycterizans (Malayan green whip snake)

Found in Malaysia and Java, the Malayan green whip snake is green to gray-green, and lighter green on its belly. It reaches four feet in length. The long-nosed tree snake (*A. nasuta*) of Sri Lanka and India east to Thailand is usually a bright green with a reddish tail.

Further Reading:
Cox, *The Snakes of Thailand.*
Mattison, *The Encyclopedia of Snakes.*
Tweedie, *The Snakes of Malaya.*

Boiga spp.

Boiga spp. are some 30 species of fairly large tree snakes found in Africa, India, Sri Lanka, southern China, and throughout Southeast Asia, with one species in Australia (*B. irregularis*). Their heads are distinct from their bodies, and they have large eyes. They are nocturnal, and all but one species (*B. trigonota*) are arboreal. Prey includes frogs, reptiles, birds, and small mammals.

B. dendrophila, the mangrove snake, is perhaps the most handsome of the species, with narrow, bright yellow bands on its glossy black back. Its belly is slate-gray, and its throat and lips, yellow. Mangrove snakes are about six feet long and are found in lowland forests and wetlands of Thailand through Malaysia and Indonesia to the Philippines. They are often spotted draped about a tree branch, hanging over water. They deliver a venomous bite, but also kill via constriction. Since they are easily tamed, they are a favorite species of snake charmers.

B. blandingii (Blanding's tree snake) and *B. pulverulenta* (blotched tree snake) are the two species found in sub-Saharan Africa. The former is a large snake, reaching up to eight feet in length, and is found in eastern, central and western Africa. In addition to eating birds and bats, it dines on the occasional chameleon. The smaller blotched tree snake is found in the same range, but also as far south as Angola.

In addition to living in Australia, the brown tree snake (*B. irregularis*) inhabits gardens, mangroves, and rainforests of Papua New Guinea, the Philippines, and Indonesia. A large snake at about six feet long, it is reddish brown with dark crossbars on its back and has a cream-colored belly. The brown tree snake was accidentally introduced on the island of Guam, probably because it hid in ship cargo from Papua New Guinea. Because they have no natural enemies on Guam, they multiplied quickly, and by the early 1970s, the snakes were widely dispersed on the island. They have caused serious ecological and economic problems, virtually wiping out the native forest birds on the island. Twelve species of endemic birds have disappeared completely; others are near extinction. The populous snakes crawl about electrical lines and frequently cause power outages and damage and they invade poultry houses and homes. An estimated 13,000 brown tree snakes per square mile reside in Guam's forests.

Further Reading:
Coborn, *The Atlas of Snakes of the World.*
Cox, *The Snakes of Thailand.*
O'Shea, *A Guide to the Snakes of Papua New Guinea.*
Phelps, *Poisonous Snakes.*
Tweedie, *The Snakes of Malaya.*

Chrysopelea spp.

The *Chrysopelea spp.* are three species of slender, arboreal snakes of Southeast Asia and Indonesia that have the ability to "fly." They straighten their bodies, draw up their ventral scales to make their undersides concave—like really sucking in your stomach—and launch themselves. Air is trapped in their hollowed bellies, like a parachute, so their fall becomes a controlled glide. In that way, they can easily float from one level of the forest to another. Diurnal, their prey includes rodents, lizards, birds, and bats. They have fangs at the back of their mouths, which deliver a slow-acting poison. *C. ornata* (golden tree snake), found in India, Sri Lanka, and Malaysia, is a black or greenish flying snake up to three feet long, with yellow or red markings. *C. paradisi* (paradise tree snake) has a similar range; it is black with a green spot on each scale and reaches three to four feet in length. *C. pelias* (twin-barred tree snake) has a brick-red or orange back with pairs of black cross-bars that

surround a greenish-yellow bar and averages two feet in length.

Further Reading:
Coborn, *The Atlas of Snakes of the World.*
Phelps, *Poisonous Snakes.*
Tweedie, *The Snakes of Malaya.*

Clelia clelia

Clelia clelia, an eight-foot snake, is found in lowland forests from southern Mexico, Central America, and much of South America. The young hatch from eggs and are bright red with black heads and yellow collars, but as they grow they turn black or gray. *C. clelia*, called the musurana, feeds mostly on other snakes, including quite venomous vipers, whose poisons have no effect on it. The musurana is one of the few snakes that has grooved fangs on the rear of its upper jaws, but it also constricts its prey. Since the typical prey is a strong and slippery snake, evolution provided both abilities. They are active both night and day.

Further Reading:
Janzen, *Costa Rican Natural History.*
Mattison, *The Encyclopedia of Snakes.*

Drymarchon corais (indigo snake)

The indigo snake is a widespread species that lives along rivers and in marshes throughout lowland tropical America, from southern Florida and Texas all the way south to Argentina. A lustrous blue-black or brown and black snake, its throat and sides have dashes of red, orange, or cream, with a short, lateral, dark bar just behind its head. The large-scaled indigo snake can reach 12 feet in length and is a diurnal, aquatic hunter with a voracious appetite for a wide variety of prey: frogs, lizards, fish, other snakes, small mammals, turtles, and birds. It does not have venom nor does it constrict; it kills with its unusually powerful jaws.

Further Reading:
Coburn, *The Atlas of Snakes of the World.*
Garel and Matola, *A Field Guide to the Snakes of Belize.*
Janzen, *Costa Rican Natural History.*

Imantodes cenchoa (chunk-headed snake)

With a relatively wide and blocky head, the chunk-headed snake is otherwise quite slim, with a very narrow neck. It is less than three feet long, has large scales, and is light brown with darker brown, irregular blotches down its back. The chunk-headed snake has bulging eyes with vertically elliptical pupils, like those of a cat. Its range is tropical Mexico to Bolivia and Paraguay, where it is well adapted to its arboreal habitat—the small twigs and leaves of shrubs and trees. It can easily bridge gaps in the foliage that are up to one-half its length, and with its thin neck, it can thrust its head out to the branch tips without startling prey. It hunts at night for small frogs and lizards dozing in the trees.

Further Reading:
Coburn, *The Atlas of Snakes of the World.*
Garel and Matola, *A Field Guide to the Snakes of Belize.*
Janzen, *Costa Rican Natural History.*

Oxybelis spp.

Oxybelis contains four or five species that range from the southwestern United States south to Brazil, in rainforest, secondary forest, or thick brush. *O. fulgidus* reaches nearly six feet in length and is slim and bright green, making it almost impossible to detect among foliage. Most feed on small frogs and lizards, but larger snakes may take birds and mice.

Further Reading:
Coburn, *The Atlas of Snakes of the World.*
Garel and Matola, *A Field Guide to the Snakes of Belize.*
Janzen, *Costa Rican Natural History.*

Oxybelis aeneus (vine snake)

The vine snake has one of the largest distributions of any Neotropical snake; it inhabits dry forests from extreme southern Arizona, through Mexico and Central America, and south to central Bolivia and southeastern Brazil. This is a very slim and long snake, reaching five feet in length, with an elongated head and sharply pointed snout. With its slender body and brown or gray coloring, it can drape itself in a tree and look every bit like a vine. The camouflage is a good defense and also allows it to capture unsuspecting meals, usually lizards. When threatened, the vine snake opens its mouth wide to display its bright, blue-black lining, and strikes repeatedly. It has large fangs in the back of its mouth. Its venom is mild, but strong enough to subdue its prey, which it hunts during the day.

Philothamnus spp. (green bush snakes)

Green bush snakes include seven species that live in rainforests of Africa. They are all slender snakes, averaging four feet in length, with oval heads set off from their necks. They have large eyes and are bright green above, with white or yellowish bellies. They hunt during the day in trees, although they may descend to the ground in search of prey, mostly lizards and frogs. *P. irregularis* (Western green snake), which is a shiny green, also feeds on nestling birds. Green bush snakes are common in riverside forests and can swim well; the Western green snake is largely aquatic. All bush snakes are alert, quick, and will dive into water after prey or if alarmed. They also puff up when bothered.

Further Reading:
Branch, *Field Guide to Snakes and Other Reptiles of Southern Africa.*
Coburn, *The Atlas of Snakes of the World.*
Isemonger, *Snakes of Africa.*

Elapidae

Elapidae includes about 200 species of snakes, all of them poisonous. The family is best represented in Australia, which is the only country where venomous snakes outnumber the nonvenomous species. They also occur in southern North America, throughout Central America and most of South America, most of Africa except the Sahara Desert, and southern and Southeast Asia.

Dendroaspis spp. (mambas)

Mambas are four species found throughout tropical Africa, in both forest and brushland. They are quick, diurnal snakes, from 8 to 13 feet in length, and highly venomous. Three of the species are arboreal and usually green, feeding on birds and small mammals. The heads of all the mambas are aptly, if somewhat dramatically, described as coffin-shaped. The black mamba (*D. polylepis*) is mainly terrestrial, dark olive to brown or gray above and grayish-white on its belly. It's the longest mamba, reaching 12 feet in length. Although all mambas are feared in Africa, only the black mamba commonly bites and is particularly aggressive. It kills its prey, fledgling birds and small mammals, with stabs from its fangs, which inject poison until the victim collapses. Its favorite homes are termite nests, hollow logs, or rock crevices. Usually it will retreat when disturbed, but if it's cornered, it raises the front third of its body, spreads a narrow hood, hisses, and opens its mouth, whose lining is black. The black mamba is believed to be the fastest-moving snake in the world, slithering along at 12 miles per hour at top speed.

Further Reading:
Branch, *Field Guide to Snakes and Other Reptiles of Southern Africa.*
Ditmars, *Snakes of the World.*
Mattison, *The Encyclopedia of Snakes.*
Phelps, *Poisonous Snakes.*

Naja spp.

Naja spp. are about 15 species of cobras, one of which, *N. melanoleuca*, inhabits rainforests. Found in most of Africa, the forest cobra reaches nearly nine feet—it is the largest African cobra—and is a deep, glossy black, with a brown head and neck. Its underside is mainly white with black patches. It is alert and swift to retreat when disturbed. Like other cobras, it raises more than two-thirds of its body from the ground when alarmed, so a nine-foot forest cobra might quite suddenly stare a human intruder directly in the eye. If this fails to frighten the perceived threat, the cobra also has a long and narrow hood that it spreads when it rises in defense. The hoods have elongated ribs that are hinged with the vertebrae in the neck. The forest cobra is nocturnal, a strong swimmer, and is often found near rivers and lakes, where it inhabits deserted termite nests, holes, hollow trees, and dense underbrush. Fish and amphibians are its principal prey, but it will also eat lizards.

Further Reading:
Branch, *Field Guide to Snakes and Other Reptiles of Southern Africa.*
Cox, *The Snakes of Thailand and Their Husbandry.*
Isemonger, *Snakes of Africa.*
Phelps, *Poisonous Snakes.*

Ophiophagus hannah (king cobra)

The world's largest venomous snake, the king cobra attains a maximum length of more than 16 feet, although few are longer than 13 feet. The brown or olive snakes prefer wooded areas in India, Southeast Asia, and the Philippines, and they feed almost entirely on other snakes, including other cobras in the *Naja* genus. They also prey on the occasional lizard and amphibian. Strong swimmers, they are common near water. The female lays a large clutch of up to 40 eggs, and both males and females linger near the nest made in leaf litter, guarding the eggs until they hatch. King cobras are relatively rare and particularly aggressive.

Further Reading:
Deoras, *Snakes of India.*
Mattison, *The Encyclopedia of Snakes.*
Phelps, *Poisonous Snakes.*

Oxyuranus scutellatus (taipan)

The taipan is the most dangerous snake in its range, northern Australia and New Guinea. It is large, fast, and highly venomous. Most often found in wet forests, it also inhabits nearly every other type of habitat, including savannas and scrubland. It is about 13 feet long, slim, with a narrow head that's distinct from its body. Its back is light to dark brown; its belly is cream-colored with orange spots. The taipan is mainly diurnal but will hunt at night, particularly during hot weather. It feeds on small mammals, which it bites, releases, then tracks down. It tends to strike its prey with numerous and rapid bites, injecting large amounts of venom with its large fangs. Before an antivenin was developed in 1955, more than 90 percent of people bitten by taipans died.

Further Reading:
Cogger, *Reptiles and Amphibians of Australia.*
O'Shea, *A Guide to the Snakes of Papua New Guinea.*
Phelps, *Poisonous Snakes.*

Tropidopheidae (wood snakes, West Indian boas)

Wood snakes, also known as West Indian boas, include about 20 species, all with prehensile, though rather short, tails.

Tropidophis spp.

The *Tropidophis spp.* range in size from one to three feet in length. Of the 16 species, 13 are found on several West Indian islands in the Caribbean, including 8 species that are endemic to Cuba; the other 3 inhabit South America as far south as the Brazilian Amazon. All are nocturnal hunters that prey on frogs, lizards, and small rodents. Some species have brightly colored tail tips, which are perhaps used as lures. Others force blood from their eyes and mouth when bothered or alarmed, defensive behavior not seen in any other snake species.

Further Reading:
Mattison, *The Encyclopedia of Snakes.*

Viperidae

Viperidae live almost everywhere on Earth except Madagascar and Australia, and in a variety of ecosystems, from forests to deserts. The more than 200 species are generally short and stocky, with broad heads. They are usually colored to blend into their surroundings, and they lie quietly coiled, waiting for unwary prey to pass by. Some species lure prey by wiggling the tip of their tails, which have evolved to look like those of invertebrates. Their most distinguishing characteristic is a pair of short maxillae, or upper jaws, to which are attached long fangs. The fangs rest along the roof of the mouth; the snakes bring them forward to strike by rotating the maxillae. Most prey are bitten and released; the snake follows the trail to devour its victim. This family includes the pit vipers, subfamily *Crotalinae*, which are the only snakes that have a pair of large, heat-sensitive pits between their eyes and nostrils. These infrared organs can register temperature differences as small as two-tenths of one degree. The strong toxins of vipers are dangerous to human beings.

Further Reading:
Coburn, *The Atlas of Snakes of the World.*
Cogger and Zweifel, *Encyclopedia of Reptiles and Amphibians.*
Mattison, *The Encyclopedia of Snakes.*

Bitis gabonica (gaboon viper)

The gaboon viper is a five-foot, nocturnal snake that inhabits forests of central and eastern Africa. Its massive body, which can measure at least three inches in diameter, is patterned in a range of deep and striking earth colors. It bears medium-size horns above its nostrils, but the horns' function is a mystery. The snake's thick body tapers at the tail and neck, so its head appears particularly wide. The gaboon viper is the largest viper in Africa, and probably the world. The fangs of the largest specimens measure some two inches. It feeds on small mammals and birds.

Further Reading:
Coburn, *The Atlas of Snakes of the World.*
Isemonger, *Snakes of Africa.*
Phelps, *Poisonous Snakes.*

Bothriechis schlegelii (eyelash viper)

An arboreal snake, the eyelash viper is found from southern Mexico to Ecuador and Venezuela; it's the most widespread tree viper in Central America. Its common name comes from an enlarged flap that extends outward above its eye. Herpetologists aren't sure what purpose the "eyelash" serves, but some suggest it may protect the snakes' eyes as the viper slithers through vines and twigs. Eyelash vipers feed mainly on frogs and lizards. They come in a range of colors, with variants of green, brown, rust, and blue patterned with darker-colored diamond shapes. A variety that lives in Costa Rica is bright gold with red flecks on the tips of a few scales.

Further Reading:
Janzen, *Costa Rican Natural History.*
Villa, "The Venomous Snakes of Nicaragua."

Bothrops atrox (fer-de-lance)

The fer-de-lance, like all pit vipers, has depressions located between its nostrils and eyes that are heat-sensitive, allowing the snake to locate warmblooded prey. Fer-de-lance are the most feared snakes in their range of Mexico south to Peru. They have triangular heads, are tan with dark brown diamond patterns on their back and yellow bellies, and can grow to seven feet in length. They are pugnacious snakes, and their venom is painful, quite fast-acting, and potentially lethal to human beings. The venom includes enzymes that begin to break down blood and tissues as soon as they are injected. Fer-de-lance are terrestrial, feeding on small mammals and sometimes birds. Like other pit vipers, they do not hunt for prey, but coil on the ground and wait. Since it may be some time before a potential meal passes by, they feed on one large animal rather than several small ones; they can swallow a creature that weighs 150 percent more than they do.

Smaller species of pit vipers or juvenile fer-de-lance strike and hang on to prey, but the adult fer-de-lance withdraws immediately after delivering its venom. The victim may travel for a distance, but soon can't move. The fer-de-lance then follows the trail of its next meal, picking up the scent by flicking its tongue while an organ in its mouth reads the scent trail.

The young tend to be arboreal. They have a yellow tail tip that may attract lizards and frogs. As they grow older, they become terrestrial and feed on mammals and sometimes birds. Fer-de-lance litters can be quite large, with as many as 75 baby snakes.

Further Reading:
Janzen, *Costa Rican Natural History.*
Kricher, *A Neotropical Companion.*

Crotalus durissus (tropical rattlesnake)

The tropical rattlesnake inhabits dry forest or open areas, from southern Mexico to Brazil and Paraguay, but for reasons not understood, not in Panama or northern Brazil. It is the only rattlesnake found in Latin America, and its venom is more powerful than that of its North American relatives.

It reaches nearly six feet in length and has a pair of stripes that extend down its neck, becoming diamonds on the rest of its body. Its tail is usually black, and although it has a fully developed rattle, it seldom brandishes it. Its prey are mainly mammals and large lizards.

Further Reading:
Janzen, *Costa Rican Natural History.*

Lachesis mutus (bushmaster)

At 10 to 12 feet long, the bushmaster is the world's third-largest venomous snake, after the king cobra (*Ophiophagus hannah*) and taipan (*Oxyuranus scutellatus*). Its habitat is forests of Nicaragua, Costa Rica, Panama, and the northern Amazon basin and coastal Brazil, along with Trinidad. A nocturnal and terrestrial snake, it is the only Neotropical pit viper that lays eggs. It feeds on small mammals and birds. It has rough, knobby scales, is tan with a pinkish tinge, and has large, dark brown or black diamond markings. The bushmaster has an unusual tail that looks like a spiny fur cone; it vibrates when the snake is bothered.

Further Reading:
Coborn, *The Atlas of Snakes of the World.*
Phelps, *Poisonous Snakes.*

TESTUDINATA (turtles)

Turtles are ancient animals that evolved into a shelled form more than 200 million years ago. The shell, which is divided into two halves, is the turtle's most notable characteristic. The upper part is called the carapace, and the lower part, the plastron. The two parts are joined on each side by a bridge. All living shelled reptiles are turtles, but some species are called tortoises or terrapins, which have different meanings in different parts of the world. Tortoise usually refers to terrestrial turtles, while terrapin usually means edible, aquatic, and hard-shelled turtles. Turtles are found on all continents except Antarctica. There are 12 families and 257 living species.

Chelidae (Austro-American side-necked turtles)

Austro-American side-necked turtles are semiaquatic side-necked turtles found in South America, Australia, and New Guinea. There are about 36 species.

Chelus fimbriatus (matamata)

Matamata inhabits slow-moving streams, muddy lakes, and wetlands of northern Bolivia, eastern Peru, Ecuador, eastern Colombia, Venezuela, the Guianas, and northern and central Brazil. It has an oblong black or brown carapace, about 17 inches long with almost parallel sides, a straight anterior rim, and a sharply serrated, round posterior rim. It spends most of its time submerged in shallow waters, just deep enough so it can raise its snout to the surface for a breath of air. Its head is triangular in shape, with a long and slender snout.

Along its head and long, muscular neck are loose flaps of skin that drift with any movement of the water, looking just like water vegetation. Scientists suspect that the nerves in the skin flaps also relay information about approaching predators and prey. Unlike most turtles, those in the *Chelidae* family draw their head in sideways, rather than straight back. To escape detection, *C. fimbriatus* can hold its breath for long periods of time and remain motionless, posing like an algae-covered rock. When it does move, it crawls on the muddy bottom rather than swims. The matamata is carnivorous, feeding on aquatic invertebrates and fish. When a fish ventures too close, the turtle opens its mouth wide and sucks in prey and water. It then snaps its mouth shout and expels the water, swallowing the fish whole. It can grow to about 16 inches, and its carapace is black or brown with some orange markings. Prior to mating, a male matamata approaches a female with its head extended, opening and closing its mouth. Females lay between one and two dozen eggs.

Further Reading:
Ernst and Barbour, *Turtles of the World.*
Kirkpatrick, "The Matamata."

Kinosternidae (mud or musk turtles)

Musk turtles are found only in the New World, from Canada to South America. They generally prefer still or slow-moving bodies of water.

Claudius angustatus (narrow-bridged musk turtle)

The narrow-bridged musk turtle lives in lakes, wetlands, and lagoons of eastern Mexico, from Veracruz state through the Yucatán Peninsula, Belize, northeastern Guatemala, and western Honduras. Its oval, dark brown, or yellowish carapace reaches about six inches; its plastron is yellow. Its large head has a sharply hooked upper jaw, with a pair of cusps on the upper jaw and a long hook on the lower jaw.

Further Reading:
Ernst and Barbour, *Turtles of the World.*

Kinosternon scorpioides (scorpion mud turtle)

The scorpion mud turtle is fairly small, reaching only eight inches in length, with a high-domed, oval, brown or black carapace. It has a large, grayish brown head and a hooked upper jaw that makes it resemble a parrot. Its scientific and common name refers to the male's spurlike tail tip, which looks a bit like a scorpion's, and is used to hold females in place during copulation. Its plastron has a hinge, so it can be closed against the carapace to protect limbs and tail. Scorpion mud turtles are found in lowland streams, rivers, lakes, and ponds, from Mexico to Peru, and eastward to the Guianas, and Brazil. They are the most common turtle of the eastern Amazon. They lay several clutches of one to five eggs during a single season and are omnivorous, eating fish, snails, frogs, insects, algae, and other plants.

If their watery home evaporates during the dry season, they bury themselves in the mud and await the next rain. When their swampy homes in the Brazilian Amazon dry up, they must seek refuge in tall grasses. Indigenous people no doubt started the tradition of setting fires to drive the turtles out of hiding, but farmers have continued it. Ground is cleared downwind of where the reptiles are, and the dry grass is set aflame, forcing the turtles to flee to open areas, where they are easily captured. Although it is illegal to capture them, scorpion mud turtles are popular delicacies, sold in markets and in restaurants in Belém and other Brazilian towns.

Further Reading:
Ernst and Barbour, *Turtles of the World.*
Goulding, *The Flooded Forest.*

Staurotypus triporcatus

Staurotypus triporcatus lives in lakes, wetlands, and lagoons of eastern Mexico, from Veracruz state through the Yucatan Peninsula, Belize, northeastern Guatemala, and western Honduras. Its carapace is brown with yellow seams, spots and dark lines. Called "guau" in Mexico, this turtle is fairly large, growing to about one foot in length. Its black head is also large, just over 3 inches wide on a 12-inch-long turtle. Its snout is protruding, and its chin bears two barbels. *S. triporcatus* has a voracious appetite, feeding on copious quantities of small crabs, snails, worms, insects, fish, and amphibians. It even feeds on other, smaller species of mud turtles. The guau's shell is particularly thick, probably as protection against its chief enemy, the Morelet's crocodile (*Crocodylus moreleii*). Its other enemies are human beings, who eat the turtle's meat.

Further Reading:
Ernst and Barbour, *Turtles of the World.*
Pritchard, *Encyclopedia of Turtles.*

Pelomedusidae (Afro-American side-necked turtle)

Afro-American side-necked turtles live in watery habitats in Africa, including Madagascar, and South America.

Pelomedusa subrufa (African helmeted turtle)

The African helmeted turtle has a brown to olive, eight-inch-long, broad, and oval carapace. The top of its head, limbs, and tail are brown to olive, while underneath, it is yellow to cream-colored. An eater of worms, insects, snails, fish, and frogs, the African helmeted turtle ranges from northern Africa south to Ghana, Senegal, Mali, Nigeria, and Cameroon, as well as South Africa and Madagascar.

Further Reading:
Ernst and Barbour, *Turtles of the World.*

Podocnemis expansa (giant Amazon river turtle)

One of seven species in the genus *Podocnemis*, the giant Amazon river turtle has a brown, olive, or black carapace. Their heads are usually yellow, with orange or red markings. *P. expansa* is about three feet long (males are usually smaller than females) and is found in the Orinoco and Amazon River basins in Venezuela, Colombia, Peru, Bolivia, and Brazil. Their front legs have five claws, and their rear legs have four; the back of each limb is covered with rows of enlarged scales. They eat aquatic vegetation and fruit.

Like sea turtles, giant Amazon river turtles migrate to specific beaches each year to lay their eggs in the sand. They nest during the dry season, which is September to November in Brazil and March to April in Venezuela. The female digs a pit about three feet wide and 24 inches deep in the sand. She lays an average of 80 spherical and soft-shelled eggs. The digging, egg-laying, and burial process takes about four hours. The eggs incubate about 45 days, then the hatchlings dig their way to the surface of the nest, usually emerging at night or at dawn. They head directly to the water. Since the nests are so deep, human beings are probably the only egg predators, but that's sufficient—for centuries, people have been digging up millions of eggs annually. As a result, this ancient species is highly endangered. Although it is illegal to hunt the giant river turtle, its meat is prized and quite popular, so poaching continues to be a problem. Urban and industrial development near nesting sites is another threat.

Further Reading:
Goulding, *The Flooded Forest*
Pritchard, *Encyclopedia of Turtles*.

Podocnemis unifilis (yellow-spotted Amazon river turtle)

The yellow-spotted Amazon river turtle is found in ponds, lakes, lagoons, rivers, and flooded forests of northern South America. It is about 20 inches long and has a pale olive or gray domed shell and yellowish-orange spots on its head. It usually bears a barbel, or feeler, on its chin. It feeds on vegetation, fruits, and flowers that fall into the water. Like the giant Amazon river turtle, the survival of the yellow-spotted river turtles is in doubt because people consume both the animal and its eggs.

Further Reading:
Pritchard, *Encyclopedia of Turtles*.

Testudinidae (terrestrial turtles, or tortoises)

Except for one other species of the *Emydidae* family, terrestrial turtles, or tortoises, are the only turtles fully adapted for life on land. They have hard, horny shells, feed mainly on vegetation, and have slightly flattened front legs with very thick scales, and quite thick hind limbs. When they retract, they leave very little of themselves exposed.

Kinixys erosa (serrated hinged-back tortoise)

The serrated hinged-back tortoise has a hinged plastron, which exists in a number of tortoise families. Usually, the hinge allows the plastron to be lowered to protect vulnerable limbs. In the genus *Kinixys*, however, the hinge is in a unique position, at the junction of the second and third back plates, so the carapace can be lowered to protect the 13-inch tortoise's hindquarters and thick, spikelike tail. When threatened, it retracts its head, while its front legs seal the anterior opening in the carapace. Enlarged scales on its forelegs face outward when pulled up, providing a defense, while it clamps down the rear of the carapace and seals itself off from the would-be threat. The carapace of the serrated hinged-back tortoise, which is found in western Africa, is usually a dark rusty brown with pale yellow blotches, and the rear edge is jagged. The head is yellow, marked with brown. Male hinged-backs are territorial and will battle intruders by butting them with their heads.

Other African hinged-backs can be found in savanna regions, but serrated hinged-backs prefer moist forest, particularly wetlands and river banks. Unlike many tortoise species, they can swim. Generally, they hide out beneath vegetation, digging themselves in. They are omnivorous, feeding on vegetation, fallen fruit, fungi, invertebrates, and carrion. The species is threatened because many are illegally exported and sold as pets.

Further Reading:
Pritchard, *Encyclopedia of Turtles*.
Tortoise Trust, USA <http://www.tortoisetrust.org/>.

PART II

Tropical Forest Plants

In the Kingdom of Green

Botanists divide the plant kingdom into "lower plants"—bryophytes (mosses and liverworts), lichens, and algae—and "higher plants," which have vascular tissue and reproduce via spores, cones, or flowers. This section concentrates on the higher plants found in tropical forests. These are divided into three groups, or classes: angiosperms, or flowering plants; gymnosperms, which are mainly conifers and cycads that bear exposed seeds; and pteridophytes, or ferns.

Most of the estimated 12,000 species of ferns are native to the tropics. The greatest diversity of conifers are in the countries of Asia and the Pacific Ocean, while cycads, or tree ferns, belong to an ancient plant group found mostly in Central and South America, South Africa, and from southeast Asia to Australia. There are at least 250,000 species of angiosperms, which, because they first appeared on earth some 135 million years ago, have come to dominate all other land plants in most ecosystems. Nearly two-thirds of all flowering plants are found in tropical countries. Angiosperms also dominate this section of the encyclopedia.

Until late 1999, scientists were perplexed about how flowering plants, the most diverse and important group of plants on Earth, first evolved. Charles Darwin called it the "abominable mystery." By studying plant DNA, scientists were recently able to identify the three oldest groups of flowering plants: water lilies, relatives of the spice star anise, and the oldest group, whose sole survivor today is the aborella, a bush found only on New Caledonia, an island in the South Pacific. But biologists still aren't sure exactly which nonflowering plant produced the first flower.

Further Reading:

World Conservation Monitoring Centre, *Global Biodiversity: Status of the Earth's Living Resources.*

Yoon, "Biologists Find Progenitors of Earth's Flowering Plants."

MEDICINAL PLANTS

Plants have long played an important role in the research and development of pharmaceutical drugs; one in four prescriptions dispensed from drugstores in the United States contains substances derived from angiosperms. About one-third of these plants are native to tropical forests. Estimated sales of plant-derived medications in the United States are about $12 billion annually. The market for herbal medicines is growing quickly: the European Union countries spend about $4 billion each year on herbal medicines.

Plants can't fly or run to escape animal predators or fungal attacks, but they can fight back. Many species protect themselves with thorns or barbs. Others contain toxic oils, latexes, or other chemical compounds. Until the nineteenth century, medicines used by human beings were derived from these active plant ingredients. Plant medicine recipes were passed from generation to generation. In Africa, up to 80 percent of the population go to traditional medicine doctors, who frequently prescribe extracts of local plants. About 35,000 to 70,000 species, more than 1 in 10 of the world's plant species, are used in traditional medicine. Surveys done by the conservation group World Wildlife Fund reveal that nearly all species used for herbal medicines, whether they are used locally or traded internationally, are collected from the wild. In Africa, people are harvesting medicinal plants faster than they can grow to satisfy this demand.

In the nineteenth century, morphine was extracted from the opium poppy (*Papaver somniferum*), the first pure substance derived from plants. Although chemical synthesis has made many drugs of natural origin nearly obsolete, sometimes the synthetic compounds are less effective than the plant-derived chemical, while many effective plant compounds simply can't be synthesized. Development of all plant-derived drugs would not have been possible unless chemists had the natural model from which to manufacture effective synthetic copies. Because less than 1 percent of tropical rainforest plants have been thoroughly researched for their potential useful properties, a number of drug companies and medical research organizations are testing tropical flora, in hopes of finding new and valuable cures. Conservation groups such as the Rainforest Alliance are working with these researchers to ensure that the countries from which the plant samples were taken will get a percentage of profits from any new medicines. Researchers are also paying tropical countries for the privilege of testing their forests' biodiversity, which has increased the value of standing forests. The search for promising new plant-derived drugs is called "bioprospecting."

Conservationists are also working in local communities in the tropics to ensure that medicinal plants are collected sustainably, so species and incomes can be safeguarded.

GOOD TO EAT

About one in every three plant species has buds, fruits, nuts, seeds, roots, leaves, or stems that can safely and deliciously provide nutrition and calories to human beings. Hundreds of plants are now cultivated to feed people on a small or grand scale, but wild plants continue to supplement the diet of millions of people worldwide. In rural northeastern Thailand, for example, plant parts gathered from the forest make up half of all the food eaten by villagers during the rainy months. Fruits of nearly 60 species of rainforest plants are sold in the marketplaces in Iquitos, Peru.

Many of our most common foods grow wild in tropical forests. Today, most are cultivated, but farmers still depend on the gene pool provided by wild seeds. Hybrid crops grown on huge farms are easily overtaken by pests unaffected by agrochemicals; genetic contributions from wild seeds have saved many farmers from billion-dollar disasters. Cultivated sug-

arcane, strawberries, tomatoes, and wine grapes were rescued from devastating losses by genetic material provided by their wild ancestors, which luckily still survived in the world's remaining forests.

INTERNATIONALLY TRADED PLANTS

Valuable plant products include timber, resins, oils, and latexes. Hundreds of plants have value as ornamentals, planted to adorn gardens and yards; scores of the potted plants common in temperate zone offices and homes have their origins in tropical forests. Naturally shade loving, they grow well under dim interior light, but seldom reach the giant sizes they do in their natural habitat. Nearly all ornamentals are grown in farms throughout tropical countries.

MATERIAL GOODS

Plants also provide people with housing, fibers, oils, tools, and resins. About one-quarter of all people worldwide depend on plants for such goods, including firewood.

Further Reading:
Myers, *A Wealth of Wild Species.*
Schultes, *The Healing Forest: Medicinal and Toxic Plants of the Northwest Amazonia.*
Sheldon, Balick, and Laird, *Medicinal Plants: Can Utilization and Conservation Coexist?*
ten Kate and Laird, *The Commercial Use of Biodiversity.*
Tuxill, *Nature's Cornucopia: Our Stake in Plant Diversity.*
World Wide Fund for Nature, Switzerland <http://www.panda.org/resources/factsheets/general/frame.htm?fct_medicinal.htm>.

A SHORT GLOSSARY OF TERMS

Bract—A specialized leaf associated with an inflorescence. Bracts may completely enclose an inflorescence and are often very colorful to help attract pollinators.

Cauliforous—A tree that bears its fruits directly from its trunk.

Deciduous—A tree that drops its leaves at least once a year; in the tropics, usually during the dry season.

Dioecious—Separate trees of the same species produce just male or female flowers, so they must be pollinated with each other in order to produce fruit.

Inflorescence—The arrangement and order of development of flowers on an axis, or stem.

Plants in Part Two are organized first by the three classes, with angiosperms first, then cycads and ferns.

Plants that have medicinal properties are marked with an asterisk (*).

Plants that provide food eaten or sold locally and/or internationally, including crops, are marked with #.

Plants that yield a nonedible product, such as timber or resins, that is traded internationally as well as locally are marked with √.

Plants that provide people with goods (other than medicines) that they can use, sell, and/or trade are marked with ‡.

Plants that are commonly grown and sold as ornamentals are marked with §.

Angiosperms: The Flowering Plants

Most of the plants described in this section are listed alphabetically by name of genus and species, except for a few large and important families: *Arecaceae*, the palm family; *Bromeliaceae*, the bromeliads; *Dipterocarpaceae*, the dipterocarps; *Graminaea*, the family of grasses; *Meliaceae*, the family of mahogany trees; and *Orchidaceae*, the orchid family.

ACACIA SPP.

Acacia spp. is one of the largest genera of trees and shrubs, with nearly 800 species. The genera is divided into two groups. One has simple leaves; most of these species are found in Australia and the islands of the Pacific, including the *koa* tree (*A. koa*) that grows on the mountainsides of Hawaii. The leaves of the second group are fernlike, called bipinnate—their leaves are arranged on opposite sides of a spine, and then subdivided into many tiny leaflets. These species are generally found in Australia, Africa, and the Neotropics, and many have spines, or thorns, at the base of their leaves.

Certain species of acacias with spines in Central America (*A. collinsii* and *A. cornigera*) have an extraordinary relationship with stinging ants (*Pseudomyrmex ferruginea*) that inhabit the hollow thorns. Unlike other acacias, *A. collinsii* and *A. cornigera* do not produce alkaloids to discourage destructive insects. Instead, the stinging ants attack any animal that lights upon or brushes up against the tree. In exchange, the acacias provide the ants with nectar produced at the tips of their leaflets.

* *A. albida*, the Ana tree, is found in much of Africa, where it is widely planted to increase the yields of crops grown under it. A tea made from boiling the bark is a folk remedy for diarrhea in many countries; in the Ivory Coast, it's used to treat leprosy.

√ *A. farnesiana* is probably native to the Americas, but has been introduced throughout the tropics.

A distillation from the flowers is often used in the perfume industry; in fact, the plant is extensively cultivated in southern France, the center of the perfume business. This acacia's gummy sap is considered more suitable for paints than the more traditional gum arabic. The bark, gum, and pods are also popularly used for folk remedies.

Further Reading:
Armstrong, "The Unforgettable Acacias."
Purdue University, USA <http://www.hort.purdue.edu/newcrop/duke_energy/Acacia_auriculiformis.html>.

§ ACALYPHA HISPIDA (chenille)

Chenille is native to southeast Asia and widely grown as an ornamental, thanks to its long, pendulous, red flowers that dangle like a velvety tail of chenille. Each tail is composed of many staminate flowers, but no petals. The plant grows to about 10 feet in height, and its leaves are pointed and have conspicuous veins.

Further Reading:
Lennox and Seddon, *Flowers of the Caribbean.*
Pierot and Hall, *Easy Guide to Tropical Plants.*

* ACHYROCLINE SATUREOIDES (macela)

The macela is a medium-size herb with small white flowers that have yellow centers. It is native to most of Central and South America, where it has traditionally been used as a medicinal plant. The flowers and leaves are prepared as a tea to treat colic, gastric problems, and epilepsy and to help regulate menstrual periods. Scientists have recently been studying its anti-inflammatory and muscle relaxant properties, especially as a treatment for salmonella poisoning. Extracts from macela have also been shown to inhibit the growth of cancer cells and to have antiviral properties against HIV-infected cells.

Further Reading:
Raintree Nutrition, USA <http://rain-tree.com/macela.htm>.

‡ *ADANSONIA DIGITATA* (baobab)

Native to east Africa, the baobab is one of the largest and longest-living trees in the world. Its thick trunk can reach a diameter of 100 feet and a height of nearly 60 feet. The tree can live for thousands of years. The trunk is sometimes hollowed out and used to hold water or provide shelter. The tree has handsome dark green foliage and 6-inch hanging white flowers. The long, oval, edible fruit hangs from a long stalk and holds about 30 seeds. Strong fiber from the baobab trunk is woven into rope and cloth.

Further Reading:
Hargreaves and Hargreaves, *Tropical Trees.*

√ *ADENANTHERA PAVONINA* (bead tree)

The bead tree is a 40-foot tree, whose spreading branches bear fine, feathery foliage high on a smooth, gray trunk. Native to southeast China and India, and now grown throughout Southeast Asia and tropical Africa as well as island nations of the Pacific and Caribbean, the deciduous tree's strong, durable wood is sometimes substituted for sandalwood. The bead tree's flowers are small, star-shaped, fragrant, and creamy yellow, borne on spikes at branch ends. The fruit are curved pods that are used in floral arrangements, while the bright red, hard, round seeds they hold are used to make necklaces and ornaments. In Asia and the Pacific, the high-protein seeds are roasted over a fire and eaten. The tree is also used for firewood and furniture.

Further Reading:
Hargreaves and Hargreaves, *Tropical Trees.*
Winrock International, USA <http://www.winrock.org/forestry/factpub/FACTSH/a_pavonina.html>.

§ *ALLAMANDA CATHARTICA*

Native to Brazil, *Allamanda cathartica* is a climbing vine with soft, yellow 4-inch trumpet-shaped flowers. At the base of each petal are several orangish lines, called honey-guides, which run down to the center of the trumpet. The vine's pointed leaves are shiny and thick, with a waxy covering. Often grown as an ornamental, *Allamanda* bears spike-covered seedpods the size of Ping-Pong balls. When ripe, they burst open and release feathery seeds that are dispersed by the wind.

Further Reading:
Lennox and Seddon, *Flowers of the Caribbean.*
Pierot and Hall, *Easy Guide to Tropical Plants.*

§ *ALPINIA PURPURATA* (red ginger)

Like other members of the *Zingerberaceae*, or ginger family, *Alpinia purpurata* is indigenous to Southeast Asia, although it is widely grown as an ornamental in the tropics and is popular with florists the world over. Red ginger has 8- to 10-inch inflorescences, about 3 inches in diameter, that grow at the end of a stalk that often reaches 10 feet in height. Large, light green leaves grow on either side of the central stem. Small, white flowers are hidden within the overlapping, boatshaped scarlet bracts.

Further Reading:
Lennox and Seddon, *Flowers of the Caribbean.*
Pierot and Hall, *Easy Guide to Tropical Plants.*

* *ALYXIA*

Alyxia is a woody vine of Indonesia and Malaysia, which grows anywhere from 3 to 85 feet long. All of the approximately 55 species bear four linked fruits and a crystalline substance in the bark that smells a bit like fresh-mown hay. It is an important ingredient of local medicines, as it hides the bitter tastes or smells of other ingredients.

Further Reading:
Veevers-Carter, *Riches of the Rain Forest.*

* # *ANACARDIUM OCCIDENTALE* (cashew)

The cashew is in the same family as mangos, pistachios, and poison ivy. The tree is grown in farms throughout the tropics, but is native to Mexico, Central America, and south to Brazil. India is the leading producer of the cashew nut, which is found inside a hard, kidney-shaped and oil-laden fruit. The nut is usually boiled or roasted. Cashew nut oil is caustic and toxic but can—cautiously—be applied to corns, warts, and calluses. Beneath the nut is an apple-shaped flower stalk, which also has a pleasant taste when ripe; its juice is often used to make wine. Unlike Brazil nuts, virtually all the cashews eaten around the world are collected from farmed trees.

Further Reading:
Harvard University Herbaria, USA.

* √ *ANIBA ROSAEODORA* (pau rosa)

Pau rosa has been nearly eliminated from its range in Colombia, Ecuador, Venezuela, French Guiana, Guyana, Suriname, Peru, and Brazil, as the tree exudes fragrant—and valuable—rosewood oil. Although synthetics are now often substituted, natural rosewood oil has long been used in preparation of expensive perfumes and fragrant soaps. At the height of its popularity, Brazil alone exported some 500 tons of rosewood a year. The entire tree and its roots are destroyed in the extraction process, and people have shown little restraint in harvesting and demolishing even quite young trees. Only Brazil continues to produce rosewood oil, but mobile distillation machinery has been moved deeper and deeper into the Amazon, so it's likely the species faces extinction there. The tree, which has reddish bark and produces yellow flowers, is considered threatened in Colombia, Brazil, and Suriname.

The oil is also used in traditional medicine to treat colds, coughs, dermatitis, fevers, and headaches.

Further Reading:
Raintree Nutrition, USA <http://rain-tree.com/rosewood. htm>.
World Conservation Monitoring Centre, England <http:// www.wcmc.org.uk/trees/ani_ros.htm>.

* # *ANNONA MURICATA* (soursop)

Soursop is a small and bushy tree, reaching about 25 feet in height, with large, dark green, fragrant, and glossy leaves and long, yellow-green, conical flowers that may emerge anywhere on the trunk, branches, or twigs. Native to the Neotropics, it produces a large (6 to 9 inches long and weighing an average 5.5 pounds), edible, heart-shaped fruit that has green, reticulated skin from which protrude pliable spines. The segmented flesh is white, fibrous, and juicy. Called "guanábana" in Spanish, the fruit is used to make drinks, yogurt, and ice cream. *A. muricata* is one of at least 60 species of *Annona* and was one of the first to be carried from the New World to the Old.

Further Reading:
Hargreaves and Hargreaves, *Tropical Trees.*
Purdue University, USA <http://www.hort.purdue.edu/ newcrop/morton/soursop.html>.
Sancho and Baraona, *Fruits of the Tropics.*

§ *ANTHURIUM ANDRAEANUM*

Native to Latin America, *Anthurium andraeanum* is now a very popular ornamental plant. In the wild, anthurium grows to about two feet in height. Each flower is a red, heart-shaped, waxy bract, between six and nine inches long. From the center of the bract there extends a long, tubelike structure called a spadix, which is white or pink. The tiny, reproductive flowers cover the spadix.

Further Reading:
Lennox and Seddon, *Flowers of the Caribbean.*

ARECACEAE (palm)

Palm trees are ancient land plants, whose fossils date as far back as the late Cretaceous era, some 85 million years ago. More than two-thirds of the world's 2,800 palm species grow in rainforests. With 1,385, the Neotropics have the greatest number of species; Africa has only 116. Africa's dearth of palms is thought to be the result of desiccation during the Pleistocene era, which greatly diminished moist ecosystems. Most palms are perennials, with a distinctive crown of leaves at their aerial growing end. The leaves are a distinctive shape as well, usually long fronds. The

African raffia palm (*Raphia regalis*), has the longest leaves in the plant kingdom, extending some 80 feet from its trunk.

The large palm family also claims the species with the largest flower: the talipot palm (*Corypha umbraculifera*), whose inflorescence rises more than 30 feet from the top of the plant and bears several million flowers. Another palm, the double coconut (*Lodoicea maldavica*), bears the plant kingdom's largest seed; it can weigh some 5 pounds. Palms range in size from the 5-inch-tall lilliput palm (*Syagrus lilliputiana*) of Paraguay, which is likely extinct in the wild, to the nearly 200-foot-tall wax palms of the Andes (*Ceroxylon quindinense* and *C. alpinum*) and the 165-foot-tall wanga palm of New Guinea (*Pigafetta filaris*).

Palms are of huge economic importance, providing food, shelter, fuel, building materials, oils, wines, and waxes. Although palms themselves are not usually logged for the international timber trade, they are often killed when other trees are cut down, as most cannot survive in the drastically altered and drier ecosystem that remains after logging. Other species are destroyed to harvest heart of palm, the tender delicacy in the center of the stem, or for their leaves, used as roof thatch or to feed livestock. At least 100 species of palm are endangered.

‡ *Arenga pinnata* (sugar palm)

The sugar palm has flower buds that yield a sweet sap that is extracted, dried, and used as a sugar. Originally found in India, Bangladesh, and Southeast Asia, it is now cultivated throughout tropical Asia. The sugar palm's trunk is thick, black, and fibrous, its pinnate leaves are long and upright, and its large inflorescences produce great clusters of fruit. The leaves provide a moisture-resistant fiber used to weave baskets and fishnets.

* # *Bactris gasipaes* (peach palm, or pejibaye)

The peach palm, or pejibaye, ranges from Nicaragua to Bolivia, but only in Costa Rica is the palm fruit widely eaten. When ripe, the fruit is peachy yellow. The pejibayes, which grow in dangling clusters and are rich in calories and vitamin A, are boiled in saltwater, peeled, and served cold with a bit of mayonnaise. The starchy pulp tastes something like a fibrous sweet potato. The peach palm trunk is lined with spines and prominently ringed; the palm leaves also have spines. In addition to people, birds are also fond of the fruit. The profusion of spines discourages mammals from taking them from the tree, but fruit that falls to the ground is gobbled up by agoutis (*Dasyprocta spp.*) and other rodents.

Many rural families grow *B. gasipaes* on their farms, both for the fruit and palm heart, which is sold locally and exported. Wild peach palms were the original source of palm heart, or *palmito*, and because the entire tree must be sacrificed to yield only a bit of palm heart, some palm species were cut into near extinction. Peach palm wood is quite hard and often used to make tool handles. Researchers at the Veterans' Administration Hospital in Minneapolis are testing the effectiveness of an extract made from the peach palm fruit against symptoms of benign prostatic hypertrophy, an enlargement of the prostate gland. Results are preliminary, but the extract apparently reduces symptoms without the side effects of current treatments.

‡ *Borassus flabellifer* (palmyra)

Native to India and Malaysia, the palmyra is usually found near the sea or coastal valleys. Its hard, black trunk is often cut for timber and tapped for the sweet sap. Its wide, fan-shaped, blue-green leaves are used as paper and to make brooms and mats. The large, black fruit contains one to three seeds that are surrounded by a layer of sweet and juicy fibrous flesh.

√ *Calamus, Daemonorops, Korthalsia spp.* (rattan)

Rattans include about 500 species of climbing palms, found mainly in India and Southeast Asia, with a few species native to Africa. The palms have long and slender stems, ranging from one-half to almost 3 inches in diameter and from 3 inches to 600 feet in length. The stems maintain the same diameter their entire length. The outer part of the stem is quite hard and durable, yet pliant, while the inside is much softer and somewhat porous. The trees' stems are used globally to make cane furniture and locally to make baskets, mats, and fish traps. The fruits of some rattan species are edible and sold in local markets. Rattans grow with the help of a long appendage lined with thorns, which firmly anchor the palm in place wherever the appendage lands.

‡ *Korthalsia spp.*

Korthalsia spp. are found in the Malay Islands to the Philippines and New Guinea. The 26 species are spiny, climbing palms that are locally called Rotang. Their long, strong, and flexible stems are widely used as rope. The leaves of *Korthalsia* palms do not form a crown, but grow along the stems. Some species have hollow extensions beyond the leaf stalk that harbor ants. In exchange for shelter, the ants defend the plant against attackers.

§ *Caryota urens* (fishtail, toddy)

The fishtail, or toddy, is a popular ornamental palm, due to its double pinnate leaves that resemble a salmon's tail. Native to India, Sri Lanka, and Southeast Asia, it has a gray trunk that grows to about 25 feet and is covered by evenly spaced leaf scar rings, marks that are left when old leaves fall from the trunk. When it reaches its maximum height, the palm produces flowers at the top of the trunk and subsequent flowers bloom lower and lower. When the lowest flower blooms, the palm dies. The red palm fruits have a high acid content that is toxic and can burn the skin on contact. The seed, however, is edible. Fishtail palm sap is high in sugars, so in India and other Asian countries, it is tapped, made into a syrup, and fermented into an alcoholic drink called a toddy. The syrup is also processed into a sugar called jaggery.

Ceroxylon spp.

Ceroxylon spp. are among the tallest of palms, reaching some 195 feet. The 17 species are found in Colombia, Ecuador, Peru, and Bolivia, in moist forest slopes and high valleys in the Andes. Their smooth and distinctly ringed trunks have a whitish, waxy covering.

§ *Chamaedorea spp.*

Chamaedorea spp. range from Mexico to Bolivia and Brazil. In Guatemala and Belize, the attractive palm fronds, called xate, have become an important export product. People gather the fronds from the forest and sell them to the international floral industry, which uses them in arrangements. *C. tepejilote*, found from southern Mexico to northern Colombia, is one of the most common of the palm species. The palms bear graceful, pinnate leaves and small black fruit. "Te-pejilote" means "mountain maize" in the Nahuatl indigenous language, as the palm's unopened inflorescence looks like an ear of corn. *C. tepejilote*, also called the pacaya, is the only palm that yields an edible inflorescence that is eaten traditionally and sold commercially. Inside the bracts is a white, edible fiber that tastes a bit like asparagus or palm heart. The usual way to prepare them is to dip them in egg batter and fry them, or chop them up for salads. Pacaya is also used to feed domestic livestock.

√ ‡ *Cocos nucifera* (coconut)

Coconuts are native to Malaysia but the durable, floating seed has allowed the palm to colonize most tropical coastlines. The palms have also been widely cultivated, so they are "native" throughout the tropics. The tree can grow nearly 100 feet tall and has feather-shaped, long leaves. Annually, coconut palms produce 40 to 100 nuts, which are about 10 inches in diameter. The hard nut is protected by a fibrous husk and contains a sweet and refreshing liquid. The nut's white meat is delicious. Palm wine and vinegar are made from the sap, baskets and mats woven from the leaves, and the coconut palm's trunk is often used for timber. Oil pressed from the nut's dried kernel is widely used in cooking and baking, and in a variety of household, cosmetic, and industrial products. Coconuts are sold in markets throughout the tropics— this one tree sustains many thousands of people.

√ *Copernicia prunifera* (carnauba wax palm)

Native to Brazil, though widely grown in tropical South America in plantations, the carnauba wax palm has leaves that yield a wax that is traded internationally and used in polishes, varnishes, and candles.

√ *Elaeis guineensis* (African oil palm)

The African oil palm originated in western Africa, but is now widely grown in plantations for the high-quality oil that can be extracted from the fruit pulp and kernel of the palm's seeds. Only the soybean is a more important edible oil source. The palm's trunk is thick and solid, and the large crown has shiny, wide, green fronds. Commercially, the oil is extracted by steam and pressure. Once refined, it is used for cooking; in margarine and baked goods; and in soaps, candles, and plastics. In many tropical countries, the number of oil-palm plantations is growing and actually displacing natural rainforest. Oil-palm companies have converted some 250,000 acres of native forests to oil-palm plantation in northern Ecuador, in an area called the Choc.

Euterpe edulis (assai palm)

Assai palms include three species that are found throughout northern and eastern South America. The tree produces a delicious fruit and is the leading palm whose heart, in the center of its trunk, is used for the palm-heart industry in Latin America. Assai palms have multiple trunks and sprout readily, so it's possible to harvest the heart without destroying the tree. These palms are tall and thin, with crowns of long, outreaching pinnate leaves that have drooping leaflets. They produce clusters of purple fruit, whose pulp is pounded and pressed, and the extracted purple liquid is mixed with cassava flour and dried into a candy called *vinho* in Brazil. Assai palms require a good deal of water and are commonly found in the Amazon River's flooded forest.

Guilielma insignis (chonta)

Chonta is a tall, slender palm with a crown of pinnate fronds and rings of short spines on its trunk. Found along large rivers in Brazil, its yellow, plum-sized fruit is sweet and widely harvested.

Lodoicea maldivica (double coconut)

The double coconut bears the largest seed in the plant kingdom. The slow-growing palm is endemic to only a few islands of the Seychelles, although it was once thought to come from the Maldives Islands, an error preserved in the plant's scientific name. The base of the tree's trunk fits into a round bowl; the tree's roots grow through holes in the bowl. Scientists aren't sure what the bowl's purpose is, but it may serve as a support, holding the tree upright during high-wind storms. The bowl is quite rot-resistant, as is the tree itself. The palm takes a year to germinate, takes another year to form its first leaf, and grows 30 to 60 years before flowering. The tree can grow to about 100 feet high, with leaf blades that are 20 feet long and 12 feet in diameter. Its fruit may weigh 45 pounds. *L. maldivica*'s common name comes from the appearance of these hefty coconuts, which, when sliced crossways, look like two coconuts joined at the middle.

‡ *Mauritia flexuosa*

Mauritia flexuosa palms are found in the Amazon and northern South America and produce a tasty fruit pulp, called aguajé, that is used to make sweets and drinks. The large kernel of the nut contains a good deal of useful oil. But because the trunk is extremely hard and slippery, the only way to gather the fruit is by cutting down the tree. Not surprisingly, the tree is rapidly disappearing. Some 15 metric tons of aguajé fruit arrives daily in the market in Iquito, Brazil, and hundreds of people have jobs making foods from aguajé. Fibers from the palm's young leaves are used to make rope, hammocks, and baskets. In the rainforest, the fruit is an important part of the diets of monkeys, rodents, tapir, and peccaries.

‡ *Metroxylon sagus* (sago palm)

Sago palms grow in Indonesia, Malaysia, the Philippines, and New Guinea, usually in dense stands in swamps. Its tall heavy trunk holds large quantities of an edible starch, called sago, which is used as a food staple. To obtain the starch, the entire trunk is cut just before flowering. Young shoots quickly sprout from the old trunk. The palm's pinnate leaves are used for thatching.

* # ‡ *Nypa fruticans* (nipa palm)

Found from India to Sri Lanka, the Philippines, and several other Pacific islands, the nipa palm usually grows along rivers and in mangrove forests. It has an underground stem, which branches above ground to form new plants. The palm's pinnate leaves arise from a stemless rosette. The inflorescences are quite close to the ground, so it is easy to tap them for their sweet sap. Because the trunk is underground, the palms regrow quickly if damaged during storms. Nipa palm seeds can float and even germinate in the water and will quickly take root if they wash up on muddy banks. The young seeds are eaten raw or preserved in a syrup. The mature seed coat is used as a vegetable ivory, often carved into buttons. The fronds, which have air-filled cavities that keep them upright, are widely used in weaving household goods and for thatching. Nipa palm shoots, bark, roots, and leaves are used to treat herpes, toothaches, and headaches, while the tree's sap is used to make vinegar and a popular fermented drink.

 # √ ‡ *Orbignya barbosiana* (babassu palm)

The babassu palm grows in Brazil and is the third most important oil palm in the world, after the African oil palm and coconut. The tree has a tall, strong, and thick trunk, long and pinnate leaves, and large inflorescences that produce heavy clusters of fruits rich in a yellow oil, which is an export product. The leaves are used in weaving, the trunk to make cellulose and paper pulp, and the fruit pulp ground into a flour. Wild stands are disappearing, so the babassu is beginning to be cultivated.

 # √ ‡ *Phytelephas aequatorialis* (tagua nut palm)

The tagua nut palm ranges from Panama to Peru. It produces hard, white seeds that were once the basis of a large button-making industry, particularly in Ecuador. Tagua nuts are a creamy white, resembling ivory; *Phytelephas* is Latin for "elephant plant." Another ivory-nut palm, *Metroxylon amicorum*, grows in the Micronesian island of Caroline, while *Hyphaene ventricosa* grows along the Zambezi River in Zambia. Tagua nut palms are dioecious; female palms can produce up to 50 pounds of large, brown, melon-sized fruits annually. Locals eat the sweet pulp of the immature seeds. Buttons and crafts made of carved tagua ivory still provide local people, particularly in Ecuador, with an important source of income, which encourages them to conserve the palms in their natural habitat.

‡ § *Roystonea regia* (royal palm)

The national tree of Cuba, the royal palm grows to about 70 feet and has a smooth, light gray trunk. With its graceful, round crown and oblong fruit, the royal palm is often grown as an ornamental, planted in rows. In Cuba the palms are cut for timber, and the fruits are used as pig feed. *Roystonea regia* is named for Gen. Roy Stone, a U.S. engineer.

‡ *Sabal spp.*

Sabal palms can live for more than 100 years, reaching nearly 80 feet and surviving hurricanes and poor soils in their habitat in North Carolina, Texas, Florida, Mexico, Guatemala, Venezuela, and the West Indies. The sturdy trunks grow to about 3 feet in diameter. Sabals are topped by 15 to 30 fronds, which stick out like punk hairdos, each frond more than 6 feet long. Birds, lizards, and bats seek shelter and build nests in the palm crowns. The palms bear small, black fruit that is eaten by resident birds along with migrant birds from North America that head south during the cold winter months. When the palms finally die, wildlife use their hollows as nesting sites. The palm fronds are also important to local residents, who use them as roofing thatch—about 1000 leaves are needed to thatch one medium-size house. A growing demand for sabal palm fronds is jeopardizing the many wildlife species that also depend on them.

 # *Salacca spp.*

Salacca include 18 species of palm native to China, Myanmar, Sumatra, Thailand, Malaysia, Java, Borneo, and the Philippines. They all have short trunks, while the crowns have long, arched leaves that crowd at the base. The leaf stalks are long and quite spiny. *S. zalacca*, found on the islands of Java and Sumatra, has a pear-shaped, purplish fruit, with an edible layer of pale yellow or pink flesh surrounding the leaves. The fruit is sold in markets in parts of Asia. *S. zalacca*'s trunk reaches only about 20 feet in height and is topped by a crown of long, erect fronds. Sharp spines cover most of this dioecious palm, so the harvesting of the fruit is a complicated matter. The palms are now grown on plantations.

§ *Verschaffeltia splendida* (Seychelles stilt palm)

The Seychelles stilt palm is a slender and feathery palm of the Seychelles, where it grows on forested hillsides. Sharp spines emerge from the rings on its tall (up to 80 feet) trunk. Black stilt roots support the tree, emerging from the trunk about 3 feet above the ground, and growing outward and downward into the earth, forming a conical cluster. The palm's crown is compact with broad, bright green, pleated leaves that curve downward. The inflorescences are about 2 to 3 feet long, each enclosed by thin bracts. There are separate, yellow, male and female flowers on the same plant. This attractive palm is grown throughout the tropics as an ornamental.

Further Reading:
Balick and Beck, *Useful Palms of the World.*
Blombery and Rodd, *An Informative, Practical Guide to Palms of the World.*
Haynes, *Virtual Palm Encyclopedia.*

Janzen, *Costa Rican Natural History.*
Jones, *Palms Throughout the World.*
Tuxill, *Nature's Cornucopia.*
Veevers-Carter, *Riches of the Rainforest.*

* √ *AQUILARIA MALACCENSIS*

Aquilaria malaccensis trees are found in the Malaysian peninsula, where they are rapidly disappearing due to demand for their fragrant and valuable heartwood, called gaharu. This resinous wood is highly prized as an ingredient in perfume, and is burned as incense in Shinto Buddhist temples in Japan. The resin can be obtained only by cutting down *Aquilaria* trees, and the oldest trees contain the most valuable resin. A kilogram of gaharu may fetch five or even ten thousand dollars in the Middle East. The incense supposedly has therapeutic properties and the resins, leaves, or roots are used to soothe rheumatism, smallpox, stomachaches, and illnesses during and after childbirth. Not every tree contains the resins, nor is it easy to tell which trees have gaharu and which don't, so many are felled needlessly.

Further Reading:
Forest Research Institute, Malaysia <http://www.frim.gov. my/tu/Aquilaria.html>.

ARTOCARPUS spp.

Artocarpus spp. includes two globally important food species, the jackfruit and the breadfruit. Both probably originated in tropical Asia, but are now grown throughout much of the tropics. The trees produce nutritious fruits all year, so they are an important food source to millions of people, especially on the islands of the South Pacific. There are about 50 species of *Artocarpus*, and about half are cauliflorous, bearing their large fruits directly from their trunks or larger branches. The trees have miniscule flowers, but some species produce hefty fruits, about the size of a small melon. The fruit has a rough rind and a white, mealy pulp that becomes soft and slightly sweet when baked. The pulp is also dried and ground into a flour. The tree reaches about 50 feet and the leaves are long, leathery, shiny, deeply lobed, and, in some species, change color slightly and drop from the trees periodically. When cut, the trees exude a sticky white sap, similar to the latex of the rubber tree, which in some species is poisonous. *Artocarpus* lumber is also popular, as it darkens after cutting to a deep mahogany shade and is resistant to insects.

Further Reading:
Hargreaves and Hargreaves, *Tropical Trees.*
Veevers-Carter, *Riches of the Rain Forest.*

* # *AVERRHOA BILIMBI* (mimbro, cucumber)

The mimbro, or cucumber, is a tree native to Indonesia and Malaysia, but is now common throughout the tropics. Its short trunk reaches from 16 to 33 feet and divides into a number of upright branches. The long, oblong leaves have a pointed tip, and the small, fragrant flowers are yellowish-green or purple and emerge directly from the trunk. The fruit is 1.5 to 4 inches long and cylindrical, and the skin turns from bright green to ivory when ripe. The flesh inside is green, juicy, and quite acidic. The fruit is commonly added to curries in the Far East and used in place of mango for chutneys; the fruit juice is a popular drink, tasting something like lemonade. Mimbro leaves are mashed up and used on rashes or to treat rheumatism. In Malaysia, a tea made from boiling the leaves is drunk as a remedy for venereal disease. In Java, a fruit conserve or syrup is used to treat coughs and fevers.

Further Reading:
Purdue University, USA <http://www.hort.purdue.edu/ newcrop/morton/bilimbi.html>.
Sancho and Baraona, *Fruits of the Tropics.*

AVERRHOA CARAMBOLA (carambola, star fruit)

Native to Sri Lanka and the Moluccas, carambola, or star fruit, is a slow-growing tree that reaches about 20 feet in height. Its spirally arranged leaflets are sensitive to light and fold inward at night, and its red and white flowers blossom on bare branches or at leaf bases. The carambola fruit is yellow, oblong, and has five deep ribs, so when you cut it in half, it looks like a five-pointed star. The fruit is sweet and watery and is eaten raw or in jams, pies, drinks, and curries. It's grown throughout the tropics. Carambola flowers and fruits several times a year.

Further Reading:
Sancho and Baraona, *Fruits of the Tropics.*
University of Connecticut, Ecology and Evolutionary Biol-

ogy Conservatory, USA <http://florawww.eeb.uconn. edu/acc_num/198500347. html>.

* # √ *AZADIRACHTA INDICA* (neem)

The neem is a medium-size tree with dark brown to gray bark, a dense rounded crown, and pinnate leaves. It is native to Myanmar and India, where it has long been recognized for its myriad of benefits that are only now being investigated more widely. Its natural habitat is dry, evergreen forests, and it can grow in poor soils and harsh climates, since it has an extensive and deep root system. The tree is fast-growing, able to reach 20 feet in only three years; it can eventually grow as tall as 50 feet. Neem's white flowers bloom in clusters and have a sweet, jasmine-like scent. The seeds of the olive-size, golden, and edible fruit are about 45 percent oil, which can be used as a natural insect repellent. Azadirachtin is one of the many compounds found in neem, and is now used as a natural and safe pesticide. Since it is termite resistant, the timber is used for fenceposts, in construction, and for furniture. In the culture of India, the neem tree has always played an important role, providing people with shelter, food, fuel, and medicine.

Planting neem was considered a sacred duty, encouraged by the Hindu religion; Hindu belief holds that anyone who plants three neem trees will enjoy life after death and never go to hell. Various parts of the tree are used to treat cough, fever, worm infestation, boils, jaundice, leprosy, and stomach ulcers. Neem is now widely planted in Southeast Asia and in Africa to help halt desertification, or the creeping spread of the Sahara desert.

Further Reading:
Neem Foundation, India.

* *BANISTERIOPSIS CAAPI* (ayahuasca)

A large woody vine, ayahuasca grows throughout the Amazon forest and is used by scores of indigenous tribes as a hallucinogen. Although the vine is known by many names, ayahuasca is the most common; in the Quechua language, "aya" means "spirit" or "ancestor" and "huasca" means "vine" or "rope." The potion has been taken for centuries: An ayahuasca ceremonial cup made of engraved stone and dating from 500 B.C. to A.D. 50 was found in the Ecua-

doran Amazon. Potions are prepared by pounding on the vine's stems, then steeping them in cold or hot water—recipes vary from tribe to tribe. The plant material is strained off and the potion drunk. Chemicals that occur naturally in the vine cause the drinker to be visited by fanciful visions, and it is believed that in this state, they have the ability to speak with spirits, predict the future, diagnose illnesses, and treat diseases. Shamans, or traditional medicine men, drink ayahuasca potions to help them better diagnose and treat patients.

In 1986, the U.S. Patent and Trademark Office (PTO) gave a patent to a U.S. businessman for 17 years of exclusive use of *B. caapi*. When indigenous and indigenous-rights groups learned of the patent in 1999, they formally petitioned the PTO, asking that the patent be canceled. The PTO finally agreed, noting that published descriptions of *B. caapi* were "known and available" prior to the filing of the patent application. According to U.S. patent law, no invention can be patented if described in printed publications more than one year before the date of the patent application. PTO regulations recognize only knowledge that is published, disregarding ancient wisdom that has been passed verbally from generation to generation, so many traditional, indigenous, medicinal-cure recipes could actually be patented in the United States.

Further Reading:
Davies, "Patent Dispute Germinates Around Plant Used in Amazon Rituals."
International Biopark Foundation, USA <http://www. biopark.org/ayahuasca.html>.

BARRINGTONIA ASIATICA

Barringtonia Asiatica is a tree native to tropical Asia that is usually found along the coast. The oval leaves form at the ends of the tree's branches, where the fragrant five- to seven-inch flowers cluster. They open in the evening, revealing brushlike white stamens tipped with deep pink, which fall to the ground by morning light. The seeds of the barringtonia's fruits contain a substance that is toxic to fish. Rural people sometimes grate and toss the seeds into streams in order to quickly harvest fish for dinner or the market. These heart-shaped fruits are buoyant and easily carried by the sea to other shorelines. A waxy coating protects the seeds from saltwater damage.

Further Reading:
Hargreaves and Hargreaves, *Tropical Trees*.

§ *BELOPERONE GUTTATA* (shrimp plant)

The shrimp plant is native to Mexico. A small shrub that can reach 6 feet, it has rough leaves that are hairy on both sides. The inflorescences are borne on curved spikes, with brick-red, overlapping bracts that grow in curved rows and resemble boiled shrimp tails. The flowers that hang from the bracts are small, white, and tubular.

Further Reading:
Armitano, *Garden Plants of the Tropics*.

* # √ *BERTHOLLETIA EXCELSA* (Brazil nut tree)

A rainforest giant, the Brazil nut tree grows to heights of 160 feet or more, with leathery leaves that can measure 15 inches long by 6 inches wide. In addition to Brazil, it's found in Colombia, Venezuela, Ecuador, and Peru. The baseball-sized fruits can weigh up to 5 pounds. They grow at the ends of thick branches, ripening and then falling from the trees in January and February. The 10 to 21 nuts inside are arranged like tangerine segments. The tree is dependent on agoutis for dispersal, as these large rodents are the only animal with teeth sharp and strong enough to gnaw open the outer shell and yank out the nuts, which, like acorn-burying squirrels, they tote off to widely scattered hoards. They bury the seeds for later consumption, but those seeds they forget will sprout. Each year, a Brazil nut tree may produce as many as 300 fruits.

Brazil exports thousands of tons of these tasty nuts annually, all gathered from the rainforest, since the trees grow quite slowly and are difficult to cultivate. Brazil nuts are a source of food and income for hundreds of thousands of Amazon residents who work in this industry, which generated more than $65 million worldwide in 1998.

About 70 percent of the nut is oil, which burns like a candle when lit. The oils are increasingly popular in natural beauty products, especially hair conditioners, soaps, and skin creams. In the Amazon of Brazil, the tree's bark is brewed as a tea to treat liver ailments. Brazil nuts are a valuable source of fat and protein for many people of the Amazon, both in the forest and in rural villages. The oils are used to make soaps and insect repellants. The empty seedpods have many practical uses: from collecting latex from tapped rubber trees to serving as drinking cups.

Further Reading:
Missouri Botanical Gardens, USA <http://www.mobot.org./MOBOT/education/feast/nuts.html>.
Taylor, "Strange Bedfellows: The Agouti's Nutty Friend."
Wolfe and Prance, *Rainforests of the World*.

* # √ *BIXA ORELLANA* (achiote)

Achiote is a large shrub—or small tree—that reaches about 20 feet in height and grows throughout the Neotropics. In Brazil, the tree is called annatto. Achiote produces large pink or white flowers that open before dawn and draw swarms of bees. The flowers wilt by midday and fall by dusk, but during their short lifetimes they produce copious amounts of pollen. Achiote has a spiny, bright red, heart-shaped fruit, whose seed contains an oily, yellowish-orange dye, used for centuries throughout Latin America as a body paint, fabric dye, and food coloring. The dye is commonly used in *arroz con pollo*, chicken with rice. One achiote tree can produce 600 pounds of seeds. A paste made from the seeds is exported to North America and Europe and used as a food coloring for margarine, cheeses, and lipsticks. In Brazil, annatto is used to treat heartburn and stomach pains.

Further Reading:
Janzen, *Costa Rican Natural History*.
Raintree Nutrition, USA <http://www.rain-tree.com/annato.htm>.
Schultes and Raffauf, *The Healing Forest*.

* # *BLIGHIA SAPIDA* (akee, or vegetable brain)

Akee, or vegetable brain, is originally from western Africa. Its scientific name comes from Captain Bligh of *Mutiny on the Bounty* fame, who introduced the species to the Americas, where it was used to feed slaves from Africa. He also introduced akee to England. The tree grows to about 40 feet in height, usually with a short trunk and a dense crown of spreading branches. Its bark is smooth and gray and its flowers, white, fragrant, and borne on racemes three to seven inches long. When the akee's leathery, deep-red fruit is ripe, its 3.5-inch, 3-sectioned pods burst open. The flesh inside is toxic unless the pods open naturally. Usually the fruit is parboiled in salt water or milk, then used in stews or fried lightly in butter. In the West Indies, "akee" means "attractive to bees." The fruit's other common name refers to the appearance of the yellowish flesh.

In western Africa, the fruit's seeds are used as a fish poison, and a mixture of the pulverized bark and ground hot peppers is rubbed on the body as a stimulant or to relieve pain. New leaves are crushed and applied to the forehead to treat a headache, and juice squeezed from the leaves is used as eyedrops to cure conjunctivitis. In Brazil, doses of an extract of the fruit's seeds are used to kill parasites.

Further Reading:
Hargreaves and Hargreaves, *Tropical Trees*.
Purdue University, USA <http://newcrop.hort.purdue.edu/newcrop/morton/akee.html>.
Sancho and Baraona, *Fruits of the Tropics*.

§ *BOUGAINVILLEA SPP.*

Originally from Brazil, *Bougainvillea spp.* is now widely grown as an ornamental vine throughout the tropics. The small, shiny leaves are triangular, and the woody stem is covered in thorns that help the plant climb. The bougainvillea's flowers are small and pale, but the bracts are large, papery and brightly hued, usually in shades of purple and red. The plant is named for a French navigator.

Further Reading:
Lennox and Seddon, *Flowers of the Caribbean*.

§ *BROMELIACEAE* (bromeliads)

A family of plants with 2,000 species, bromeliads are found in the Neotropics, except for one species that is native to western Africa—it must have found its way there through some strange dispersal accident. Many are epiphytes, or plants that live on other plants. They are not parasites, but take moisture and nutrients from the surrounding air, rather than from the soil via a root system. Hundreds may festoon the branches of tropical trees, sometimes causing a branch to break under their weight. Some bromeliad species do root in the ground, while others form water tanks with their tightly overlapping leaf rosettes so they can catch rainfall, like a bucket left outdoors. The leaves inside the rosettes have umbrella-shaped

scales that open to collect water but close during rainless periods to avoid water loss. Leaves and debris that fall in the water tanks foster growth of algae and single-celled animals that feed mosquitoes, insect larvae, and other wildlife—the bromeliad water tanks become a mini-ecosystem all their own, plus provide the host plant with water and nutrients during dry seasons. Some bromeliads can hold as much as two gallons of water. Tree frogs, flatworms, snails, mosquitoes, salamanders, and tiny crabs may spend their entire lives inside bromeliad tanks.

Bromeliad flowers grow on a spike-like stem from the plant's center. They are usually flaming red to attract hummingbirds, a principal pollinator.

Ananas spp. are found in Latin America, principally in South America. Although some wild species with small fruits grow in the forest, the domesticated pineapple, *A. comosus*, is not found in the wild. Its ancestors probably originated in the Amazon. By the time Columbus arrived, it was already widely cultivated throughout the Americas, and he brought it back to the Old World. Now the plant is grown in huge plantations in the tropics, with Hawaii and Malaysia among the principal producing areas.

Tillandsia spp., found throughout forests of Central America, have narrow, tapering leaves that are bushy or spiky. They have just a few flimsy and wiry roots with which they cling to their host branch, and collect water and nutrients through the tiny silver scales that cover the plant. The scales allow moisture to enter the leaves but prevent its escape. Like other bromeliads, the bright red, orange, purple, or blue flowers are borne on stiff inflorescences that rise from the center of the plant.

The inflorescence of *T. fasciculata*, which grows in dry tropical forests and has narrow, spiky gray-green leaves, is scarlet on the bottom half and sunny yellow at the top.

Several *Tillandsia* species provide shelter to ants. The rosette of leaves in these species is closed near the top, creating a dry, hollow space the ants can colonize.

Further Reading:
Janzen, *Costa Rican Natural History*.
Kricher, *A Neotropical Companion*.
Wolfe and Prance, *Rainforests of the World*.

BROSIMUM UTILE (cow tree)

The cow tree attains a height of nearly 100 feet and gets its common name from the white latex it produces, which is actually drinkable, if not necessarily very tasty. The pulp surrounding the tree's small, round fruit is sweetish and edible. Indigenous people once used the bark of the cow tree for blankets by soaking pieces in water for leaching, then drying and beating the bark to soften it.

Further Reading:
Janzen, *Costa Rican Natural History.*
United States Department of Agriculture Forest Products Laboratory <http://www2.fpl.fs.fed.us/TechSheets/ Chudnoff/TropAmerican/html%20files/brosim3new. html>.

§ *BRUGMANSIA SUAVEOLENS* (angel's trumpet)

A shrub native to Latin America, angel's trumpet is a popular ornamental because its striking blooms are 8 to 12 inches long, white, and trumpet-shaped. The plant itself reaches about 15 feet, and its bluish green leaves are large and fuzzy. In the evenings, an angel's trumpet plant covered with dangling flowers emits a strong and heady perfume. In spite of its divine name, the leaves and flowers are toxic. Amazonian indigenous people drink a brew made from the leaves as a hallucinogen.

Further Reading:
Lennox and Seddon, *Flowers of the Caribbean.*
Pierot and Hall, *Easy Guide to Tropical Plants.*

* ‡ *BURSERA SIMARUBA, B. GRAVEOLENS* (gumbo limbo)

Gumbo limbo trees are native to Haiti, Central America, and Brazil, and are known for their unusual bark, which is red and shaggy and peels off of the tree in papery thin strips. The tree grows to a height of about 80 feet and is called "Indio desnudo," or "naked Indian," in Spanish. They are dioecious and produce small flowers that are yellow-green to white, growing in clusters at the end of the branches. The flowers last only one day but produce a good supply of nectar, which draws stingless bees, flies, ants, and beetles. The fruits need a good eight months before they are ripe and are favorite foods of white-faced and spider monkeys. The unusual bark of gumbo limbo trees is a common remedy for skin afflictions, like sores, insect bites, rashes, and sunburn. Taken as a brewed tea, it is supposed to cure urinary tract infections, soothe colds, break fevers, and purify blood.

The tree yields an aromatic resin that is used as an incense and in varnish.

Further Reading:
Janzen, *Costa Rican Natural History.*
Raintree Nutrition, USA <http://www.rain-tree.com/ gumbo.htm>.
United States Department of Agriculture Forest Products Laboratory <http://www2.fpl.fs.fed.us/TechSheets/ Chudnoff/TropAmerican/htmlDocs%20tropamerican/ Burserasimaruba.html>.

§ *CAESALPINIA PULCHERRIMA* (dwarf poinciana, red bird-of-paradise)

The dwarf poinciana, or red bird-of-paradise, is a small tree that reaches about 12 feet in height. Native to the Neotropics, it is grown widely as an ornamental. The tree flowers nearly year-round; the vermilion, yellow-edged blossoms cluster at the tips of the branches and have five petals. The red stamens and pistils are about six inches long and extend well beyond the petals. The dwarf poinciana's nectar is a favorite with hummingbirds. The fernlike, bipinnate leaves are divided into numerous, paired leaflets, about three-fourths of an inch long, and the branches bear soft prickles. The tree's leathery, beanlike seedpods hang from stems and, once dry, burst open to project seeds, which are toxic.

Further Reading:
Lennox and Seddon, *Flowers of the Caribbean.*
University of Nevada Las Vegas Arboretum, USA <http:// hiddenvegas.com/arboretum/potm/caesalpinia_ pulcherrima.html>.

§ *CALATHEA INSIGNIS*

Calathea insignis is found in forest understories from Mexico to Ecuador, as well as gardens throughout the tropics. The inflorescences are bright yellow and are a series of folded bracts that hold 8 to 19 pairs of flowers. Flowers open first at the base of the inflorescence and then proceed upward. Each day the next-highest bract reveals the flowers within. Six to nine days later, the second pair of flower buds in the bottom bract matures, so there are continual waves of flowering and constant rewards for pollinators. *Euglossa* bees are principal pollinators, frequently working the inflorescences systematically, starting with the lower flowers and proceeding upward. The plant's wavy-edged leaves can grow up to 18 inches long and are pale green on top with blotches of dark green, while the underside is dark purple, which helps the plant capture the greenish light that is reflected up from the forest floor.

Further Reading:
Botany.com <http://www.botany.com/calathea.htm>.
Janzen, *Costa Rican Natural History.*

§ *CALATHEA MAKOYANA* (peacock plant)

With broad, oval leaves to capture the limited light that reaches the forest floor, the undersides of the peacock plant's leaves are purplish, while the leaves' uppersides are pale green with oblong, dark green blotches edged with creamy white. In nurseries, it is also called "zebra plant." The flowers are tubular, white with purple lobes, and produced on short, thick inflorescences. Native to the Neotropics, it is a member of the maranta family like the prayer plant, and also closes its leaves at night.

Further Reading:
Missouri Botanical Gardens, USA <http://www.mobot. org/MOBOT/education/tropics>.

* √ *CALYCOPHYLLUM SPRUCEANUM* (mulateiro)

Mulateiro grows in the Amazon and is a popular source of high-density lumber. But it's the tree's bark that really attracts attention. Every year, mulateiro, like other species of this genus, completely sheds and then regenerates its bark. The shedding action probably evolved to discourage vines from climbing the tree. As the bark grows older during the year, it changes color, from a greenish to a very smooth and shiny brown. Indigenous people understandably assumed such bark contained special qualities, so it's long been used in rituals and in folk medicine. A dressing made from the bark is used to treat cuts and burns and a variety of skin ailments such as fungal infections.

Further Reading:
Raintree Nutrition, USA <http://rain-tree.com/mulateiro. htm>.
Wolfe and Prance, *Rainforests of the World.*

√ § *CANANGA ODORATA* (ilang-ilang, or ylang-ylang)

Native to Southeast Asia, the ilang-ilang, or ylang-ylang, can grow 60 feet high and is cultivated throughout the tropics as an ornamental. Its leaves grow along one side of the flat branches, and the tree's flowers are yellow and drooping, with large and narrow petals. The fruits are black and the size of berries. Ilang-ilang is in flower and fruits nearly year-round. Oil extracted from the highly fragrant flowers is used in some of the world's most expensive perfumes.

Further Reading:
Missouri Botanical Gardens, USA <http://www.mobot. org/MOBOT/education/tropics>.
University of Guam, College of Agriculture and Life Sciences <http://uog2.uog.edu/cals/POG/cananga.html>.

* # √ *CARICA PAPAYA* (papaya)

Papaya is native to the West Indies and Central America. A fast-growing, short-lived tree, it bears fruit when one year old. The large leaves are deeply lobed and cluster in a spiral near the apex of the trunk, which reaches from 7 to 30 feet high. The small white or yellow flowers are quite fragrant and bloom in abundance near the trunk apex. The tree is usually dioecious—some trees produce only pollen-bearing flowers, some fruit-bearing flowers—but some are hermaphroditic, producing both. The oblong fruit can grow up to 11 inches in length and weigh almost 20 pounds. The thin skin turns from green to yellowish orange when ripe, and the sweet and buttery flesh is yellow to reddish orange. Papain is prepared from the dried latex of immature fruits and is used in meat tenderizers, chewing gum, and cosmetics; as a drug for stomach ailments; and to bathe hides in the tanning industry. The trees are grown on fruit farms for local markets and export.

Further Reading:
Janzen, *Costa Rican Natural History.*
Sancho and Baraona, *Fruits of the Tropics.*

§ *CASSIA GRANDIS* (pink shower tree)

The pink shower tree is one of many *Cassia* trees that bloom throughout the tropics. The tree originated in Java, but is now grown widely as a shade-giving ornamental. The 30- to 50-foot-tall tree produces clusters of flowers with five coral pink petals. It also bears a 3 inch black pod, with flat, round, yellow seeds buried in a dark, sticky, odiferous pulp.

Further Reading:
Hargreaves and Hargreaves, *Tropical Trees.*
Lennox and Seddon, *Flowers of the Caribbean.*

√§ *CASUARINA EQUISETIFOLIA* (Australian pine)

With its small, scalelike leaves, the Australian pine, although evergreen, is not a true pine. Found along seacoasts from Malaysia to Australia, Micronesia, the Philippines, and the Polynesian Islands, it is named for the cassowary (*Casuarius casuarius*), a large, flightless bird found in Australia's rainforests. The tree's long, drooping bottom branches with dull green needlelike leaves look a bit like the black, sagging feathers of the bird. The tree reaches about 80 feet in height and bears tiny, red, tufted flowers and conelike fruits. Australian pines are widely grown in the tropics as windbreaks and shoreline stabilizers and, in fact, seem to grow best along the coasts, where sea spray supplements moisture in arid climates. In some areas, the tree has become an invasive plant species. *C. equisetifolia* is an excellent source of fuel and charcoal. The wood burns slowly, evenly, and hotly. The bark is used for tanning.

Further Reading:
Hargreaves and Hargreaves, *Tropical Trees.*
Purdue University, USA <http://www.hort.purdue.edu/newcrop/duke_energy/Casuarinaequisetifolia.html>.
Winrock International, USA <http://www.winrock.org/forestry/factpub/FACTSH/C_equisetifolia.html>.

* *CATHARANTHUS ROSEUS* (rosy periwinkle)

Native to Madagascar, the rosy periwinkle is now widespread. The plant is a perennial herb with deep green, shiny, opposite leaves and white or pink flowers; it grows to a height of about one foot. It is used to treat such diverse ailments as eye inflammations, rheumatism, and diabetes, but the plant's fame comes from two alkaloids it contains: vincristine and vinblastine. The former is used to treat childhood leukemia, while vinblastine is used as an effective treatment for Hodgkin's disease and testicular can-

cer. Other periwinkle alkaloids are used to treat breast cancer. The rosy periwinkle is often cited during debates over who should benefit from profits that result in sales of medicines derived from medicinal plants. Villagers in Madagascar told researchers with the pharmaceutical company Eli Lilly about the powers of *C. roseus* in the 1950s. While the drug company continues to earn millions from the lifesaving drugs derived from the plant, no one in Madagascar profited. On the other hand, Eli Lilly learned to isolate the alkaloids and use them to concoct the lifesaving drugs, investing millions of dollars in the process. Alkaloids from the rosy periwinkle are still produced by isolation from raw materials; the plant is grown in plantations in Texas, Spain, and elsewhere specifically for this purpose.

Further Reading:
Blackwell, *Poisonous and Medicinal Plants.*
National Wildlife Federation, USA <http://www.nwf.org/wildalive/periwinkle/sciencefacts.html>.
ten Kate and Laird, *The Commercial Use of Biodiversity.*

CECROPIA SPP.

Cecropia spp. are among the first trees to sprout in natural or manmade clearings in Neotropical rainforests. The trees' wood is light and has large hollow chambers, like bamboo. They have large, platter-shaped leaves and long fruit clusters that hang down beneath the leaves and so are easily accessible to birds and bats. The tree invests much energy in growth rather than chemical defenses against pests, but has evolved a handy relationship with certain species of *Azteca* ants. The ants live in the hollows of the tree trunk; in addition to providing housing, cecropias produce tiny capsules filled with high-energy carbohydrates, whose sole purpose is to feed the ants. In exchange, the feisty ants stand ready to attack any pest or vine that dares to touch the home tree. Not all cecropias have protective ant populations; those that do not tend to grow in higher elevations, where there are fewer leaf munchers. Cecropias are a favorite food of sloths, whose coarse fur protects them from biting ants.

Further Reading:
Forsyth and Miyata, *Tropical Nature.*
Wolfe and Prance, *Rainforests of the World.*

‡ *CEIBA PENTANDRA* (ceiba tree, kapok tree)

The ceiba, or kapok, tree is unusual because it is native to not only the Neotropics, from southern Mexico to the southern Amazon, but also to West Africa. A towering canopy emergent that can reach 200 feet, it grows rapidly, with young trees vaulting some 13 feet per year. Ceibas are among the first trees to colonize clearings, when its seeds are blown into open areas. The large ceiba flowers are white or pink on the outside and dark brown within. They produce a good amount of nectar and have a musty odor that attracts bats. While feeding, the bats become dusted with pollen. Each ceiba produces only a few flowers per night over an extended period of time, which guarantees that a not-quite-satiated and pollen-covered bat will find its way to another tree and, thus, pass on the pollen. A mature tree can produce from 500 to 4,000 fruits at once, each holding some 200 seeds. The fruits open while they are on the tree, and gray-white, silky fibers, with seeds lightly embedded in them, are carried away by the breeze. The silk, called kapok, is too small to be woven into thread, but is used to stuff pillows, mattresses, and life preservers.

Unopened fruit can float, which may be how the seeds reached western Africa from Latin America. Ceiba wood is very light and absorbent but not very durable, so it usually isn't logged extensively. The wood is used to make coffins, carvings, and dugout canoes. Ceiba is supposedly an ancient Caribbean word for "canoe." Maya indigenous people revered the tree and believed that upon death, souls rose to the heavens cradled in a kapok tree.

Further Reading:
Hargreaves and Hargreaves, *Tropical Trees.*
Janzen, *Costa Rican Natural History.*

* *CEPHAELIS IPECACUANHA* (ipecac)

Native to the rainforests of Brazil and Bolivia, ipecac is a perennial herb with opposite leaves and small, funnel-shaped white flowers that grow in bracket clusters. The plant's underground stem and roots contain several alkaloids, used to treat amoebic dysentery and chronic respiratory congestion. But its main use is as an emetic agent. Ipecac syrup is commonly used to induce vomiting in children, especially in cases of poisoning. This unassuming, shade-loving plant has saved countless lives.

Further Reading:
Blackwell, *Poisonous and Medicinal Plants.*

* *CHONDRODENDRON TOMENTOSUM*

Chondrodendron tomentosum is a chief source of curare, along with another broad-leaved, woody vine of Neotropical, lowland forests: *Strychnos toxiferal.* The vine grows long enough to reach the forest canopy, and may get as thick as three inches at its base. Its leaves are about 6 inches long, smooth above and hairy below, and the inflorescences consist of separate clusters of male and female, small, greenish-white flowers. Indigenous people have used the plants to make a potion extracted from the bark, into which they dip the tips of their arrows, spears, or blowgun darts. The name curare comes from indigenous words for "poison." Before synthetic drugs replaced it, curare was also used in surgery, because it causes muscles to quickly relax. Today synthetic products based on the plant's effective compound, d-tubocurarine, are still used in surgery.

Further Reading:
Blackwell, *Poisonous and Medicinal Plants.*
Raintree Nutrition, USA <http://www.rain-tree.com/curare.htm>.
ten Kate and Laird, *The Commercial Use of Biodiversity.*

§ *CHORISIA SPECIOSA* (floss-silk tree)

The floss-silk tree is native to Argentina and Brazil, but is now widely grown as an ornamental for its pretty, pale pink, crimson, or white five-petaled blooms, which are 3 inches long. The fruits are pear-shaped capsules that burst open to send forth silky white floss seeds into the wind. The tree grows to about 40 feet high, and the smooth and stocky trunk is usually covered with substantial cone-shaped thorns, about 1.5 inches tall and one-half inch across at the base.

Further Reading:
Hargreaves and Hargreaves, *Tropical Trees.*

* # CINCHONA SPP. (quinine)

Quinines include 40 trees that reach a height of 60 feet and grow at medium to high elevations in the Andes of South America. The tree's corky bark yields alkaloids such as quinine, a standard treatment for malaria. The tree was supposedly named for the Countess of Chinchon of Peru, who in 1638 was cured of a malarial fever with a medicine made from the bark of a cinchona tree. The bark was used to treat malaria throughout the 1600s to mid-1800s. In 1820, scientists were able to isolate the quinoline alkaloid in the bark and develop an extract that more effectively treated malaria, which is transmitted by mosquitoes.

Until the end of the nineteenth century, South American countries were the principal exporters of quinine bark. The Dutch managed to smuggle *Cinchona* seeds out of South America and establish plantations in Java, which soon dominated world production. When Java was occupied by the Japanese in 1942, South American sources were again used by the Allies, and new plantations were established in Africa. In 1944, scientists successfully synthesized the quinine alkaloid, and bark extracts were seldom used to treat malaria. Today, the bark is still harvested and exported for use in the beverage industry to flavor quinine water and tonic water, as an herbal medicine, and in a prescription drug that treats irregular heart rhythms.

Further Reading:
Raintree Nutrition, USA <http://www.rain-tree.com/quinine.htm>.
ten Kate and Laird, *The Commercial Use of Biodiversity.*

√ CITRUS SPP. (citrus)

Citrus is the general name used for the evergreen trees and shrubs of the *Rutaceae* (rue) family, which includes grapefruit, lemon, lime, orange, and tangerine. These fruit trees originated in Southeast Asia, but are now grown in tropical and subtropical regions worldwide. The early orange and mandarin trees spread north into China from northeast India and Myanmar, while the limes and lemons spread south into Malaysia. Oranges, lemons, and limes were carried to Latin America by early Spanish and Portuguese explorers. Lime is the citrus tree most widely grown in Latin America, with Mexico being the world's largest producer of the fruit.

Citrus trees are generally grown as monocultures on farms. They have white or purple flowers and bear a fruit with a leathery rind and a juicy pulp that's divided into sections. The leaves, flowers, and fruit of citrus trees are replete with fragrant oils.

C. aurantifolia (lime) is widely cultivated in tropical and subtropical regions. Lime trees are fairly small, reaching no more than 15 feet, and have irregular, crooked trunks. The round green fruit produces an acidic juice that was long used to prevent scurvy. Because English sailors used limes as a scurvy preventative, they were given the nickname "limey."

C. aurantium is the sour orange; *C. sinensis*, the sweet orange; and *C. reticulata*, the tangerine, or mandarin orange. The white flowers have an enchanting scent that is used in flavorings and perfumes, as are the oils of the fruit and leaves. The majority of cultivated oranges are used to produce orange juice, extracts, and preserves.

Further Reading:
Janzen, *Costa Rican Natural History.*
Plant Explorer, USA <http://www.iversonsoftware.com/business/plant/Citrus.htm>.

‡ § COCHLOSPERMUM VITIFOLIUM (silk tree, cotton tree)

The silk, or cotton, tree is a small (averaging 30 feet in height) deciduous tree with bright yellow blossoms that look like large buttercups. Found from Mexico to northern South America and a popular ornamental tree, it drops its leaves before producing the cup-shaped flowers that are about four inches across. The fruit capsule contains kidney-shaped seeds that are covered with long white floss, like cotton, which is easily dispersed by the wind. This soft substance is sometimes used to stuff pillows and mattresses.

Further Reading:
Hargreaves and Hargreaves, *Tropical Trees.*
Janzen, *Costa Rican Natural History.*

√ COFFEA SPP. (coffee)

Coffee was originally an understory shrub native to Ethiopia and Sudan; today it is the second-most important product in international commerce, after oil, on the basis of volume traded, and first on the basis of value—about $2 billion a year, or 1 percent of total, legal world trade. In the wild, the coffee bush can reach 30 feet in height. Its five-inch-long leaves are dark green and glossy on top, duller green underneath. Coffee flowers are small, white, and fragrant. After they fall from the plant, the berries develop, ripening from dark green to bright crimson. Beneath the berry skin and pulp is the green bean; each berry usually contains two beans.

Legend holds that an Abyssinian goatherd noticed that after his goats ate the bright red berries from a small tree growing wild in the pasture, they became quite frisky. The goatherd shared his discovery with the monks at a nearby monastery. They developed a drink made by brewing the berries and found they had no problem staying wide awake during the evening prayers. Traders brought the beans across the Red Sea to Arabia, where the plant was cultivated as early as A.D. 600. The beans became lucrative trade items by the thirteenth century. Brewed coffee spread from Arabia to Turkey in 1554, from Turkey to Italy in 1615, to France in 1644, and then quickly to the rest of Europe. The Arabs closely guarded their valuable plants, but in 1690, the Dutch managed to obtain a few seedlings, planted them in farms in Java, and gave seeds to several botanical gardens in Europe. In 1723, a Frenchman stole a plant from a botanical garden in France and brought it to the island of Martinique, where it flourished. The offspring of that kidnapped plant spread throughout the West Indies and eventually South America, reaching Brazil in 1727. Today, more than two-thirds of the world's coffee is grown in Latin America, with Brazil producing more than 25 percent of world production. Coffee is an important economic crop in at least 50 countries and is the third-largest import in the United States, after oil and steel.

The two species of coffee that are most important economically are *C. arabica* and *C. robusta*. The former is more delicate and is usually sold at higher prices. Since coffee is naturally a shade-loving plant, for centuries it was grown under hardwood and fruit trees. But in the 1970s, coffee farmers started planting dwarf coffee plants in cleared land that required no shade and produced higher yields. Without the diversity of the forest, though, these full-sun farms required fertilizers, pesticides, and constant care. A shaded farm produces less coffee, but is nearly a self-sustaining ecosystem that needs little input from people.

Sun farms, like any monoculture, support almost no wildlife. Biologists realized that migratory songbirds in particular were suffering as sun coffee farms replaced shade in Latin America. To survive, the birds that fly south to spend the winter months in the tropics must have food, water, and trees—readily available in shade coffee farms but nonexistent in the full-sun plantations. Now conservationists are urging consumers to drink only coffee that is certified as shade-grown. Organizations like the Rainforest Alliance are inspecting and giving green seals of approval to these bird-friendly coffee farms.

Further Reading:
Paige, *Coffee and Power.*
Rice and Ward, *Coffee, Conservation, and Commerce in the Western Hemisphere.*
Salvesen, "The Grind over Sun coffee."

§ *CORDIA SEBESTENA* (Geiger tree, anaconda, geranium tree)

Variously known as the Geiger tree, the anaconda, and the geranium tree, this popular ornamental tree is native to the Caribbean. It blooms almost constantly. It grows to about 20 or 30 feet and produces tubular, orange flowers and plumlike fruit that have a pleasant fragrance but a not very appealing taste. The flowers grow in large bunches of about 15 blossoms per cluster. Each blossom is about one inch in diameter, with frilly-edged, papery petals. Ornithologist John James Audubon named the tree in memory of John Geiger, a nineteenth-century ship pilot from Florida.

Further Reading:
Hargreaves and Hargreaves, *Tropical Trees.*
Lennox and Seddon, *Flowers of the Caribbean.*

‡ *COUROUPITA GUIANENSIS* (cannonball tree)

Native to northern South America, the cannonball tree is an unusually constructed large tree with thick, tough bark. Short and crooked branches jut out from the main trunk and support flowers and fruit, but the spear-shaped leaves grow at the end of twigs at the crown of the tree. The 3-inch, saucer-shaped, waxy, red or yellow flowers have a strong, sweet smell. The 7-inch, brown, round fruits do indeed look like cannonballs. The fruits are inedible, but the shells are sometimes used to make bowls and other household implements, and in crafts.

Further Reading:
Hargreaves and Hargreaves, *Tropical Trees.*
Lennox and Seddon, *Flowers of the Caribbean.*

√ *DALBERGIA SPP.* (rosewood)

Rosewood includes about 200 species found throughout the tropics, where its timber is highly valued and used to make furniture, tools, and sporting goods. *D. retusa*, called "cocobola" in Spanish, is an endangered species found in the Pacific lowlands of Mexico south through Panama. So few trees remain that it's difficult to know how tall cocobola can grow. It appears to be a sub-canopy species, reaching about 65 feet. The heartwood is a rich dark brown, woven with black lines; other varieties are reddish, yellow, or black. The tree's flowers are small and white, and its fruits are thin and long like pea pods and contain about five seeds.

Further Reading:
Janzen, *Costa Rican Natural History.*

§ *DELONIX REGIA* (flamboyant, royal poinciana, flame tree)

The flamboyant, royal poinciana, or flame tree is a popular tropical ornamental tree. Although it originated in Madagascar, it is grown widely in the Caribbean and is, in fact, the national flower of Puerto Rico. It reaches a height of about 50 feet, branches widely, and blooms with dense clusters of flame-red flowers. Each flower has five petals; the uppermost petal is striped white. The seedpods remain—sometimes for months—after the flowers and feathery leaves drop off, and may eventually grow two feet long. Each pod holds up to 50 cylindrical seeds, which are sometimes fashioned into necklaces.

Further Reading:
Hargreaves and Hargreaves, *Tropical Trees.*
Lennox and Seddon, *Flowers of the Caribbean.*

**DIOSCOREA COMPOSITA*

Dioscorea composita is a vine native to southern Mexico, with heart-shaped leaves and white-fleshed tubers, often called the Mexican yam. The vine also produces a chemical called diosgenin, from which oral contraceptives and sex hormones are synthesized. The yam's chemicals are used to produce anti-inflammatory compounds, including cortisone.

Further Reading:
Blackwell, *Poisonous and Medicinal Plants.*

√ *DIOSPYROS SPP.* (ebony)

Diospyros spp. include 200 species of trees and shrubs native to western Africa but now grown in many tropical regions worldwide. Although the sapwood of *Diospyros* is white and soft, the heartwood is hard and jet-black. The heartwood is used to make piano keys, cutlery handles, musical instruments, and carvings, but good-quality timber trees are becoming rare due to overharvesting. The fruit of *D. kaki*, a nontropical tree called the oriental persimmon, is one of the most popular fruits in the world, in terms of the quantity consumed.

Further Reading:
Purdue University, USA <http://www.hort.purdue.edu/newcrop/morton/japanese_persimmon.html>.
The Royal Botanic Gardens, Kew, England <http://www.rbgkew.org.uk/herbarium/ftea/eben.html>.

√ *DIPTEROCARPACEAE* (dipterocarps)

Dipterocarps is a large family of towering trees, the dominant tree species in the forests of Southeast Asia. The word "diptserocarp" comes from the Greek words for "two-winged seeds," although the seeds look more like badminton shuttlecocks. Of the 500 species of dipterocarps, 380 are found in Malaysia, Indonesia, Papua New Guinea, the Philippines, and Borneo, and the majority of these only in low-lying forests. This region has frequent cyclones, so counting on the wind to disperse seeds is a good strategy. No single species dominates the forests; rather, the entire family is well represented. Dipterocarps are usually forest emergents, with tree crowns that are elevated above other trees. In some parts of Borneo, dipterocarps make up 70 percent of the forest canopy.

An average dipterocarp rises 165 feet high and typically weighs at least 50 tons, about as much as an average redwood. But unlike the few remaining stands of ancient redwoods, dipterocarp forests cover some 200 million acres in Southeast Asia. An average-size tree is at least 100 years old, and some of the most massive may have stood for 400 to 600 years. *Shorea javanica* (damar) trees are giant dipterocarps, canopy emergents that grow to some 120 feet high in the forests of Indonesia. The trees produce a valuable resin that can be refined for industrial uses, such as paint varnishes. The resin is clear and strong, easily tapped from the trees, and considered superior to synthetic products. Thousands of villagers earn their livings tapping damar trees in natural forests.

Dipterocarps flower rarely, usually when they are at least 60 years old, and perhaps only once a decade. Many species reproduce by "mast-fruiting": the synchronous production of vast numbers of single-seeded fruits every few years, with almost no seed production in the intervening years. Not surprisingly, no wildlife species depend solely on the fruits or flowers of a dipterocarp. The hard wood of dipterocarps is highly prized, and the tree is rapidly being lost to the international timber trade—already gone from forests in Sumatra and Java.

Logging is also damaging the dipterocarp population in Borneo. Intensive logging leaves behind piles of debris that easily catch fire during drought years, so the trees have failed to produce in many areas. But even in protected, 220,000-acre Gunung Palung National Park, the dipterocarps are in trouble. The mammals in the park are devouring masting dipterocarp seeds, not allowing any to reproduce. In the past, seed-eating wildlife could move outside the park to other food sources, so at least some seeds had a chance to sprout. But now the park is completely surrounded by heavily logged and burned areas, so animals are trapped. Studies show that there have been almost no new dipterocarp seedlings in Gunung Palung in eight years.

Further Reading:
Collins, Sayer, and Whitmore, *The Conservation Atlas of Tropical Forests: Asia and the Pacific.*
Hartshorn and Bynum, "Tropical Forest Synergies."
Myers, *The Primary Source.*
Veevers-Carter, *Riches of the Rain Forest.*

§ *DOMBEYA WALLICHII* (hydrangea tree, pink ball tree)

The hydrangea, or pink ball, tree is native to eastern Africa and Madagascar and now widely grown in the tropics as an ornamental. The tree grows to about 30 feet in height and bears huge, heart-shaped, and toothed leaves, about 10 inches wide, and bright pink flowers. Each flower has five wide petals and is crowded into round clusters. The clusters hang on foot-long, hairy stalks like Christmas tree ornaments.

Further Reading:
Hargreaves and Hargreaves, *Tropical Trees.*

DURIO ZIBETHINUS (durian)

Durian bears a delicious or odious fruit, depending on your tastebuds. About the size of a football, the durian tastes something like a mixture of "rancid garlic and best strawberries," according to the biologist Norman Myers, who adds that consuming it is like "eating dessert in a run-down public toilet." Tigers (*Panthera tigris*) have a particular passion for durian fruit, although they are mainly meat-eaters. The tree grows in the Malay Peninsula and has so many fans that it generates a crop worth many millions of dollars. Most durian trees are now grown in plantations. The durian is pollinated by just one species of bat, *Eonycteris spelaea*, whose populations are falling due to development—particularly of mangrove swamps, where another flower favored by this bat species blooms. Although development of the mangroves may mean short-term profits for some, it also negatively affects the profitable durian industry.

Further Reading:
Myers, *The Primary Source.*
Newman, *Tropical Rainforests.*

√ *DYERA COSTULATA* (jelutong)

Jelutong is scattered in forests of Thailand, Malaysia, Sumatra, and Borneo, where it can reach a height of 200 feet, with a trunk diameter of eight feet. A popular timber tree and commonly tapped for its latex, it is also grown in plantations. Although it regenerates fairly well, it is a threatened species in Malaysia, where its export is now banned. The wood is used for pencils, picture frames, dowels, carvings, blackboards, wooden toys, furniture, doorknobs, and matchboxes. Jelutong latex is used to make chewing gum and in paints. Most jelutong gum is exported from Indonesia to the United States, Japan, and Italy.

Further Reading:
United States Department of Agriculture Forest Products Laboratory <http://www2.fpl.fs.fed.us/TechSheets/

Chudnoff/SEAsian%26Oceanic/htmlDocs%20seasian/ Dyeracostulata>.
World Conservation Monitoring Centre, England <http:// www.wcmc.org.uk/trees/dye_cos.htm>.

*‡ ENTADA GIGAS (monkey ladders)

Monkey ladders are Neotropical rainforest vines, or lianas, which twist through the forest canopy. A similar species, *E. phaseoloides*, is a monkey ladder found in the Old World tropics. Their braids and turns make them look like ladders, and indeed, monkeys—along with lizards, snakes, and sloths—use them as arboreal conduits. The fast-growing vine, which can reach 100 feet in 18 months, produces the world's largest legume pod, which hangs three to six feet from the vine. The pods are not very durable and usually break apart into 15 or so compartments that each contain a shiny, brown, heart-shaped seed. The seeds can float for many miles, washing ashore far from the tree that produced them. Since so many are carried to land by the surf, it was once thought that the seeds belonged to some strange underwater plant. The seeds have long been used to make jewelry and other objects by cutting them in half, hollowing out the insides, and hinging the two halves together. The seeds are also ground into a plaster to sooth inflammations or ground and swallowed to relieve constipation or, on the other end of the scale, to serve as an aphrodisiac or—conversely—as a contraceptive.

Further Reading:
Palomar College, USA <http://daphne.palomar.edu/wayne/ plmay97.htm#draped>.

√ ENTANDROPHRAGMA ANGOLENSE (gedu nohor tiama)

Gedu nohor tiama is a principal export tree species across tropical Africa and one of the main sources of African mahogany. A deciduous tree, it can reach a height of 160 feet, with a trunk diameter of four to seven feet, over large buttresses. The pinkish-brown or dull red hardwood, which dries to a deep reddish-brown, is used for decoration and cabinets, boat-building, flooring, and furniture. It is the principal export wood of Democratic Republic of Congo, where it is called kalungi, and one of the major exports in Congo. Because it has been heavily logged, it is a threatened species in several countries of West Africa. Ghana and Liberia are considering a ban on export of gedu nohor tiama in log form, while Cameroon, Ghana, and Gabon have imposed minimum diameter limits, so loggers do not cut young trees. The slow-growing tree does not thrive in plantations, and regeneration is poor at a distance from the parent trees.

Further Reading:
Tropical Forest Foundation, USA, "Species Bulletin # 1."
World Conservation Monitoring Centre, England <http:// www.wcmc.org.uk/reference/footer.html>.

√ ENTEROLOBIUM CYCLOCARPUM (elephant's ear, monkey ear)

The elephant's ear, or monkey ear, is a huge tree indigenous to Latin America, from Mexico to Brazil. Its gray trunk can be as large as 10 feet in diameter. It drops its feathery foliage during the dry season, and a few months later, as new leaves appear, it blooms with small, white flowers, while fruit from the previous year's flowering begin to mature and turn brown. The fruit begin as small and green, eventually reaching five inches in diameter and packed with as many as 22 seeds. Parrots are fond of the green seeds, digging them out of the fruit and seed coat. When mature, the ear-shaped fruits turn dry and brown and eventually fall to the ground. Peccaries break into *E. cyclocarpum* seeds with their sharp molars and eat the pulp, so the seeds are dispersed throughout the forest when the peccaries defecate. The seed won't germinate unless the seed coat has been cracked. Seeds are often made into necklaces, and the tree's dark red-brown heartwood is used to make cabinets and boats. The bark is used for tannin, soap, and traditional medicines.

Further Reading:
Hargreaves and Hargreaves, *Tropical Trees*.
Janzen, *Costa Rican Natural History*.

§ EPIPHYLLUM SPP.

Epiphyllum spp. are epiphytic cacti that grow on tree limbs high in the forest canopy of the Neotropics. Like other epiphytes, they draw their nourishment and moisture from the air, dust, rain, and debris that fall and decay near their skinny, aerial roots. The environment in the canopy can be harshly sunny and windy, not unlike the severe desert habitat where cacti are common. *Epiphyllum* have no leaves, but rather flat or triangular thickened stems or branches. Unlike desert cacti, *Epiphyllum* species are not covered with spines, but have tiny bristles in the areolas. Many are night-blooming species, with large white or tawny cup-shaped or tubular blossoms that are pollinated by bats or moths.

E. anguliger, sold as an ornamental called the "fishbone cactus," is native to southern Mexico's forests. It has fragrant pale yellow flowers and blooms just a few nights a year. *Epiphyllum* hybrids are commonly sold as ornamental plants.

Further Reading:
Newman, *Tropical Rainforests.*
Pierot and Hall, *Easy Guide to Tropical Plants.*
San Diego Epiphyllum Society, USA <http://www.surfnfax.com/epi/epiwhat.htm>.

* ‡ *ERYTHRINA SPP.* (mulungu, coral trees)

Mulungu, or coral trees, bloom with a profusion of brightly colored flowers. Their black seedpods hold large red and black seeds often used to make jewelry. The tree reaches about 30 feet in height and is native to Panama and south to Bolivia. There are more than 100 different species of *Erythrina*, many with blooms the color of coral, and several are used for medicines, insecticides, and fish poisons. Drunk as a tea made by boiling the tree's bark, it's a natural sedative popular among indigenous people in Brazil. It is also used as a tranquilizer, as a sleeping potion, and to soothe asthma, bronchitis, hepatitis, and liver and spleen inflammations. Because the tree is a nitrogen-fixer, it's frequently used in agroforestry throughout the tropics to improve soil quality and, since it grows rapidly, on coffee and cacao plantations as a shade tree.

Further Reading:
Raintree Nutrition, USA <http://rain-tree.com/mulungu.htm>.
Rodale Institute, USA <http://fadr.msu.ru/rodale/agsieve/txt/vol4/issue1/3.html>.

* *ERYTHROXYLUM COCA* (coca)

Coca is a shrub or small tree that grows on the eastern slopes of the Andes and in the Amazon valley. For centuries, the Incas chewed coca leaves (and then so did the invading Spaniards and the descendents of

both) to combat fatigue while working in the thin, mountain air. In 1860, the leaves' active ingredient, cocaine, was isolated in Germany, and the drug was then used as a local anesthetic. Its molecular structure served as a model for the synthesis of novocaine and xylocaine. Another coca derivative, lidocaine, is used to prevent and treat the most common cause of death from heart attack: ventricular fibrillation and ventricular tachycardia.

Further Reading:
Blackwell, *Poisonous and Medicinal Plants.*

√ *EUCALYPTUS DEGLUPTA*

One of the few eucalyptus trees that can thrive in rainforests, *Eucalyptus deglupta* is native to the Philippines and other western Pacific Islands. It grows tall—to 100 feet—and straight, and its distinctive bark has a camouflage pattern. A fast-growing tree, it's a popular timber plantation species, grown throughout the tropics.

Further Reading:
United States Department of Agriculture Forest Products Laboratory <http://www2.fpl.fs.fed.us/TechSheets/Chudnoff/SEAsian%26Oceanic/htmlDocs%20seasian/Eucalyptusdeglupta>.
Veevers-Carter, *Riches of the Rain Forest.*

* # √ *EUGENIA AROMATICA* (clove tree)

The clove tree is one of 700 species of the genus *Eugenia* found throughout Indonesia, Malaysia, and the western Pacific Islands. Like the clove, all *Eugenia* trees have tiny oil glands on their leaves' undersides, which release a strong odor when they are crushed. The clove tree is native to the Moluccas, the group of Indonesian islands once called the "Spice Islands," and along with nutmeg, was in large part responsible for creating many wealthy people, turf wars, and political upheavals for some 17 centuries. Cloves are actually the immature, unopened flower buds of the tree, which are picked just when they turn red. They need only be dried or steamed to distill their fragrant oils. Clove oils are commonly used as an anesthetic and antiseptic. Most cloves are today grown in plantations.

Further Reading:
Hargreaves and Hargreaves, *Tropical Trees.*
United States Department of Agriculture Forest Prod-

ucts Laboratory <http://www2.fpl.fs.fed.us/TechSheets/ Chudnoff/SEAsian%26Oceanic/htmlDocs%20seasian/ Eugenia%20spp.htm>.
Veevers-Carter, *Riches of the Rain Forest.*

* # √ § *EUGENIA JAMBOS* (rose apple)

According to the legend, the rose apple is the tree that bore the golden fruit of immortality. Grown throughout the tropics as an ornamental, this 30-foot tree has narrow, pointed leaves and large, pale yellow, hairy flowers that look like pompons. They bear oval, white, or pale yellow fruits that have a pink blush and smell and taste faintly of rose water. The fruit is sometimes eaten raw, but usually stewed or made into jams. The strong dark red or brown heartwood is used to make furniture, wheel spokes, frames for musical instruments, and packing cases. In India, the fruit is used as a brain and liver tonic. The seeds are roasted and powdered in Nicaragua to treat diabetes, while in Cuba, the root is used to treat epilepsy.

Further Reading:
Hargreaves and Hargreaves, *Tropical Trees.*
Purdue University, USA <http://www.hort.purdue.edu/ newcrop/morton/rose_apple.html>.
Veevers-Carter, *Riches of the Rain Forest.*

* # § *EUGENIA MALACCENSIS* (Malay apple)

The Malay apple is presumed to have originated in Malaysia, although it is now commonly cultivated from Java to the Philippines and Vietnam, and in southern India. From there, the history of its worldwide dissemination can be partly traced. It was introduced into East Africa by Portuguese travelers and then to the Pacific Islands centuries ago, since the wood was used by Hawaiians to carve idols. By some historic accounts, the only fruits in Hawaii, before the arrival of missionaries, were bananas, coconuts, and the Malay apple. In 1793, three small Malay apple trees were brought to Jamaica by Captain Bligh, aboard the *Providence*, and the plant soon reached other Caribbean islands. It may be that the Portuguese also brought the plant to Brazil, for it is now grown there, as well as in Suriname, Venezuela, and Panama. In 1921, the botanist Dr. David Fairchild sent seeds from Panama to the United Sates Department of Agriculture, and in 1929, young trees from Panama were also brought to a botanical garden in

northern Honduras. The tree is now grown in most of Central America. Not surprisingly, it is known by dozens of different names, in dozens of different languages and dialects, from jambu merah in some parts of Malaysia to ohia in Hawaii and maraon japons in El Salvador. The tree grows to 50 feet in height and has large, shiny green leaves that are some 10 inches long. Its deep rose-colored blossoms look like small brushes. It blooms abundantly in clusters on the upper trunk and along leafless parts of mature branches. The glossy, pink and white fruit is about 2.5 inches, pear-shaped, and has a mild, applelike taste. The fruit are eaten raw or cooked or turned into wine. In Indonesia, the flowers are eaten in salads, and young leaves and shoots are eaten raw with rice. All parts of the tree are used in a variety of medicinal remedies wherever it is grown, e.g., to heal cuts, and to treat dysentery, constipation, diabetes, coughs, and headaches.

Further Reading:
Hargreaves and Hargreaves, *Tropical Trees.*
Purdue University, USA <http://www.hort.purdue.edu/ newcrop/morton/malay_apple.html>.

√ *EUSIDEROXYLON ZWAGERI* (Borneo ironwood, belian)

The Borneo ironwood is typical of a number of tree species that bear the nickname "ironwood" in that it has extremely hard and dense wood. Found in Sabah, Malaysia, Borneo, several other Indonesian islands, and the Philippines, it is extremely durable and resistant to insect attack. The trees are moderately tall, between 50 and 100 feet high, with large and long leaves and one of the largest seeds in the world, measuring about 5 inches in length and 1.5 inches in diameter. The seed sizes vary from tree to tree—and even on the same tree. Seeds are protected by a bone-hard covering marked with long furrows, which surrounds an only slightly less resistant fibrous coating. It takes months for the coating to rot away, but the decay provides a stockpile of nutrition when a seed finally begins to germinate. Seedlings may grow 3 feet or more before they expend the energy to produce leaves. Borneo ironwoods have been logged for more than a century, so few stands remain.

Further Reading:
Veevers-Carter, *Riches of the Rain Forest.*

§ *FICUS SPP.* (fig)

Fig trees are found throughout the tropics; nearly every tropical country and island group (except the Hawaiian Islands) has at least one indigenous species of fig. About 200 species of fig trees are known as "strangler figs." Stranglers, like all figs, produce fruit three times a year or more, so there's always a supply of fruits available to hundreds of species of wildlife, including parrots, bats, toucans, and monkeys. The small fig seeds pass through animal digestive tracts and are deposited on treetops throughout a forest. If a seed sprouts, it sends a single root from the canopy to the forest floor, a lifeline for the strangler fig seedling. The root sends up water and nutrients. More roots grow downward, gradually fusing and wrapping around the host tree's trunk, while fig branches grow toward the canopy, eventually overshadowing the host tree.

Over many decades, strangler figs will grow so large as to effectively cut off water, sunlight, and nutrients from the host tree, as its swelling roots twist around the doomed tree like a massive wooden weaving, finally killing it. They have an advantage over other forest trees, since they start life in the bright sunlight, and don't have to invest energy and nutrients growing a trunk to reach the light.

Each of the nearly 900 species of fig trees—800 of which are found in the tropics—is pollinated by a different species of wasp, which lays its eggs in the tree's fruit. Should one wasp species disappear, so will the tree that depends upon it, and if the female wasp cannot find her particular fig tree, she and her eggs will die. Figs are a favored fruit for scores of rainforest animals, from tiny ants to two-ton elephants. They are plentiful, available year-round, and an essential source of calcium. An average fig has three times more calcium than other fruits.

In India and the South Pacific islands, strangler figs are known as banyans (*F. bengalensis*), an English adaptation of the word "banians," which refers to the Hindu merchants who set up shop under the shade of immense fig trees. Because it is believed that Buddha meditated under a banyan, Hindus consider the tree to be sacred. Supposedly, Alexander the Great was able to camp with 7,000 soldiers under the shade of one banyan. One tree in India measured 2,000 feet across by 85 feet high. The tree's aerial roots grow down to the earth from horizontal branches. *F. bengalensis* and *F. religiosa* are widely planted in the tropics and subtropics as ornamentals. The former was a popular ornamental in Florida, but is now considered a nuisance tree in some areas. After nearly a century of celibacy in this subtropical locale, they are now reproducing. Sometime in the 1980s, the banyan's pollinator, a small gall wasp, arrived in south Florida, probably accidentally transported by ship. The wasp quickly spread and reproduced, and Florida's *F. bengalensis* began producing fertile seeds. The tree's fruit is a favorite among birds, which dispersed seeds all around south Florida. Baby banyans began budding in every nook and cranny. Floridians fear that the widely spreading roots of banyan trees will damage roads and building foundations.

Further Reading:
Bannan, "The Cornucopia Tree."
Forsyth and Miyata, *Tropical Nature*.
Kinnaird, "Big on Figs."
Laman, "Borneo's Strangler Fig Trees."
Veevers-Carter, *Riches of the Rain Forest*.

§ *FITTONIA VERSCHAFFELTII*

Fittonia verschaffeltii is a creeping herb and popular ornamental known as the red nerve plant. Native to Colombia and Peru, the plant has leaves that are short stemmed and heart shaped, a dark green with a network of narrow, red veins. The small yellow, tubular flowers are borne on erect, narrow spikes with overlapping, green bracts. Their taxonomic name is a tribute to Elizabeth and Sarah Mary Fitton, the English authors of *Conversation about Botany*, published in the early 1800s. Indigenous women of the Peruvian Amazon may chop up *Fittonia* leaves and apply them to their bodies, believing that this will enhance their artistic skills. The lines and patterns on the plant's leaves resemble designs found on some Amazonian clay pottery.

Further Reading:
Armitano, *Garden Plants of the Tropics*.
Forsyth, *Portraits of the Rainforest*.

* √ *GMELINA ARBOREA* (melina)

Native to Southeast Asia, *Gmelina arborea* grows from India east to Vietnam. It is a popular plantation species, cultivated throughout the tropics. A fast-growing tree, on good soils, *Gmelina arborea* can reach 100 feet in 20 years, with a trunk diameter of 2 feet. The tree has straw-yellow wood that is mainly used for paper pulp, particle board, construction, and matches. The leaves are oval and about 6 inches long, and the flowers are bell-shaped and bright yellow. The smooth, oblong fruits turn yellow when ripe. In India, a concoction made from the roots is used to treat abdominal tumors, blood disorders, cholera, colic, epilepsy, headaches, and a variety of other ailments.

Further Reading:
Purdue University, USA <http://www.hort.purdue.edu/ newcrop/duke_energy/Gmelina_arborea.html>.

* # √ *GRAMINAEA*

The *Graminaea* family of grasses includes the fastest-growing plant on Earth: bamboo. Some species gain 2 inches a day in height. *Dendrocalamus giganteus,* a Southeast Asian bamboo, can reach more than 100 feet in height and 12 inches in diameter. It was famous during World War II for its ability to promptly engulf parked jeeps. A particularly useful plant to human beings, about 1,200 species are found worldwide in the humid tropics. There are two main types of growth among bamboo species, depending on the species' root structure. Most of the tropical species are "clumpers," while "runners" are more common in temperate climates; these can be quite invasive.

Bamboo stems have segmented, hollow chambers and flower only once every 7 to 120 years, depending on the species. Once it blooms, the parent plant often dies. Particularly in Asia, bamboo provides food (tender, fresh bamboo shoots), housing, musical instruments, cooking and eating utensils, medicines, baskets, ropes, furniture, and hunting weapons. *Bambusa balcooa,* native to India, is widely eaten and used in construction. In the Neotropics, the largest genus of bamboo is *Chusquea*, found mainly at high elevations from Mexico to Chile and in the Caribbean.

Further Reading:
American Bamboo Society, USA <http://www.tropical bamboo.org/about.htm>.
Myers, *The Primary Source.*
Rainforest Information Center, Australia <http://forests. org/ric/good_wood/bamboo.htm>.

* √ *GUAIACUM OFFICINALE* (lignum vitae, tree of life)

The lignum vitae, or tree of life, is the heaviest of all commercial woods and produces a resin that has been used for centuries. It grows slowly, reaching about 30 feet at its peak, and has a broad, rounded, dense crown. The leaves are small and round, and blue flowers bloom in large clusters at the branch tips. The flower of the lignum vitae, which means "wood of life" in Latin, is the national flower of Jamaica; the tree is native to the West Indies and from southern Mexico south to Venezuela. The wood has been traded since the sixteenth century, when its sap was used as a medicine, and its dense, olive-brown heartwood was used for ship propeller shafts and wagon-wheel bearings. Today the wood is used to make gavels, mallets, and bowls. The resin is still used to treat arthritis. Until more effective drugs were manufactured, the sap was also used in a treatment for syphilis.

Further Reading:
Hargreaves and Hargreaves, *Tropical Trees.*
Lennox and Seddon, *Flowers of the Caribbean.*

§ *HELICONIA SPP.*

Heliconia spp. include about 40 species, most of which are found in the Neotropics, with only three or four species in the Pacific basin. The inflorescences are colorful bracts, usually flame red or orange, that arise alternately from the stem and enclose and protect the many tiny flowers that bloom at different times. In some species, the stems hang down, and in others, the stem is erect. The plants are related to the banana; the leaves are paddle-shaped, which is characteristic of the banana family. Those that have upright bracts collect water, which provide homes to dozens of species of aquatic, minute fauna. The miniature ponds also protect the flower base from stingless bees, which like to chew their way into nectar centers.

Heliconias are rich in nectar, and the shapes, lengths, and curves of the inflorescences are adapted to fit the shapes, lengths, and curves of various species of hummingbirds, the plant's principal pollinators. Heliconias have no scent, but hummingbirds' sense of smell, like that of most birds, has atrophied. The plants tend to produce a surge of nectar at dawn, just when hummingbirds are making their rounds. About 500 species of flower mites, members of the *Ascidae* family, are also involved in the hummingbird-Heliconia relationship. For their size, these tiny bugs run as fast as a cheetah. They sprint from the Heliconia into a hummingbird's nose as it pauses to sip nectar. When the hummer moves on to the next Heliconia blossom, it dashes out of the nasal passage and onto the flower, where it feeds on nectar and pollen and also breeds.

H. pegonantha is the species frequently seen in Neotropical rainforests or along the banks of a forest stream. The red inflorescence is pendulous and often 3 feet long. The stalk can grow as high as 25 feet.

Further Reading:
Forsyth, *Portraits of the Rainforest.*
Pierot and Hall, *Easy Guide to Tropical Plants.*

√ *HEVEA BRASILIENSIS* (rubber trees)

Rubber trees evolved in the Amazon basin and are the primary source of natural rubber. The 100-foot-high trees are in the *Euphorbiaceae* family, which includes manioc and poinsettias. They have waxy leaves and very soft bark, which exudes a milky white sap when scratched. Latex, which has natural insecticide properties, probably evolved to provide rubber trees protection against termites and other insects that can easily bore through the tree's yielding bark. When any object permeates the bark, the tree begins to bleed the gummy latex—no fun for a wood-seeking insect. Latex has remarkable elasticity and for centuries was used by indigenous people, but it was Westerners who turned latex into a billion-dollar industry.

The English word "rubber" is derived from one of the first uses of latex—to erase, or rub off, pencil marks on paper. Latex was exported to Europe and the United States in the late 1700s and early 1800s to make hot-air balloons, syringes, rubber boots, and bottles. When a U.S. entrepreneur, Charles Goodyear, and an English inventor, Thomas Hancock, discovered that heated latex vastly improved the durability of the product, the rubber industry took off. The advent of the bicycle and then the automobile created huge demand for the product for tires. Although synthetic rubber was invented during World War II, there is still a demand for natural rubber, which has superior resilience. The tire industry uses two-thirds of natural rubber supplies. The tires on the U.S. space shuttle fleet are totally natural rubber. Although the rubber plantations of Malaysia and several other Asian countries provide millions of tons of rubber, demand outstrips supply.

Although other latex-bearing trees live in the forests, *H. brasiliensis* provides the best quality. Supplying the demand for rubber completely altered the Amazon forest, as rubber tappers poured into the region at the turn of the nineteenth century. To collect rubber, tappers score the bark of a tree, and the latex bleeds into a cup propped against the tree, much the same way maple sugar is collected. Wealthy businessmen laid claim to vast stretches of Brazilian forest, then hired cheap labor to tap the rubber trees.

The Brazilian rubber boom came to an end in the early 1900s, when the tree was successfully cultivated and grown in plantations in Singapore. The entire Asian rubber industry is based on a mere 26 seedlings of *H. brasiliensis* that were brought from Brazil. The Amazon saw another boom during World War II, when Asian supplies became uncertain. During that time, the grandfather of Francisco "Chico" Alves Mendes moved his family to the Amazon to tap rubber.

In 1988 Chico Mendes, a rubber tapper and labor organizer who tried to defend the forest from exploitation by ranchers, was murdered. His brutal death caused an international outcry and helped launch the rainforest conservation movement. To many, his death became a symbol for the struggle to conserve the rainforest for those who directly depend upon it.

Further Reading:
Myers, *The Primary Source.*
Revkin, *The Burning Season.*

§ *HIBISCUS SPP.*

Hibiscus spp. include about 230 species of plants widely grown in the tropics. Some are shrubs with delicate blossoms, and others are hardy trees that can grow nearly anywhere. The most common form, *H. rosa-sinensis*, originated in Hawaii. The plant has serrated, alternate leaves, and blooms throughout the year. The petals of each flower form a trumpet shape that can be as long as 5 inches across. A stamen protrudes from the center, with female stigmas at its tip. The most common hues of hibiscus flowers, which last only one day, are red, white, pink, and yellow.

H. esculentus produces the edible capsules, or seedpods, known as okra or gumbo, a popular food. *H. esculentus* is believed to be native to Africa or Southeast Asia; it was used by the Egyptians as early as the twelfth century. It arrived in the United States in the eighteenth century, probably on slave ships, and became a key ingredient in Cajun cooking. The plant bears the typical hibiscus flowers. The seedpods can reach lengths of 10 inches, but are less fibrous when they are younger; the optimal eating size is about 3 inches.

Further Reading:
Lennox and Seddon, *Flowers of the Caribbean.*
Purdue University, USA <http://www.hort.purdue.edu/rhodcv/hort410/okra/ok00001.htm>.

≠ HURA CREPITANS (sandbox tree)

The sandbox tree is native to Central America, south to the Amazon, and to the West Indies. The tree's milky sap is caustic, and some fishermen in the Neotropics use the sap to stun fish. The sandbox tree bears heart-shaped leaves and a dark maroon flower. Its 3-inch, woody fruits look like miniature pumpkins. Each of these seed capsules comprises 15 to 20 divisions that hold the round, flat seeds. The capsules explode when ripe, spurting seeds, which are sometimes fashioned into jewelry. In the past, the fruit was hollowed out, filled with sand, and used for blotting ink or as paperweights. *H. crepitans* saplings need sunlight and can grow only in light gaps in the forest canopy.

Further Reading:
Hargreaves and Hargreaves, *Tropical Trees.*
Janzen, *Costa Rican Natural History.*

√ HYMENAEA COURBARIL

Hymenaea courbaril is a large legume that grows throughout the Neotropics, where it's often called "stinking toe" tree. Reaching 130 feet or more in height, the trees' crowns are part of the rainforest canopy from western Mexico to central South America and on the Caribbean islands. The trunk and roots of the tree exude a yellow resin that hardens into globs that, when they drop from a tree or an ancient tree falls, become buried in the soil. The hardened resin, called copal, fossilizes over millions of years into a glowing amber, used to make beads and jewelry. A relative in east Africa, *H. verrucosum*, produces amber.

Further Reading:
Janzen, *Costa Rican Natural History.*

§ HYPOESTES SANGUINOLENTA

Known as the "pink polka-dot plant" or "spatter plant," *Hypoestes sanguinolenta* is native to Madagascar. A popular ornamental, the plant has oval, heart-shaped leaves that are speckled with spots of bright pink. The tiny flowers are lilac-colored. Pink polka-dot was brought to the Neotropics as a cultivated pot plant and escaped into the wild.

Further Reading:
Armitano, *Garden Plants of the Tropics.*
University of Vermont, Dept. of Plant and Soil Sciences <http://pss.uvm.edu/pss123/folhypo.html>.

* √ ILEX PARAGUAYIENSIS (yerba maté)

Yerba maté is related to holly and has similarly shaped, leathery leaves. There are about 350 species in the genus *Ilex*, found on all continents except Antarctica. *I. paraguayiensis* is native to Paraguay, Brazil, and Argentina; has an average height of 23 feet; and bears small white flowers and red, black, or yellow berries that are about the size of peppercorns. Throughout South America, its leaves are brewed to make a tea that is a popular herbal medicine, used as a diuretic and stimulant to the nervous system. It is also used to reduce appetite and fatigue. Traditionally, dried, minced yerba maté leaves are placed inside a gourd, hot water is added, and the brew is sipped through a metal pipe fitted into the gourd, with a strainer at its lower end. The tree is cultivated to meet demand for its leaves, although some production still comes from forest trees in Brazil and Paraguay, the leading exporting nations. Some 300,000 tons of maté leaves are harvested each year. Conservationists promote cultivation of yerba maté because it provides jobs and income and promotes reforestation, since the crop must grow under shade. One wild tree can yield up to 85 pounds of dried leaves annually. Harvesting is done between May and October, when the tree is in full leaf, and leaves are collected from the same tree only every third year.

Further Reading:
Raintree Nutrition, USA <http://www.rain-tree.com/yerbamate.htm>.
University of California at Davis, USA <http://agronomy.ucdavis.edu/gepts/pb143/crop/mate/mate.htm>.

§ *JACARANDA ACUTIFOLIA* (fern tree)

The fern tree grows throughout the Neotropics but originated in Brazil. Trees grow to heights of about 30 or 40 feet, and the foliage is fernlike. The trees burst with clusters of blue, bell-shaped blossoms. When mature, the fern tree's fruits are hard and dark brown and release winged seeds that are dispersed by the wind. It is widely grown as an ornamental, and in its natural forest habitat, is a gap colonizer—one of the first species to sprout and grow quickly when an opening occurs in the forest canopy.

Further Reading:
Hargreaves and Hargreaves, *Tropical Trees.*
Lennox and Seddon, *Flowers of the Caribbean.*

√ *JUGLANS* SPP. (walnut)

The walnut includes 23 species native to North, Central, and South America; Eastern Europe; and Asia. *J. regia*, the Persian or English walnut, produces the popular nut, while *J. nigra*, the eastern black walnut, is grown primarily for its timber. There are about 11 species native to Central and South America. *J. neotropica*, the South American walnut, is a popular timber species, used in furniture and cabinets, musical instruments, and sporting goods. It reaches a height of 60 feet and is found in southern Mexico, most of Central America, Colombia, Ecuador, Peru, and the mountains of Argentina.

Further Reading:
Germplasm Resources Information Network, USA <http://www.ars-grin.gov/npgs/cgc_reports/jugcgc.html>.

§ *LANTANA CAMARA*

Lantana camara is a weedy shrub native to the West Indies but now distributed worldwide. The plant reaches a height of four to five feet, and the leaves are small, dark green, and serrated. Individual lantana flowers are small and flat, but bloom in dense groups and are particularly attractive to butterflies. The flowers are yellow when they first open, then gradually over 24 hours change to orange, then red. Only the young, yellow flowers produce nectar. The small, round, berry-sized fruits are green when young, then turn purple. They hold two seeds and are popular food for manakins (*Pipridae*) and tanagers (*Thraupidae*). Lantana leaves contain toxic alkaloids, which

explains why insects avoid them and why the species has spread so widely and become a pest for farmers. Cattle and sheep are sometimes poisoned after munching lantana.

Further Reading:
Janzen, *Costa Rican Natural History.*
Lennox and Seddon, *Flowers of the Caribbean.*

√ *LEUCAENA LEUCOCEPHALA* (leadtree or white epopinac)

Leadtree is native to the Neotropics but is such a versatile tree that it is grown worldwide. A legume, it can grow 13 feet in just six months and reach 50 feet in six years. The wood is used for fuel, lumber, pulpwood, crafts, and charcoal, and the foliage is an excellent animal fodder, probably the highest quality of any tropical legume due to its high protein content. The pods of this multipurpose tree can be eaten raw or dried and ground into flour, and the gum can be used as a commercial thickener for sauces and ice creams, like gum arabic.

Further Reading:
Myers, *The Primary Source.*
Purdue University, USA <http://www.hort.purdue.edu/newcrop/cropfactsheets/leucaena.html>.

* # √ *MAMMEA AMERICANA* (mammee apple)

The mammee apple is native to the West Indies and northern South America. It is a 70-foot tree with a short trunk and ascending branches that form a densely foliaged crown. The leaves are glossy, dark green, and leathery; the tree looks a bit like a southern magnolia. The white, fragrant flowers open to 1.5 inches and have orange stamens or pistils or both. The mammee apple is grown throughout the Neotropics for the brown, thick-skinned edible fruits, which have an orangish, sweet pulp that tastes a bit like apricots or raspberries. The fruit is eaten raw or stewed, cooked in pies, or made into a spiced marmalade. Various parts of the tree contain chemicals and have been used in traditional medicines and as insecticides. The heartwood, which is reddish or purplish-brown, is used for cabinets, pillars, rafters, or fence posts because it is fairly resistant to rot. In the French West Indies, a liquor called Eau de Creole, or Crem de Creole, is distilled from the flowers.

Further Reading:
Hargreaves and Hargreaves, *Tropical Trees.*
Purdue University, USA <http://www.hort.purdue.edu/newcrop/morton/mamey.html>.

√ *MANGIFERA INDICA* (mango)

The mango tree originated in India, and although it is now grown widely throughout the tropics, more mangos—some 400 varieties—are grown and consumed in India than anywhere else. Mango trees grow to about 65 feet high, and they live long lives, some surviving 200 or 300 years. Hundreds or even thousands of tiny, pink or yellow, hairy flowers bloom at once in clusters. Shiny, green, and leathery mango leaves have a smell like turpentine when crushed. One of the world's most popular fruits, the oval-shaped, smooth-skinned mango may be pale yellow, yellowish-orange, reddish-pink, or deep pink. Mango flesh is pale yellow to deep orange. The flesh is quite juicy, with the consistency of a peach but more fibrous. Buddhist monks likely brought mangos on voyages to Malaya in eastern Asia in the fourth and fifth centuries B.C., and the Persians are believed to have carried the fruit to East Africa in the tenth century A.D. The Portuguese introduced the tree to West Africa and Brazil in the sixteenth century. From Brazil, it was carried to the West Indies, and from there, reached Mexico in the nineteenth century. The first mango was planted in Florida in the 1860s.

Mango trees 10 years or younger may flower and fruit regularly each year, but after 10 years, they usually flower and fruit only every two to four years. The fruits take from two to five months to develop. Mangos are now grown in plantations, and a healthy tree aged 10 to 20 years will produce about 250 fruits per year. India has more than two million acres of mango trees, producing some 65 percent of the world's crop, or nearly 10 million tons. Other leading exporters of fresh mangos are the Philippines, Thailand, Mexico, and Indonesia, while England and France import more mangos than any other country.

Mangifera indica is in the same family as cashews, poison oak, and other poisonous plants. Its sap can irritate human skin, even causing blisters. Some people can't touch mangos or eat them without a skin reaction or swelling around the eyes and lips. Wherever they are grown, mangos are used to relieve a variety of human ills. People make a tea of dried mango flowers or leaves to treat diarrhea, fever, hypertension, or diabetes, or sip a brew made from the bark to treat rheumatism and diphtheria.

Further Reading:
Hargreaves and Hargreaves, *Tropical Trees.*
Janzen, *Costa Rican Natural History.*
Purdue University, USA <http://www.hort.purdue.edu/newcrop/morton/mango_ars.html>.
Sancho and Baraona, *Fruits of the Tropics.*

* # *MANIHOT ESCULENTA* (cassava)

Native to Latin America, cassava has become an extremely important food crop for millions of subsistence farmers throughout the tropics. Cassava grows as a shrub about 9 feet tall, with edible, fleshy, tuberous roots. It has an erect stalk and reddish leaves. Because of the root's high starch content, cassava is a major source of calories for poor families; it is easily grown and cultivated within six months to three years after planting. In Africa, the leaves provide an inexpensive source of protein and vitamins. In the Philippines, cassava root is boiled to a pulp, wrapped with coconut and sugar in banana leaves, and boiled to make a popular dessert called suman. Cassava starch is used in the manufacture of paper and textiles and as monosodium glutamate, an important ingredient in Asian cooking. The tapioca used in puddings is cassava starch. In traditional medicine, different parts of the plant are used to treat tumors, boils, conjunctivitis, diarrhea, flu, snakebite, and hernias. Before it can be eaten, cassava must be boiled to break down the natural cyanide that accumulates in its roots.

Manihot probably originated from two regions: northeastern Brazil and from Mexico to Guatemala. Sweet cassava was cultivated by indigenous people at least 4,000 years ago in Peru and 2,000 years ago in Mexico. At least 150 varieties are cultivated today, classified as bitter—due to their higher cyanide content—or sweet. The largest producers of cassava are Nigeria, Brazil, Congo, Madagascar, Thailand, and Indonesia.

Further Reading:
Consultative Group on International Agricultural Research, USA <http://www.cgiar.org/areas/cassava.htm>.
Purdue University, USA <http://newcrop.hort.purdue.edu/newcrop/duke_energy/Manihot_esculenta.html>
University of Connecticut, Ecology and Evolutionary Biology Conservatory, USA <http://florawww.eeb.uconn.edu/acc_num/199200534.html>.

* # √ *MANILKARA ZAPOTA* (chicle, sapodilla)

Chicle, or sapodilla, is native to southern Mexico, Belize, and northeastern Guatemala. It now grows throughout Central America and the West Indies, where apparently it was cultivated centuries ago, as well as in the Old World tropics. It's a slow-growing tree that reaches a height of 60 feet and bears glossy green leaves and small, bell-like, cream-colored flowers. The fruit's yellowish flesh becomes very soft, sweet, and juicy, with a grainy texture and taste something like a pear. Sapodilla is eaten raw and used to make custards and ice cream. The fruit is also popular with wild mammals, including howler monkeys, kinkajous, tapir, peccaries, and bats.

The sap of the tree is chicle gum, the original base of chewing gum. The sap is tapped from wild and cultivated trees in Mexico, Guatemala, and Belize, coagulated by stirring it in large pots set over fires and pouring the mixture into molds. In 1866, a former president of Mexico brought a sample of chicle to businessman Thomas Adams in New York, thinking that the chicle latex might be used as rubber. But chicle latex doesn't produce an elastic rubber as does the latex of *Hevea brasiliensis,* the rubber tree. Recalling that people in Mexico chew on chicle sap, Adams thought of adding sugar to the sap of the *M. zapota* tree—at the time, most chewing gum was made from paraffin. As a result, another new industry from a natural forest product was invented. At the peak of the chicle-based chewing-gum industry, in 1930, some 14 million pounds of chicle were imported to the United States. Latex from other trees and synthetic gums have largely replaced chicle, although there is a small cottage industry for natural chewing gum. Wild Things, based in Florida, sells JungleGum made of natural chicle. The product has received certification from the Rainforest Alliance, verifying that the sap has been harvested sustainably.

The tree's fruit and leaves are used in folk medicine to treat diarrhea, coughs, and colds. Sapodilla wood is strong and durable—it was used for lintels and beams in Maya temples that are still intact among the ruins of the great Maya buildings. The timber is used for railway crossties, flooring, tool handles, furniture, and cabinets, but it is illegal to harvest the tree in Mexico because of its value as a chicle source.

Further Reading:
Missouri Botanical Gardens, USA <http://www.mobot.org/MOBOT/education/tropics/>.

Purdue University, USA <http://newcrop.hort.purdue.edu/newcrop/morton/sapodilla.html>.
Wild Things, USA <http://www.junglegum.com/Chicle/chicle.html>.

MARANTA SPP.

Maranta spp. includes about 400 species found throughout the tropics. Marantid flowers are well adapted for bee pollination: Some species have evolved a "trigger" that shoots a load of pollen on any bee that enters the flower. All marantid species have a structure called a pulvinus at the base of their leaves that regulates the flow of fluids and causes the plants to fold their leaves at night. The leaves point upward in the evening to avoid contact with invading plants, then lower in the daylight to shade out competitors. Marantid leaves have colored patterns that look like hand-painted stripes, and emerge in tight rolls that gradually unfurl as they mature. Prayer plants (*Calathea spp.)* also do this and belong to the same family (*Marantaceae*) as the marantids.

M. arundinacea (arrowroot) originated in the Neotropics, where indigenous groups prized it for the starchy food they made from its tubers. The Arawak indigenous people of the Caribbean Islands and coasts used arrowroot tubers to draw poison from wounds inflicted by poisoned arrows. Europeans who learned this custom gave the plant its common name in English. The plant's flowering stem grows up to six feet tall and bears creamy flowers at the ends of slender branches. The numerous leaves are from 2 to 10 inches long, with long sheaths enveloping the stem.

The plants, which occur in pairs, grow six months to a year before they are harvested for their roots. The tubers are first soaked in hot water, peeled, cut into small pieces, mashed to a pulp, and further macerated to break down the cells surrounding the starch. The pulp is then washed on screens to separate the starch from fibers, and the separated starch is dried and ground to a powder. Arrowroot is commonly used as a sauce thickener and is also obtained from other species of the genus.

Further Reading:
Graf, *Exotica.*

* √ *MELIACEAE*

Meliaceae is a family of trees whose members are commonly called mahogany. Many species are logged for their valuable hardwoods. The Spanish began exporting mahogany from Cuba in the late sixteenth century, while mahogany from the West Indies was supposedly first brought to England by Sir Walter Raleigh, who had a mahogany table made for Queen Elizabeth I. The popularity of the mahogany wood grew steadily, and today mahogany has been nearly logged out of its natural range. The wood ranges in color from a light golden, to a rich red, to a deep golden brown, and is quite strong, hard, and durable. Most mahogany trees can reach at least 150 feet in height and 6 feet in diameter.

The most common sources of mahogany are the Neotropical genus *Swietenia* (called *caoba* in Spanish) and the West African genus *Khaya*, particularly *K. ivorensis*. *S. macrophylla* is found from Mexico south to Brazil and Peru, while *S. mahagani*, found in the Caribbean, is no longer logged, due to its scarcity. *S. macrophylla* is the only mahogany species still widely available, with most exports coming from Brazil. Its small flowers are pale yellow and its mature fruit are woody capsules that hold about 40 wind-dispersed seeds.

S. humilis is found only in the dry Pacific forests of Mexico south to northern Costa Rica, while *S. mahagoni* is restricted to Caribbean islands. *K. ivorensis*, whose heartwood is a pale pink to dark reddish brown, isn't quite as strong as the Neotropical species.

Cedrela odorata (Spanish cedar)

Spanish cedar is another member of the *Meliaceae* family that is commercially important. Found in the Caribbean, Mexico, and Central America south to Argentina, it most often grows in humid or dry lowland forest. It colonizes secondary forest, abandoned pastures, and agricultural lands. The tree can reach a height of about 130 feet and bears numerous small, greenish white flowers. The fruits are large, woody capsules that ripen, split open, and shed seeds while still attached to the tree. Though still abundant, Spanish cedar has been widely exploited for timber throughout its range for 200 years, so it is difficult to find large, well-formed trees. It is threatened with extinction in Panama and the Dominican Republic. The wood has a pleasant scent, is durable, and is termite- and rot-resistant, and therefore often used to construct closets for storing clothes, boats, cabinets, and cigar boxes. The bark and leaves are used to relieve headaches.

Further Reading:

Janzen, *Costa Rican Natural History*.
United States Department of Agriculture Forest Products Laboratory <http://www2.fpl.fs.fed.us/TechSheets/Chudnoff/African/htmlDocs%20africa/khayaivor> and <http://www2.fpl.fs.fed.us/TechSheets/Chudnoff/TropAmerican/htmlDocs%20tropamerican/swieteniamacroph.html>.
World Conservation Monitoring Centre, England <http://www.wcmc.org.uk/trees/ced_odo.htm>.

MUSA SPP. (banana)

The most popular fruit in the United States, the banana is indigenous to tropical Asia, from India and Malaysia south to northern Australia. A domesticated hybrid is grown throughout the tropics as a major export fruit. Worldwide, about 11 million tons of bananas are exported, totaling more than $55 billion annually. Ninety percent of the total exports come from Latin American and the Caribbean. The leading banana-exporting country is Ecuador, followed by Costa Rica, where banana plantations total about 100,000 acres. Around the world, some 25 million acres are under cultivation. Most bananas are grown in small gardens for local consumption.

The banana plant is herbaceous, with annual stems. After producing fruit, the stem dies and a new one grows from the parent root. The plant is fast-growing, reaching 10 to 40 feet and producing a crown of quite large, thick, oval leaves. The flowers grow from spikes at the center of the leaf crown. The fruit needs about 10 months to ripen. The larger and starchier plantain (*M. paradisiaca*) is also an important local food crop in the tropics, particularly in Africa.

Further Reading:

Food and Agriculture Organization, Italy <http://www.fao.org>.
Janzen, *Costa Rican Natural History*.
Plant Explorer, USA <http://www.iversonsoftware.com/business/plant/Banana.htm>.
Sancho and Baraona, *Fruits of the Tropics*.

MYRISTICA FRAGRANS (nutmeg)

Nutmeg trees grow in Indonesia, Sri Lanka, the Philippines, the western Pacific Islands, and the West Indies, but probably originated in the Banda Islands in Indonesia. The genus grows best at low altitudes and includes at least 350 different species that all have aromatic fruit and nuts. The tree reaches a height of about 75 feet and has small, pale yellow flowers with no petals. Nutmeg trees are dioecious; only the female bears the oval, yellow, 2-inch fruits. About six months after flowering, the fruit ripens and splits open to reveal the glossy, dark-brown seed. The Malay name for the tree is *pendarah; darah* means "blood." When the thin bark of the tree is cut, the sap that oozes out is pink but dries to a deep brownish red. *Pendarah* also signifies magic; with its blood-colored sap and fragrant nuts, *Myristica* was assumed to hold powerful cures. The bright orange or red covering of the nutmeg seed is called mace, also used as a seasoning. Mace and nutmeg are the only spices that are obtained from the same fruit. The seed is also red, to attract seed dispersers. On some Indonesian islands, "nutmeg pigeons" (*Ducula spp.*) perform this service. They eat the mace covering, but drop the hard-shelled nut to the ground.

Further Reading:
Hargreaves and Hargreaves, *Tropical Trees.*
Missouri Botanical Gardens, USA <http://www.mobot.org/MOBOT/education/tropics/>.
United States Department of Agriculture Forest Products Laboratory <http://www2.fpl.fs.fed.us/TechSheets/Chudnoff/SEAsian%26Oceanic/htmlDocs%20seasian/Myristicaspp>.
Veevers-Carter, *Riches of the Rain Forest.*

* √ *MYROXYLON BALSAMUM* (balsam of Peru)

The balsam of Peru is a massive tree, reaching up to 100 feet in height and some 18 to 36 inches in diameter. It ranges from Mexico to Argentina. It has a straight, smooth gray trunk, and the heartwood, which becomes deep red when exposed, is used for flooring and furniture. The spicy-scented wood yields a gum used to flavor baked goods, candy, chewing gum, ice cream, soft drinks, and syrups. Balsam gum is collected by cutting v-shaped slashes into the bark; harvesting can begin when trees are about 20 years old. One tree yields about 6.5 pounds of gum annually and, if handled carefully, can continue producing gum for 40 years. In 1993, a half-pound of gum in El Salvador, a leading exporter, fetched about $2 to $3. Balsam oil is extracted from the gum by steam distillation and is used in the perfume, cosmetic, and soap industries. In traditional medicines, the gum is used as an antiseptic ointment and for skin diseases.

The common name arose during the seventeenth century, when the resin was shipped to Europe from ports in Peru. A similar species, *M. pereirae*, is more commonly found in Mexico and Central America, where indigenous tribes used the leaves and fruit to treat asthma, rheumatism, and wounds. One legend holds that when the Aztec king Moctezuma decided to conquer what is now Central America, he sent scouts south from Mexico to cultivate the tree so his army would have a way to treat their battle wounds.

Further Reading:
Winrock International, USA <http://www.winrock.org/forestry/factpub/factsh/myrox.htm>.

NEPENTHES SPP. (pitcher plants)

Pitcher plants are epiphytic vines found in the Far East, from India to northern Australia, plus the islands of Madagascar, the Seychelles, and New Caledonia. The approximately 70 species of pitcher plants tend to grow in nutrient-poor areas, such as swamps, so they've evolved a clever, if somewhat brutal, way to get the sustenance they need. The vines climb with the help of tendrils that form at the tip of the leaves. Some of the tendrils develop hollow buckets at their ends. As these pitchers develop, they become inflated with air and in a few days, a lid at the top pops open. The usually brightly colored pitchers secrete nectar that attracts insects, which slip on the waxy surface at the pitcher's rim and tumble into a pool of digestive fluid. There are other insects that live in the buckets and are unaffected by the enzymes. They provide nitrogen to the plant by breaking down the detritus that collects at the bottom of each pitcher. One species of pitcher plant, *N. ampullaria*, hosts a half-inch crab to do this cleanup detail.

N. bicalcarata, found in Borneo, has hollow tendrils that house small ants that freely run up and down the walls of the pitchers and swim in the digestive fluid to no ill effect. Apparently the ants help the pitcher plant by removing some of the more sizeable prey that fall into the deadly well water, insects so large that they would rot before the plant could digest them, and thus cause the plant to decay. The ants ignore the small dead insects, but haul up the larger ones. The ants benefit from this labor-intensive service because once they drag up the prey, they break it into pieces and eat them. Pieces that might slip back into the pitcher are small enough that the plant can quickly digest them. The ants also sip nectar from the pitcher plant's rim. If an insect arrives to nibble on their hospitable pitcher plant, the guardian ants dash under the rim of the pitcher and make loud, frenzied clicking sounds to scare away the intruder.

Further Reading:
Duke University Medical Center <http://www.duke.edu/nplummer/Nepenthes.html>.
Wolfe and Prance, *Rainforests of the World.*

NEPHELIUM SPP.

Nephelium spp. trees are native to the lowland rainforests in Malaya, Borneo, Java, Sumatra, and southern China. There are about 35 species; 25 of them are in Borneo. Rambutan (*N. lappaceum*), kapulasan, mata kucing, and litchi are all domesticated species of *Nephelium*, grown for their delicious fruits. Rambutan fruits are red and yellow and have soft spines. Like all the fruits from domesticated species, the pulp surrounding the seeds is milky white (*Nephelium* means "little cloud" in Greek), juicy, and sweet. Since the kapulasan fruits don't have spines, they are the ones most often exported.

Further Reading:
Veevers-Carter, *Riches of the Rain Forest.*

√ OCHROMA SPP. (balsa)

Balsa is a fast growing tree of the Neotropics. A lover of light, it grows in forest gaps and larger clearings; the tree is an important colonizer of deforested areas. It's common to see the leaves of a balsa tree nearly eaten away by insects. The tree puts its energy into growing fast, not in chemical defenses against predation. Because balsa concentrates on height rather than density, its wood is quite light, even lighter than cork. Balsa wood has long been used by indigenous people of Central and South America to construct rafts and canoes. Balsa wood, the lightest commercially available timber, is used to make life preservers, surfboards, and toys such as model airplanes. Because it is a good insulator, it is used on airplanes to reduce vibrations and noise. In addition to growing in clearings and pastureland, it thrives in lowland forests, especially along rivers; it is also grown in plantations.

Further Reading:
Forsyth and Miyata, *Tropical Nature.*
Janulewicz, *The International Book of the Forest.*

ORCHIDACEAE (orchids)

Orchids are the largest family of plants, with about 20,000 species. They grow nearly everywhere except the most frigid and dry environments, but more species—at least 10,000—are found in the tropics than anywhere else on earth. The Neotropics are particularly rich in orchid species, but even there, they are not evenly distributed. Southern Central America, northwest South America, and the countries that lie along the Andes hold the greatest number of species. Ecuador (3,270 species) and Colombia (2,899) have the most, but the much smaller countries of Costa Rica and El Salvador have the highest number of species per square mile.

Many orchids are epiphytes, growing on other plants, especially in Neotropical forests. One tree in a Venezuelan rainforest was found to hold 47 different species of epiphytic orchid. Most orchids obtain nourishment from the air, rain, or moisture in the soil. To survive an often-dry environment, many epiphytic species have thick, waxy leaves that lose less water than do terrestrial species with thinner leaves. They also may have bulbous stems or leaf bases that store water for dry times. Other species have roots that can soak up moisture like sponges. All orchids depend on mycorrhizae fungi during some phase of their lives. In a mutually beneficial relationship, the fungi grow partly inside orchid roots and help the plant take in water and minerals, while the fungi ingest some of the nutrients the orchids produce through photosynthesis.

Although the shape and size of flowers in the orchid family vary widely, the general floral arrangement is about the same in all species. There are six parts to an orchid flower. The outer three are sepals, and are usually green, while the inner three are petal-like and often beautifully colored. In the middle of the sepals and petals is a column that contains the orchid's reproductive parts. The tip of the column has a cap containing the pollen, which is transferred to other orchids by hummingbirds, bees, moths, and other insects. Bees are the primary pollinators of Neotropical orchids. Many orchid species depend on only one species of bird, bee, or other insect for pollination, which means that if the pollinator disappears from its habitat, so does the orchid.

Because they are so lovely and delicate, orchids have historically been a favorite flower of human beings, who have managed to overcollect many species into extinction or near extinction. Of course, orchid fans also have demanded laws to protect the species that remain. The Orchid Specialist Group of the World Conservation Union (IUCN) has called on orchid collectors to ensure that they do not buy or sell wild orchids. Species such as slipper orchids (*Paphiopedilum spp.*) of Southeast Asia have long been targets of orchid collectors; IUCN estimates that 26 of the 60 species of this genus are seriously endangered in the wild. Overcollecting is a problem for some of the showy species, but thousands of less glamorous orchids are threatened by the continued loss of rainforest.

Further Reading:
Cullen, *The Orchid Book.*
Forsyth, *Portraits of the Rainforest.*
IUCN Species Survival Commission, Switzerland <http://dlp.CS.Berkeley.EDU:8080/iucn/docs/orchids/>.
Kricher, *A Neotropical Companion.*
Pijl and Dodson, *Orchid Flowers: Their Pollination and Evolution.*
Wolfe and Prance, *Rainforests of the World.*

Angraecum spp.

Angraecum spp. include more than 200 orchid species found mainly in tropical Africa but also extending to Madagascar and Sri Lanka. *A. sesquipedale* is found primarily in Madagascar. Charles Darwin described it in 1850. Because the white orchid has a 16-inch spur with a sweet liquid at its base, Darwin surmised that a moth with a 16-inch proboscis was its pollinator and that this long-tongued moth was nocturnal because the orchid is fragrant only at night. Such a moth was unknown at the time of the evolutionary biologist's prediction, which was dismissed. But 50 years later, Darwin's imagined moth, *Xanthopan morgani predicta*, was found in Madagascar. The moth is indeed nocturnal and has a coil on the front of its head, which it unfurls to a length of 16 inches so that it can dip into the sweet liquid of *A. sesquipedale*. The tree-dwelling plant, also called Star of Bethlehem Orchid, is about 14 inches in height when mature, and its white blossoms are often up to 4 inches across.

§ *Cattleya spp.*

Cattleya spp. is one of the smaller genera of orchids, with about 60 species, all epiphytes found in Central and South America, usually high in trees in mountain forests. *Cattleya* is the orchid most often used in corsages.

Further Reading:
Cullen, *The Orchid Book.*

Coryanthes spp. (bucket orchid)

The bucket orchid includes about 30 species found from Guatemala to Bolivia. *Coryanthes* has one of the most complex flowers of all orchids and are often found in conjunction with nests of the ants in the *Camponotus* and *Azteca* genera. They are pollinated solely by male bees of the *Euglossa*, *Eulaema*, and *Euplusia* genus. Attracted by the orchid's fragrance, the bee enters the large flower, slips on the waxy surface, and tumbles into the orchid's lower, hollow lip, which is filled with water secreted by the plant's faucet gland, called the pleuridia. The only escape route is through a narrow spout in the flower, which the bee reaches by brushing against the orchid's sticky stigma and the pollen glands. Pollen is thus securely glued to the bee. The same pollen-painted bee must fall into another *Coryanthes* to complete pollination. If a visiting bee already has pollen on its back, a hook strategically located on the roof of the escape tunnel collects it. Amazingly enough after such an ordeal, the bees don't hesitate to visit another *Coryanthes* orchid.

Further Reading:
Cullen, *The Orchid Book.*
Wolfe and Prance, *Rainforests of the World.*

§ *Encyclia cordigera* (Easter orchid)

A medium-size epiphytic orchid found from southern Mexico to Venezuela, the Easter orchid has olive brown flowers with a white and lavender lip. Pollinated by bees, the orchid flowers in March, around Easter time.

Further Reading:
Janzen, *Costa Rican Natural History.*

Vanilla planifolia

Vanilla planifolia is one of more than 100 varieties of vanilla orchids and the only one that produces fruit. It is also the only orchid grown for something other than its lovely flowers. Native to Mexico, it was brought back to Europe by Spanish invaders. The vines were planted, but the small, yellowish flowers yielded no fruit, which, when produced, are elongated and bean-shaped. The main pollinator of the plant is a tiny bee (*Melipona*), which is not found in Europe. In 1841, a 12-year-old African slave named Edmond Albium developed a method of artificially pollinating vanilla that is now widely used in producing countries. Pollination must be done during the short period of the day that the vanilla flower is open. Flowering season lasts just about a month. The main ingredient in a cured vanilla pod is called vanillin, which, among several other substances produced by the orchid, yields the delicate vanilla flavor that chemists have been unable to duplicate. Because vanilla orchids require painstaking care, vanilla beans and extract are expensive. In 1780, a pound of Mexican vanilla beans fetched $17—a fortune at the time. Wild vanilla vines grow around trees; in farms they must be grown under trees to protect the plant from winds and rains, since the vanilla beans stop growing if the flower is knocked off the vine. Most of the vanilla today is grown in Madagascar, Central America, and the Seychelles.

Further Reading:
Four-K Vanilla <http://members.home.net/abid/backg. html>.
Missouri Botanical Gardens, USA <http://www.mobot. org./MOBOT/education/>.

§ *PACHYSTACHYS LUTEA* (gold candle)

Native to Peru, gold candle is a popular ornamental in tropical and subtropical gardens worldwide. The small, shrubby plant forms clumps of upright stems that have bright, golden-yellow, and overlapping bracts at the tip of each stem. Once the bracts develop, white flowers emerge from their sides. The leaves are a deep green and have prominent veins.

Further Reading:
Armitano, *Garden Plants of the Tropics.*
University of Connecticut, Ecology and Evolutionary Biology Conservatory, USA <http://florawww.eeb.uconn. edu/acc_num/198500167.html>.

§ *PASSIFLORA SPP.* (passion flowers)

Passion flowers include between 350 and 400 species that are native to the Neotropics. Most passion flower plants produce blossoms that have five sepals, five petals, five anthers, and three stigmas. Supposedly, when Spanish missionaries first saw a scarlet passion flower in the New World, they associated it with the passion of Jesus Christ; hence the popular and scientific names. Passion flowers are primarily pollinated by hummingbirds. The nectar is secreted within a doughnut-shaped chamber that is usually surrounded by erect filaments that prevent most insects from entering, but allow easy entry for the long bills and tongues of hummingbirds. Nectar is also stored outside the flower, in parts of the plant called nectaries, which draw ants and other insects. The nectaries protect the plant by attracting insects that prey on plant-eating bugs. Passion flowers are cultivated as climbing vines that cling to walls and fences with aid from long, coiled tendrils.

Some species produce round, edible fruits, such as *Passiflora ligularis,* whose orange skin is easily broken to reveal a sweet pulp that surrounds small, black seeds. *Passiflora edulis* bears what is commonly called "passion fruit." The purple or yellow fruit has a tart golden pulp inside that is blended into a luscious juice.

Further Reading:
Janzen, *Costa Rican Natural History.*
Lennox and Seddon, *Flowers of the Caribbean.*
Sancho and Baraona, *Fruits of the Tropics.*

√ *PERSEA SPP.* (avocado)

Avocados are native to the Neotropics, found in the West Indies and from southern Mexico south to Chile. *P. americana* is the species grown in plantations for its soft-fleshed, subtle-tasting, nutritious, pale green fruit. The tree can grow up to 65 feet high. Other *Persea* species are grown for their timber, which is used in furniture, joinery, construction, crates, plywood, and flooring. The bark of *P. lingue* is used for tanning.

Further Reading:
Sancho and Baraona, *Fruits of the Tropics.*

PHYSOSTIGMA VENENOSUM (calabar bean)

The calabar bean is a sizeable woody vine found in West Africa. It has three-parted leaves, pink or purple budlike flowers, and brown pods that contain large reddish-brown seeds. Eating just a few beans can be poisonous, as they contain several toxic alkaloids. Once the bean was used in "trials" in Nigeria. If the accused merely vomited after drinking a brew made from calabar, then he or she was innocent. If the brew caused more serious poisonings, or death, then the accused was declared guilty. Nigerian law now prohibits trial by calabar bean.

Further Reading:
Blackwell, *Poisonous and Medicinal Plants.*

* *PILOCARPUS JABORANDI*

Pilocarpus jaborandi is a small shrub, a member of the citrus family, which grows in northeast Brazil. The leaves of the jaborandi contain an alkaloid called pilocarpine, commonly used in the treatment of glaucoma—a pressure buildup in the eye that often causes blindness if not treated. When drops or ointment containing pilocarpine are administered, the drug relieves pressure in the eyeball. Pilocarpine is also used to treat xerostomia, or dry mouth, a condition caused by radiation treatments. The Tupi indigenous people of Brazil have long known this treatment; in the Tupi language, "jaborandi" means "what causes salivation." It is cheaper to produce pilocarpine from jaborandi leaves than to produce it synthetically, so Brazilians collect leaves from the wild. Although it's generally less damaging to a plant to collect its leaves than its bark or roots, jaborandi leaf harvesting has begun to have a detrimental impact on the plant's population in the wild. A German pharmaceutical company, E. Merck, spent 10 years developing a healthy jaborandi plantation in Brazil; at the same time, it has developed a system to ensure that wild leaves are collected sustainably.

Further Reading:
Blackwell, *Poisonous and Medicinal Plants*.
ten Kate and Laird, *The Commercial Use of Biodiversity*.

* # *PIMENTA DIOICA* (allspice)

Allspice is a 35-foot tree native to the Caribbean. Its gray bark; dark green, large, and leathery leaves; and greenish white flowers all have a strong and pleasant scent. The allspice fruit is a berry that ripens from green to dark purple and contains two seeds. Allspice berries are harvested by hand when they are unripe and strongly flavored, and are ground to produce allspice powder. Christopher Columbus carried allspice back to Europe; because it was thought to be a kind of pepper, it was given the scientific name *pimenta*, as *pimienta* means "pepper" in Spanish. The common name comes from the perception that the ground spice has the flavor of a mix of cloves, nutmeg, cinnamon, and pepper. Historically and currently, Jamaica is the largest producer of allspice, which is used in catsup, pickles, baking, and men's aftershave lotions. Oil extracted from the seeds is used in liqueurs, such as Chartreuse. In folk medicine, allspice is used to relieve arthritis pain.

Further Reading:
Purdue University, USA <http://newcrop.hort.purdue.edu/newcrop/Crops/Allspice.html>.
University of the West Indies <http://wwwchem.uwimona.edu.jm:1104/lectures/pimento.html>.

PIPER SPP.

Piper spp. are found throughout the tropics, but most of the 500 species are Neotropical. Most are shrubs that reach a height of about 8 feet. *Piper* has tiny flowers that are tightly packed in a long, slender, white, arching or drooping spike, called a spadix, that is borne opposite the leaves. Pollinators are tiny bees and small beetles. The single-seeded fruits are also quite small and packed on the spike. They are commonly eaten by bats, the major seed disperser, which nab them with their mouths while in flight.

P. auritum ranges from Mexico to Colombia. These plants grow to about 20 feet tall and have large, round leaves that smell like sarsaparilla when crushed. They are most commonly found in light gaps in the wet rainforest, rather than in deeply shaded areas.

P. nigrum, the source of black pepper used to flavor foods, originated in India and Cambodia, and as with other spices of Asia, was once worth its weight in gold and still accounts for about 35 percent of the total world trade in spices today. The use of pepper is documented in records from ancient Rome and the Middle Ages. In the twelfth century, one of the most influential guilds in London was the Guild of Pepperers. Pepper was used to pay taxes, rents, and ransoms. The quest for pepper sent explorers out to sea to find new lands and new sources. As with other *Piper* species, the plant's tiny flowers grow on the spadix and mature into peppercorns. When the berries are ripe, they are red. The fruits are harvested when they are immature and green in color. They turn black and wrinkly when they are dried. Milder, white pepper is the inner husk of the berry, obtained by removing the outer husk from ripe, red fruits. Several alkaloids—piperine, piperidine, and chavicin—are responsible for the pepper berry's tangy taste.

Further Reading:
Janzen, *Costa Rican Natural History*.
Missouri Botanical Gardens, USA <http://www.mobot.org./MOBOT/education/>.
Pierot and Hall, *Easy Guide to Tropical Plants*.

* § *PLUMERIA SPP.* (frangipani)

Native to southern Mexico and Central America, frangipani is now grown throughout the tropics as ornamentals. The trees can reach about 15 feet and bear enchantingly sweet smelling blossoms, usually reddish orange, pink, yellow, or white. The 2-inch, waxy flowers cluster together at branch ends and are a source of perfume. The flowers are used in Hawaii for making leis. The dark-green leaves are about 10 inches long and sprout near the ends of the branches. Most species shed their leaves during the dry season. The tree's trunk is rather short and begins to branch close to the ground. Each new shoot branches into two, so the frangipani quickly becomes a sprawling, multibranched tree. The milky sap is poisonous and used in traditional medicines, as are the flowers.

Further Reading:
Lennox and Seddon, *Flowers of the Caribbean.*
Pierot and Hall, *Easy Guide to Tropical Plants.*

‡ *PONGAMIA PINNATA* (pongamia)

Pongamia is native to India, east to Fiji and Australia, and produces a seed that contains an oil that can be substituted for diesel fuel. The oil is driving generators in rural Indian villages, part of a project called Sustainable Transformation of Rural Areas (SU-TRA), directed by the Indian Institute of Science and funded with $10 million from the United Nations and World Bank. Two pounds of pongamia seeds yield slightly more than a half-pound of oil; one tree produces between 20 and 200 pounds of seeds annually. Two pounds of oil generates about three units of electricity, at a cheaper rate than the current price of diesel. After the seeds are crushed to produce the oil, the residuals are a good source of mulch.

Further Reading:
Acharya, "Tree Provides Biodiesel For India."
Purdue University, USA <http://www.hort.purdue.edu/newcrop/duke_energy/Pongamia_pinnata.html>.

* *PRUNUS AFRICANA* (African cherry)

The African cherry is a hardwood tree related to the European cherry that grows in montane forests in Africa, from Cameroon east to Kenya and Madagascar, and as far south as South Africa. It's a light-demanding gap colonizer, a species that grows quickly whenever an opening is created naturally by storms or by people with chainsaws. Extracts from the tree's bark are used to treat benign prostatic hyperplasia, a swelling of the prostate gland. Harvested

from wild populations, the extract is exported to Europe; annual sales are around $150 million annually. As long as a tree is not completely girdled, the bark can be harvested repeatedly. Cameroon has been the leading supplier of African cherry bark to European countries. With as much as 4,900 tons of bark from the trees being exported to Europe for medicinal use, the tree is endangered in several countries and no longer commercially profitable. In Madagascar, people may trek through the forest for three days to search for the bark. They are stripping trees at eight times the rate of regrowth. In order to stave off extinction, at the April 2000 meeting of the Convention on Trade in Endangered Species (CITES), delegates agreed to new controls on trade of *P. Africana* bark.

African cherry doesn't produce seeds until it is 15 to 20 years old, but researchers with the International Centre for Research in Agroforestry in Nairobi have discovered a way to trick the tree into producing more quickly using a process called "air layering." After the bark is removed from a branch, the branch is coated with peat and covered with plastic. Under the right conditions, this induces the branch to produce roots that can eventually be planted directly, cutting the regeneration time to just three years. The center is also helping bark harvesters get paid more for their arduous work; in 2000 their payment stood at less than fifty cents per pound of bark they collected; *P. africana* bark is eventually sold at 200 times that price.

Further Reading:
Acharya, "Prostate Demand Strips Curative Trees."
The Economist, "All Bark, No Bite."
New Scientist Online, "Plant Poachers" <http://www.newscientist.com/news/news_223627.html>.
Tuxill, *Nature's Cornucopia: Our Stake in Plant Diversity.*

* # *PSIDIUM GUAJAVA* (guava)

Guava originated in Latin America. Historians believe it may have been domesticated in Peru several thousands of years ago because at archaeological sites, they have uncovered the seeds of the guava along with beans, corn, squash, and other cultivated plants. European explorers, traders, and missionaries carried the tree to the Old World, and it is now cultivated throughout the tropics. The trees have smooth, copper-colored bark and are fairly small, reaching about 30 feet in height. They produce veined, oblong leaves; fragrant white flowers; and a yellow, lemon-shaped, thin-skinned fruit. Inside, pink or white sweet pulp surrounds a mass of small, hard, white seeds. Called "guayaba" in Spanish, the fruits are used to make jams, drinks, and ice cream. The fruit is also a favorite among birds and monkeys, which disperse the seeds throughout the rainforest. Tea made from the bark and leaves is used in folk medicine to treat diarrhea and dysentery and to regulate menstrual periods.

Further Reading:
Hargreaves and Hargreaves, *Tropical Trees.*
Purdue University, USA <http://www.hort.purdue.edu/newcrop/morton/guava.html>.
Sancho and Baraona, *Fruits of the Tropics.*

QUERCUS OLEOIDES (encino)

Encino is the only lowland oak in the Neotropics. Hundreds of species of *Quercus* exist on Earth, but most occur in the northern, temperate region, with only a few species reaching as far south as Colombia, none in Africa south of the Sahara desert, and none in the region of Papua New Guinea and Australia. Except for *Q. oleoides,* all other *Quercus* species of the tropics grow only in evergreen montane forests above 2,000 feet. Encino ranges from the Gulf coast of Mexico to northwestern Costa Rica. It is unusual among tropical trees because it grows in high densities, dominating all the forests in which it occurs, like mangrove trees. But unlike mangroves, other tree species share encino habitat as well. It can grow to about 150 feet and, like its closest relative—the live oaks of the southeastern United States—does not drop its shiny green, oval leaves. The tree bark is thick, rough, and fire-resistant, which is lucky because most of its habitat has been converted to pastureland that frequently burns. Male flowers are yellow and pendulous and are borne on year-old twigs, while the female flowers lack petals and bloom on new twigs. The thin-walled acorns are popular food for squirrels, pacas (*Agouti paca*), deer (*Odocoileus virginianus*), peccaries (*Tayassu*), and white-faced monkeys. In protected areas where there are substantial mammal populations but isolated encino stands, the trees cannot produce enough acorns both to provide food and to reproduce, so they are effectively sterile. Enough acorns fall in natural (and increasingly rare) encino forests to allow for germination.

Further Reading:
Janzen, *Costa Rican Natural History.*

RAFFLESIA SPP.

Rafflesia spp. are famed for two reasons—they are the world's largest flower, with a blossom that measures some three feet across, and they really stink. As *Rafflesia* matures, it gives off a foul odor like rotting flesh, which makes many species back off, but attracts the bluebottle flies that pollinate it—the same flies that are first on the scene when an animal dies. The leafless flower is a parasite that grows inside the roots or lower stems of only a few species of vines, from which it draws nourishment for up to two years before it erupts into enormous bloom. The cup in the center of the plant can hold more than a gallon of water. The plant is named for Sir Stamford Raffles, an English explorer who stumbled upon it during an 1818 expedition to Indonesia. *Rafflesia* is found only in the rainforests of the Malay islands and is quite rare because most of its habitats have been heavily deforested.

Further Reading:
Line, "The Biggest, Stinkiest Flower."
Newman, *Tropical Rainforest.*

* RAUVOLFIA SERPENTINA (Indian snakeroot)

For centuries, the Sadhus, or holy men, of India chewed the root of Indian snakeroot to achieve a state of detachment while they meditated. It is native to deciduous forests in India, Myanmar, Thailand, Malaysia, Java, and Indonesia. The plant is a low-growing shrub, with red, pink, or white flowers. In some regions, the plants grew in thick tangles, but global demand for Indian snakeroot over the past few decades has greatly diminished wild populations in forests where it was once copious. Roots are usually harvested in December, when the plants have dropped their leaves and the alkaloid content is highest. The most effective alkaloid in the root is reserpine, which in the 1950s became the first Western tranquilizer, revolutionizing treatment for hypertension and mental illness. Resulting demand exhausted wild populations in Java and Indonesia, once major exporters of Indian snakeroot. Efforts to farm the plant have been only partially successful. There is a continuing demand for reserpine—estimated sales of the alkaloid in 1989 were $42 million in the United States alone. But the plant is listed by CITES in Appendix 2 as "threatened with extinction."

Further Reading:
Sheldon, Balick, and Laird, *Medicinal Plants.*

* √ *RHIZOPHORA MANGLE* (red mangrove)

The red mangrove is a principal tree species of the world's mangrove forests, which are found throughout the tropics but particularly in Southeast Asia. Mangrove forests ring shorelines and are flooded daily by the rising tide. Like other mangrove trees, red mangrove can grow in saltwater because its succulent leaves secrete excess sodium and chloride. Red mangrove holds itself above the waterline with arching prop roots that anchor the tree securely in the mud. The tree reaches 65 feet in height with stilt roots of about 13 feet in height. The bark is gray and the leaves oval-shaped, thick, and leathery. The pale yellow, bell-shaped flowers are borne on forked stalks. Inside the oval, dark brown, berry-sized fruit is one seed that germinates on the tree. It forms a cigar-shaped, buoyant structure so that the developed seedling can float on sea currents before washing ashore and taking root. Red mangrove timber is used for construction, poles, shipbuilding, and wharves. Bark extracts are used to stain floors and furniture. The bark is also used in a wide assortment of medicinal cures, to treat everything from backache, diarrhea, and sore throat to leprosy, tuberculosis, and syphilis.

Further Reading:
Purdue University, USA <http://newcrop.hort.purdue.edu/newcrop/duke_energy/Rhizophora_mangle.html>.
Wolfe and Prance, *Rainforests of the World.*

* √ § *RICINUS COMMUNIS* (castor plants)

Castor plants are native to east Africa, but now grow in most tropical or subtropical regions of the world, particularly along river banks. In one season, the castor plant can grow 6 to 15 feet high, so it's become a pest in some regions of the world. Its leaves are pointed and have slightly serrated edges and prominent central veins. The flowers are usually small and green, although pink or red in some varieties, growing in long inflorescences, with male flowers at the base and female flowers at the tips. The fruits have soft spines. When mature, the seedpods become dry and split open, and the seeds are ejected as if shot from cannons. The poison in castor seeds is called ricin, and it is deadly—ounce for ounce, 6,000 times more poisonous than cyanide. It takes only four seeds to kill an adult human being. In 1978, ricin was used to murder a Bulgarian journalist named Georgi Markov, who had written articles critical of the government. While waiting at a bus stop in London, he was stabbed with the point of an umbrella that inserted a ricin-laden, perforated metallic pellet into his leg.

A positive use of the toxic seeds is in treating malignant tumors—a cancer treatment using castor seeds attacks a tumor without damaging other body cells. Castor beans are pressed to extract castor oil, a common health tonic. Ricin is not present in the oil, which is also commonly used in paints and varnishes and the production of nylon. Because it coats well to moving hot metal parts, it is also an important ingredient of motor oil for high-speed cars and motorcycles. The oil has long been used medicinally to treat everything from constipation to dermatitis. Castor plants are frequently grown as ornamentals.

Further Reading:
Cornell University, Poisonous Plants Homepage <http://www.ansci.cornell.edu/plants/castorbean.html>.
Purdue University, USA <http://newcrop.hort.purdue.edu/newcrop/duke_energy/Ricinus_communis.html>.

§ *SAINTPAULIA IONANTHA* (African violets)

African violets include about 20 species of small plants with fuzzy, fleshy leaves, and violet-shaped flowers. They are indigenous to the mountain forests of Tanzania and Kenya, and are widely grown as ornamental houseplants, but are endangered in their native habitats.

Further Reading:
Plant Explorer, USA <http://www.iversonsoftware.com/business/plant/African%20Violet.htm>.

* √ § *SAMANEA SAMAN* (rain tree)

One of the most beautifully symmetrical trees in the world, the rain tree grows to nearly 200 feet in height and has a thick, spreading canopy that can stretch to

250 feet. Simon de Bolivar supposedly parked his entire liberation army under a rain tree in Maracay, Venezuela. It produces fernlike, bipinnate leaflets that fold up at night. Legend said the saman—which is native to Latin America, from Mexico south to Bolivia and Brazil—produced rain at night, probably because the closed up leaflets allowed showers to fall directly to the ground. The tree's flowers are small, delicate white tufts with pink tips that cluster at the end of twigs, each cluster on a green, hairy stem. The long black bean pods have a sweet pulp. The pods provide fodder for cattle, and the durable yellow-brown wood is used for posts, furniture, cabinets, oxcart wheels, and moldings. In Hawaii, the wood is used for building boats and "monkeypod bowls." Different parts of the tree are used in traditional medicines to treat colds, diarrhea, headaches, sore throats, and stomachaches.

Further Reading:
Hargreaves and Hargreaves, *Tropical Trees.*

√ *SANTALUM SPP.* (sandalwood)

Sandalwood is a partially parasitic plant that grows off the roots of host plants. It is found in Southeast Asia and Hawaii, and the pleasantly aromatic wood, with a cedarlike scent, is used to make chests and boxes, while oils from the wood are used in perfumes. The plant's use in Asia, particularly *S. album*, dates back many thousands of years. Because it is parasitic, sandalwood is difficult to propagate, so it cannot be grown commercially. Not surprisingly, there are few sandalwood stands left in the world. Sandalwood trees in India were wiped out centuries ago, and the remaining trees and shrubs in Hawaii are also scarce.

Sandalwood was the first profitable export from the Hawaiian islands, in the late eighteenth century. Throughout the 1800s, merchants bargained with Hawaiian chiefs for the right to log the sandalwood trees, called 'Iliahi in the native language, and shipped many thousands of tons back to Asia and Europe. Several species of sandalwood are endemic to Hawaii, such as Haleakala sandalwood (*S. haleak-alae*), which grows on the slopes of Haleakala volcano on the island of Maui. Lanai sandalwood (*S. freyci-netianum*), a small, gnarled tree that bears small clusters of vibrant red flowers, is found on the island of Lanai and is an endangered species, with perhaps 40 surviving plants.

Further Reading:
Endangered Species Information System <http://fwie.fw.vt.edu/WWW/esis/lists/e701044.htm>.
Southern Illinois University at Carbondale <http://www.science.siu.edu/parasitic-plants/Santalaceae/>.

* ‡ *SAPINDUS SAPONARIA* (jaboncillo, soapberry)

Jaboncillo, or soapberry, is commonly found in dry forests from Mexico to Brazil, and traditionally played an important cultural role in the region. The tree grows to about 80 feet in height and produces amber-colored berries that were once widely used as soap to wash clothes. The fruits contain a high content of lather-producing saponin, which is used in traditional medicine to treat epilepsy, cataracts, and ulcers. Brewed, the oblong leaves are used to treat snakebite. People also tossed jaboncillo leaves into pools to stupefy fish, so that they could easily be caught. Species of the genus *Sapindus* are also found in eastern Asia and Europe, where the berries were also used as detergents.

Further Reading:
USDA Forest Service <http://www.fs.fed.us/database/feis/plants/tree/sapsapd/>.

* # *SCHINUS MOLLE* (Brazilian peppertree)

A small tree that grows to about 20 feet, the Brazilian peppertree is native to Central and South America. It has narrow, spiky leaves and produces many tiny flowers and copious amounts of berrylike fruits. Nearly every part of the tree exudes a spicy-smelling oil, and has long been used in folk remedies. The sap is used as an astringent, purgative, and diuretic. Other parts of the Brazilian peppertree are used to treat fevers, coughs, upper respiratory infections, diarrhea, premenstrual syndrome, and problems related to menopause, and as an antiseptic. The peppery-flavored berries are used in syrups, vinegars, wines, and other beverages, or are ground up and used as a spice.

Further Reading:
Raintree Nutrition, USA <http://www.rain-tree.com/peppertree.htm>.

SELENICEREUS WITTII (moonflower)

The moonflower is an epiphytic cactus of the Amazon. The Amazon River and its tributaries overflow their banks each year, flooding thousands of acres of forest, and the flat, spiny stems of the moonflower keep it tightly attached to tree trunks above the high-water area. The moonflower has bright white flowers that open only in the evening. The flowers have a tube with nectar at the base, which means it is likely pollinated by moths. Scientists are unsure because it isn't easy to catch the moonflower in full flower. Each cactus produces just a few flowers, and the blooms last just one night.

Further Reading:
Wolfe and Prance, *Rainforests of the World.*

SOLANUM QUITOENSE (naranjilla)

Naranjilla is a thorny bush native to the Andean highlands of Colombia, Ecuador, and Peru. It grows about 8 feet high and has thick, spiny stems and soft, wide, fuzzy leaves. The fragrant flowers are white on the upper surface, with purple hairs beneath. Naranjilla is Spanish for "little orange"; the plant's round fruit is bright orange when ripe. Naranjilla, however, is not a citrus, but belongs to the nightshade family, as do the tomato and eggplant. Until the fruit is ripe, it is covered with a brown, hairy, protective coat. Inside, the fruit has four compartments containing a slightly acidic pulp that tastes something like a lemony pineapple.

Further Reading:
Purdue University, USA <http://www.hort.purdue.edu/ newcrop/morton/naranjilla_ars.html>.
Sancho and Baraona, *Fruits of the Tropics.*

§ *SPATHODEA CAMPANULATA* (African tulip tree, flame of the forest, fountain tree)

The fiery red flowers give the name to the African tulip tree, known also as flame of the forest and fountain tree. It is a tall (40 to 50 feet), full tree that continually produces the flowers. Originating from western Africa, it was first identified in Ghana, and is now grown throughout the tropics as an ornamental. The tulip-shaped, scallop-edged flowers grow in a circle, at the center of which is a crowd of buds.

Only a few flowers in the circular formation open at any one time. The buds are often full of water, which can be squirted out one end with a pinch—hence the common name fountain tree. The tree's boat-shaped seedpods may be as long as 2 feet.

Further Reading:
Hargreaves and Hargreaves, *Tropical Trees.*
Lennox and Seddon, *Flowers of the Caribbean.*

STAPELIA GIGANTA

Stapelia giganta are fleshy, cactuslike, and leafless plants native to Africa. They have thick, pale green stems, which form four notched angles. Their flowers are pink and star-shaped, and have numerous horizontal purplish margins edged with white, wooly hairs. They also have a distinctively fetid odor, which attracts flies and other pollinating insects and earns the plant the common name, "carrion flower."

Further Reading:
Armitano, *Garden Plants of the Tropics.*

* *STROPHANTHUS GRATUS*

A large vine found in western tropical Africa, *Strophanthus gratus* has funnel-shaped purple, pink, or white flowers. Its seeds contain a rapidly acting toxin; indigenous people mashed them to a thick fluid, into which they dipped their arrow tips. The chemical ouabain has been isolated from the seeds and is used to quickly stimulate the heart during acute cardiac failure.

Further Reading:
Blackwell, *Poisonous and Medicinal Plants.*

* √ *TABEBUIA SPP.* (Pau D'Arco)

Pau D'Arco trees, found in Central America and tropical South America, have unusually resilient wood. During World War II, the Navy experimented with using *Tabebuia* wood for ball bearings. Extracts from the tree's inner bark have been used for centuries by indigenous people to treat a variety of ailments. Few studies have been done to see how effective the extracts are, but they are still marketed as treatments for cancer, leukemia, baldness, allergies, diabetes, dysentery, malaria, herpes, and acne. Based on reputation, if not scientific fact, *Tabebuia* extract's popularity has caused overharvesting, a common problem with plant-derived medicines when their value suddenly increases. When trees are stripped of too much of their bark, they die. In Peru, *Tabebuia* has been so overharvested that a similar-looking bark, from the *Cariniana* tree, is being passed off as the magical *Tabebuia* in some marketplaces. The ruse seems to be effective, as *Cariniana* trees have also become scarce. *Tabebuia* is also cut for timber, since its high-quality wood is quite heavy and durable.

§ *T. pentaphylla*, also called pink poui or pink cedar, is an emergent that can reach 150 feet, often planted as an ornamental or to shade coffee and cocoa plantations. It is not evergreen, but drops all its leaves before producing clusters of dazzling, deep pink flowers—a rosy blush in the green carpet of the canopy and the national flower of El Salvador.

Further Reading:
Hargreaves and Hargreaves, *Tropical Trees.*
Lennox and Seddon, *Flowers of the Caribbean.*
Sheldon, Balick, and Laird, *Medicinal Plants: Can Utilization and Conservation Coexist?*
USDA Forest Service <http://www2.fpl.fs.fed.us/TechSheets/Chudnoff/TropAmerican/htmlDocs%20tropamerican/tabebuialapach.html>.

* √ *TAMARINDUS INDICA* (tamarind)

Native to Africa and Southeast Asia, the tamarind grows to about 40 feet, with a dense, rounded crown of feathery leaves. Its timber is commonly used in cabinetmaking. The pale yellow flowers have dark red buds, which yield long gray pods. Inside is a dark brown, tart fruit, commonly made into drinks, jams, and candy. The tree's bark, which contains tannin, is used to dye clothes, while the roots and leaves are used to cure a variety of ills, from stomachaches to coughs to morning sickness.

Further Reading:
Hargreaves and Hargreaves, *Tropical Trees.*
National Center for Biotechnology Information, USA <http://www.ncbi.nlm.nih.gov/Taxonomy/tax.html>.

√ *TECTONA GRANDIS* (teak)

Teak, native to India, Myanmar, Thailand, Indonesia, and Java, is one of the most durable timbers in the world—and now one of the most expensive. Native stands have nearly been logged out, as demand far exceeds supply. Teak is farmed in plantations throughout the tropics. Its resilient and beautiful dark golden wood, which turns to a dark brown when exposed, is resistant to decay and termites. The oil in teak's heartwood is what gives the tree its ability to resist the elements. The wood is used in cabinets, flooring, furniture, and boats. The tree can reach 150 feet.

Further Reading:
Food and Agriculture Organization, Italy <http://www.fao.org/WAICENT/FAOINFO/FORESTRY/forestry.htm>.
Tewari, *A Monograph on Teak*; also <http://www.vedamsbooks.com/no7501.htm>.
United States Department of Agriculture Forest Products Laboratory <http://www2.fpl.fs.fed.us/TechSheets/Chudnoff/SEAsian%26Oceanic/htmlDocs%20seasian/tectonagrandis.html>.

THEOBROMA CACAO (cacao)

Cacao probably originated in the lowland rainforests of the upper Orinoco and Amazon River basins in South America. Today it is found as far north as southern Mexico; seeds probably migrated north with people, as cacao is one of the oldest cultivated plants in history and one of the great food discoveries by human beings. Many indigenous tribes believed that cacao was planted by the gods. After the Spanish conquest of Mexico, cacao became an important food crop, and cacao seeds continued to be used as currency until the 1800s. The tree is cauliflorous; the flowers and fruits grow directly from the trunk. The flowers are pollinated by midges and probably by other small, crawling insects. The oblong fruits are about 7 inches long and need about a month to ripen. Inside are five rows of large, dark brown seeds surrounded by a sugary pulp. Monkeys and small mammals gobble up the pulp, leaving the seeds behind. But some early visionary human being in Mexico—probably as far back as 600 to 200 B.C.—had the wisdom to mash the seeds, to mix them with water, chili peppers, vanilla, and other spices, and so concoct a tasty beverage—the first hot chocolate drink.

Although the plant may have originated in South America, its introduction to the human palate occurred in Mexico and Central America. Eventually, it became a major cultivated food crop, although it is very susceptible to fungal diseases. Cacao seeds contain more than 300 chemicals, including two mildly addictive caffeine-like substances, so they have a true pharmacological impact on the brain, as any choco-holic will attest.

Further Reading:

Janzen, *Costa Rican Natural History.*
Young, *The Chocolate Tree.*

* UNCARIA TOMENTOSA (cat's claw)

Cat's claw is a large, woody vine that ranges from southern Mexico to the Amazon basin. The vines can stretch more than 100 feet, literally clawing their way up tree trunks to the canopy via curved thorns that grow near their leaves. Their flowers are small and yellowish white. Indigenous people have used cat's claw as medicine for centuries, and many Latin Americans still commonly use it to treat a wide range of illnesses: urinary tract infections, gastric ulcers, asthma, tumors, diabetes, depression, and viral infections. The vine's bark or root is usually boiled, and the decoction is drunk like a tea. Scientists in the north have been studying the alkaloids in the bark and roots to better understand their properties, which seem particularly effective against stomach and bowel disorders. Researchers in Colombia have found that a mixture of 143 Amazonian plants, including *U. tomentosa*, seems to be effective against the virus that causes AIDS.

Further Reading:

Nutrition Science News <http://www.nutritionsciencenews.com/NSN_backs/Oct_95_NSN/understanding_herbs.html>.
Raintree Nutrition, USA <http://www.rain-tree.com/catclaw.htm>.

VICTORIA AMAZONICA (giant water lily)

Giant water lilies produce the world's largest floating leaf, a saucer-shaped platform measuring nearly eight feet in diameter. Giant water lilies are found in the muddy backwaters of Amazon lakes. When the rains abate and water levels drop, the lilies are often stranded on dry ground and decompose. The root base may live, however, and a new plant sprouts when the Amazon basin floods with the next season's rains. The lilies' white flowers, as large as soccer balls, are stimulated to open when darkness falls. They release a strong, sweet scent, like pineapple and butterscotch, which attracts brown scarab beetles (*Cyclocephala hardyi*) that spend the night inside the flowers, feeding on starchy appendages. When morning comes, the flowers close with the beetles trapped inside. With the appendages eaten away, pollen is released and strewn all over the captive beetles. The lilies' petals gradually change from white to purple, and when they open the second night, the pollen-coated beetles stagger off to find another fragrant white blossom in which they can pass a pleasant night and, not coincidentally, brush off the pollen they carry on the stigma of the new flower.

Many Amazon wildlife species make excellent use of the giant water lily's vast floating leaves. One is the jacana (*Jacana jacana*), a rail-sized, black bird with bright yellow feathers on the underside of its wings. Jacanas have long toes and nails that allow them to distribute their weight in such a way that they can dash across the water's surface or walk on vegetation. They often lay their eggs atop giant water lily leaves.

Further Reading:

Goulding, *The Flooded Forest.*
Wolfe and Prance, *Rainforests of the World.*

Gymnosperms: Conifers, Cycads, and Gnetophytes

> The gymnosperms are the most ancient seed plants in the forest. Only five of the original eight orders still exist, totaling about 700 species. Gymnosperms are nonflowering plants with vascular systems. Their seeds are always borne upon leaves that are arranged in cones and exposed to the air during all stages of development, the opposite of angiosperm seeds, which are protected and hidden during at least some stage of development.

CONIFERS

Conifers include 550 of the 700 known species of gymnosperms. They are more often found in temperate zones than in the tropics, where there are no firs, spruces, true cedars, or cypresses. Southern forests do, however, have coniferlike trees. Most occur south of the Tropic of Capricorn, but a few extend north into tropical areas. Conifer pollination counts on the wind to blow the abundant yellow pollen from the male cones to the female cones. Leaves are usually needle-shaped or scalelike, and nearly all are evergreen. The trees generally have straight trunks with horizontal branches that vary regularly in length from top and bottom, so they have a conical outline.

Further Reading:
Janulewicz, *The International Book of the Forest.*

§ *Abies guatemalensis* (pinabete)

Pinabete grows in the humid highlands of Guatemala, with some stands as well in Mexico, Honduras, and El Salvador. It reaches 160 feet in height and bears 4-inch cones. The pinabete grows in a graceful conical shape, with deep green, needle-shaped leaves. Its pale wood is used for furniture and construction. In Guatemala, the tree's shape and fragrant aroma has made it the most popular Christmas tree. As a result, it is endangered in Guatemala; each December, about 150,000 trees are cut and sold.

Further Reading:
Gymnosperm Database <http://www.geocities.com/Rain Forest/Canopy/2285/pi/ab/guatemalensis.htm>.

√ *Araucaria spp.*

Araucaria spp. are native to most of the tropics but Africa. One of the most important timber trees in South America is the paraná pine (*A. angustifolia)*, or candelabra tree, found in the hills of southern Brazil, extending into Paraguay and Argentina. Its height reaches 150 feet, usually with a clean and straight trunk with branches that extend like a candelabra and end with a flat-topped crown. It is a particularly important timber species in Brazil, but has nearly been logged out. The original extent of the tree has declined by more than 80 percent in the past century, and it is now an endangered species. In Argentina, *Araucaria* covers less than 2,500 acres—all that remains of a stand that in 1960 carpeted some 500,000 acres. The seeds are fed on and dispersed by a number of wildlife species, in particular the red spectacled Amazon parrot (*Amazona pretrei)*. Brazilians also harvest some 3,400 tons per year of fruit and seeds for local consumption. The timber is mainly used in construction, furniture, and veneer. It is also used in Brazil to make musical instruments, boxes, and matches. The Brazilian government will permit only licensed cutting of paraná pine, and Paraguay, where only a small stand of the pine remains, has declared the species protected.

√ *A. bidwillii* (bunya pines)

Bunya pines are located in Australia's Queensland coastal rainforest. The pines produce a tasty, hard-shelled nut. Bunya pines were logged nearly to extinction; now they are protected in 23,000-acre Bunya Mountains National Park. The conifer is still grown in plantations, and its cream-colored wood is used for veneers, plywoods, and boxes. Another conifer common to Queensland rainforests is the hoop pine (*A. cunninghmaii*).

Further Reading:

Gymnosperm Database <http://www.geocities.com/Rain Forest/Canopy/2285/ar/ag/robusta.htm>.

World Conservation Monitoring Centre, England <http://www.wcmc.org.uk/trees/ara_ang.htm>.

√ *Agathis*

Agathis trees are found in Malaysia, some Pacific islands, Australia, and New Zealand. Most have been cut for timber, as the tree grows tall and straight, and its wood is fine-grained and dense. In Indonesia and the Philippines, *A. dammara* is tapped for its sticky resin, which is made into copal, used in the manufacture of paints and varnish. The kauri pine (*A. robusta*) is one of the dominant and emergent trees of Australia's Queensland coastal rainforest. It towers up to 140 feet, with a heavily branched crown. The timber is used for plywood, cabinets, furniture, and flooring, but it has nearly been completely logged out and doesn't grow well in tree farms.

Further Reading:

Janulewicz, *The International Book of the Forest.*
Veevers-Carter, *Riches of the Rain Forest.*

§ CYCADS

Cycads were the most prevalent plants when dinosaurs ruled the earth. They are called "living fossils," as they've changed very little in the last 280 million years. They look like palms with full and feathery, fernlike leaves, but are actually more closely related to conifers. They are dioecious; females produce the seeds, and males produce cones that contain pollen. The cones of some species can be as long as 2 feet and weigh as much as 60 pounds. Their seeds are rather large and often brightly colored, in shades of red, purple, and yellow. Some species reach 80 feet high; these taller species may be pollinated by the wind, but cycads that grow in the understory are more likely to be pollinated by beetles, bees, and other insects. They grow slowly and reproduce infrequently. Cycads contain chemicals that are toxic to people and most other mammals.

There are about 200 species of cycads. They are most prevalent and diverse in Central America. Many species have only recently become extinct due to deforestation and overzealous collectors. Dozens of other species are endangered or threatened. Some species can live for more than two centuries, and collectors particularly value these, and will pay top price for a plant. *Encephalartos spp.*, native to Africa and Southeast Asia, can be sold for thousands of dollars; most poached plants are taken from the wild. People of several Asian Pacific countries eat the pith of the *Encephalartos* stem, so the cycad's extraction has an effect on their food supply.

The most widely distributed genus is *Zamia*, which ranges from the southern United States to northern Chile and grows in lowland rainforest up to about 3,000 feet. *Zamia fairchildiana* is found only in southwestern Costa Rica and neighboring Panama. Similarly, many cycads species are endemic to small regions.

Bowenia spp.

Bowenia spp., restricted to Australia's rainforests, particularly near streams, are the only cycads with bipinnate leaves. Their young parts bear short, curved, colored hairs. When they were more populous, their adult fronds were sold commercially for their long-lasting characteristics and deep green color. The seeds and leaves of *Bowenia* contain a toxin that can be harmful to people and animals.

* # ‡ *Cycas spp.*

The 40 species of *Cycas spp.* are widely distributed in tropical regions, but concentrated in Southeast Asia and Australia. Australian species do well in the dry tropical climate in full sun. They have a bluish shade that fades in *Cycas* that grow in shadier, cooler areas. *Cycas* stems bear sago that is used as a starch, while the plant's woolly hairs are used to stuff linens, as roof thatch, and to weave baskets. Their tall stems are used as house supports. Their leaves provide a gum used to soothe snake and insect bites.

Further Reading:

Janulewicz, *The International Book of the Forest.*
Janzen, *Costa Rican Natural History.*

Jones, *Cycads of the World.*
Kubitzki, *The Families and Genes of Vascular Plants.*
University of California, Museum of Paleontology, USA <http://www.ucmp.berkeley.edu/seedplants/cycadophyta/cycadlh.html>.

GNETOPHYTES

Gnetophytes include only three orders, Ephedrales, Welwitschiales, and Gnetales, each of which contains only a single genus, for a total of some 70 species.

One is *Gnetum,* composed of 30 species. *Gnetum* grows in tropical and subtropical regions, usually as lianas, although some may grow as shrubs or small trees. They have large and flat leaves. *G. gnemon* is a small tree of Indonesia that is cultivated for its edible seeds.

Further Reading:

Gymnosperm Database <http://www.geocities.com/Rain Forest/Canopy/2285/gn/gn/leyboldii.htm>.
University of Georgia, Department of Botany, USA <http://www.botany.uga.edu/Tour3.html>.

Pteridophytes: The Ferns

There are three orders in Filicopsida, the class of ferns: Ophioglossales, Marattiales, and Filicales. Ophioglossales and Marattiales contain primitive ferns whereas Filicales include the most abundant and complex fern species. Ferns do not produce flowers and differ from algae and mosses primarily because they possess internal vascular structures.

During the Carboniferous era, 350 million years ago, *Pteridophytes* were the most prevalent plants on Earth. The decomposed and compressed remains of these ancient fern ancestors today provide us with coal and natural gas. About 12,000 fern species are found worldwide; 65 percent of them flourish in the lowland tropics. The Malaysian Peninsula alone holds 500 fern species, while tiny Costa Rica has 900 species, twice as many as in the United States and Canada combined.

Ferns have fronds that radiate out from the center of the plant. Individual fronds may be miniscule one-cell structures or as complex and thick as 13 feet. Young fronds are usually coiled in an arrangement known as circinate vernation. Ferns reproduce via their spores, thousands of which sit on the undersides of their fronds and are dispersed by the wind. They are self-fertilizing, producing both male and female sex organs. Most fern stems are rhizomes that grow horizontally along the ground or just underneath the surface of the soil. Some stems are as narrow as threads, particularly in tropical, epiphytic ferns.

Ferns grow on the forest floor, along riverbanks, under rocks, on tree branches, or in mangroves. In dryer regions, they proliferate around marshes, streams, lakes, and lagoons. Ferns can be epiphytes, growing on other plants, or lithophytes, growing on rocks. Both receive nourishment from decomposing matter.

A few ferns are true aquatics that are able to flourish while fully and permanently immersed in freshwater. Tiny *Azolla pinnata* (mosquito fern) grows on the surface of rice-paddy fields in Southeast Asia. It hosts an alga that converts nitrogen to a form that's usable by other plants, so its presence increases rice-paddy productivity. *Stenoclaena palustris* thrives in freshwater swamps. People pick and eat their young, rosy-colored fronds. *Acrostichum spp.*, which include the mangrove fern, are among the few species that can grow in salty, muddy inlets and mangroves.

Another water-loving fern, *Salvinia molesta*, has become a pest in Papua New Guinea and other tropical countries, where it proliferates on the surface of polluted ponds and lakes, making transportation nearly impossible and suffocating fish. It is virtually impossible to remove, as each reproductive leaf forms a new plant.

Where forests have been replaced by cropland and pastures, the ferns that thrive in shady and humid climates are displaced by weedy, sun-loving ferns. *Gleichenia, Dicranopteris,* and *Pteridium* can grow in poor soils and help prevent soil erosion with their rigorous and extensive system of rhizomes, but they also crowd out other fern species. *Pteridium spp.* (bracken) spreads rapidly via its ropelike, underground rhizomes that rapidly invade fields in both temperate and tropical regions. *Pityrogramma calomelanos*, or silver-backed fern, is a common weed that has colonized the oil-palm plantations in central and West Africa. The fern originated in South America, but over the past two centuries, it has become a nuisance in nearly every tropical region in the world. It produces tiny, abundant, and easily dispersed spores.

People use ferns as medicines and food. Edible young fronds, known as crosiers or fiddleheads, possess carcinogens and so must be eaten in moderation.

In many countries, however, the poisonous *Pteridium spp.* is considered to be a tasty delicacy, as long as it's eaten in small doses. In marketplaces in New Guinea, Malaysia, and the Philippines, the young, uncurled fronds of *Diplazium esculentum* and *Pneumatopteris sogerensis* are regularly sold. The fronds are commonly boiled in coconut milk with sweet potato or banana.

Fronds of a few species, such as those of genus *Pteridium*, and rhizomes of the species *Dryopteris filix-mas* have been used as an alternative to tea and a substitute for hops in beer. In Costa Rica, *Equisetum giganteum* is used to treat kidney failure. Various fern species are also used as remedies for parasitic worms, rheumatism, bowel disorders, and bleeding. To treat burns, sprains, bites, and stings, a paste made from mashed ferns is applied to the affected area.

ADIANTUM SPP. (maidenhair ferns)

Maidenhair ferns can grow under slight shade, but they typically flourish in sunny areas. They are most common next to waterfalls and rivers along the more open edges of the forests of Central and northern South America. *A. capillus-veneris*, also known as Venus's-hair fern or southern maidenhair, is found in tropical forests worldwide. It reaches about 14 inches high, with leaves that are up to 20 inches long. In the Peruvian Amazon, where it's called "culantrillo," it is used as a diuretic to treat coughs, urinary disorders, and rheumatism. In Brazil, it's known as "avenca" and is used to cure respiratory infections. Worldwide, it's used to remedy menstrual problems.

ANGIOPTERIS EVECTA (elephant fern)

The elephant fern is a majestic plant with enormous fronds that emerge from a thick brownish protrusion—its stem. The stalks are generally swollen with water, and the pressure causes the fronds to stand erect. When the climate is dry, the fronds droop. *A. evecta* is found in lowland forests. People often harvest its edible, young shoots.

§ *ASPLENIUM SPP.*

Asplenium spp. is a large and highly varied genus, with leaves that may either be shaped like straps or be extremely fine and delicate. Some species have threadlike rhizomes and are high climbing, while others have erect crowns that may form short trunks. The tropical species *A. antiquum* and *A. bulbiferum* are favorites among fern enthusiasts. *A. nidus* is an epiphyte that dominates the trees, rock faces, and boulders of the humid, tropical rainforests. Their shiny green fronds naturally form a basket that traps organic debris.

§ *NEPHROLEPIS BISERRATA* (sword fern)

Native throughout the tropics, the sword fern is commonly grown as an ornamental. Fronds are bright green and can each span one foot in width and four feet in length.

Further Reading:

Australian National Herbarium <http://www.anbg.gov.au/projects/fern/taxa/index.html>.
Chin, *Ferns of the Tropics.*
Croft, "Ferns and Man in New Guinea."
Flora of North America <http://www.fna.org/Libraries/plib/WWW/Pteridophytes/T40032854.html#26602671>.
Jones, *Encyclopedia of Ferns.*
University of Connecticut, Ecology and Evolutionary Biology Conservatory, USA <http://florawww.eeb.uconn.edu/acc_num/198500923.html>.

Top: During the annual rainy season in South America, the downpours spill over the banks of the Amazon River and its hundreds of tributaries, flowing into surrounding floodplains. The Amazon's flooded forest encompasses at least 25 million acres, an area larger than England. © Michael Goulding

Bottom: The view from 150 feet high in the canopy of a Costa Rican rainforest is spectacular. Scientists estimate that the little-explored forest canopies are home to more than half the world's total biodiversity. © Gary Braasch

Top Left: Tropical forests at mid-elevations on mountain ranges and volcanoes are called cloud forests. Plant growth is slower here, since the nearly constant cloud cover makes photosynthesis more difficult, but mosses, ferns, orchids, and bromeliads flourish. Shown here is the cloud forest of Mombacho volcano in Nicaragua. © Gerald Bauer

Top Right: The layer of forest beneath the canopy, or tallest treetops, is called the understory. Shade-tolerant species abound here; palms often dominate, particularly in the Neotropics, where there are more than 1,000 palm species. © Gary Braasch

Bottom Left: Most tropical trees have buttressed roots. Scientists are unsure what purpose the buttresses serve, but guess that they provide support to the towering trees, which have shallow roots. © Gary Braasch

Bottom Right: Commercial logging of tropical forests accounts for some 14 million acres of forest lost each year. © Chris M. Wille/Rainforest Alliance

Top Left: Cassowaries (*Casuariidae*) are huge, flightless birds that stand about five feet tall and weigh as much as 130 pounds. All three species are found in Papua New Guinea. © Art Wolfe

Top Right: The strange-looking hoatzin (*Opisthocomus hoazin*) lives in the flooded and mangrove forests of the Amazon, the Orinoco River in Venezuela, and the Guianas. It can barely fly and makes only short treks from roosting trees to feeding trees. © Michael Goulding

Bottom Left: Small and local populations of spectacled owls (*Pulsatrix perspicillata*) are found from southern Mexico to western Ecuador, Bolivia, and northern Argentina, where deforestation has greatly reduced their habitat. © Gary Braasch

Bottom Right: Considered by many to be the most beautiful bird on Earth, the resplendent quetzal (*Pharomachrus mocinno*) nests in cloud forests from southern Mexico to Panama. © Art Wolfe

Top Left: Hornbills (*Bucerotidae*) are important seed dispensers in their forest habitats of sub-Saharan Africa and southeast Asia. They may fly more than 100 miles, scanning the canopy for ripe fruit. © Gerry Ellis/GerryEllis.com

Top Right: The Andean cock-of-the-rock (*Rupicola peruviana*) is found in the Andes Mountains from Colombia to Bolivia. © Art Wolfe

Bottom: Assassin bugs (*Reduviidae*) swiftly snatch their prey, which can be nearly any kind of insect, insert their beaks, and inject paralyzing saliva. © David Julian

Top: Worker leaf-cutter ants (*Acromyrmex spp.* and *Atta spp.*) of the Neotropical forests snip off leaf pieces and carry them back to the nest, where they are cut into smaller pieces, chewed, spit out, and planted. The ants feed on the fungus that grows in the leaf garden. © Gerry Ellis/GerryEllis.com

Bottom: Found from Mexico to Colombia, the morpho butterfly (*Morpho peleides*) has a slow and loping flight. Its dazzling blue wings may serve as a defense mechanism, blinding predators with blazing flashes of color. © David Julian

Top Left: The babirusa (*Babyrousa babyrussa*) is one of three wild pig species found in tropical Asia; the endangered babirusa inhabits forested riverbanks in just a few Indonesian islands: Sulawesi, Togian, Sulu, and Buru. © Art Wolfe

Top Right: The clouded leopard (*Neofelis nebulosa*) is an endangered wild cat that inhabits dense tropical forests of southeast Asia. © Art Wolfe

Center: The largest American wild cat, the jaguar (*Panthera onca*) is shy and solitary. © Gerry Ellis/GerryEllis.com

Bottom: The tayra (*Eira barbara*) is a member of the weazel family, found from southern Mexico to northern Argentina. Diurnal, it hunts for rodents and other small mammals, nesting birds, lizards, eggs, fruits, and honey. © Art Wolfe

Top: At home in the trees, the binturong (*Arctictis binturong*) is the only carnivore in the Old World that has a prehensile tail. © Art Wolfe

Center: Biologists believe that just 50 to 60 Java rhinoceros (*Rhinoceros sondaicus*) survived in Java, Indonesia, until 1999, when a tiny population was found in Vietnam. The Java rhinoceros is one of the most endangered animals on Earth. © Art Wolfe

Bottom: Found from Panama to northern Argentina, the capybara (*Hydrochaeris hydrochaeris*) is the world's largest rodent. It lives in small groups of about a dozen animals, grazing on grasses along rivers and flooded meadows. © Michael Goulding

Top Left: Endemic to Colombia, the cotton-top tamarin (*Saguinus oedipus*) is highly endangered due to deforestation, hunting, and capture for the local and international pet trade and the biomedical industry. © Art Wolfe

Top Right: The black-and-white colobus monkey (*Colobus guereza*) is disappearing from its forest habitat in west central and eastern Africa as the primate is hunted for meat and its glossy black coat, used to trim coats and dresses. © Art Wolfe

Bottom Left: Found only in Madagascar, the aye-aye (*Daubentonia madagascariensis*) lives a solitary life in the treetops. © Art Wolfe

Bottom Right: A picky eater, the white-handed gibbon (*Hylobates lar*) swings through the trees in its southeast Asian habitat, searching for the fruits it prefers. © Gerry Ellis/GerryEllis.com

Top Left: Lemurs spend virtually all their lives in the treetops of Madagascar; only the ring-tailed (*Lemur catta*) occasionally walks along the forest floor. Because Madagascar has just 10 percent of its forests left, all lemurs are extremely endangered. © Gerry Ellis/ GerryEllis.com

Top Right: A slow-moving arboreal primate, the potto (*Perodicticus potto*) lives in tropical forests of Africa, from Sierra Leone to western Kenya. © Art Wolfe

Bottom Left: The mountain gorilla (*Gorilla gorilla beringei*) is found only in the high altitude forests along the borders of Congo, Uganda, and Rwanda. Probably fewer than 200 remain in the war-torn region. © Gerry Ellis/GerryEllis.com

Bottom Right: The orangutan *(Pongo pygmaeus)* was once found throughout southeast Asia, but now is restricted to Borneo and Sumatra. Fewer than 20,000 orangutans survive—or some 30 to 50 percent fewer than just a decade ago. © Gerry Ellis/ GerryEllis.com

Top: This Brazilian fisherman's boat is laden with rock-baçu (*Lithodoras dorsalis*), a large catfish that moves in and out of estuarine forests in the Amazon. © Michael Goulding

Center: Eighty million years ago, ancestors of the spectacled caiman (*Caiman crocodilus*) looked very much the same as the reptiles do today. Spectacled caiman live in the rivers and wetlands of the Amazon River basin, where they are heavily poached. © Art Wolfe

Bottom: Found from Mexico to Argentina, the insect-eating basilisk (*Basiliscus basiliscus*) is famous for its uncanny ability to dash across the water. © Gary Braasch

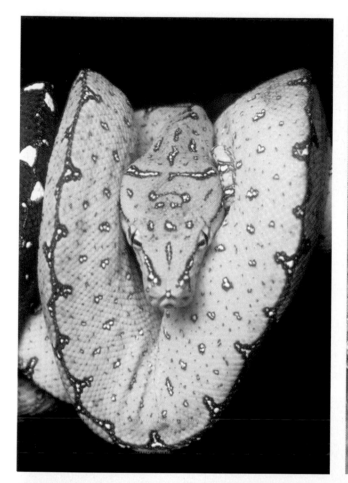

Top Left: Completely arboreal, the green tree python (*Chondropython viridis*) is endangered in its range of Papua New Guinea and far northeastern Australia. The female python descends to the ground only to lay eggs. © Gerry Ellis/GerryEllis.com

Top Right: A kind of pit viper, the fer-de-lance *(Bothrops atrox)* is feared throughout its range from Mexico south to Peru, as its venom is fast-acting and potentially lethal to human beings. © Gary Braasch

Bottom: About three feet long, the giant Amazon river turtle (*Podocnemis expansa*) is found in the Orinoco and Amazon river basins in Venezuela, Colombia, Peru, Bolivia, and Brazil. They are an endangered species because after the females lay eggs in the sand, people dig up millions of the eggs to eat and to sell. © Michael Goulding

Top: Poison-arrow frogs (*Dendrobates spp.*) have toxins in their skin that can cause convulsions, paralysis, and eventually death if they enter an animal's bloodstream. They are usually bright hues, to warn off would-be predators. © David Julian

Bottom: Red-eyed leaf or tree frogs (*Agalychnis callidryas*), found from Mexico to Panama, live in the forest canopy, where they hide in leaf foliage during the day and hunt for insects at night. © David Julian

Top Left: Native to Brazil, Colombia, Venezuela, Ecuador, and Peru, the Brazil nut tree (*Bertholletia excelsa*) is an important source of food and income for hundreds of thousands of Amazon residents who work in the Brazil-nut industry, which generated more than $65 million worldwide in 1998. © Gary Braasch

Top Right: Epiphytes are plants that grow on other plants. In cloud forests, epiphytes comprise some 40 percent of the total plant biomass. © Gary Braasch

Bottom Left: Ceiba or kapok trees (*Ceiba pentandra*), like this one in Costa Rica, can reach 200 feet. They are native to the Neotropics as well as West Africa, a strong indication that these two continents were at one time connected. © Gary Braasch

Bottom Right: Monkey ladder vines (*Entada gigas*) twist through the Neotropical rainforest canopy and are used by monkeys and other animals as footbridges through the treetops. © Gary Braasch

Top Left: While there are other latex-bearing trees, *Hevea brasiliensis* provides the best quality. Supplying the demand for rubber completely altered the Amazon forest, as rubber tappers poured into the region at the turn of the twentieth century. To collect rubber, tappers score the bark of a tree and harvest the latex that oozes out. © Michael Goulding

Top Right: Once the most populous plants on Earth, tree ferns, or cycads, have changed very little in the past 280 million years. The 200 or so species are found throughout the tropics but are most diverse in Central America. © Gary Braasch

Bottom: Red mangrove (*Rhizophora mangle*) is a principal species in the world's mangrove forests, particularly in southeast Asia. Like other mangrove trees, red mangrove can tolerate saltwater because its succulent leaves secrete excess sodium and chloride. © Gerald Bauer

Top: The Yanomami live in the Amazon rainforest on the border of Brazil and Venezuela. © Art Wolfe

Bottom: Fuel wood collecting is a leading cause of deforestation. Wood provides nearly 60 percent of all the energy used in Africa, 15 percent in Latin America, and 11 percent in Asia. This young boy is hauling wood home in El Salvador, a country that has lost 98 percent of its original forest cover. © Chris M. Wille/Rainforest Alliance

Top: Deforested hillsides, like this one outside Guatemala City, pose particular dangers during the frequent tropical rainstorms. Without trees to hold soil in place, rain crashes onto the earth, often causing destructive, sometimes fatal, mudslides and filling once-clear rivers with dirt and debris. © Gerald Bauer

Bottom Left: Throughout the tropics, communities are establishing tree nurseries, like this one in Costa Rica, so they can reforest degraded land to protect their watersheds or to provide a future source of fuel wood, timber, and food. © Chris M. Wille/ Rainforest Alliance

Bottom Right: Scientists use such technology as radio telemetry to learn about wildlife to improve their understanding of how to save endangered species. These researchers are tracking the movements of the resplendent quetzal in a Guatemalan cloud forest. © Chris M. Wille/Rainforest Alliance

PART III
People and Tropical Forests

No species has a greater impact on tropical forests than human beings. For millennia, people have lived in forests, finding all they needed to survive from the plants and animals that share their habitat. Some 50 million people continue to live in tropical forests, but they all face uncertain futures. More than 6 billion more people live outside of forests, and no matter how far they dwell from rainforests, the decisions they make about how they live have an impact on whether forest dwellers— whether human, plant, or animal—will survive.

Forest-Dwelling Indigenous Groups

Indigenous people are the descendants of the original inhabitants of land. More than 1,000 rainforest cultures exist today, usually living in small, scattered villages, so the forest can support their numbers, just as it did their ancestors. They use forest resources, but a majority also grow crops. When the size of their territory permits, they grow food on cleared land for a few years, then move on to new land, allowing the original croplands to recover. This system of farming works well in the rainforest, where soils are generally poor. But often, governments have restricted indigenous people to territories that are too small to allow for shifting cultivation or to provide enough wildlife for food or the other natural resources on which the forest dwellers have traditionally depended.

Many indigenous groups have struggled to keep their lands and maintain their traditional languages, social values, and cultures. Most groups that depend on forests are threatened by nonindigenous ranchers and farmers, logging, mining, dams, and other development projects, which all encroach on their territories. Only a few indigenous groups have been granted secure title to the lands on which they have lived for generations. Because they have no land titles, their homes can be taken away at the whim of politicians, with a penstroke of a lawyer, or at gunpoint by ranchers. Contact with non-indigenous people can have a devastating effect on tropical-forest people, who have not developed resistance to contagious diseases common among non-indigenous people, nor do they have access to modern medicines. Contact with outsiders can also be damaging to native traditions and cultures. Millions of indigenous people have been forced or have chosen to abandon their traditions, language and territory. Millions more continue to live in or close to the forests on which they have always depended for resources. There are an estimated 50 million tribal people living in tropical forests, out of a total of 300 million indigenous people worldwide. Virtually all have at least some contact with non-indigenous people. The list below describes 32 indigenous groups, some of the true people of the rainforest. They are grouped alphabetically by region.

Further Reading:
Alcorn, "Indigenous Peoples and Conservation."
Chapin, "Losing the Way of the Great Father."
Rainforest Information Centre, Australia <http://forests.org/ric/Background/people.htm>.

AFRICA

The Aka

The Aka number about 40,000 and live in forested areas of southwestern Africa, where they survive by hunting and gathering. They provide produce and bush meat to nearby villages. The men may also work as day laborers, but are usually paid less than non-indigenous men for the same work. Catholic missionaries have attempted to establish pilot villages to encourage the Aka to abandon their nomadic existence, but without much success. Their future is closely linked to the survival of the forest. They have no voice when deals are made between the government and logging companies.

The "Pygmies" of Africa

The "Pygmies" never refer to themselves by this name. It is a collective term sometimes used to describe the approximately 200,000 indigenous groups of the forested areas of Burundi, Cameroon, Central African Republic, Democratic Republic of Congo, Congo, Equatorial Guinea, Gabon, Rwanda, and Uganda. The word *pygmy* derives from the Greek *pyme*, which refers to a unit of measure equivalent to the distance from the elbow to the knuckle. There is no pygmy culture or solidarity. A group of forest dwellers in one part of Africa, whom Europeans called "pygmies," were unaware of a group of forest dwellers in another part of Africa, although Europeans referred to them as "pygmies" as well.

Indigenous tribes of Africa were nomadic hunters and gatherers. Today governments and loss of forest resources have forced most to live in permanent settlements, although they still hunt and gather what forest resources remain. Virtually all have been affected by logging, farming, gold and diamond mining, and other development, which have destroyed their homes and the forests. Few groups receive any health, education, or other services from governments. They frequently trade with nearby farmers, exchanging honey, medicinal plants, and meat for grains and other goods. A growing population in Central Africa has caused a stiff demand for bush meat, which is wild game meat, especially monkeys, so overhunting is becoming a serious problem, as indigenous groups sell the meat to outsiders, one of the few ways they have to earn currency.

Further Reading:

Living Earth, England <http://www.gn.apc.org/LivingEarth/RainforestDB/People/>.

Rainforest Information Centre, Australia <http://forests.org/ric/Background/people.htm>.

Further Reading:

Gonen, *The Encyclopedia of the Peoples of the World*.

The Akan

The Akan form distinct kingdoms in southern and central Ghana and southeastern Cote d'Ivoire and total between five and six million people. They have in common a tonal language known as Twi, which has many dialects. Nearly all Akan are forest dwellers. They also farm yams, sweet potatoes, and plantains—all forest plants—along with food-bearing trees such as palms and cacao. Cotton and silk weavings and woodcarvings are important sources of income, as are minerals, such as gold and bauxite. Men and women share in labor, and both may own farms and houses. Until the end of the nineteenth century, the Akan practiced slavery, obtaining slaves from dealers to the north, usually Muslims. Each Akan kingdoms is ruled by a different king, who is elected by various officials, chosen from a royal line. While missionaries converted most Akan to Christianity in the nineteenth century, the kings maintain the indigenous religious beliefs, which is based upon worship of a high god, various spirits, and ancestors.

Further Reading:

Levinson, *Encyclopedia of World Cultures*.

The Baka

The Baka of southeast Cameroon depend on the forests for nearly everything they need to survive. They live in villages scattered throughout 31,000 square miles. A very rough estimate of their total population is 30,000. During the three-month rainy season, when the forest is at its most productive, the Baka pack a few belongings and travel, building camps for a week or so, hunting and gathering in each spot. At the end of the rainy season, the trees flower, and bee-hives are brimming with honey, a particularly prized food among the Baka. During this time of year, most of their activities center around honey gathering. During the dry season, when forest streams are low, they concentrate on fishing and collecting crusta-ceans. The Baka trade with the neighboring Bantu farmers, exchanging forest products or labor for plan-tains and other crops, and such items as machetes, metal cooking pots, and clothes.

Further Reading:
Living Earth, England <http://www.gn.apc.org/Living Earth/RainforestDB/People/>.

The Mbuti

The Mbuti live in the Ituri forest of Democratic Re-public of Congo, where they lead a nomadic exis-tence, hunting and gathering and trading forest resources with farmers. Blue duikers (*Cephalophus monticola*) are the most commonly hunted mammal, while termites and honey are other valuable forest commodities. The Ituri forest covers about 17.3 mil-lion acres; an estimated 20,000 Mbuti live within them. They traditionally look on the forest as their parents, calling themselves "BaMiki BaNdula": "chil-dren of the forest."

Further Reading:
Duffy, *Children of the Forest.*
Hart and Hart, *The Ecological Basis of Hunter-Gatherer Sub-sistence in African Rain Forests: The Mbuti of Eastern Zaire.*

The Ogiek

The Ogiek live in Kenya's central Great Rift Valley, where they are engaged in a battle for their territory. With a population of about 5,000 families, they hunt and gather honey and farm bees in the Tinet Forest, about 155 miles west of Nairobi. In 1991, the gov-ernment had allocated five acres of Tinet Forest per family to the Ogiek, who began farming and building schools, in addition to gathering honey from the for-est. But in March 2000 the Kenyan government evicted the Ogiek from Tinet Forest, on the grounds that the forest was made a national protected area in 1961. Conservationists and human-rights advocates believe that powerful Kenyans pushed through the recent eviction ruling so they could log the forest without interference.

Further Reading:
Okoko, "Kenya's Indigenous Honey Hunters Lose Their Forest Home."
World Rainforest Movement, Uruguay, "The Future of the Ogiek and Their Forests" <http://www.wrm.org.uy>.

The Twa

The Twa are the original inhabitants of the small nations of Rwanda and Burundi in central Africa; today about 10,000 Twa live in each country. A smaller population, which likely migrated from the forests of Democratic Republic of Congo, is also found in Uganda. They work as potters and musi-cians—their cultural tradition is centered on songs, dance, and music. Some groups live as traditional forest dwellers; others work as day laborers and pot-ters. The Twa are caught in the middle in the con-tinuing battle between the Hutus and Tutsis. Both these groups consider the Twa to be ignorant and uncivilized. As a result, the Twa are a disadvantaged and marginalized caste with little power. Because a very small number of Twa had privileged positions at the Tutsi royal court, the Hutu perceive all Twa as Tutsi sympathizers. In 1994 thousands of Twa were slaughtered by the Hutu, while others were forced to flee from their homes.

Further Reading:
Gonen, *The Encyclopedia of the Peoples of the World.*
Living Earth, England; <http://www.gn.apc.org/Living Earth/RainforestDB/People/>.

ASIA

The Akha

The Akha live in villages surrounded by forests in the mountains of southwest China, eastern Myanmar, western Laos, northwestern Vietnam, and northern Thailand. Their total population is estimated to be around 430,000, with more located in Myanmar (180,000) than the other countries. Villages may have as many as 200 houses, or as few as 10. Since the 1930s, Akha communities in Thailand have grown smaller as the mountain environment has de-teriorated, causing worsening economic conditions for the indigenous people.

The entrance to an Akha village has two wooden gateways, one positioned upslope and the other downslope, flanked by carved female and male figures. The gates mark the separation between the domain of human beings and their livestock and the domain of spirits and wild animals. Homes are built of logs, bamboo, and thatch, either on the ground or on stilts. The Akhas divide their homes into a female side and a male side, paralleling that between the village and the surrounding forest. Their main crop is rice, grown in plots for a few years then moved to new ground, which allows the former rice field to be reclaimed, and renourished, by the forest. They also grow vegetables and supplement these with wild fruits, mushrooms, and other edible forest plants. They hunt the forests for wild boar, deer, and birds, but game is less plentiful now, due to deforestation.

Further Reading:
Levinson, *Encyclopedia of World Cultures.*

The Dani

The Dani are an ethnic group numbering about 110,000 who live in the mountains of Irian Jaya, the part of the island of New Guinea that belongs to Indonesia. Today they survive by growing sweet potatoes and raising pigs. While most have been converted to Christianity, they maintain many of their traditional religious beliefs, which center around the spirits of the dead. Many of their rituals and ceremonies are related to war.

Further Reading:
Gonen, *The Encyclopedia of the Peoples of the World.*

The Dayaks

The Dayaks are the most populous racial group in Sarawak, on the island of Borneo. About half of Sarawak's population of 1.5 million are Dayaks, including the forest tribes of the Kayan, Kenyah, Kejaman, Kelabit, Punan Bah, Tanjong, Sekapan, and Lahanan. Anthropologists believe these tribes originated from the mainland of Southeast Asia. Traditionally, the people who lived along the coasts were fishermen, while inland, they practiced shifting agriculture—mainly growing rice—supplemented by hunting and gathering. Some Dayak groups live in longhouses, which can be several hundred yards long and serve as home to many families. They are usually built about 15 feet from the ground on massive stilts and have thatched roofs and a communal veranda.

Most Dayaks farm communal land and also depend on swamp areas and forests for hunting and gathering plants for food and medicines. Traditionally, tribe boundaries were marked by streams or ridges. Within each group, individual families have rights to plots of land that they clear for their gardens, and these plots are passed down through generations.

Loir Botor Dingit is the chief of the Bentian people, a Dayak tribe, who has mobilized his people to fight unsustainable deforestation of rattan forests. Rattan is a main source of income for the Bentian. Bowing to pressure from the Bentian, the Indonesian government eventually gave the tribe rights to manage a 24,700-acre rattan forest. Dingit was awarded a Goldman Environmental Prize in 1997. With the prize money, he established a scholarship fund for young Bentian students and provided rice for Bentian villages after the severe drought of 1998.

Further Reading:
Environmental Defense Fund, USA, "Indonesian Environmental Hero Acquitted in Landmark Case" <http://www.edf.org/pubs/newsreleases/1998/nov/d_hero.html>.
Goldman Foundation <http://www.goldmanprize.org/recipients/recipient.html>.
Living Earth, England <http://www.gn.apc.org/LivingEarth/RainforestDB/People/>.

The Iban

The Iban are a riverine people who practice shifting agriculture in western Borneo, on the island of Sarawak. In 1960, their population was estimated to be 238,000. Historically, they moved across the island in search of land. An aggressive people, they practiced headhunting and enslavement of women and children. Each Iban community has a longhouse, with an average of 14 doors for that many families. They grow rice and tap rubber as a source of income and cultivate vegetables such as pumpkin, cucumber, and gourds along with the rice. After 2 years, fields are abandoned and allowed to recover for 10 to 20 years. Although the Iban hunt deer and pigs, fish are their most important source of protein.

Further Reading:
Lebar, *Ethnic Groups of Insular Southeast Asia.*

The Ifugao

The Ifugao have lived in the same area—the slopes of Mount Data, in northern Luzon, a Philippine island—for several hundred years. About 150,000 Ifugao live in small villages, growing rice in terraces on the hills and practicing slash-and-burn farming of tubers. They also hunt and gather forest foods. They are experts in rice terracing and have an intricate ritual and legal organization. The Ifugao believe that the universe consists of five regions: the Earth, the heavens, the underworld, the downstream region, and the upstream region.

Further Reading:
Gonen, *The Encyclopedia of the Peoples of the World.*

The Korowai

The Korowai number about 3,000. They are another indigenous group of Irian Jaya, where 250 different tribes live. Because their territory apparently holds no minerals, oil, or commercially valuable trees, the government and outsiders largely ignored them until the late 1970s. That's when two Dutch missionaries arrived and stayed for 15 years. They learned a lot about the Korowai, but did not manage to convert a single person to Christianity. The Korowai have a fundamental belief that a *laleo*, an evil spirit that might be disguised as a white man wearing clothes, could arrive to destroy their world. They live in close family and neighbor units, hunting game like the cassowary (*Casuariidae*) and farming such crops as taro, tobacco, sweet potatoes, and bananas. The sago palm (*Metroxylon sagus*) provides them with building material and is a source of food, since the tree yields an edible starch that can be pounded into flour. The Korowai build their homes on stilts or in treetops, sometimes as high as 150 feet.

Further Reading:
Steinmetz, "Irian Jaya's People of the Trees."

The Maisin

The Maisin of Papua New Guinea are fighting developers who are clearcutting forests with plans to replace them with plantations of African oil palm (*Elaeis guineensis*) on what the Maisin claim is their traditional territory. The forests are inland from the coast on an eastern island in the Papua New Guinean archipelago (PNG). The Maisin have taken the developers to the nation's highest court. Under Papua New Guinea's constitution, the Maisin and other indigenous groups supposedly are the legal owners of their traditional lands. The developers, however, say they have valid permits to clear forests and establish the plantation.

In addition to farming and hunting, the Maisin support themselves by selling betel nuts and pounded bark art called tapa cloth. Their territory encompasses nearly one million acres that stretch from Mount Suckling, one of the PNG's tallest mountains, to the coast. About 3,000 Maisin live in nine villages. They won their lawsuit in a lower court, but the developers appealed. Conservationists are watching closely to see what the outcome of this landmark suit will be.

Further Reading:
Fair Trade Zone, "Painting a Sustainable Future" <http://www.fairtradezone.com/products/tapa/tapashow/index.html>
Strieker, "Rainforest Tribe Takes Battle with Developers to Court."

The Orang-Asli

The Orang-Asli live in the forested hills of the Malaysian Peninsula, spread out in different villages. The total population of about 60,000 is divided into three ethnic groups: the Semang, Senoi, and Proto-Malays. Traditionally, the Semang are nomadic foragers, the Senoi are shifting cultivators, and the Proto Malay grow mainly tree crops. All depend on forest resources for goods they can use and sell, such as rattan. They have traded these products with their nonindigenous neighbors for centuries. The Orang-Asli have always used bamboo from the forest to construct their houses, cooking utensils, blowpipes, and dozens of other tools. They hunt wildlife in the forest for food and also to sell and trade.

Further Reading:
Living Earth, England <http://www.gn.apc.org/LivingEarth/RainforestDB/People/>.

The Penan

The Penan live in the forests of Sarawak, a Malaysian state that shares the island of Borneo, the third-largest island in the world. Deforestation has ended the nomadic way of life for most Penan, who number about 7,000. Perhaps no more than 300 continue to live as nomads. The Penan society is egalitarian, with no leaders, and they have always been dependent on forest resources, including wild pigs, birds, fish, and primates. The Penan are renowned hunters, masters at the blowpipe. Their forest territory is crisscrossed by a maze of hunting paths and trading paths, with specific rocks, streams, and hills marking land that belongs to one family or another. In the Amazon, indigenous groups often stay in one place to tend gardens that can provide them with a source of carbohydrates. To fill this diet requirement, the Penan depend on the Sago palm (*Metroxylon sagus*), a fast-growing tree whose trunk is loaded with starch. The Penan harvest a palm, cut it into sections, and roll it to a source of water, where it is pounded with mallets. The pulp is placed on a fine-meshed rattan mat on a wooden frame, and water is poured over the mash. The white starch filters through the rattan onto a mat beneath the frame and is dried over a fire to produce the sago flour.

The Penan gained worldwide attention in 1987 when, after their pleas to the government to end deforestation failed, they formed human barricades across timber trails to halt invading loggers. They were assisted by the Kelabit and Kayan, two other indigenous groups. Their protests were unheeded, and most Penan today have been forced to abandon their nomadic way of life and live in desolate resettlement camps.

Further Reading:
Davis, "Vanishing Cultures."
Linden, "Lost Tribes, Lost Knowledge."
Rainforest Information Centre, Australia <http://forests.org/ric/Background/people.htm>.

The Tuboy Subanon

The Tuboy Subanon live on Mindanao, the second-largest island of the Philippines. They had little contact with nonindigenous people until the twentieth century, except with Muslim merchants who lived along the coast, who tended to exploit them. The Subanon, who believe in many gods, are often caught in the middle of clashes between Christians and Muslims on the islands. About 13,500 Tuboy Subanon live in scattered villages in the mountainous interior of the Zamboanga Peninsula of Mindanao, where they plant rice and other grains and raise livestock. Every few years, they move to a new location and clear more forest to plant their crops. The Tuboy Subanon have no division of labor by gender—both men and women work the fields, do the cooking, and care for their children.

Further Reading:
Humboldt University <http://www.humboldt.edu/~lfs1/ethnog/mike/research.html>

AUSTRALIA

The Kuku-Yalanji

The Kuku-Yalanji live in the Queensland forests of the northeast and are the only tribal forest people in Australia who maintain their own culture and language. They live in closely connected communities and extended families, where they are cared for by the Rrunyuji, or "clever man," who cures illnesses with plants from the forest, and can predict the future and stop and start the rain. The Kuku-Yalanji wear Western-style clothing, but no shoes; they may use aluminum boats and outboard motors, but still fish with spears, considered much more effective than fishing lines. In other words, they have chosen to adopt tools from other cultures if they are advantageous.

Further Reading:
Levinson, *Encyclopedia of World Cultures.*
Ministry for the Environment, "Launch of the Kuku Yalanji Conservation and Management Plan" <http://www.environment.gov.au/minister/ps/98/mcsp19feb98.html>.
Rainforest Information Centre, Australia <http://forests.org/ric/Background/people.htm>.

CENTRAL AMERICA AND MEXICO

Central America's indigenous population is about 5.5 million people, spread out among 43 different groups, or about 22 percent of the total population of the isthmus. The majority live in the volcanic highlands of Guatemala and the Caribbean coastal plane, from Belize to the border of Panama and Colombia, an area that is still blessed with forests. During the Spanish conquest, indigenous people fled into these more remote areas. Over the intervening centuries, they have been pushed further into the mountains or deeper into the lowland Caribbean forests by the newcomers.

The Emberá and the Wounaan

These groups, the Emberá and the Wounaan, live in eastern Panama and adjacent Colombia. Eight groups of Ember and Wounaan, totaling about 20,000, live in the Chocó of Colombia, while about 15,000 Emberá and Wounaan live in Panama, most in some 80 villages of the still largely forested Darién province. The two groups have similar cultures—they live in dwellings raised on poles, women traditionally wear short wraparound skirts of brightly colored cloth (once bark cloth), while men, who once wore loin clothes called *guayuco*, wear western clothes. Both men and women paint geometric designs on their skin with vegetable dyes. They were once feared for their skillful use of poison blow darts.

In the 1700s, the Spanish brought the Emberá and the Wounaan to the Chocó to fight the Kuna indigenous people of the region, who were violently resisting Spanish efforts to force them to work in gold mines. Most of the Kuna were pushed to the Pacific coast, while the Emberá and the Wounaan settled in the Darién, where in 1983 the Panamanian government gave them rights to a 740,000-acre reserve. They cultivate crops such as bananas, fruit trees, yams, corn, and rice and also hunt and fish. They now work with the Kuna, once their bitter enemies, in a fight for land rights and against encroaching colonists and illegal loggers.

Further Reading:
Levinson, *Encyclopedia of World Cultures.*
Minority Rights Group International, *World Directory of Minorities.*

The Kuna

The Kuna have a long tradition of political organization and self-rule in Panama, where they have control over their territory, called the Comarca of Kuna Yala. The Comarca, a semiautonomous state, is a band of rainforest that extends for 125 miles along the Caribbean coast nearly to the border of Colombia and encompasses more than 300 tiny islands off the coast, called the San Blas Islands. The Kuna population of about 40,000 lives on some 40 of the islands, plus in 12 mainland villages along the coast.

The Kuna have integrated ecotourism into their society, as more and more foreign travelers visit the beautiful San Blas Islands, stay in modest lodgings there, and learn about the Kuna culture. Although Western society has greatly influenced and altered Kuna traditions, they still have managed to adapt and retain their identity. The traditional dress of Kuna women is particularly striking, as they wear gold nose rings, bracelets of beads around their arms and legs, and colorful woven blouses, called *molas.*

Further Reading:
Chapin, "Losing the Way of the Great Father."
Institute for the Development of Kuna Yala <http://www.ecouncil.ac.cr/rio/focus/report/english/kuna.htm>.

The Maya

The Maya lived in south central Mexico between 7000 and 5000 B.C., where they were among the earliest farmers. By 2500 B.C., they had moved south to Chiapas, Mexico, and Guatemala. By 2000 B.C., the hillsides of Guatemala were dotted with their small farms, planted in maize, squash, and avocados. When the soils of their garden plots, or *milpas*, became depleted, they abandoned the land and moved on, cutting down trees to plant new gardens. The ceremonial village of Tikal was settled as early as 600 B.C. and by 300 B.C., the Maya had constructed buildings of limestone, which housed as many as 10,000 people. The Maya reached their height of power from A.D. 300 to 900, having built more than 200 cities in southern Mexico, Guatemala, Belize, and parts of Honduras and El Salvador.

The collapse of the ancient Maya civilization has long tantalized historians. What could have happened to bring about the end of such an advanced culture? Anthropologists believe the cause could have been civil war, changing trade routes, or a top-heavy aristocracy. A more recent theory holds that ecological deterioration was responsible. Because the Maya needed intense fires to convert limestone into lime stucco for building their magnificent tombs, temples, and other structures, deforestation and resulting erosion brought about an end to fertile farming.

Today the Maya are concentrated in the northern highlands of Guatemala, where there are about 4.5 million people of some 22 different Maya language groups. Maya also live in Belize, Honduras, El Salvador, and the Mexican states of Tabasco, Chiapas, Campeche, Yucatan, and Quintana Roo. Many Maya live in the cities, dress in western clothing, and are part of the Guatemalan mainstream. Many more live in traditional communities, and the women in particular continue to wear the traditional and colorful woven clothing. Using traditional techniques, they continue to fashion brightly colored textiles, baskets, pottery, and wood carvings. They have been persecuted by the Guatemalan government, and until a peace accord was signed in December 1996, thousands were killed in a brutal civil war.

The 450 Maya of the Selva Lacandona forest in Chiapas state, Mexico, have retained many of their traditions and culture. Until the 1960s, the Lacandona Maya were isolated, then bulldozers began to carve roads throughout the region and prompted the immigration of some 150,000 colonists and cattle ranchers, who burned the forest for pastureland. By 1987, nearly half of the Lacandon forest, which once covered 5,000 square miles, had been destroyed. Today the Maya live in three villages dispersed in the remaining forest. They continue to depend on the forest for food and products, but also practice agroforestry: cultivating a mix of food, timber, and medicinal plants.

Further Reading:
Carpenter, "Faces in the Forest."
Honan, "Did the Maya Doom Themselves by Felling Trees?"
Jacobson, *Great Indian Tribes.*
Menchú, *I, Rigoberta Menchú: An Indian Woman in Guatemala.*

The Mayangna

The Mayangna were once called Sumu by outsiders, referring to several groups of related indigenous people living in eastern Nicaragua and Honduras. Once they were the most widespread population on the Caribbean slope of Central America; today about 13,000 Mayangnas live in Nicaragua and 1,000 in Honduras. The government of Nicaragua has granted them and the Miskito indigenous people territorial rights and self-government along the Caribbean coast, while Honduras has established indigenous reserves. Their culture is closely tied to river systems, where they have established villages and planted subsistence crops in the rich, alluvial soils of river beds. They also hunt and fish. When Hurricane Mitch hit Central America in 1998, bringing floods and landslides that killed about 11,000 people throughout the region, the Mayangna in Nicaragua and Honduras lost virtually all their crops and many homes. The devastation was due to deforestation upriver; that is, not in indigenous territory but in lands that had been cleared by colonists—although it was the indigenous that suffered as a result. Without trees to break the flow of water, the rains crashed onto hillsides and carried away tons of soil and debris, which flowed downstream and washed away the Mayangna's crops and belongings.

Further Reading:
Jukofsky, "Unnatural Disaster: Conservation Lessons from Hurricane Mitch."
Levinson, *Encyclopedia of World Cultures.*
Minority Rights Group International, *World Directory of Minorities.*

The Pech

The Pech originally inhabited the coasts of northern Honduras, but after the Spanish invaded, they fled to the mountains. Today some 1,500 Pech live on and farm the land in north central Honduras, near the Tinto and Patuca Rivers. While most are nominal Christians, they continue to revere the community shamans, the medicinal-plants healers. Their political leader, who is elected, is called a *cacique.*

Further Reading:
Lentz, "Medicinal and Other Economic Plants of the Paya of Honduras."

SOUTH AMERICA

The Awá

The Awá live along the border of Ecuador and Colombia, in an area known as the Chocó. The forests of the Chocó are among the wettest on Earth, averaging 400 inches of rainfall each year. Little was known about the 2,200 Awá in Ecuador and 5,000 in Colombia until the early 1980s. To defend their land from colonists, the Awá of Ecuador established a government and built schools in the mid-1980s. They also carved out a 150-mile cleared swath around their territory, a boundary they call the *manga.* In 1988, the Ecuadorian government declared the territory within the manga an "ethnic forest reserve," the first such designated land in South America.

Further Reading:
Cultural Survival, USA <http://www.cs.org/AVoices/articles/LevelFour-Levy.htm>.
Schwartz, "Drawing the Line in a Vanishing Jungle."

The Embera Katio

The Embera Katio live along the Caribbean coast of Colombia, where they hunt and fish. They are waging a battle to stop construction of a huge hydroelectric dam project planned for the Sinu River, within their territory. The Urra dam would flood more than 17,000 acres of forests. Although Colombia's constitutional court ruled in the Embera Katio's favor, construction is under way. As a result of the drop in water levels, riverbanks are eroding, homes are washing away, and the population of freshwater fish, an important food source for the indigenous group, is plummeting. To bring world attention to their plight, about 200 Embera Katio people marched 700 miles to the capital city of Bogotá in late 1999. After camping out in the yard of the environment ministry for four months, representatives of the Embera Katio signed an agreement with the government on April 26, 2000. The accord provides economic compensation to allow the group to expand their reserve, bars construction of a second proposed hydroelectric project, and includes provisions to protect human rights in the zone.

Indigenous People of the Amazon

Anthropologists believe that the South American Amazon is home to the world's largest remaining number of indigenous people who have had no contact with the outside world. Most tribes, however, have been obliterated. When Europeans first reached Brazil 500 years ago, the indigenous population was around five million. It is now less than 200,000. Anthropologists estimate that one forest-dwelling tribe in Brazil has been lost every year since 1900.

The story of the decline of the Waimiri-Atroari tribe is typical of the experience of many indigenous groups in Brazil. In 1903 around 6,000 members of Waimiri-Atroari lived in the Amazon, but by 1973, the population was only 3,500. Many had died violently, trying to protect their land from invading colonists. By 1985, only 374 remained, most of them children. Hundreds died during an epidemic of measles, a disease brought by outsiders to a vulnerable group with no access to health care. Others were killed by assassins hired by wealthy landowners who usurped lands abandoned by the Waimiri-Atroari.

Although most indigenous groups of the Amazon are in contact with non-indigenous people or trade with other tribes, many maintain their traditional lifestyles, practicing shifting agriculture and depending on the rainforest for food and products, including medicines. The medicine man, called the shaman, continues to treat illnesses with cures derived from forest plants.

Many tribes of the Amazon have joined the Coordinating Body for Indigenous Peoples' Organizations of the Amazon Basin and are fighting, with some success, for titles to their territories. The Brazilian government has demarcated hundreds of indigenous territories by cutting a 6-foot swath through the forest to alert outsiders of the boundaries, but hundreds more groups have yet to receive official recognition. Even though several South American countries have set aside reserves specifically for indigenous groups—such as the Xingu National Park in Pará, Brazil—these reservations can be vulnerable to political whims. In 1971, for example, the Brazilian government confiscated nearly two million acres from the Pará indigenous people to build highways. While land was added to the reservation, it was savanna—useless to a forest-dwelling people. In spite of Brazil's efforts to expel them, gold miners, called *garimpeiros*, continue to be one of the biggest threats to indigenous groups in Brazil's Amazon.

In the Amazon of Colombia, indigenous people control more territory than any other tribal people in the world. Nearly 15 million acres of land belong to some 70,000 indigenous people of more than 50 ethnic groups. With the creation of Chiribiquete Park in 1989, 2.5 million acres in the Amazon are protected; some of this land overlaps with the indigenous peoples' territory. Colombian law gives indigenous people the right to follow their own customs and traditions, develop their own education programs, and organize their own health services.

Further Reading:

Chatterjee, "Indigenous Communities Fend Off Gold Miners." Living Earth, England <http://www.gn.apc.org/LivingEarth/RainforestDB/People/>.

Native Web, "Waimiri-Atroari Fight Paranapanema For Their Rights" <http://www.nativeweb.org/abyayala/brazil/cimi/232.html>.

Park, *Tropical Rainforests*.

Schemo, "The Last Tribal Battle."

Further Reading:
Weekly News Update on the Americas, "Embera Win Agreement, End Occupation" <http://home.earthlink.net/~dbwilson/wnuhome.html>.
World Rainforest Movement, Uruguay, "Colombia: the Embera Katio's Struggle for Life" <http://www.wrm.org.uy>.

The Huaorani

The Huaorani live in the Amazon of Ecuador, called "El Oriente." They have fiercely defended their territory from intruders, but have a long history of persecution and intervention at the hands of rubber barons during the late nineteenth and early twentieth centuries, from missionaries, and most recently from the petroleum industry, whose activities in El Oriente have forever changed the ecosystem there. In the 1990s, representatives of the Huaorani, whose population is about 1,300, brought their vocal protests against development by the U.S. petroleum company Maxus to Washington, D.C. In January 1999, the government of Ecuador set aside 2.7 million acres as the Cuyabeno-Imuya and Yasuni National Parks, declaring them off-limits to oil drilling, mining, lumbering, and settlement by outsiders. The parks are open to some ecotourism development. The Huaorani live in Yasuni Park, where they are still superb hunters, particularly adept with the blowgun.

Further Reading:
Kane, *Savages*.

The Kayapo

The Kayapo are one of 17 tribes that live in Xingu National Park in the heart of the Brazilian Amazon. They are descendents of people who migrated south from Mexico and Central America more than 10,000 years ago and were among the first human beings to live in the Amazon. Their territory today is about the size of Austria, and consists mostly of rainforests and some open grasslands. Since the invasion of the Portuguese centuries ago, the Kayapo have fiercely defended their territories and thus escaped the fate of many indigenous groups that came in contact with trespassing Europeans. Dozens of squatters and gold miners have disappeared after venturing onto Kayapo lands.

Interaction with gold miners has led many Kayapo to abandon their farm plots and traditions, although they are also using modern technology to record many of their ceremonies as a way of preserving their tribal knowledge. Each Kayapo village has a communal garden, where the women grow such crops as sweet potatoes, manioc, fruit, tobacco, and cotton. They also transplant useful plants from the forest that can be used for construction, medicines, and foods. They cultivate plants that are favored by wildlife that are in turn eaten by the Kayapo, so that their garden can eventually mature into a productive forest. As of 1993, there were about 14 Kayapo villages, with a total population of about 4,000.

Further Reading:
Park, *Tropical Rainforests*.
Rainforest Foundation, England <http://www.rainforestfoundationuk.org/rainhome.html>.

The Makuna

The Makuna live in Colombia near the headwaters of the Comena River and the border with Brazil. Nearby indigenous groups include Maku, Tukano, Banyanin, Tutuyi, Tanimuka, and Yauna. The last two are traditional enemies of the Makuna. Most of the Makuna's land is communally owned and allocated according to clan affiliation. The Makuna are hunters, gatherers, and slash-and-burn farmers. They use canoes and have elaborate fishing techniques. There are about 600 surviving Makuna. Threats to their traditional way of life include gold mining and resultant river contamination and the cocaine trade, which has introduced currency into their culture.

Further Reading:
Cultural Survival, "The Peoples of 'Millennium'."

The Pemon

The Pemon of Venezuela are opposing construction of a high-voltage, 470-mile-long power line that would cross their territory and Canaima National Park. They have pulled down newly erected electricity towers and blockaded a key highway that links Venezuela with Brazil. Much of six-million-acre Canaima National Park is covered by table mountain formations, sheer cliffs, and waterfalls, including Angel Falls, the world's highest, at more than 3,000 feet. In November 1999, the Venezuelan National Constituent Assembly voted to include in the country's constitution a law that establishes legal rights for indigenous communities. The new law should help the Pemon, the third-largest of Venezuela's 25 indigenous groups, in their fight.

Further Reading:
Johnson, "Energy Projects Opposed: Latin Indians Using Sabotage."
World Rainforest Movement, Uruguay, "The Pemons' Struggle" <http://www.wrm.org.uy>.

The Shuar

The Shuar live in southeastern Ecuador, along the border with Peru. The Shuar always fiercely defended their lands and once were headhunters. When the invading Spaniards tried to enslave the Shuar and force them to mine for gold in their own territory, the indigenous group successfully revolted, destroying most of the Spanish settlements. In a switch to more pacific techniques of preservation in the mid-1960s, they formed the first ethnic organization in the Ecuadorian Amazon, which used the media to safeguard their way of life. They traditionally live in isolated homes along rivers and streams, which they navigate in dugout canoes. They hunt, fish, and grow a few crops. Shaman have always had a central role among the Shuar.

Further Reading:
Native Web <http://www.nativeweb.org/abyayala/ecuador/shuar/>.

The U'wa

The U'wa are engaged in a fight for their land in Colombia, by blocking access at the construction site of an oil well constructed by Occidental Petroleum of Los Angeles. The U'wa reservation is 544,000 acres of cloud forest in the Sierra Nevada de Cocuy Mountains near the border with Venezuela. In 1997, the U'wa, a seminomadic tribe of about 6,000, threatened to commit suicide if drilling began on their reservation. Occidental backed down, but in September 1999, the government of Colombia granted a license to the U.S. company to drill for oil just outside the reserve. Occidental has predicted that the U'wa's land could sit on top of the country's largest oil field, potentially holding 1.4 billion barrels of oil. The U'wa, who believe oil is the blood of the Earth and should not be extracted, say that the drilling will cause environmental damage and social disruption.

A tribal leader, Berito Kuwaru'wa, has become the spokesman for the U'wa in their ongoing fight, traveling widely to explain why the group is fighting the drilling; he even spoke at Occidental's 1998 Annual Meeting. That year Kuwaru'wa won both the Goldman Environmental Prize and Spain's Bartolome de las Casas Award.

Further Reading:
Goldman Foundation, USA <http://www.goldmanprize.org/recipients/recipients.html>.
Runyan, "Colombia Opts for Oil Over Indigenous Rights."
World Rainforest Movement, Uruguay, "The U'wa People Do Not Surrender" <http://www.wrm.org.uy>.

The Xavante

The Xavante of Brazil became known to the western world as warriors during the 1930s and 1940s, when they defended their lands against encroachment from outsiders. They succeeded in keeping invaders out until the 1960s, when newly constructed roads brought floods of colonists. They were forced to seek refuge in Mato Grasso, in central Brazil. They then began appealing to government officials to help them protect the natural resources on which they depend. In 1980 they led a protest that led to the reorganization of the government's ineffective indigenous affairs agency. Eventually, the Xavante won rights to their lands. Today, about 6,000 Xavante live on six reserves in Mato Grasso. Many produce rice as a source of income.

Further Reading:
Cultural Survival, USA <http://www.cs.org/sprojects/SPXavante.html>.

The Yanomami

The Yanomami control 23 million acres in Brazil, near the border with Venezuela, granted to them by the Brazilian government in 1992. Numbering about 11,000, they live in communal villages called "Yanos," which are scattered in the Amazon rainforest. The Yanos are doughnut-shaped houses where the whole community of 25 to 400 people live. Each family has its own hearth that opens onto the center of the Yano. Each family also has its own plot of land where they cultivate bananas, plantains, sweet potatoes, cassava, palm, papaya, and other fruits and vegetables. They also grow plants used for medicines, tools, household goods, and ceremonies. Their crops provide most of their food, but they also fish, hunt, and gather fruits, nuts, and honey from the forest. Every 5 to 10 years, they move to a different area and build a new Yano.

A Yanomami hunter will not eat an animal he has

killed but rather distribute it among friends and relatives. Although Yanomami villages may have male leaders who make helpful suggestions to his neighbors, there are no designated "chiefs." Invasion by gold miners has brought fatal diseases to the Yanomami and mercury poisoning of their rivers. The Brazilian government has been able to remove most gold miners from Yanomami territory, but many are slowly returning. Yanomami continue to die from malaria, and dozens have been murdered by the invaders. Missionaries have encouraged some populations of Yanomami to live near missions rather than move from place to place in the forest. As a result, they have nearly hunted out howler and spider monkeys near the stationary population centers. Illegal gold miners also hire Yanomami to hunt spider monkeys, exchanging gold dust for the meat. Yanomami also live in the forested highlands of southern Venezuela.

Further Reading:

Amanaka'a Amazon Network, USA, "Goldminers Expelled From Yanomami Reserve" <http://amanakaa.org/ccpy.htm>.

Rainforest Information Centre, Australia <http://forests.org/ric/Background/people.htm>.

Naturalists, Explorers, Scientists, and Activists

The 54 people whose contributions to rainforest conservation are described below are obviously only a small fraction of the thousands of smart, brave, and dedicated people who have devoted themselves to understanding and protecting this rare and priceless ecosystem. Many of them are recipients of prestigious international honors, such as the Global 500 Award, given by the United Nations Environmental Programme, and the Goldman Environmental Prize, given by the Goldman Foundation to honor grassroots conservationists. The selection below gives an idea of the wide range of people, from many different countries, cultures, and disciplines, who deserve praise and gratitude for their achievements.

Amooti, Ndyakira (1956–1999) was an environmental journalist from Kampala, Uganda. As a reporter and editor for *The New Vision,* an independent newspaper, Amooti was the only journalist in the country whose primary beat was environmental news. He exposed wildlife smugglers, particularly those who captured and sold endangered chimpanzees and gray parrots. His reporting also disclosed news of illegal poaching, mining, and logging, particularly in upland forests—habitat for the nearly extinct mountain gorilla. Amooti's reports spurred the Ugandan Parliament to upgrade the upland forests from a reserve to a national park and establish several other forest reserves. Amooti won the Global 500 Award in 1993 and the Goldman Prize in 1996. Amooti was still reporting for *The New Vision* when he died of leukemia in August 1999.

Further Reading:
Dam, "The Green Team: Defender of the Pearl."
Environmental Law Alliance <http://pan.cedar.univie.ac. at/>.
Goldman Foundation, USA <http:www.goldmanprize.org/ recipients/ recipient.htm>.
Leach, "Don Quixotes of the Environment."

Attenborough, Sir David (1927–) has been making natural history documentaries since the 1950s and is one of only a few from a generation of English broadcasting pioneers that still presents on television. Attenborough documentaries showed scientists' images on film of the animals they were researching. He is an innovator of camera technology and has been able to capture phenomena such as how hummingbirds beat their wings, how cheetahs catch gazelles, and how marsupials give birth. Attenborough writes the scripts for his television series on wildlife and has the knack of explaining to the public complex ecological concepts in accessible and appealing language. His most ambitious endeavor was a trilogy documenting the Earth's wildlife from 300 million years ago to the present. Attenborough's television series and books have been used as study aids for students around the world. He was a Global 500 laureate in 1987.

Further Reading:
Public Broadcasting System <http://www.pbs.org>/lifeof birds/sirdavid/ index.html>.
Royal Television Society <http://www-royaltv.pp.asu.edu/ Cooke98.htm>.
World Wide Fund for Nature, England <http://www.wwf-uk.org/news /attenbor.htm>.

Banks, Sir Joseph (1743–1820) was a self-taught botanist born into a prosperous English family. In 1768 he set sail on the *Endeavour,* under the command of Lt. James Cook, with biologist Daniel Solander, artists Sydney Parkinson and Alexander Buchan, and four assistant collectors. The expedition sailed to South America, stopping at Rio de Janeiro, then on to Tahiti and other Pacific Islands, New Zealand, Australia, and the East Indies, returning to England three years later. Along the way, Banks, Parkinson, and Buchan documented, drew, and collected 30,000 plant and animal specimens, at least 1,400 of which were plant species new to scientists. To Tahiti he brought citrus from Brazil, encouraging an exchange of plants among Europe, the Americas, and the Pacific Islands that helps account for the pantropical range of many plants today.

Further Reading:
Philadelphia Academy of Natural Sciences <http:// www.acnatsci.org/exhibits/banks/banks.html>.
Watkins, T.H. "The Greening of the Empire: Sir Joseph Banks."

Bates, Henry Walter (1825–1892) was an English naturalist and explorer who, after introducing Alfred Russel Wallace to entomology, traveled with his colleague to the Amazon River in 1848. Wallace left the Amazon in 1852, but Bates remained for 11 years, exploring and, above all, collecting specimens. He amassed nearly 15,000 species, mainly insects, 8,000 of which were previously unknown to Europeans.

He was greatly weakened by malaria when he returned to England, and it was Charles Darwin who convinced him to write up his Amazon notes. In 1861, he published a now-seminal paper, "Contributions to an Insect Fauna of the Amazon Valley," and in 1863, his two-volume *The Naturalist on the Rivers Amazon*. He developed the theory of mimicry to explain why two unrelated species of butterfly look alike. Some butterflies of the family *Heliconiinae* have black, yellow, and orange patterns and are poisonous or noxious; other species that are not distasteful to predators have evolved the same bright patterns. This misleading and beneficial impersonation is referred to as "Batesian mimicry."

Further Reading:
Maslow, *Footsteps in the Jungle.*

Beebe, William (1877–1952), born in Brooklyn, N.Y., was director of tropical research for the New York Zoological Society, a position he held for 30 years until his death. He traveled the world and did groundbreaking biological research wherever he went. He wrote 21 books for general audiences about his experiences, including *Jungle Days* and *The Edge of the Jungle*, which described the Neotropical rainforest. He taught the general public about the Earth's flora and fauna and, for many, his books were their first introduction to the tropical forest. In 1930, he became the first person to see the deep ocean floor by lowering himself in a 5,000-pound steel diving chamber he had designed with his partner, Otis Barton.

Further Reading:
Maslow, *Footsteps in the Jungle.*

Belt, Thomas (1832–1878) was an English mining engineer, naturalist, and author of dozens of papers on geology, entomology, and paleontology. His engineering background aided him in writing brilliant descriptions of nature's designs and systems, evolution, and earth sciences. From 1868 to 1872, he traveled in Nicaragua while working for the Chontales Gold-Mining Company. His absorbing account of this Central American country's flora and fauna are detailed in an influential book, *The Naturalist in Nicaragua*, published in 1874. Along with his observations of nature, the book describes his theories of ancient indigenous life, the effects of the Ice Age, and the origin of mineral veins. He realized, for example, that the glaciers that passed through the Central American isthmus must have caused mass extinctions, but that many species must have survived in lowland forests. Some of his conjectures were dismissed at the time of the book's publication, but have since been confirmed. *The Naturalist in Nicaragua* is a memorable record of Neotropical biodiversity of more than a century ago.

Further Reading:
Belt, *The Naturalist in Nicaragua.*
Maslow, *Footsteps in the Jungle.*

Brundtland, Gro Harlem (1940–) is a Norwegian physician who was prime minister of Norway for 10 years. In 1983, the secretary-general of the United Nations invited her to set up and chair the World Commission on Environment and Development. In 1987, the commission published its report, called "Our Common Future," although it is often referred to as the Brundtland Commission Report. The report coined the phrase "sustainable development," meaning the use of natural resources for economic gain without depleting them. The report led to the 1992 United Nations Conference on Environment and Development, or the Earth Summit, held in Rio de Janeiro, Brazil.

Further Reading:
Bland, "Giving WHO Moral Clout."

Buchori, Damayanti (1960-) is an Indonesian entomologist and university lecturer who specializes in pest and plant diseases. She is the director of the Center for Integrated Pest Management and program coordinator for Wildlife Preservation Trust International. Buchori focuses on the link between conservation and agriculture, promotes insect diversity, and is concerned with increased pesticide use as Indonesia modernizes its agricultural system. Buchori centers much of her work in the agro-ecosystems and the rainforests of Java's Gunung Haliman National Park, the last remaining rainforest in Java, and also in Karawang, an area of intensive rice production. Buchori is studying such factors as biodiversity at the forests' margins, the impacts of land-use systems on biodiversity, and the distribution of botanical pesticides.

Buchori received awards in 1997 and 1999 from the Ecological Society of America, and in 2000, she received the Iris Darnton Award for International Nature Conservation from the Royal Geographical Society of London.

Further Reading:

The Whitley Awards Foundation, England <http://www.whitleyaward.org>.

Wildlife Preservation Trust International, USA <http://www.thewildones.org/Scientists/askDami.html>.

Cassells, David (1961–) is a forester from Australia who in 1996 left his position at the World Bank environmental department to be the first director general of the Iwokrama International Centre for Rain Forest Conservation and Development in Guyana. The center comprises nearly 900,000 acres of rainforest, or 2 percent of Guyana, and its goal is to prove that tropical forest can be sustainably managed to benefit the people living in it, near it, and all Guyanans. The center is supported by the United Nations' and World Bank's Global Environmental Facility and the International Development Research Centre in Canada, plus donations from dozens of other countries and international organizations. Cassells plans to make the center self-sufficient include certified sustainable logging, ecotourism, and bioprospecting, or the search for new medicines and other marketable products derived from flora and fauna.

Further Reading:

Gilmour, "Rainforest Pioneer."

Cox, Paul (1953–), a U.S ethnobotanist, moved to a remote village on Savai'i Island in Western Samoa in 1984 to conduct research. He found that villagers had recently sold a piece of their forest to loggers to pay for a new schoolhouse. With help from High Chief Fuiono Senio, Cox began a fund-raising campaign so villagers could keep their forest and still have their schoolhouse. They designed a covenant stipulating that donations would pay for the schoolhouse, while the villagers agreed to preserve their rainforest—one of the largest, intact, lowland forests in all of Polynesia—for 50 years. For their efforts, Cox and Senio won the Goldman Prize in 1997. Cox also helped establish a 20,000-acre rainforest preserve elsewhere on the island and the 10,000-acre National Park of American Samoa. His research has led to the development of five drugs derived from Samoan forest plants. Any profits eventually made from the drugs will be shared with the people of Samoa. In 1998 he became director of National Tropical Botanical Garden in Hawai'i and Florida.

Further Reading:

Goldman Foundation, USA <http://www.goldmanprize.org/recipients/recipients.html>.

Levy, "Ethnobotanist Paul Cox Helps Save Samoan Rainforest."

Seacology <http://www.seacology.org/prize/index.html>.

Sonne, "Paul Cox."

Darwin, Charles (1809–1882) had a profound effect on modern ecological thinking. Before he published *On the Origin of Species*, at age 50, most naturalists believed that flora and fauna were stable parts of an ideal design, and in Christian Europe and the Americas, the designer was God. This view meant that human beings could do little permanent damage to the design; all they could do was to rearrange the parts. All species, it was believed, existed for the good of other species as much as itself, and all worked to benefit human beings.

Darwin changed the perception of nature after his travels from his native England to South America and the Pacific Islands, particularly after his observations in the Galapagos Islands. He embarked on the HMS *Beagle* in 1831, when he was only 22 years old. The *Beagle*'s goal was to survey and map the east coast of South America, from Bahia, Brazil, south to Rio de Janeiro, from Montevideo south to Tierra del Fuego. Darwin traveled to the Argentine pampas and the plains of Patagonia, where he found the fossilized remains of gigantic prehistoric land mammals.

The *Beagle* arrived in the Galapagos in 1835. He spent five weeks exploring at least five of the islands, where he was struck by the number of plant and animal species that were found nowhere else on Earth and often in a very confined range. Of 25 new bird species he collected, 13 formed what he called a "most singular group of finches" that were found only on the islands and were related to one another in the structure of their beaks, short tails, stature, and plumage. The finches, which were later named for Darwin, formed the basis of the theory of evolution that Darwin presented in his influential book *On the Origin of Species*, which was an overnight bestseller.

After Darwin first saw a tropical rainforest, he wrote: "It is easy to specify the individual objects of admiration in these grand scenes; but it is not possible to give an adequate idea of the higher feelings of wonder, astonishment, and devotion, which fill and elevate the mind."

Further Reading:
Maslow, *Footsteps in the Jungle.*
Paehlke, ed., *Conservation and Environmentalism.*

Ehrlich, Paul (1939–) is a U.S. entomologist whose scientific research has focused on the evolutionary biology of insects and butterflies. He has also studied and written extensively about the interaction among human population, resource exploitation, and the environment and has long been an advocate of tropical biodiversity conservation. He is the author of *The Population Bomb*, published in 1968, which documents the environmental impacts of overpopulation. He also founded the U.S.-based nonprofit organization Zero Population Growth. He was awarded the Crafoord Prize in 1993, and with his wife, Anne Ehrlich, the United Nations Environmental Programme Sasakawa Environment Prize in 1994. The Bing Professor of Population Studies at Stanford University in California, he has authored 37 books and more than 700 technical papers, many written in close conjunction with Anne Ehrlich. In 1996, they co-authored *Betrayal of Science and Reason: How Anti-Environmental Rhetoric Threatens Our Future.*

Further Reading:
Becher, *Biodiversity: A Reference Handbook.*
Public Broadcasting Service <http://www.pbs.org/kqed/population_bomb/theshow/bio.htm>.
Toth, "Stanford's Nuttiest Tenured Turkey."

Eisner, Thomas (1929–) is a U.S. entomologist named by his contemporaries as the father of chemical ecology, a science that explores chemical interactions of organisms and a subject that Eisner has studied for the last four decades. As a field biologist on four continents, Eisner has made scores of discoveries about the chemical defense and communications systems in insects, other arthropods, and plants. His insect research encompasses the biological sciences, chemistry, ecology, evolution, behavior, morphology, and basic entomology, highlighting details that were previously unexplored. His studies are coupled with activism for conservation and animal rights. By introducing and promoting "chemical prospecting," or the search for potential medicines and other useful products in plants and animals, Eisner has linked corporate interests, scientific research, environmental protection, and sustainable development in tropical countries. Among the numerous international awards and fellowships Eisner has won are the Tyler Prize for Environmental Achievement in 1990, the National Medal of Science in 1994, the National Conservation Achievement Award in 1996, and the Rainforest Alliance's Green Globe Award in 1997. Since 1957, Eisner has taught at Cornell University.

Further Reading:
Cornell University <http://www.nbb.cornell.edu/neurobio/eisner/svita.html>.
Rawat, "Nature as Laboratory."
Wildlife Preservation Trust International, USA <http://www.thewildones.org/Scientists/eisner.html>.

Fossey, Dian (1932–1985) was a biologist from the United States who studied the mountain gorillas in Rwanda and Democratic Republic of Congo. Her reports led to a much better understanding of these primates, dispeling myths about their aggressive behavior, and to the protection of the gorillas by the government of Rwanda. Unfortunately, civil wars have once again put the gorillas in jeopardy. Fossey was murdered in 1985, some believe because of her outspoken complaints against gorilla and other wildlife poachers.

Further Reading:
Watkins, "One Hundred Champions of Conservation."

Galdikas, Biruté (1946–) is an Indonesian scientist and professor, who since 1971 has been studying and fighting to conserve the orangutan (*Pongo pygmaeus*) of Borneo's forests, first as a student of the anthropologist Louis Leakey and later on her own. She is an activist battling the destruction of orangutan habitat and Borneo's rainforests, which are steadily being lost to fires and illegal logging. Galdikas founded Orangutan Foundation International in 1986 with the Indonesian government's permission and endorsement. She created its Orangutan Care Center and Quarantine in 1999, and has implemented local patrols to curtail illegal logging. In 1998, she negotiated with the Indonesian government to secure nearly 190,000 acres as protected wildlife reserves after logging concessions on the land had expired. Galdikas has been honored with awards from many international conservation organizations, including the Global 500 award from UNEP in 1993 and the prestigious "Kalpataru" award in 1997, the highest award given by the Republic of Indonesia for environmental leadership.

Further Reading:
Dreifus, "Saving the Orangutan, Preserving Paradise."
Orangutan Foundation International <http://www.orangutan.org>.

Gámez, Rodrigo (1936–) is a Costa Rican scientist and agronomist. Before becoming active in biodiversity conservation, Gámez worked as the director at the University of Costa Rica's Institute of Cellular and Molecular Biology and spent three decades in the field studying plant pathology and plant virology. Between 1987 and 1990, he was named advisor in biodiversity to the president of Costa Rica, a position that launched a career in conservation. He created a new, well-rounded approach to the development and management of the nation's extensive national park system and founded the Neotropical Foundation, a nonprofit conservation group. Gámez is the director of the National Institute for Biodiversity (INBio) a research center that he founded in 1989. Under his leadership, INBio is cataloging the plants and insects in Costa Rica, a feat that has not been attempted in any other tropical country. INBio is also pioneering in the creation of agreements with businesses to sustainably utilize Costa Rica's natural resources. The Ministry of Natural Resources in Costa Rica receives donations from INBio as a percentage of the payments the group receives from its negotiations with businesses. Gámez received the Bernardo Houssey Inter-American Prize in Sciences from the Organization of American States in 1983, and in 1995, he accepted the Prince of Asturias Prize on INBio's behalf.

Further Reading:
Becher, Anne. *Biodiversity: A Reference Handbook.*
Tranberg, "Unique Biodiversity Program in Costa Rica."

Goodall, Jane (1934–) is an English naturalist and scientist whose career investigating chimpanzees (*Pan* spp.) began in 1960 in the Gombe Stream National Park of Tanzania, where she was an assistant to the anthropologist Louis Leakey. In 1967, Goodall established and directed the Gombe Stream Research Center. Her research revealed that chimpanzees were smart toolmakers, capable of feeling happiness, pain, and anger, and that they formed societies nearly as complex as humans'. This new view demonstrated that the connection between chimps and people was closer than anyone thought. Goodall teaches, travels, and campaigns in promotion of chimpanzee research, conservation, and funding to establish primate protected areas. She helped secure six reserves in Africa and has established Jane Goodall Institutes for Wildlife Research, Education and Conservation in nine countries. She received the J. Paul Getty Wildlife Conservation Prize in 1984, the Green Globe Award

from the Rainforest Alliance in 1993, and the Global 500 award in 1997.

Further Reading:
Golden, "A Century of Heroes."
Jane Goodall Institute, USA <http://www.janegoodall.org/jane/jane_bio_day.html>.

Hafild, Emmy (1958–) is a trained agronomist who was raised in Sumatra, Indonesia, and is now one of that country's most active environmentalists. She is chief executive of the Indonesian Forum for the Environment (WALHI), which coordinates activities of more than 300 nongovernmental organizations. In 1994 and 1996, WALHI filed lawsuits against the government for siphoning money away from budgeted reforestation projects. She has protested against logging, burning, mining, and environmental degradation from the timber and palm oil industries.

Further Reading:
Liebhold, "Crusader for Indonesia's Enchanted Forests."
Ressa, "Gold and Blood in the Wilderness: Indonesia Mine a Blessing and a Curse."
Tesoro, "Millennium Politics and Power."

Hasbun, Carlos Roberto (1963–) is a wildlife biologist from El Salvador who helped author his country's first wildlife conservation law. He is the founder of the Sea Turtle Conservation Programme, the National Convention on International Trade in Endangered Species Commission, the Zoological Foundation of El Salvador, and the co-founder of the Wildlife Rescue and Rehabilitation Centre and the Environmental Association. Working closely with villagers, he developed management schemes for the sustainable use of mangroves, the green iguana (*Iguana iguana*), and sea turtles. He organized the Salvadoran Environmental Association and its project "Friends of the Trees," which sponsors the planting of mangrove seedlings for each adult mangrove tree cut down. Hasbun is studying geographic variations of spiny tailed iguanas (*Ctenosaura similis*). A winner of the Global 500 award in 1996, he serves as director of the National Parks and Wildlife Service in El Salvador.

Further Reading:
BioNews, University of Hull, "PhD Student in the Aftermath of Hurricane Mitch" <http://www.hull.ac.uk/biosci/html/feb99.html>.
Environmental Research and Wildlife Development Agency <http://www.erwda.gov.ae/turtles.html>.
United Nations Environment Programme, Global 500 <http://www.unep.org/newdraft/unep/per/ipa/g500/roll.htm.>

Janzen, Daniel (1939–), a U.S. conservation biologist, has studied the ecology and biodiversity of ecosystems, particularly the dry tropical forests in Costa Rica, for more than 40 years. To secure protection for Costa Rica's remaining dry forests, he spearheaded a fund-raising effort to purchase land adjacent to a national park. Thanks in large part to his and biologist Winnie Hallwach's efforts, 300 square miles of land were added to the park, called the Guanacaste Conservation Area, and a plan to reforest and restore degraded lands in the new reserve is well underway. Janzen and Hallwach serve as scientific advisors for the Guanacaste Conservation Area, where trees have been protected from forest fires and repopulated from seeds of native species dispersed by the region's high winds. Thanks to their fund-raising efforts, some $30 million have been donated to Costa Rica's national parks. Janzen donated half of his $100,000 award from the 1984 Crafoord Prize to the parks system. He edited *Costa Rican Natural History* (1983), a comprehensive encyclopedia of the Central American country's flora and fauna, and has been a professor at the University of Pennsylvania since 1976.

Further Reading:
Allen, "Biocultural Restoration of a Tropical Forest."
University of Pennsylvania <http://www.sas.upenn.edu/biology/faculty/janzen/>.

Kahumbu, Paula (1966–) is a Nairobian ecologist and evolutionary biologist. It was while she was studying elephants (*Loxodonta africana*) in the Shimba Hills and the Diani forest of Kenya in 1996 that Kahumbu decided to establish the Colobus Trust, a primate-conservation organization. The Colobus Trust, of which Kahumbu remains the director, is the first privately funded organization in the southern coast of Kenya. The group looks for ways to ease pressures on colobus monkeys (*Colobus* spp.) from human encroachment on their habitat. The organization builds treetop-to-treetop bridges that allow the arboreal primates to safely cross roads, and raises funds for forest conservation, education, and awareness activities. Kahumbu was winner of the Whitely Award for her project to conserve the threatened black and white colobus monkeys (*Colobus guereza*) that live in the Kenyan and Tanzanian forests. She continues to research elephants in tropical rainforests of the Shimba Hills.

Further Reading:
Philadelphia Zoo, <primatereserve. philly.com/conservationists. html>#Paula>.
The Whitley Awards Foundation, England <http://www. whitleyaward.org.>

Kuroda, Yoichi (1954–), born in Tokyo, has intensely lobbied the Japanese government and logging industry to stop irresponsible timber imports. In 1987, Kuroda founded the Japan Tropical Forest Action Network (JATAN) and has since become its coordinator. Japan imports more tropical timber than any other country in the world, buying wood from rapidly dwindling logged forests of South America, Africa, and, particularly, Southeast Asia. After researching and reporting the first data collected on Japan's timber-consumption patterns in 1989, Kuroda influenced foreign policy and has had international success in motivating Japanese companies to reduce their exploitation of timber in Southeast Asia and the Amazon. He received the Goldman Prize in 1991.

Further Reading:
Goldman Foundation, USA <http://www.goldmanprize. org/recipients/recipients.html>.

Lovejoy, Thomas (1941–) began studying in the Amazon in 1965 as a conservation biologist. Since then, he has held prominent positions at the Smithsonian Institution, the World Wildlife Fund, the Society for Conservation Biology, and the President's Council of Advisors on Science and Technology. He is credited with coining the term "biological diversity" and is responsible for first designing debt-for-nature swaps, which allows countries to reduce their international debt in exchange for protecting their environment. Lovejoy has spent decades working with the Biological Dynamics of Forest Fragments Project, an experiment that studies the minimum size that a national park or biological reserve must be to remain viable and sustain wildlife. The project's influential research conclusions showed that biodiversity declines proportionally as the size of a forest reserve decreases. Lovejoy works as lead specialist for the environment for Latin America and the Caribbean and chief advisor on biodiversity to the president of the World Bank. He is the founder of the Public Broadcasting System's television series *Nature*. In 1992, he received the Global 500 award.

Further Reading:
Australian Broadcasting Corporation <http://www.abc.net. au/rn/science/earth/stories/s48203.htm.
Becher, *Biodiversity: A Reference Handbook*.

Lutzenberger, José (1926–) is an agronomist who grew up in Porto Alegré, Brazil. In 1970 he left his position at a multinational chemical corporation and in 1971 founded the first environmental organization in Brazil, the Association for the Protection of the Natural Environment of Rio Grande do Sul. Lutzenberger is popularly known in Brazil as the father of the environmental movement and is a recognized expert on plant health, soil science, and organic fertilizers. He served as special secretary for the environment to the president of Brazil between 1990 and 1992. He presided over the Earth Summit Conference in Rio de Janeiro in 1992 and actively voiced support for preservation of the Brazilian Amazon's remaining rainforests. He outlawed iron mining and suspended granting of fiscal incentives that encouraged agricultural development in the Amazon, and began closely monitoring rates of deforestation. Because of his outspokenness against the government's ineffective environmental policies, he was dismissed. Lutzenberger lectures widely and is a 1988 winner of the Right Livelihood Award.

Further Reading:
The Right Livelihood Award <http://www.rightlivelihood. se/recip1988_2. html>.

Maathai, Wangari (1940–), an anatomy professor at the University of Nairobi in Kenya, is the first Kenyan woman to receive a Ph.D. In order to mitigate the effects of deforestation, desertification, and soil erosion, in 1977 she launched the Green Belt Movement, whose members have planted some seven million trees in the east African nation. In 1998 she and fellow activists took on wealthy developers who tried to raze hundreds of acres of the Karura Forest for luxury housing. After holding off the bulldozers, the Green Belt Movement began planting tree seedlings in the portion of the Karura Forest that had already been cleared. In January 1999, police beat and teargassed protestors who were marching to Karura to plant trees, which instigated three days of rioting in Nairobi and brought cries of protest from the international environmental community. Maathai is copresident of Jubilee 2000, which is lobbying for cancellation of Third World debt. She won the Goldman Prize in 1991.

Further Reading:
Goldman Foundation, USA <http://www.goldmanprize. org/recipients/recipients.html>.
Mutiso, "Wangari Maathai: Her Women's Army Defies an Iron Regime."
Salmon, "Mobilising the Mothers."

Matola, Sharon (1954–) is a U.S.-born zoologist who became enamored with tropical plants and ani-

mals when she enlisted in the U.S. Air Force and was stationed in Panama. A former lion tamer, she established the Belize Zoo in 1983, when she began to care for animals that were once used in a nature documentary filmed in Belize. As director, she turned the small collection into a world-class zoo and environmental education center; her efforts have allowed thousands of Belizean children to learn about their nation's wildlife. Only native wildlife is displayed in the zoo, and all animals are in natural surroundings, with no enclosed cages. As a natural resources advisor to the government of Belize, she was able to focus policymakers' attention on the need to protect the country's forests. She was the first chair of the World Conservation Union's Tapir Specialist Group. In 1998 she received the Whitley Award.

Further Reading:
Hoffman, "Wonder Woman of Belize."
The Whitley Awards Foundation, England <http://www. whitleyaward.org>.

Mee, Margaret (1909–1988) was a botanical artist from the United States who began traveling extensively in the Brazilian Amazon when she was 46. For three decades she chronicled the flora of the Amazonian forest. She identified several new plant species; several bear her name. Hers was an early voice of protest against the destruction of the Amazon rainforest. One of her ambitions was to paint a moonflower (*Selenicereus wittii*) in full bloom in the Amazon. This required precise timing since the flowers last for only one night, and each plant produces just a few flowers. In 1988, she achieved her dream and painted a lovely composition by perching on top of a river launch for an entire night. The moonflower was her last painting, as she died later in the year in a car accident.

Further Reading:
Maslow, *Footsteps in the Jungle.*
Wolfe and Prance, *Rainforests of the World.*

Mendes, Francisco "Chico" (1944–1988) was a labor activist and grassroots environmentalist. He organized a union of fellow rubber tappers in the Amazon town of Xapuri, Brazil, and demanded that the government protect the forest on which they depended for their livelihoods from the annihilating fires set by powerful cattle ranchers. In 1988, the government complied, declaring a 61,000-acre tract as an "extractive reserve," meaning that it could not be cut or burned but used only in a sustainable way—

for the harvest of rubber, Brazil nuts, and other forest resources. With the establishment of this and three other extractive reserves, Mendes had accomplished an amazing feat for someone who first heard the word "environment" three years earlier. On December 22, 1988, Mendes was murdered by local landowners angered that the extractive reserves prevented them from expanding their cattle ranches. His death sparked international outrage and new interest in the plight of rainforests worldwide.

Further Reading:
Revkin, *The Burning Season.*

Mittermeier, Russell (1949–) is a U.S. primatologist and president of Conservation International (CI), a leading conservation organization working to save tropical biodiversity worldwide. He has also worked with the World Conservation Union, the World Health Organization, the World Wildlife Fund, and the World Bank's Task Force on Biodiversity. In 1997, Mittermeier and CI convinced the government of Suriname to set aside one-tenth of the country as a protected reserve. He is the author of several books, including *Megadiversity*, which promotes a plan to establish protection zones in regions worldwide that hold a high level of biodiversity.

Further Reading:
Becher, *Biodiversity: A Reference Handbook.*
Rosenblatt, "Russell Mittermeier: Hero of the Week."

Montiel, Rodolfo (1955–) is a Mexican farmer who in 1998 created a grassroots ecological movement called the Organization of Campesino Ecologists of the Sierra de Petatlán and Coyuca de Catalán. Composed of subsistence farmers and environmentalists, the group originally united to express their concerns about logging in the mountains outside the village of Zihuatanejo. They then turned their focus to the forests of Sierra of Petatlán in the state of Guerrero, a mountain once carpeted with ancient white pine and fir trees. Montiel convinced more than 100 members to launch a letter-writing campaign that reported environmental degradation, corruption, and human-rights violations in the area to Guerrero's governor, Mexico's environmental minister, and other authorities. The group of farmers reported that local rivers were drying up and fish were dying as a result of environmental neglect and government indifference. In 1995, when the group protested against logging companies that were granted concessions in Guerrero's ancient forests, the Mexican government began

to aggressively denounce the farmers' activities, accusing them of being a guerilla movement and trafficking in drugs. Several members of Montiel's organization have been murdered.

In May 1999, after Montiel led protests against deforestation and pollution, he and a cohort were detained and imprisoned by the Mexican Army. Only then did Montiel's activities attract attention and support from the international environmental community, which began to pressure the Mexican government to release him from prison. In another effort to increase awareness of Montiel's situation, he was awarded the Goldman Environmental Prize for the year 2000; his award was announced in advance of other prize winners. Montiel plans to use the award money to purchase irrigation equipment for his fellow farmers.

Further Reading:
Dillon, "Jailed Mexican Wins Environmental Prize."
Environment News Service, "Goldman Prize Awarded Early to Jailed Mexican Farmer-Ecologist" <http://www.ens.lycos.com/ens/apr2000/2000L-04–05–02.html>.

Myers, Norman (1936–) is a British ecologist and a critical figure in biodiversity conservation. In the 1960s Myers worked for the British government in Kenya, where he began an innovative community-development program in a region inhabited by two indigenous tribes. He remained in Kenya after the country gained independence and became a professional photographer, which allowed him to travel and explore the country's wildlands. He has provided the intellectual and scientific foundation for many natural-resources conservation concepts, including "biodiversity hotspots." Myers has been a scientific consultant and policy advisor to the White House and an advisor to the World Bank's Global Environment Facility. He has also worked with the U.S. Departments of State and Defense, NASA, the Smithsonian Institution, the European Commission, and seven United Nations agencies. He was the first British scientist to receive the Volvo Environmental Prize and the second to receive a Pew Fellowship in Conservation and Environment. In 1988, Myers won the Global 500 Award, and in 1998, the Queen's Honour for his dedication to the global environment. A Fellow of Green College in Oxford, England, Myers is the author of 15 books, including *The Primary Source: Tropical Forests and Our Future* and *Environmental Exodus: An Emergent Crisis in the Global Arena.*

Further Reading:

Foo, "Instruments for Change: An Electronic Conversation with Norman Myers" <http://iisd.ca/susprod/new consumers.htm>.

Kreisler, "The Journey of an Environmental Scientist: Conversation with Norman Myers."

Myers, *The Primary Source.*

Nakhasathien, Seub (1949–1990) was born into a farming family in the Prachin Buri Province of the Muang District, Thailand. During his twenties, Nakhasathien studied silviculture and worked as a junior government official in the government wildlife agency. In the 1980s, he served as chief of the Bang Phra nonhunting area, where he managed the training of forest rangers and also directed the Wild Animals Relocation Project. In 1987, he became chief of a wildlife sanctuary in Surat Thani Province. Throughout his career in public service, he stressed the need to educate the public about the environment. In 1988, he was honored Forest Official of the Year, and in 1991 was posthumously awarded a Global 500 laureate. In 1990, the year he died, Nakhasathien established a fund in order to protect two of Thailand's wildlife sanctuaries. Named in his honor, the Seub Nakhasathien Foundation has become one of the most effective and popular conservation organizations in Thailand.

Further Reading:

The Seub Nakhasathien Foundation, Thailand <http://www.seub.ksc.net/aboutus/introseub/his1life-e.htm.

United Nations Environment Programme, Global 500 <http://www.unep.org/newdraft/unep/per/ipa/g500/roll.htm>.

Nguiffo, Samuel (1966–) is a 1999 Goldman Prize recipient from Cameroon and founder of the Center for Environment and Development (CED). Nguiffo won his prize in recognition of his dedicated efforts to curtail rampant deforestation in Cameroon. An attorney, he travels throughout his country's forests to inform indigenous tribes of their land rights, specifically a provision in the forestry law that allows local people to legally manage their traditional land. CED also supports sustainable income-generating projects like beekeeping. Nguiffo is leading an international campaign to ensure that a proposed oil pipeline between Cameroon and neighboring Chad does not destroy forests, cause marine pollution, or dislocate people.

Further Reading:

Environment News Service, "Valiant Grassroots Environmentalists Rewarded with Goldman Prizes" <http://www.ens.lycos.com/ens/apr99/1999L-04–19–03.html>.

Goldman Foundation, USA <http://www.goldmanprize.org/recipients/recipients.html>.

Nugkuag, Evaristo (1950–) is the leader of the Aguaruna tribe of the Peruvian Amazon. After studying medicine in Lima in the 1970s, he shifted his attention to organizing members of his tribe to obtain better environmental, social, and economic welfare. In 1977, he helped found the Aguaruna and Huambisa Council, which represented 45,000 inhabitants of 140 communities in the tropical-forest region. Developing alternative methods of land protection, human development, health care, and education, the council became one of the most effective indigenous organizations in South America. Nugkuag realized that all indigenous people in the Peruvian Amazon faced the same threatening conditions and motivated different tribes to form an organization known as the Alliance of the Indian Peoples of the Peruvian Amazon.

In 1984, he formed and became president of COICA, a federation representing the national indigenous organizations of the five Amazon Basin countries: Peru, Bolivia, Ecuador, Brazil, and Colombia. In 1992, after Nugkuag's presidency at COICA expired, he continued to forge alliances, becoming president of the Alliance of European Cities and the Indigenous People of the Amazon for the Protection of Tropical Rainforests, Climate and Human Life. Under his leadership, Nugkuag nurtured relationships among tribes that had not communicated in more than 25 years. He also built ties between indigenous tribes and the international environmental community. Nugkuag was awarded the Right Livelihood Award in 1986 and a Goldman Environmental Prize in 1991.

Further Reading:

Goldman Foundation, USA <http://www.goldmanprize.org/recipients/recipients.html>.

Right Livelihood Award <http://www.rightlivelihood.se/recip1986_5. html>.

Padua, Claudio (1948–) and **Suzana Padua** (1950–) are Brazilian conservationists who founded the Institute for Ecological Research (IPE). The organization's conservation efforts are supported by the partnerships it has established with government bodies, other nongovernment organizations, and members of the agrarian movement. The Paduas developed a creative strategy to protect the highly endangered black lion tamarin and its vulnerable habitat, Brazil's Atlantic coastal forest. Their initiative focuses on education and involvement of local communities. C. Padua, a former businessman, began studying ecology and biology before working with endangered primates at the Rio de Janeiro Center. He is trained in primatology, the breeding of endangered species, and the integration of geographically isolated subpopulations. His work has been recognized by the American Society of Primatology, and he has won the Henry Ford Award for Conservation, the Whitley Award of the Royal Geographic Society, and the Society for Conservation Biology Achievement Award. S. Padua is an environmental educator who specializes in conservation of protected areas by raising awareness among local Brazilians. She is the director for the Brazil program of Wildlife Preservation Trust International.

Further Reading:

Iams, "Trotting Out Honors and Looking Ahead."
Wildlife Preservation Trust International <http://www.thewildones.org/Scientists/askPaduas.html>.

Peal, Alexander (1944–) of Liberia organized the Wildlife and National Parks section of Liberia's forestry department and led the agency from 1977 to 1990. In 1983, with help from the World Wide Fund for Nature and the World Conservation Union, Peal established Liberia's first national park and drafted many of the country's wildlife-management laws. He founded the country's first and private conservation organization—the Society for Nature Conservation in Liberia (SNCL). Peal left his country for the United States in the midst of political upheaval and to support SNCL and gather funds for conservation in Liberia. In 1996, Peal won the Senior Biology and Conservation Award from the American Society of Primatologists in recognition of his dedication to conserving Liberia's wildlife. He received the Goldman Prize in 2000.

Further Reading:

American Society of Primatologists, "Conservation Award Winners for 1996" <http://www.asp.org/conservation/conservation_awards96.htm>.
Goldman Foundation, USA <http://www.goldmanprize.org/recipients/recipients.html>.
Stang, "Saving Liberia's Rainforests."

Plotkin, Mark (1955–) is a U.S. tropical ethnobotanist who has been researching the healing powers of plants in Central and South America for more than 20 years. He has categorized more than 300 botanical cures and written popular books about ethnobotany, including *Tales of a Shaman's Apprentice*. His efforts focus on ensuring that rainforest cures are not destroyed by outside influences and pressures from loggers and farmers. He has worked to encourage large pharmaceutical companies that are screening tropical flora for their potential as new medicines to return a portion of any profits to forest conservation and forest inhabitants. He has served as director of plant conservation at the World Wildlife Fund and vice president of Conservation International. He is now research associate in the Department of Botany at the Smithsonian Institution and the Executive Director of Amazon Conservation Team, a nonprofit group.

Further Reading:

Hallowell, "In Search of the Shamans' Vanishing Wisdom."
Reed, "Sorcerers' Apprentice."

Prance, Ghillean (1937–), a botanist, grew up in Gloucestershire, England. He was senior vice president for science at the New York Botanical Garden, where he established the Institute of Economic Botany. From 1988 to 1999 he served as director of the Royal Botanic Gardens at Kew, England. Prance has traveled extensively in the Amazon rainforest and has published more than 300 papers on plant systematics, plant ecology, ethnobotany, and conservation. He has also written 14 books, including the text of *Rainforests of the World*, published in 1998. Prance won the Linnean Medal in 1990 and the Royal Geographical Society Patron's Medal in 1994. He was knighted in 1995. After his retirement from the Royal Botanical Gardens, he joined the Eden Project as a consultant and director of science. The Eden Project manages the largest indoor rainforest environment in the world and experiments with conservation techniques with the goal of furthering plant biodiversity in the wild.

Further Reading:

The Eden Project <www.edenproject.com/19_8_99.htm>.

Quansah, Nat (1953–) is originally from Ghana, but has worked for the last decade with the World Wide Fund for Nature in Madagascar, the world's fourth-largest island and home to 5 percent of the world's animal and plant life. Quansah launched a WWF program in 1989 that trains local researchers to study the island's natural resources. The project investigates the relationship between local communities and local plant diversity and taps the expertise of both villagers and foreign scientists. In 1994, Quansah established a clinic in a remote village of Madagascar where he created the Integrated Health Care and Conservation

Program, which aims to retain and teach traditional knowledge of natural remedies. Quansah's goals are to increase awareness of the virtues of natural medicine and decrease deforestation and loss of traditions. Quansah received a 2000 Goldman Prize.

Further Reading:

American University, "Deforestation in Madagascar" <http://www.american.edu/projects/mandala/TED/MADAGAS.HTM>.

Cunningham, "Healthcare, Conservation, and Medicinal Plants in Madagascar."

Goldman Foundation, USA <http://www.goldmanprize.org/recipients/recipients.html>.

Ranjitsinh, M. K. (1938–) started working in conservation in his native India in the late 1960s, when he began assisting families who were moving out of newly designated protected areas. As director of forests and wildlife in the Ministry of Agriculture from 1973 to 1975, Ranjitsinh spearheaded the Wildlife Preservation Act of 1972, which established national parks and sanctuaries. He was instrumental in passing laws that banned trade in snakes and lizard skins and launched projects to preserve the few surviving populations of the highly endangered Asiatic wild buffalo. While working with the forest and wildlife department, he pushed through laws to regulate sawmills, established a zoo, and instigated campaigns to prevent forest fires. He also helped establish 8 national parks and 11 wildlife sanctuaries. He served on the committee that initiated Project Tiger, a conservation initiative that identified reserves for Royal Bengal tigers (*Panthera tigris*). He is the director of the World Wide Fund for Nature's tiger conservation program. Ranjitsinh won the Global 500 award in 1991.

Further Reading:

Ridge, "India's Tiger Threatened by Land Use and Parts Demand."

Raven, Peter (1936–) is a U.S. botanist and environmental activist. In the early 1970s, he helped place tropical deforestation and the species extinction crisis on the agendas of influential international organizations, such as the United Nations Environmental Programme, the United Nations Development Programme, and the World Bank. In 1970, he and Norman Myers co-authored a report entitled "Research Priorities in the Humid Tropics," which presented information revealing that global destruction of tropical forests was far more serious than most ecologists were aware. With colleague Paul Ehrlich,

he developed a theory of coevolution to describe how species change as the species with which they interact change. Since 1971, he has been director of the Missouri Botanical Garden, the nation's oldest public botanical garden. Raven received the Global 500 prize in 1987, the Sasakawa Prize in 1995, and the Engler Medal in 1999, awarded for his lifetime service to plant taxonomy. He is a world leader in plant biotechnology, a leading researcher in plant evolution, and an ardent spokesperson against biological extinction.

Further Reading:

Jackson, "Through Politicking for Plants, He Made His Garden Grow."

Reynal, Miguel (1936–) founded the Fundación Vida Silvestre Argentina in 1970 and has served on its executive council since 1987. The group is now Argentina's leading environmental group. In 1987 he helped launch Save the Forest, an international project that promotes environmental reporting and debt-for-nature swaps. In 1994 Reynal established Fundación Ecos, a conservation organization with headquarters in Uruguay. He remains the executive director and helps provide information and resources for community leaders in South America. Reynal won the Global 500 award in 1997. His environmental achievements include helping to save the endangered pampas deer, establishing a breeding station for the pudu-pudu deer, and training more than 1,500 teachers in environmental education. Reynal has created private and state-held reserves and a bird sanctuary in Buenos Aires.

Further Reading:

United Nations Environment Programme, USA, "Global 500: Roll of Honour for Environmental Achievement."

Robbins, Chandler (1918–) is a U.S. ornithologist who led the efforts to monitor populations of non-game birds. He established the Breeding Bird Survey that since 1965 has tracked bird populations throughout North America. When the survey showed declines in the populations of migratory songbirds, Robbins began researching and brought attention to the impact of tropical deforestation on the migrants.

Further Reading:

Lipske, "American Heroes: He Wrote the Book on Birds."

Saro-Wiwa, Ken (1941–1995) was a Nigerian environmentalist, leader of the Ogoni indigenous people, and writer who was executed in 1995 by the Nigerian government. He was a newspaper columnist, novelist, poet, and author of a popular television show, but he quit his writing career in the 1990s to lead protests in favor of Ogoni self-determination and against the refusal of Shell Petroleum to clean up oil spills, bury pipelines, and build health and educational facilities in the Niger Delta. He also called for a boycott against Shell. His movement was undermined by the Nigerian military government when he and colleagues were implicated in the killings of four pro-government chiefs. He and eight other Ogonis were hanged for murder, in spite of international outrage and protests. His final words on the gallows were, "Lord, take my soul, but the struggle continues." His son continues to fight against environmental destruction in Ogoni territory.

Further Reading:
Chernos, "Ken Saro-Wiwa's Son Continues Fight Against Big Oil."
Golden, "A Century of Heroes."

Schaller, George B. (1933–) was born in Berlin, Germany, and raised in the United States. He is a biologist, mammalogist, naturalist, and zoologist who has devoted his career to understanding endangered species, such as the mountain gorilla (*Gorilla gorilla beringei*). His wildlife research showed that gorillas were not the aggressive beasts most people believed they were, underlined the importance of predators, and emphasized that there was little point in saving animals unless you can also save their habitat. He has been a research zoologist at the New York Zoological Society and was the director of its international conservation program from 1979 to 1988. In addition to the hundreds of magazine articles he has written, Schaller has also authored popular books on African and Asian mammals based on his research. He received the National Book Award in 1973 for *The Serengeti Lion: A Study of Predator-Prey Relations*; the 1980 World Wildlife Fund Gold Medal, awarded for his contributions to the understanding and protection of endangered species; the Global 500 award in 1995; and, with two other primate specialists—Jane Goodall and Biruté Galdikas—the 1997 Tyler Prize.

Further Reading:
Golden, "A Century of Heroes."
Orangutan Foundation International <http://www.orangutan.org>.

Schultes, Richard Evans (1915–) is internationally recognized as one of the greatest Amazon explorers of the century. He pioneered the study of relationships between plants and people, mainly by investigating the effects plants have on people's psyches. Born in the United States, he traveled to the Colombian Amazon in 1941 on a fellowship to study the plant-derived poisons that indigenous people use on their arrows and darts. Between 1941 and 1953, Schultes collected and identified more than 24,000 plant species, many of which are named after him. He has published more than 400 technical papers and 9 books, including a photographic essay with over 160 images of his travels prior to 1954. In the 1960s, as the pharmacological industry's interest in plants for medicinal purposes waned, Schultes continued to praise indigenous peoples' sacred rituals and medicinal practices involving plants. Well after his retirement, Schultes continued to travel to Colombia to help establish national parks in the Amazon basin. In 1994, he was awarded the Global 500 prize. He has also received the World Wildlife Fund Gold Medal, the Tyler Prize for Environmental Achievement, and the Linnean Gold Medal.

Further Reading:
Becher, *Biodiversity: A Reference Handbook*.
Schultes, *Where the Gods Reign: Plants and Peoples of the Colombian Amazon*.
Synge, "One Man, One River; A Story of Amazon Exploration."

Silva, Marina (1958–) spent her childhood tapping rubber, hunting, and fishing in the Amazon forests of Acre, Brazil. She was illiterate until the age of 16, when she left home to attend school and earned her bachelor's degree when she was 20. With Chico Mendes, Silva co-founded an independent trade union movement of rubber tappers in Acre. As a team, they became famous for mobilizing grassroots resistance to environmental destruction. Silva helped organize peaceful demonstrations for rubber tappers that protested deforestation by cattle ranchers and threats to their traditional land holdings. As a result, the government set aside more than 60,000 acres in Acre as an extractive reserve, which is sustainably managed by the villagers who live nearby. Silva was the first rubber tapper ever elected to Brazil's federal senate. As a populist senator, she increased protection of extractive reserves and promoted awareness of social justice and sustainable development in the Amazon region. Silva is a 1996 Goldman Prize winner.

Further Reading:
Goldman Foundation, USA <http://www.goldmanprize.org/recipients/recipients.html>.
Leach, "Don Quixotes of the Environment."

Sinclair, John (1939–) has been a leader in the Australian conservation movement, particularly on Fraser Island, for more than 30 years. He has fought against sand mining and logging on the narrow island, which is located off Queensland. In 1967 he founded the Maryborough Branch of the Wildlife Preservation Society of Queensland, and in 1971 he started the Fraser Island Defenders Organization (FIDO). Sinclair has been the leader of the Australian Conservation Foundation, president of the Wildlife Preservation Society of Queensland, and currently serves as director of the Australian Tropical Research Foundation. He is researching the lowland rainforest of Queensland and helping Aboriginal groups fight against a proposed dam. He was named "Australian of the Year" in 1976, won the Global 500 award in 1990, and received the Goldman Environmental Prize in 1993.

Further Reading:
Goldman Foundation, USA <http://www.goldmanprize. org/recipients/recipient>.
John Sinclair Trust for Conservation <http://www. sinclair.org.au.>.

Skutch, Alexander (1904–) is a U.S. ornithologist who for some 50 years has lived on a farm in Costa Rica—now the Los Cusingos Bird Sanctuary, a reserve he donated to a local research organization. With a doctorate in botany, Skutch sailed to Panama in 1928 to study bananas in the Neotropics but became more enthralled with the birds of the plantations than the fruit. He discovered that little was known of their behavior and became devoted to ornithology. Skutch has authored more than 30 books —on ornithology, philosophy, or a little of both. His scientific and passionate accounts of tropical bird life include studies of some 200 species. In honor of his dedication to the conservation and scientific understanding of birds, in 1997 the Association of Field Ornithologists created The Alexander F. Skutch Award for excellence in Neotropical ornithology.

Further Reading:
Association of Field Ornithologists <http://www.afonet.org/ about.html#Skutch>.
Skutch, *A Naturalist amid Tropical Splendor.*
Tropical Science Center <http://www.cct.or.cr/alex.htm>.

Tiensuu, Roland (1978–) was a nine-year-old student in teacher Eha Kern's class in Sweden when Kern invited a guest biologist from the United States to speak to her class about Costa Rica's Monteverde Cloudforest Reserve. After the presentation, Tiensuu encouraged his classmates to raise funds to help protect the reserve. What started as a campaign to buy just a few acres as part of a land-purchase program soon turned into a program that conserved more than 60,000 acres of forest in Monteverde. Kern and her husband then launched an organization called the Children's Rainforest, which is dedicated to raising funds for tropical rainforest conservation. In the group's honor, an area of the reserve in Monteverde is called the Children's Rainforest. Both Kern and Tiensuu received the Goldman Prize in 1991.

Further Reading:
Goldman Foundation, USA <http://www.goldmanprize. org/recipients/recipients.html>

Varela, Jorge Márquez (1947–) is a Honduran conservation biologist who has worked for nearly four decades to preserve biodiversity in the coastal wetlands and mangrove forests along the Gulf of Fonseca. He is founding director of the Committee for the Defense and Development of Flora and Fauna of the Gulf of Fonseca (CODDEFFAGOLF), which represents more than 10,000 subsistence fishermen, salt extractors, farmers, schoolchildren, and local residents of villages along the Gulf. Thanks to efforts of Varela and CODDEFFAGOLF, in late 1999 the government of Honduras established a Gulf of Fonseca protected area and placed a moratorium on the expansion of shrimp farms, which have destroyed thousands of acres of coastal mangrove forests. Varela won the Global 500 award in 1992 and the Goldman Prize in 1999.

Further Reading:
CODDEFFAGOLF <http://www.ben2.ucla.edu/~alexagui/ cgolf/cgolf-en.htm>.
Goldman Foundation, USA <http://www.goldmanprize. org/recipients/recipients.html>.

Vázquez-Yanes, Carlos (1945–1999) was one of the first tropical ecologists to sound the alarm at the high rate of tropical deforestation. Born in Venezuela, he moved to Mexico while young and became a citizen of that country. Vázquez co-wrote an influential paper in 1972 that focused on the nonrenewable characteristics of tropical rainforests. His pioneering research on seed germination helped scientists understand how tropical forests regenerate. He also used his knowledge of tree regeneration to restore damaged forest ecosystems. Vázquez received numerous awards from the Mexican Academy of Sciences, the Botanical Society of Mexico, and the University of Mexico.

Further Reading:
Núñez and Gómez-Pompa, "Carlos Vázquez-Yanes."

Von Humboldt, Alexander (1769–1859) began his explorations of the New World in 1799, near the mouth of the Orinoco River in Venezuela, accompanied by the French botanist Aimé Bonpland. He learned that the Orinoco shared waters with the Amazon, and before returning to England seven months later, he had collected 12,000 plant species. He was one of the first to write about the exuberance of tropical flora; before his publications, Europeans had no knowledge of the interior of tropical America. Von Humboldt and Bonpland returned in 1801, exploring Cuba, Colombia, Ecuador, Peru, and Mexico—collecting specimens, taking copious notes, and mapping. In Peru he fixed the magnetic equator at 7 degrees, 27 minutes south latitude. He shipped back samples of bird guano that launched a fertilizer industry. He was the first to survey the cold current that flows off the coast of Peru and bears his name. He was the first to document the engineering marvels of Inca ruins. Based on the ruins, he proposed the Asian origin of American indigenous people, not having any idea that there once indeed was a land bridge connecting Siberian Asia with Alaska.

After spending five years exploring the Neotropics, he passed the remainder of his life writing up the results of his studies, publishing books that had a huge influence on the next generation of explorer naturalists; Darwin decided to pursue natural science after reading von Humboldt.

Further Reading:
Maslow, *Footsteps in the Jungle.*

Wallace, Alfred Russel (1823–1914) traveled with partner Henry Walter Bates up the main channel of the Amazon River in 1848. Before their explorations, no Europeans had ever explored the upper Amazon and its main northern tributary, the Rio Negro. Wallace's expertise was zoogeography; he collected specimens to find out their variety and range. He spent eight years in the Malay Archipelago from 1854 to 1862, traveling and collecting specimens and writing scientific articles. He proposed that new species arise by the progression and continued divergence of varieties that outlive the parent species in the struggle for existence. He forwarded this paper to Charles Darwin, who was struck by the similarity of Wallace's conclusions to his own. Darwin and Wallace submitted a single and singular article on natural selection to the Linnean Society in 1858. Wallace's

Geographical Distribution of Animals (1876) and *Island Life* (1880) were the standard authorities in zoogeography and island biogeography.

Further Reading:
Maslow, *Footsteps in the Jungle.*
Severin, *The Spice Islands Voyage.*

Werikhe, Michael (1956–1999) raised worldwide awareness of the plight of the rhinoceros, while raising $1 million to conserve this endangered animal in his native Kenya. He brought the mammal's plight to the public's attention by canvassing on foot in the United States, Africa, Taiwan, and Europe from 1982 to 1993. The Kenyan government used the funds he raised to transfer rhinos from threatened habitats to protected areas where they could breed and thrive, thus significantly increasing the animal's population in Kenya. Werikhe won the Global 500 award in 1989 and the Goldman Prize in 1990. At the time of his death in 1999, he was involved in a project to develop housing for rangers and researchers in Kenya's Tsavo National Park.

Further Reading:
Goldman Foundation, USA <http://www.goldmanprize. org/recipients/recipients.html>.
McClanahan, T.R. "The Rhino Man's Final Rest."
The Wildlife Society, USA <http://www.wildlife.org/press/ rhinoman.html>.

Wilson, Edward O. (1929–), a professor at Harvard University, is one of the best known and most respected advocates of global biodiversity protection. A U.S. entomologist, his expertise is on the social behavior of ants, but early in his career it expanded to include population biology, ecology, genetics, conservation biology, sociology, philosophy, and activism. With ecologist Robert MacArthur he developed a theory in the 1960s on island biogeography, which suggested that smaller islands support a vastly lesser diversity of species than larger ones. This led biologists to realize that protected areas and parks were becoming increasingly fragmented, and unless they could be enlarged and linked to other wildlands, the isolated island parks would not prevent the extinction of numerous species. His interest in studying the social behavior of ants led to his contributions to the field of sociobiology, which links biology with social behavior. His landmark book *Sociobiology: The New Synthesis* was published in 1975, and in 1989, the Animal Behavior Society named it the most important book ever written on animal behavior. Wilson is the only person to have received both the National

Medal of Science (in 1977) and the Pulitzer Prize. He won the latter twice, first in 1978 for *On Human Nature* and again in 1991 for *The Ants*, coauthored with Bert Holldobler. He edited two books, *Biodiversity* (1988) and *Biodiversity II* (1996), that focus on the devastating loss of tropical flora and fauna. His 1992 book, *The Diversity of Life*, describes the remarkable biological richness that exists on Earth and the threats to this biodiversity. He has received the Crafoord Prize and dozens of other awards, prizes, medals, and honorary degrees from around the world.

An international poll placed Wilson in the top 100 of the most influential scientists of all time.

Further Reading:
Luoma, "The Sociobiologist: Edward O. Wilson."
Oliveira, "Edward O. Wilson, Doyen of Biodiversity's Crusade."
Quammen, *Song of the Dodo: Island Biogeography in an Age of Extinctions.*
Wade, "From Ants to Ethics: A Biologist Dreams of Unity of Knowledge."
Wilson, *Naturalist.*

PART IV

Saving Tropical Forests

Imaginative Experiments and Practical Solutions

The causes of deforestation are numerous, complicated, and overlapping, as explained in this book's introduction. Fortunately, conservationists, scientists, government officials, business leaders, and many other dedicated and creative people are searching for ways to curb the loss of tropical forests. Clearly, no single answer exists, but rather a panoply of promising responses that have undoubtedly already saved millions of acres of trees. As long as trees continue to fall, however, people with courage and expertise will continue to search for the most efficient and effective ways to keep tropical forests intact. Below are explanations of 27 of the areas in which the experts are working to find solutions to deforestation.

AGRICULTURE

Agriculture is the most common cause of deforestation throughout the tropics. As human populations grow, more people are forced to clear forested land to plant the food they need to survive. Since most tropical soils do not have enough nutrients to support crops for more than a few years, people are forced to abandon these plots and clear even more forest. This centuries-old system of farming in tropical soils is called slash-and-burn agriculture, shifting cultivation, or swidden cultivation. For the system to work, farmers must abandon farm plots after a few years, allowing them to lie fallow for at least 10 years, so they may recapture nutrients. The system did indeed provide for indigenous people of the humid tropics for hundreds of years, but now there is no longer enough land available for shifting cultivation to work effectively.

The border where farmland meets tropical forest is called the agricultural frontier. One of the toughest challenges that conservationists face is how to stop the spread of the agricultural frontier while human populations continue to grow and so much arable land is lost to urban sprawl or ruined by poor management. Many initiatives focus on helping people who live near the agricultural frontier develop alternatives to farming that will provide them with a livable income (see "Community-Based Conservation" and "Forest Resources"). Other projects focus on helping subsistence farmers make their plots more productive, so they can grow sufficient food without having to slash-and-burn more forest land.

A particularly successful program is sponsored by World Neighbors, which is based in the United States and active in 18 tropical countries. Via community groups, World Neighbors not only trains farmers how to grow crops more productively and ecologically but also trains them how to teach these same methods to their neighbors.

Further Reading:

Bunch, *Two Ears of Corn.*
Neugebauer, *Ecologically Appropriate Agri-Culture.*
Pagiola, *Mainstreaming Biodiversity in Agricultural Development.*
Thrupp, *New Partnerships for Sustainable Agriculture.*
World Neighbors, USA <http://www.wn.org>.

AGROFORESTRY

Agroforestry is the practice of growing trees in combination with perennial and annual food crops and sometimes livestock. It is an ancient practice, still fairly common in the tropics. Conservationists encourage subsistence farmers, who may be more accustomed to clearing land of trees before planting food crops, to practice agroforestry, which makes maximum use of the land and protects and even improves soils.

Trees may be planted around farm plots, protecting against erosion from wind and rain—particularly on hillsides—or interspersed with crops. Planted in closely spaced rows, frequently trimmed trees are "living fenceposts," whose sprouting branches and vegetation are used to feed cattle. Trees are often planted to provide shade for crops such as coffee or to support vine crops, such as vanilla and pepper. They also serve as a source of fuel wood, fibers, fruits, nuts, and other foods, medicines, resins, and construction material.

Further Reading:
Gliessman and Grantham, "Reshaping Agricultural Development" in Head, *Lessons of the Rainforest.*
Griffith, "Agroforestry: A Refuge for Tropical Biodiversity after Fire."
Pagiola, *Mainstreaming Biodiversity in Agricultural Development.*

BIODIVERSITY HOTSPOTS

Biodiversity hotspots is a term that refers to areas on Earth that have relatively large concentrations of flora and fauna. The British ecologist Norman Myers created the hotspot concept in the late 1980s, and since then, biologists have identified biodiversity hotspots to guide decisions about where conservation efforts should be focused. In the face of an inflated rate of flora and fauna extinction brought on by human destruction, biodiversity hotspots are a form of triage, so that the regions on Earth with the greatest and most threatened biodiversity are saved first. Agriculture, logging, and other human development activities are extinguishing tropical species at some 1,000 times the normal evolutionary rate of extinction. Scientists estimate that at the current rate of deforestation, all tropical forests will be lost by the end of this century, along with more than half of all the species on earth. By some predictions, human beings may have already destroyed so many plant and animal species that the planet will need ten million years to recover—that's 20 times as long as humans have already existed.

By concentrating on biodiversity hotspots, the grim statistics could be reversed. Myers and scientists at Conservation International have identified 25 hotspots, based on five factors: the number of native plants, number of native vertebrates, ratio of native plant numbers to land area, ratio of native vertebrate numbers to land area, and how much primary vegetation remains as compared with what originally existed. Nineteen of the 25 hotspots are in the tropics. At least half of the world's terrestrial species live within these areas.

To help catalogue the number of endemic and other species these areas contain, Conservation International dispatches expert scientific teams as part of its Rapid Assessment Program (RAP). During a RAP visit to Ecuador in 1993, two leading scientists, the ornithologist Theodore Parker and the botanist Alwyn Gentry, were killed in an airplane accident.

Part of their legacy is the hundreds of new species in numerous tropical hotspots that they identified.

In March 2000, Myers and four colleagues identified the world's eight hottest spots, based on their having the greatest concentration of endemic species while also facing exceptional habitat loss. All but one of them is in the tropics.

Further Reading:
Conservation International, USA <http://www.conservation.org/web/fieldact/hotspots/hotspots.htm>.
Myers, "Tackling Mass Extinction of Species: A Great Creative Challenge."
Wilson, *The Diversity of Life.*

BIOSPHERE RESERVES

Biosphere reserves are land and coastal ecosystems that are internationally recognized by the United Nations Educational, Scientific and Cultural Organization (UNESCO). To receive the special designation, an ecosystem must be nominated by a national government, meet a minimal set of criteria, and adhere to certain conditions. Under UNESCO rules, biosphere reserves must contribute to biodiversity conservation; encourage socially, culturally, and ecologically sustainable economic growth; and support research, monitoring, education, and training related to local, national, and global conservation and development issues.

Biosphere reserves have an established core protected area, where no development or activities that would harm biodiversity are permitted; outside that lies an area, called a buffer zone, where some ecologically sustainable development is permitted, such as grazing, tourism, or sustainable logging; and surrounding that lies a broader region where local residents, scientists, and others can work together to develop eco-friendly growth. There are 352 biosphere reserves in 87 countries; more than 100 are in the tropics.

Although the biosphere reserve designation does not afford any protection to an area under national law, international recognition and prestige wield influence. Removal by UNESCO of the biosphere reserve designation would be a global embarrassment to any country. In early 2000, UNESCO threatened to repeal the designation from the 12.3 million-acre Río Plátano Biosphere Reserve in Honduras if the country did not control illegal logging. From 1980 to 2000, some 3.5 million acres of the reserve had been deforested.

Further Reading:
Reuters, "Honduran Congress seeks troops to protect forests."
UNESCO, Man and Biosphere Reserves, USA <http://www.euromab.org/brprogram/bioreser.html>.

CARBON BONDS

Carbon bonds can be purchased by companies, such as electricity-generating utilities, that emit carbon dioxide to help mitigate the environmental damage that these emissions cause. The bonds represent a promise by the bond sellers that they will protect a specified, forested area or otherwise encourage sustainable forestry. Since trees absorb carbon dioxide, forests are considered to be "carbon sinks." Emissions like carbon dioxide cause global climate warming, so forests help moderate global temperatures. Scientists and economists are currently studying exactly how much carbon dioxide forests absorb and what the monetary value of this environmental service might be. They suggest that about 20 percent of atmospheric damaging emissions come from the logging and burning of forests.

Under a 1997 international climate-change treaty that was signed by 174 nations but not the United States, industries that emit carbon dioxide may be required either to cut back on emissions or in effect to mitigate them by purchasing carbon bonds. Trading of carbon bonds is also called "joint implementation," since two countries are often involved. Several tropical countries have already sold carbon bonds and struck joint implementation deals with utility companies. One of the first was in 1995; Tenaska, Inc., an energy company in Washington State, provided $500,000 toward the purchase of forest in Costa Rica. Additional funds came from the National Fish and Wildlife Foundation and the Rainforest Alliance, both based in the United States, and the forest was turned over to the Costa Rican government. Today it is protected as Piedras Blancas National Park.

In a 1998 joint implementation deal, the Bolivian government, the Nature Conservancy (TNC), a U.S. conservation group, and three U.S. corporations purchased for $9.5 million more than two million acres of forested land in Bolivia. TNC scientists say that by keeping the land forested, about 15.8 million metric tons of carbon will be kept from entering the atmosphere. American Electric Power, BP America, and PacifiCorp are the three U.S. corporations that helped purchase the land.

Further Reading:
International Climate Change Fund, USA <http://www.ji.org/>.
The Nature Conservancy, USA <http://www.tnc.org>.
Nilsson, *Greenhouse Earth.*
Richards and Costa, "Can Tropical Forestry Be Made Profitable?"
Zollinger, "Private Financing for Global Environmental Initiatives."

CERTIFICATION

Certification, known also as eco-labeling, is one way that consumers can contribute to rainforest conservation. If a product is "certified" or bears an "eco-label," it means the product was made in such a way as to have a relatively minimal impact on biodiversity. One of the most common certified products is timber from tropical countries, or items made from tropical woods, like furniture and musical instruments. The first and currently largest timber certification program, called SmartWood, was started by the Rainforest Alliance in 1989. SmartWood has certified 93 forestry operations as "well managed." In the tropics, 3.5 million acres of forest and tree plantations have been certified. To gain certification, a forestry operation must allow experts to inspect and verify whether or not the program's environmental regulations were followed.

Other forestry certification programs operate worldwide, and some are more restrictive than others. To maintain consumer confidence in these seals, the Forest Stewardship Council accredits the certifiers. FSC approval means that the regulations of a particular certification program at least meet, if not exceed, internationally approved environmental and social requirements. Consumers should look for the FSC seal to ensure that environmental claims on a wooden product are legitimate.

The Rainforest Alliance is also working with a coalition of Latin American conservation groups to certify farms that meet stringent guidelines to protect workers and wildlife. This coalition is certifying banana, coffee, citrus, and cocoa farms.

Further Reading:
Environmental Protection Agency, *Environmental Labeling.*
Rainforest   Alliance,   USA   <http://www.rainforest-alliance.org>.

COMMUNITY-BASED CONSERVATION

Community-based conservation is known as Integrated Conservation and Development Projects, or ICDPs, in professional conservation lingo. The objective of these projects is to protect and manage ecosystems in ways that involve and benefit residents who live nearby. Most projects link protected-area conservation with local social and economic development. The theory is that if people have alternative and reliable sources of income plus access to health care and education, they will no longer need or want to take resources unsustainably—such as wildlife, plants, and timber, as well as land—from forest reserves. So community-based conservation projects aim to improve the quality of life of people living near areas that are rich in biodiversity, while promoting the conservation and management of these ecosystems. This is an alternative to strictly protected parks that penalize local people who use forest resources. Although valuable forest resources are often marketed to improve people's incomes, they are harvested carefully and sustainably, following a management plan. In that way, residents will view the forest as a source of income on which they can depend for many years, as long as they don't exploit the ecosystem.

A project in southwestern India is a good example of a typical ICDP. In the Bilgiri Rangan Temple Sanctuary in the Western Ghats, several Indian and international groups began working in the mid-1990s with the local Soligas people who live in the sanctuary's buffer zones. Together they developed a honey collection, processing, and marketing project; a food processing plant; and an herbal medicine collection and marketing project. These initiatives have provided employment and increased income for scores of Soligas. But the project is too new to ascertain if it is helping to protect the sanctuary's wildlife. To be successful, community-based conservation projects generally need outside funding for a long period of time—at least 20 years.

Further Reading:
Biodiversity Conservation Network, *Final Stories from the Field.*
Brown & Wyckoff, *Designing Integrated Conservation & Development Projects.*

Clay, *Generating Income and Conserving Resources.*
Kramer, Schaik, and Johnson, *Last Stand.*

CORRIDORS

Corridors are vegetated landscapes that connect one natural area with another. When human activities fragment habitats—as has happened to wildlands across the globe—the animals and plants within the resulting patches become isolated by farmland, houses, or roads. Most animals need to move daily or intermittently to find food, shelter, breeding sites, or mates. Young animals must abandon the areas where they were born to avoid inbreeding and competition with other family members. Plants also need to exchange genes, via pollen and seed dispersal.

Many biologists believe that habitat fragmentation is the most serious threat to biological diversity and a main cause of extinction. As a result, conservationists are studying how protected areas could be connected with each another, via smaller greenways, or corridors. In the early 1990s, Wildlife Conservation Society, a U.S. conservation group, first proposed a green corridor that would stretch throughout the Central American isthmus. Called *Paseo Pantera*, or "Path of the Panther," the proposal was eventually adopted by the governments of Mexico and the seven Central American countries, and now the Mesoamerican Biological Corridor has millions of dollars in funding from the United Nations and World Bank, through a financing institution called the Global Environmental Facility (GEF). The corridor is envisioned to stretch along the Caribbean coast of the region, spanning borders in a true international conservation effort.

Meanwhile, Wildlife Conservation Society is working on another kind of corridor in Central America, a system of trailways that would allow hikers—and wildlife—to trek from Guatemala south to Panama without stepping out of the woods. And the Costa Rican biologist Mario Boza is tackling the most ambitious corridor project yet—he wants to establish an Ecological Corridor of the Americas that would stretch from the forests of Alaska to Tierra del Fuego on the tip of Argentina and shelter more than 50 percent of the biodiversity on Earth.

Further Reading:
Harris, *The Fragmented Forest.*
Jukofsky, "Path of the Panther."
Muñoz, "Ambitious Plan to Protect Biodiversity."

Noss, "Corridors in Real Landscapes."
Wildlife Conservation Society, Mesoamerican Trails Project, USA <http://www.mesosenderos.org/index.html>.

DEBT-FOR-NATURE SWAPS

Debt-for-nature swaps were a popular tropical forest conservation mechanism in the 1980s. First proposed in 1984 by conservation biologist Thomas Lovejoy, the innovative transactions were a response to two overlapping problems. First, many tropical countries have onerous, multimillion-dollar debt burdens to international lending organizations, foreign governments, and commercial banks. Second, indebted countries feel justified in liquidating natural resources for quick cash to service their debts. In a debt-for-nature swap, the banks holding the loans sell them at reduced rates to a conservation organization. The indebted country's central bank then acquires the debt from the conservation organization in return for a promise to issue bonds in the local currency. The interest on those bonds is used to finance environmental projects, often purchase of forested lands for protection.

Conservation International, an environmental group based in the United States, engineered the first debt-for-nature swap in 1987. CI purchased $650,000 worth of Bolivia's debt from Citicorp bank for the bargain price of $100,000. As part of the deal, the Bolivian government established a $250,000 fund to maintain a biological research station in the Beni Biosphere Reserve, three million acres of forest and grasslands at the foot of the Andes. In the following years, debt-for-nature swaps grew larger and more sophisticated. Costa Rica benefited most from the swaps, exchanging nearly $55 million in debts for conservation funding. Debt-for-nature swaps fell out of favor in the early 1990s, amid criticisms that they were inflationary and legitimized the foreign-debt situation, which these critics believe are massive transfers of wealth from poor to wealthy nations. The swaps did not significantly reduce the debt of any one country, nor did they truly protect significant natural resources from development, except perhaps in Costa Rica. They did bring a lot of attention to the issues of foreign debt and tropical conservation and did benefit small but important conservation initiatives.

Since Third World debt is still an important factor in global relations, the swaps may regain their previous popularity. During a trip to Bangladesh in March 2000, President Bill Clinton announced that the United States would forgive $6 million in the Bangladesh's external debt if the country would devote funds to preserving tropical forests and increase protection of Bengal tigers (Panthera tigris).

Further Reading:
Jakobeit, "Nonstate Actors Leading the Way: Debt-for-Nature Swaps" in Keohane, editor, Institutions for Environmental Aid.
Patterson, "Debt for Nature Swaps and the Need for Alternatives."
Wille, "The Greening of Debt."
Williams, "Banking on the Future."

ECOTOURISM

Ecotourism is a relatively new word that applies to the venerable concept of combining ecology with tourism. It is usually defined as a kind of travel to natural areas that conserves the environment and improves the well-being of local residents. To be considered ecotourism, travel should have minimum environmental impact, educate the visitor, provide direct financial benefits for conservation, be economically favorable for local communities, respect the local culture, and be sensitive to the host country's political, environmental, and social climate. Ecotourism is one of the few clear and concrete examples of conservation bringing direct and measurable financial benefits to governments and communities that protect tropical forests. In fact, the economic benefits of ecotourism total in the millions of dollars, as more tourists visit parks and reserves in the tropics, willing to pay for a chance to walk through a rainforest and perhaps spot a rare animal. During the 1990s, tourism became the number-one source of foreign income in Costa Rica, surging ahead of bananas and coffee; the country's beaches and famed national parks were the prime tourist destinations.

To ensure that ecotourism really is environmentally and culturally beneficial, several groups have established standards for tour operators, hotel owners, and travelers. The Ecotourism Society, based in Vermont, has published guidelines for nature tour operators and is dedicated to establishing standards for the profession. Costa Rica's tourism institute has an excellent program that actually grades hotels as to their sustainability—travelers need only check the institute's Web site and choose a hotel with an eco-rating <http://www.turismo-sostenible.co.cr>.

One of the biggest challenges for ecotourism advocates is ensuring that nature travel brings an economic benefit to those living in or near the scenic area. The RARE Center for Tropical Conservation, a nonprofit group based in Virginia, developed a nature-guide-training course, now active in Central America, Mexico, and the Caribbean. The intensive, three-month course trains villagers how to be naturalist-guides and how to speak English, the language most international travelers have in common. Graduates easily find jobs leading tourists on nature hikes through the parks in their own backyards, and they have an obvious incentive to help protect the reserve that provides them with a good income.

Further Reading:

Brandon, "Ecotourism and Conservation."

Costa Rica Tourism Institute, Sustainable Tourism Certification Program <http://www.turismo-sostenible.co.cr>.

Ecotourism Society, "Ecotourism Guidelines."

Jukofsky, "Guides in Their Own Backyards."

ENDANGERED SPECIES

Like biodiversity hotspots, endangered species give conservationists a way to set priorities about which species and ecosystems need immediate protection. Endangered plants and animals are also powerful environmental education tools, as the public tends to respond to the plight of particular species, especially larger animals—often called "charismatic species." Many conservation campaigns focus on just one charismatic and endangered species, such as elephants or tigers. Of course, securing habitat for these species means safer futures for hundreds of other plants and animals that share the same ecosystems.

Beyond its function as a way to engage the public, focusing on endangered species makes clear scientific sense—once a species disappears, it's gone forever, and the repercussions of removing a species from the web of life may be severe. Many biologists believe that as human beings destroy tropical forests, they are causing species loss at a rate so high that if it continues, the result will be one of the major extinctions in history, on a par with the disappearance of the dinosaurs. New research shows that the Earth will need 10 million years to recover from the impact already made to date by the disappearance of so many species. All species, including *Homo sapiens*, are linked together in a great, largely unexplored, web of interdependence. When one species declines or disappears, the survival chances of all the co-dependent organisms decline. To keep a species from extinction, scientists must first identify which ones are in trouble,

not easy when you consider the sheer numbers: somewhere between 10 and 100 million species. Only a fraction—about 1.4 million—of these existing species have actually been identified, so we're losing thousands of plants and animals that we've never even seen. Some scientists estimate that at least 100 species disappear from Earth every single day.

The World Conservation Union (IUCN) maintains a list of threatened species and categorizes them as "critically endangered," "endangered," or "vulnerable." These are compiled on the "Red List of Threatened Plants" and "Red List of Threatened Animals." Currently, 34,000 species are on the Red List of Threatened Plants. The Red List of Threatened Animals lists 5,025 species as threatened with extinction. All known species of birds have been evaluated for the list; 25 percent of mammals and 11 percent of birds are classified as endangered. Although not all reptile, amphibian, and fish species have been analyzed, of those that have, 20 percent of reptiles, 25 percent of amphibians, and 34 percent of fish (mostly freshwater species) are threatened with extinction. Each plant and animal taxa is assigned to a Species Survival Commission, composed of experts who are trying to develop strategies to save the threatened flora and fauna. Nearly 5,000 experts from 170 nations serve on the commissions. The Red Lists can be found at <http://www.redlist.org>.

The Convention on International Trade in Endangered Species of Wild Fauna and Flora (CITES) is only one of the ways that this information about species status in the wild is being used. Around the world, millions of plants and animals are bought and sold each year—for timber, leather, and fur, and as folk medicines, pets, and ornamentals plants. This global market is worth billions of dollars and involves some 350 million plants and animals annually. CITES is an international treaty, first signed by eight countries in 1975, designed to protect endangered species from international commercial trade. Nations that sign the treaty agree that they will control commercial trade in the species on the CITES list; 150 countries have signed. There are three different categories in the CITES treaty. Appendix I includes all species threatened with extinction or whose survival may be jeopardized by trade. More than 500 animal and 300 plant species are listed on Appendix I. Appendices II and III include thousands more species that scientists feel may soon be affected by trade. CITES signers meet every two or three years to discuss which species should be on the three different lists.

World Wide Fund for Nature and IUCN direct a program called TRAFFIC, which monitors CITES and provides scientific information and training to biologists in the signatory nations. TRAFFIC recently helped persuade Greece, Mexico, Vietnam, and Yemen to sign onto CITES.

Further Reading:

CITES, Switzerland <http://www.wcmc.org.uk/CITES/eng/index.shtml>.

Di Silvestro, "Rescue from the Twilight Zone."

IUCN USA Multilateral Office <http://www.iucn.org/>.

Kirchner and Weil, "Delayed Biological Recovery from Extinctions Throughout the Fossil Record."

TRAFFIC, USA <http://www.traffic.org/>.

Webster, "The Looting and Smuggling and Fencing and Hoarding of Impossibly Precious, Feathered and Scaly Wild Things."

Wilson, *The Diversity of Life*.

ENVIRONMENTAL EDUCATION

Environmental education and public awareness are key to any successful conservation campaign. There is no point in announcing regulations to control natural resource use or establishing a new forest reserve's boundaries unless people understand why conservation makes sense and is important to them. Changing human behavior is the only way to improve the global environment—or prevent its collapse. And the way to change human behavior is through education.

Most environmental campaigns are directed toward children because they can be reached at school (although not all in some particularly vulnerable countries attend school) and are more likely to adopt new behaviors than are adults, who are often inflexibly set in their ways. Large, international conservation groups and governments across the globe have invested millions of dollars in environmental education. Local grassroots groups also incorporate education in their conservation campaigns. In Veracruz state, Mexico, Pronatura Veracruz, a local conservation group, works with schoolchildren to teach them about the millions of raptors that fill the skies each fall and spring. The birds migrate from their nesting grounds in the United States to their wintering habitat in southern Veracruz. Before Pronatura began visiting schools with education materials, children didn't know where the birds originated or where they were going. The conservation group's ultimate message is that unless Veracruz's endangered forests are left standing, the raptors will lose their winter habitat and disappear. Pronatura has created a good deal of pride and awareness among local children, and the group assumes that this investment in education will mean that when the schoolchildren are adults, they will be more interested in forest conservation than their parents were. But this is a risky, long-term investment; by the time these children are adults, Veracruz's forests may already be gone. Nor is there any guarantee that the children's school lessons will affect their adult behavior.

Time is of the essence, so conservationists must also invest in environmental education programs for adults. This can be done through town meetings, posters, pamphlets, and, often most effectively, the media. On the island of St. Lucia, RARE Center for Tropical Conservation, a U.S. nonprofit group, produced a soap opera called "Apwé Plézi," from a Creole folk saying, "after the pleasure, comes the pain." The dramatic program's message was to encourage people to practice family planning. With some 156,000 people living on a 240-square-mile island, St. Lucia's small pieces of remaining forests are in jeopardy unless the population stabilizes. "Apwé Plézi" became an island-wide hit, and according to RARE's follow-up surveys, the family-planning message hit home. Whether that translates into safeguarded biodiversity remains to be seen.

It is difficult to measure the long-term effectiveness of environmental education campaigns, but conservationists agree that education and awareness are fundamental building blocks for progress toward sustainable use of the planet's natural resources.

Further Reading:

Biodiversity Project, "Life. Nature. The Public. Making the Connection."

Jukofsky, "After the Pleasure."

Jukofsky, "Raptor Rapture in Veracruz."

Wood and Wood, "How to Plan a Conservation Education Program."

ENVIRONMENTAL SERVICES

Environmental services are what nature provides for human beings at no charge. Conservationists point out that people take these services for granted, so they are undervalued, exploited, and disappearing. We rely on oceans to provide us with fish; on forests for wood, medicines, and food; on rivers and lakes for water; on plants to pump oxygen into the atmosphere and absorb carbon dioxide; on wildlife to keep pests in check and distribute seeds and pollinate plants that provide us with food and other goods. As explored in the introduction to this book, forests provide numerous other services; just their protection of our water supplies can exceed the value of the timber they provide.

If we put a price tag on these services, we might be more disposed to protecting earth's other living inhabitants. The scientists Charles Peters, Alwyn Gentry, and Robert Mendelsohn were among the first to tally a forest's total value. In 1989, they estimated that 2.5 acres of forest in Peru were worth $6,820.

More recently, the World Watch Institute has estimated the value of environmental services. For example, wetlands—which were once considered wastelands that should be drained and filled—cleanse water, recycle nutrients, recharge aquifers, control floods and storms, and serve as nurseries for fish and other wildlife. Wetlands near cities have values of more than $20,000 per acre per year for these services. In Malaysia, the value of a now-destroyed coastal mangrove wetland's services in flood control alone was estimated to be $185,000 per mile. It cost twice that to construct rock retaining walls that are less effective at providing the same service.

Some economists and environmentalists suggest that a country's wealth would be more accurately reflected if environmental costs were deducted from reported profits. "Green accounting" would depreciate from a nation's gross domestic product the value of forests that had been logged or minerals extracted. The new green math might sufficiently awaken and educate governments so that they correct customary and short-sighted overexploitation of nature and natural resources.

In an effort to more fairly demonstrate the value of forests as protectors of watersheds, a nonprofit group in Costa Rica called the Foundation for the Development of the Central Volcanic Range (FUN-DECOR) engineered an innovative deal with a hydroelectric company. The company will pay Costa Rica's National Forestry Fund $10 for each 2.5 acres of land whose owner agrees to reforest his or her property, sustainably manage the forests already on the property, or agree to permanently protect the forested land. The forestry fund then pays participating landowners about $40 per 2.5 acres annually, which is competitive with what they might make if they used the land to graze cattle or plant crops. The deal is a bargain for the utility company, which will make a profit from the hydro-generated electricity they sell only if forests remain to protect their source of income—the San Fernando River—from siltation.

Further Reading:

Abramovitz, "Valuing Nature's Services" in Brown, *State of the World 1997*.

"Agreement between Energía Global and FUNDECOR," see organizations in Costa Rica for FUNDECOR's contact information.

Chichilnisky, "Sustainable Development and North-South Trade" in Guruswamy and McNeely, *Protection of Global Biodiversity*.

Panos Media Briefing #38, "Economics For Ever: Building Sustainability into Economic Policy."

Peters, et al., "Valuation of an Amazonian Rainforest."

Tuxill, *Nature's Cornucopia*.

FIRE

Fire is a leading cause of tropical forest loss worldwide. More forest burned around the world in 1997 than at any other time in recorded history, with at least 12.3 million acres of forest lost in Indonesia and Brazil alone. In the Brazilian Amazon, forest fires increased that year by more than 50 percent over 1996. Vast areas of Papua New Guinea, Colombia, Peru, Tanzania, Kenya, Rwanda, and other parts of Africa also burned in 1997. Some 2.47 million acres of peat forests burned in Indonesia for months, throwing up a thick haze across Southeast Asia and producing more carbon dioxide in six months than the entire annual contribution from cars and power stations in all of western Europe. The 1997 fires were responsible for damages totaling an estimated $4.5 billion.

The following year wasn't much better, with 226,000 acres of forest lost in Mexico; nearly 2 million in Guatemala, Honduras, and Nicaragua; and another 2.5 million in the Amazon of Brazil. Busy airports in Honduras had to be closed due to the haze. Fires again raged across Brazil in 1999. And in early 2000, more than a thousand fires burned on the Indonesian islands of Sumatra and Kalimantan, causing choking pollution.

Most of these fires are set intentionally by ranchers and farmers who clear forested land with fires. Farmers also burn their fields at the end of each dry season to ready them for new crops. Tradition, rather than science, holds that crops grow well in mineral-laden ashes. Fires combined with continual plantings actually exhaust tropical soils, often beyond regeneration. Soils become compacted, increasingly less hospitable to seed germination. Land is parched at the end of the dry season, particularly in pastures and fields, so winds easily spread the flames. Since the disastrous fires of 1998, the governments of Mexico and Central America have met with farmers' and conservation groups to try to coordinate education campaigns and fire-fighting brigades. Mexico's natural resources agency launched a hard-hitting information campaign, but at least one group of farmers publicly announced that they had burned their fields for generations and weren't about to stop. Besides, they claimed, drug traffickers were mainly responsible for the fires.

Farmers did not set the fires that devastated Indonesia's tropical forests in 1997, killing unknown wildlife species and destroying habitat for such endangered species as orangutans *(Pongo pygmaeus)* and Asian elephants *(Elephas maximus)*. The fires were deliberately set by companies and concessionaires who wanted the trees out of the way so they could create industrial pulpwood and oil-palm plantations. When the same culprits were suspected of setting the year 2000's fires, the government announced it would make public the names of suspects and review their concession permits.

In early 2000, the World Wide Fund for Nature and World Conservation Union announced a new joint program called "Firefight," aimed at fire prevention through education about fire management, development of economic incentives that would improve fire control, elimination of skewed government incentives that actually encourage fires, and policies and legislation that would protect forest areas from fires. The five year campaign is underway in six regions of the world, including Central and South America, sub-Saharan Africa, and Asia.

Further Reading:
Hecht and Cockburn, *The Fate of the Forest.*
Jukofsky, "Same Time, Next Year: More Fires?"
Moore, "Firefight."
United Nations Environment Programme, "Going Up in Smoke" and "UNEP Welcomes Tough Stance on Forest Fires."

FOREST RESOURCES

Aside from timber, forest resources include the wildlife and plants that human beings have collected and used for centuries. People who live in and near tropical forests utilize hundreds of these resources—fruits, medicinal plants, resins, fibers, and more. Usually, the products are used and consumed by those who gather them, but increasingly, the harvesters are selling them locally and even internationally. People who obtain income from this age-old harvest are more likely to protect the forests that furnish the products for free, so conservationists promote the sustainable collection of a wide variety of marketable forest resources.

Brazil has set up "extractive reserves" where people are allowed to tap the latex from rubber trees *(Hevea brasiliensis)*; logging is illegal in these protected areas. To supplement their income, rubber tappers also gather Brazil nuts from the forest floor. The softball-sized fruits contain up to 20 nuts, which are brought to a local processing plant and, from there, dried, shelled, and sorted for shipment to foreign markets. Ben & Jerry's is just one company that purchases these valuable wild nuts. The U.S. group Cultural Survival was one of the first nonprofit organizations to broker contracts between harvesters of forest products to international companies like Ben & Jerry's.

Farmers began planting damar trees *(Shorea javanica)* in their forests on the island of Sumatra hundreds of years ago. Today, the trees grow among other forest species, and the clear resin extracted from the trees continues to provide people with a steady income. Damar resin is used to formulate paint and varnishes.

Following models provided by Brazil nuts and damar resin, conservationists are working to develop other markets in conjunction with local residents. As with any new business, problems abound. First, markets for new products are difficult to establish. Further, economies, weather, and political situations are unpredictable, so if people depend on a particular product for income, and the price suddenly drops or the product is no longer available, they may be left with burdensome debts. Sometimes demand may exceed available supply or a market may dry up completely. Still, the concept is promising both for improving the well being of people living near forests and the forests themselves.

If forest-product promoters are successful in meeting the business and marketing challenges, they must also consider the ecological restraints. How many pounds of seeds can be gathered, for example, without endangering the ability of a plant to regenerate? The Rainforest Alliance's SmartWood program has certified JungleGum, which is made from chicle collected by Maya people in Mexico from the sapodilla tree *(Manilkara zapota)*. Most chewing gum is made from an artificial base, but there is a growing market for all-natural JungleGum. The certification guarantees that the chicle harvesting does not damage the forest ecosystem. JungleGum is manufactured by a Florida company called Wild Things, Inc., which pays at least 20 percent more than the world market price for chicle to the *chicleros* in Mexico who collect it.

The indigenous Hatam people live outside the Arfak Mountains Nature Reserve in Indonesia and survive through farming and hunting. They often collected and sold endangered birdwing butterflies (*Ornithoptera*) without considering the species' long-term survival. A local chapter of the World Wildlife Fund began working with the Hatam in 1989 to establish a butterfly ranch. Today, six species of the birdwing butterflies are raised at the various stations just outside the reserve. About 1,300 Hatam are organized in 88 ranching groups. Specimens are collected, eggs are carefully watched, caterpillars are fed, and the chrysalae are legally shipped to collectors, mainly museums, that are willing to pay good prices for unusual species. The Hatam realize that their livelihood depends on the long-term survival of wild birdwings, as well as the health of the forest ecosystem. Although the project has yielded positive results, problems typical of any small business persist, from a faltering Indonesian economy to unreliable shipping methods. And since the birdwings still have a high value, poaching and a black market continue.

Further Reading:

Biodiversity Conservation Network, *Final Stories from the Field.*

Carr, Pedersen, and Ramaswamy, "Rain Forest Entrepreneurs."

Jukofsky, "Can Marketing Save the Rainforest?"

Peters, *Sustainable Harvest of Non-timber Plant Resources.*

Rainforest Alliance, SmartWood, USA <http://www.smartwood.org>.

Taylor, "Saving the Forest for the Trees."

FORESTRY MANAGEMENT

Forestry management is the controlled and regulated harvest of trees for timber, which involves specialized techniques so that the commercial value of the forest that remains after logging is the same or higher. Management usually means selective cutting in a forest— only certain, preselected trees are logged. Proper natural-forest management should be sustainable; that is, harvests should not exceed the ability of the forest to regenerate itself, should not cause soil depletion, and should not harm the genetic potential of the remaining tree species.

The science of forestry management was developed in Europe, and the techniques were transferred to tropical forests, with mixed results. Only a small portion of the world's tropical forests are being managed sustainably today, although many foresters still believe that sustainable, natural-forest management in the tropics is feasible. But conservationists and foresters alike understand that natural forest-management today should not be confined by traditional definitions. Land ethics should play a larger role. Forest management should be both socially and ecologically sustainable, so the objective goes beyond merely providing a continuous supply of timber. Local communities should receive a fair share of the benefits and not bear a disproportionate share of the costs. The changing tropical forest ecosystem and shifting needs of society must be taken into account. The goal of forest management should not only be timber production but also be conservation of other forest resources and protection of water quality and biodiversity.

To receive certification or green seals of approval, well-managed forestry operations must meet these goals, plus take into account the well being of workers and local communities.

Forestry operations that fully involve local residents often hold the best promise for long-term sustainability. In Papua New Guinea, the logging industry is rife with corruption; a former Minister of Forests was found guilty of more than 80 charges of corruption, while a predecessor lived rent-free in a house owned by the largest logging company in the country. Some three-million cubic meters of timber are cut by loggers each year, which is at least three times more than the sustainable rate. But the scene is remarkably different in the Bainings Mountains of New Britain, off the east coast of Papua New Guinea. Rather than sell out to Asian logging companies, many residents have formed a cooperative to selectively log their own forests, with guidance and training from the Pacific Heritage Foundation, a local nonprofit group. Trees are cut only after inventories and careful planning. Once an area has been selectively logged, it is left to regenerate. Local sawmills provide steady income. The Bainings Community Forest Project exports timber to Europe and has won certification from SGS Forestry.

Further Reading:

Baird, "Saying 'No' to Asian Loggers."

Buschbacher, "Ecological Analysis of Natural Forest Management in the Humid Tropics" in Goodland, *Race to Save the Tropics.*

Johnson and Cabarle, *Surviving the Cut.*

World Commission on Forests and Sustainable Development, *Our Forests, Our Future,* Canada <http://www.iisd.ca/wcfsd>.

FUEL WOOD

Fuel wood provides 5 percent of the energy consumed by people worldwide each year. Wood provides nearly 60 percent of all the energy used in Africa, 15 percent in Latin America, and 11 percent in Asia. In 40 developing countries, fuel wood accounts for more than 70 percent of all energy use. It also accounts for half of all wood consumption, or about 1.9 billion cubic meters annually. About half of all the world's fuel wood is produced in five countries, four of them tropical—India, China, Brazil, Indonesia, and Nigeria. In many tropical cultures where fuel wood is the main energy source, wood gathering is the task of women and children. As trees become scarce, women and children must spend more time trekking greater and greater distances to find wood, which they then must chop, bundle, and haul home on their backs. As a result, fuel-wood use has an impact not just on the forests but on human health and cultures. Further, smoke from wood fires pose serious human health hazards.

Agroforestry techniques that include growing tree species specifically for their potential to provide fuel wood can help alleviate the problem. Switching to more efficient stoves can also make a big difference. Most stoves deliver 20 percent or less of the fuel wood's energy content; a typical western gas stove delivers 70 percent of the fuel's energy. Improved wood stoves with better designs make firewood last longer. Solar cookers have worked well in many countries as an efficient and practical alternative to fuel wood. The cookers make good sense in tropical countries—at least during the dry season—where the sun is a ready resource. It takes practice to use the cookers, though, and women must stand directly in the hot sun. Cultures that have cooked for centuries with wood aren't always ready to switch to solar-cooked food. But many families have no choice—there is simply no more wood available within a day's walk.

Further Reading:

Abramovitz and Mattoon, "Reorienting the Forest Economy" in Brown, *State of the World 1999.*

Lankford, "Making a Difference with Solar Ovens." Myers, *The Primary Source.*

GEOGRAPHIC INFORMATION SYSTEMS (GIS)

Geographic information systems (GIS), as well as maps, satellite photos, and remote-sensing devices, are providing invaluable—and irrefutable—information for conservationists and policy makers. Information on species, population, geographic distribution, and other factors can be entered into databases and then mapped via computerized GIS. The resulting maps can show, for example, the location of mangrove swamps in relation to protected area boundaries, whether monarch butterfly staging areas are too close to logging operations, or whether coffee farms have impinged on resplendent quetzal habitat. GIS technology was used in Costa Rica to produce a map that outlined the current breeding and migration range of the endangered green macaw (*Ara ambigua*), along with recommended boundaries for a refuge that would protect green macaw nesting sites. The maps were presented to government officials and helped to convince them that a reserve was needed and would be feasible.

Satellite photos have enabled conservationists to see where forests remain on Earth. In Brazil, the government is using satellite photographs to identify illegal logging. In 1997, when fires raged in Indonesia, the politically powerful timber industry placed blame on small-scale, slash-and-burn farmers. But the environment minister, Sarwono Kusumaatmadja, downloaded satellite images of burning Indonesian rainforests from a U.S. government Web site and compared them to timber concession maps. The satellite pictures revealed that many of the fires were being set in the same areas that timber companies wanted to clear for plantations. Using his maps as evidence, Kusumaatmadja convinced his government to revoke the permits of 29 timber companies.

The U.S. nonprofit group World Resources Institute is using satellite imagery, GIS, mapping software, and on-the-ground observation to analyze threats to the world's forests. The group's Global Forest Watch has revealed extensive logging operations in the Congo Basin of Africa, the second largest contiguous tropical forest after the Amazon. Global Forest Watch, which has 75 partners in seven countries, will eventually map 21 countries and 80 percent of the world's remaining ecologically intact and natural forests.

Further Reading:
Office of International Affairs, "A Map Is Worth 1,000 Words."
Savitsky and Lacher, *GIS Methodologies for Developing Conservation Strategies.*
Scott, Csuti, Jacobi, and Estes, "Species Richness."
Sheehan, "Gaining Perspective."
World Resources Institute, USA <http://www.globalforestwatch.org>.

HUMAN POPULATION

The growth of human population has a profound affect on the environment, as people put increasing demands on the Earth's dwindling resources. More than 1.1 billion live in 25 biodiversity hotspots—threatened areas of the world that are particularly rich in flora and fauna. Some 75 million people live within three major tropical wilderness areas, which hold the largest expanses of tropical forest: the upper Amazon and forests of the Guyanas in South America; the Congo basin; and Papua New Guinea, Malaysia, and islands of the Pacific.

Many conservation groups recognize that it makes little sense to help rural communities build better, more sustainable lives without addressing population growth. Because impoverished villages often lack basic reproductive health information and facilities, conservation groups can fulfill an important need of residents by helping them get these services. Easing human pressure on the natural environment and meeting the health needs of the rural poor are mutually reinforcing goals.

A local nonprofit group in Indonesia is working with communities in East Java, bringing men and women together to discuss sustainable agriculture and reproductive health. By using plant and pest biology analogies, the project has been able to overcome villagers' resistance to discussing the sensitive topic of reproductive health and family planning. A conservation group in the Yucatán Peninsula of Mexico has worked with villagers living outside the Calakmul Biosphere Reserve for 10 years. During that time, the population in the area increased nearly threefold, due to a high birth rate and increased migration. Now the group has added health and family-planning programs to its conservation agenda.

According to a director of Population Action International in Washington, D.C., studies show that women who gain control over their reproduction become better environmental stewards, as they gain the free time needed to learn how to sustainably use natural resources.

The current world population is more than six billion people, with a growth rate of 1.3 percent annually. The United Nations projects that the world's population will reach 8.9 billion in 2050. Although the population is growing much faster in the world's poorer countries, citizens of wealthier countries are consuming far more natural resources. So conservationists know that in addition to addressing population growth, they must educate consumers in wealthy nations to make smarter, more thrifty choices—to embrace the three Rs: reduce, recycle, reuse. Decreasing human population and curbing consumption require tremendous changes in human behavior, but unless both are achieved, the planet is imperiled.

Further Reading:
Caudill, *Lessons from the Field: Integration of Population and Environment.*
Cincotta and Engelman, *Nature's Place: Human Population and the Future of Biological Diversity.*
Engelman, *Plan & Conserve.*
Ericson, Freudenberger, and Boege, *Population Dynamics, Migration, and the Future of the Calakmul Biosphere Reserve.*
Kates, "Population and Consumption: What We Know, What We Need to Know."
Vogel and Engelman, *Forging the Link.*

INDIGENOUS PEOPLE

The indigenous people who continue to live in tropical forests are losing ground to invaders just as their ancestors did, just as many indigenous groups that are now extinct did. Today they live in reservations that they did not create, and people whom they did not elect and who do not represent their interests control their existence. Indigenous forest dwellers have proven themselves to be far better land stewards than nonindigenous people. Recognizing this, conservationists are working with scores of indigenous groups in the tropics, assisting them in gaining legal rights to their land and resources, so they can at least choose their own futures.

Conservation and human-rights groups are also careful to introduce health-care and education systems that complement indigenous culture, rather than impose foreign models that can disrupt traditions. In the eastern Amazon of Ecuador, the

Shuar people wanted to avoid assimilation but educate their children so they could choose to compete with other Ecuadorians. They set up a radio education program; each remote village has its own portable receiver and a trained teacher's assistant. Children have received a formal Ecuadorian education plus have studied Shuar history, language, and resource management.

Obviously, indigenous people are not all alike. Some groups have strong connections to natural resources; others, particularly small groups, are more vulnerable to pressures to exploit their resources. Virtually all groups are now linked to the market economy, either through barter or cash exchange. They understandably want access to health care, education, and material goods that can improve the quality of their lives. Their traditional knowledge may provide for these necessities and desires in ways that are compatible with biodiversity conservation, but not always, especially as governments, colonists, and ranchers crowd indigenous populations into smaller and smaller reserves.

Conservationists cannot demand that indigenous people live in isolation and maintain traditional ways against their will, simply because this way of living may be more compatible with resource protection. Conservationists can, however, work together with native people to ensure that they have the time and opportunity to shape their own destinies and to adapt their traditional, low-impact ways of life to the pressures of the modern world.

Further Reading:

Alcorn, "Indigenous Peoples and Conservation."
Clay, "Indigenous Peoples" in Head, *Lessons of the Rainforest.*
Redford and Mansour, *Traditional Peoples and Biodiversity: Conservation in Large Tropical Landscapes.*
Redford and Sterman, "Forest-Dwelling Native Amazonians and the Conservation of Biodiversity."

MEDICINAL PLANTS

Medicinal plants have cured human beings of a wide range of maladies for centuries. Of the 121 useful drugs now on the market that were obtained from higher plants, more than one-third originated from tropical forests. Yet less than 1 percent of tropical forest plants have ever been screened for their potential curative effects. Pointing to the lifesaving cures that were derived from such plants as *Catharanthus roseus*, used to treat leukemia, and *Cinchona*, which yields quinine to treat malaria, conservationists argue that tropical forests should be preserved if for no other reason than the fact that they certainly hold countless as-yet-unknown cures.

That may not be a convincing argument for a local farmer who wants to use forested land to grow crops. Studies in Belize, however, show that people who collect medicinal plants and sell them at local markets can actually earn more per acre than if they used the land for farming or grazing cattle. As with other forest resources, medicinal-plant ecology must be thoroughly understood so plants are not overcollected. This has already happened in Africa. Extracts from the bark of *Prunus africana* is commonly used in Europe to treat prostatic hyperplasia, a swelling of the prostate gland. The bark can be harvested sustainably, but instead, people have destroyed thousands of trees to sell the bark, and the species is endangered in several countries. Ethnobotanists like Mark Plotkin are working closely with indigenous people, particularly traditional healers, to catalog their knowledge of medicinal plant properties before the plants disappear. The cures themselves are endangered; as forest dwellers have more access to western medicines, they are less inclined to use traditional medicines, and healers no longer pass the prescription recipes to younger generations.

Until the early 1990s, when pharmaceutical companies had a renewed interest in searching for new drugs in rainforests, drugs derived from plants in tropical countries never returned a profit for the countries of origin or the people who introduced the plants to western biologists. Those days of exploitation are over. Costa Rica's National Biodiversity Institute (INBio) set an important precedent in 1991, when it signed a lucrative deal with the U.S. drug company Merck. INBio granted Merck the right to analyze plants and animals that the institute collected from Costa Rica's rainforests. INBio received $1 million, which was used for forest protection and to fund an inventory of the country's flora and fauna. Similar deals were brokered between other tropical countries and western drug companies, universities, and research institutions.

The 1990s rainforest search for new drugs, called "bioprospecting," has yielded a number of promising plant compounds, such as calanolide A, isolated from leaves and twigs of *Calophyllum lanigerum*, a tree native to Malaysia. In laboratory tests, the compound has been shown effective against one strain of HIV. Another compound that shows indications of fighting AIDS has been derived from a woody vine of the genus *Ancistrocladus*, found in Cameroon.

At the same time, a debate continues over who "owns" these compounds and the drugs they may eventually produce, even when botanists have paid for the right to screen plants. Wild flora and fauna were once considered the property of no one, and were sampled, taken, and exchanged without payments or contracts. Now they are considered to have monitory value like a country's mineral and oil deposits; simply because a nation had no role in creating the resource doesn't mean it doesn't own it.

The United States has been criticized for refusing to sign the Biodiversity Convention, drawn up at the 1992 Earth Summit in Brazil. The international treaty acknowledges that developed countries should pay for genetic resources from poor nations. The United States already has a history of what some critics call "bio-piracy." The Eli Lilly Company developed two cancer-fighting drugs, vincristine and vinblastine, both derived from the rosy periwinkle (*Catharanthus roseus*), found in Madagascar. The drugs have not only saved thousands of lives but also brought millions of dollars in profits to Eli Lilly. The complaint is that neither Madagascar, nor the villagers who first showed Eli Lilly botanists the plant, ever earned a cent.

A 1999 controversy over an Amazon vine, *Banisteriopsis caapi*, further deepened mistrust toward the intentions of U.S. botanical entrepreneurs. In 1986 the U.S. Patent Office granted a patent to a California businessman that allowed him exclusive use of the vine, although for centuries, Amazon shamans have used in religious ceremonies a powerful hallucinogen brewed from the vine. The Californian had been given the plant by indigenous people in Ecuador; as a thank-you, he built a school. After protests from indigenous and human-rights groups, the Patent Office withdrew the patent.

The dispute will make it that much more difficult and expensive for foreign botanists to search for miracle cures in tropical forests. Bioprospecting is already an expensive and risky venture for research institutes and pharmaceutical companies, which are balking at the fees many tropical countries are demanding for the right to search and sample flora and fauna. Experts agree that eventually the problems will be smoothed out, and fair and equitable agreements eventually reached between nations and researchers. The problem, of course, is that potential drugs are disappearing every day as tropical forests fall.

Further Reading:

Balick and Mendelsohn, "Assessing the Economic Value of Traditional Medicines from Tropical Rain Forests."

Balick, Elisabetsky, and Laird, *Medicinal Resources of the Tropical Forest.*

Bird, "Medicines from the Rainforest."

Cox and Balick, "The Ethnobotanical Approach to Drug Discovery."

Joyce, *Earthly Goods: Medicine-Hunting in the Rainforest.*

Jukofsky, "Medicinal Plant Research Leads Scientists to Rain Forests."

Miller, "High Hopes Hanging on a 'Useless' Vine."

Plotkin, *Tales of a Shaman's Apprentice.*

Pollack, "Biological Products Raise Genetic Ownership Issues."

ten Kate and Laird, *The Commercial Use of Biodiversity.*

MONITORING AND EVALUATION

Monitoring and evaluation help conservationists understand if their efforts are yielding results. They have various ways to measure the success of a project whose goal is to save forest biodiversity by improving the quality of life of people living. Comparing residents' incomes before the project with their incomes one or two years later is one measure. The biodiversity of the forest must also be measured. That means that before a project begins, conservationists must do an ecological assessment of the protected forest, so that they can compare results with another survey after the project has run its course.

Collecting this kind of data can be expensive and time-consuming, and in the face of an emergency situation—a forest or species that's about to be lost—may seem unnecessary. But many conservationists believe more monitoring and evaluation is needed so successful projects can be duplicated and less successful projects can be adjusted.

In 1990, The Nature Conservancy launched a program called Parks in Peril, whose goal was to strengthen park management in the Neotropics. To help the organization assess whether this ambitious effort was working, it developed a scorecard that aided project directors in determining how far a particular park had progressed in 16 areas, such as physical infrastructure, land tenure issues, long-term financing, and environmental education programs.

The Biodiversity Action Network and World Conservation Union published an extensive list of biodiversity indicators to help measure the effectiveness of the Convention on Biological Diversity, signed in 1992 at the Earth Summit in Brazil. The World Bank has published a comprehensive collection of indicators for each country in the world. *World Development Indicators* covers such development-related topics as people, environment, economy, and trade, and includes, when available, figures from previous years for comparison. One of the environmental indicators is number of endangered mammals, which is listed with total land area. The most recent numbers are used, so when the next edition of *World Development Indicators* is published, new figures will allow for comparison.

The Biodiversity Support Program, a consortium of three U.S. nonprofit groups—the World Wildlife Fund, The Nature Conservancy, and World Resources Institute—had as its mission biodiversity protection worldwide. All initiatives of this ambitious project included a carefully designed monitoring and evaluation component, so when the initiative concluded in 1999, everyone could understand what methods worked best and why, and which obstacles were the most difficult to overcome. These results were published so that other conservationists could learn from the program's successes and frustrations. The program also published a step-by-step explanation of how to design and implement an effective monitoring and evaluation component.

Monitoring and evaluation of conservation projects are particularly important because the results help the public understand whether or not progress is being made. After all, most conservation projects are supported by public funds, so the public should be informed about their investments. Unfortunately, it is nearly impossible to measure "what might have been"; that is, while deforestation may be continuing, we cannot measure how much worse the situation would surely be without tropical forest conservation projects and public support of these efforts.

Further Reading:
Biodiversity Action Network and World Conservation Union, "Exploring Biodiversity Indicators and Targets Under the Convention on Biological Diversity."
Biodiversity Conservation Network, *Evaluating Linkages Between Business, the Environment, and Local Communities.*
Brandon, Redford, and Sanderson, *Parks in Peril.*
Margoluis and Salafsky, *Measures of Success.*
Rainforest Alliance, USA, "Consensus Statement on Monitoring and Evaluation."
Ross-Larson, *World Development Indicators.*

PROTECTED-AREA MANAGEMENT

Protected-area management is a misnomer in many parts of the world. Merely because an area is designated as "protected" doesn't mean it is being managed for conservation. These are the so-called "paper parks"—their boundaries and protective regulations exist only on maps in government offices, not in reality. That an area is designated as a national park or reserve may help protect it from development, but throughout the tropics, hundreds of parks are under threat, regardless of their official status.

The world's tropical parks and reserves protect just 5 percent of all the rainforests on Earth, probably not enough to forestall massive species extinction. Most rainforest reserves are seriously affected by an average of three of the following most common threats, all of which are usually illegal, at least on paper: agricultural expansion, hunting and fishing, logging and fuel-wood gathering, livestock grazing, mining, fires, road building, and hydropower development. Those who perpetrate these threats may be impoverished, wealthy, or politically powerful people, or government institutions. Governments may choose to ignore the threats to the forests in the interest of political and economic gain or due to a lack of the funds needed to deter them. Worldwide, the average rainforest park is inadequately staffed; its guards are untrained and lack necessary equipment; local police do not support park guards by arresting violators; and judges do not impose fines or punishments, or they are too meaningless to be deterrents.

A 1998 study by the World Wildlife Fund–World Bank Alliance revealed that less than one-quarter of forested protected areas in 10 assessed countries were well managed with solid infrastructures. Only 1 percent was considered completely secure, while another 1 percent had lost all conservation values. Another 22 percent are suffering from some form of degradation. Of the countries surveyed, 8 of the 10 are in the tropics: Brazil, Gabon, Indonesia, Mexico, Papua New Guinea, Peru, Tanzania, and Vietnam. The World Wildlife Fund–World Bank Alliance has pledged to convert 125,000,000 acres of currently threatened, forested protected areas into well-managed and socially responsible reserves by 2005.

Recognizing the troublesome situation of scores of protected areas in Latin America, in 1990 The Nature Conservancy (TNC) launched a parks rescue program called "Parks in Peril." Sixty parks in 18 countries are part of the program; they cover more than 75 million acres in the region. Parks were selected based on their biological significance, socioeconomic and cultural value, endangerment, and management capacity and opportunity to resolve problems. TNC works with partner groups and government agencies in each country to identify and reduce the threats these parks face.

In Costa Rica, Bolivia, and several other tropical countries, private reserves are becoming more important in the movement to conserve biodiversity. One of the most famous and visited rainforest parks in the world, the Monteverde Cloud Forest Preserve in Costa Rica, is privately owned and managed, and quite successfully. The Environmental and Natural Resources Law Center, a nonprofit group in Costa Rica, is leading the Central American Private Conservation Initiative, whose goal is to make it easier and more attractive for landowners to protect the forests on their property. Conservation easements are one tool being promoted by the center. A property owner can put restrictions on the land deed that dictates how the land will be used—what will be permitted and what prohibited—and that covenant stays with the land even after it is sold or inherited.

A group called the Tropical Science Center, which manages the Monteverde Reserve in Costa Rica, and the U.S.-based Nature Conservancy agreed on a conservation easement on land owned by TNC that lies outside the reserve. The easement prohibits TNC from cutting any trees on one part of the land, although trails can be built for educational purposes, while on other parts of the property the conservancy may construct limited number of low-impact buildings, as long as they are used only for educational or scientific purposes. The legal restrictions stay with the land even if TNC should decide to sell it. One benefactor from this conservation easement is the resplendent quetzal (*Pharomachrus mocinno*), an endangered and beautiful bird that nests in the reserve but during certain times of the year looks for food in forests on TNC's land.

Where protected tropical forest reserves have been created, planned, and managed with care, as in the case of Monteverde, they are successful in meeting their principal goal of conserving tropical biodiversity.

Further Reading:

Brandon, Redford, and Sanderson, *Parks in Peril: People, Politics, and Protected Areas.*

Dudley and Stolton, "Management Effectiveness of Protected Areas: An International Workshop."

Environmental and Natural Resources Law Center, Costa Rica, "Conservation Easements: The Private Landowner Contributing to Sustainable Development."

Kramer, van Schaik, and Johnson, *Last Stand: Protected Areas & the Defense of Tropical Biodiversity.*

REFORESTATION, RESTORATION, AND TREE PLANTATIONS

Reforestation, restoration, and tree plantations can help heal degraded land and take pressure off existing forests, but the results will never duplicate the forest habitat that was lost. Restoration is a natural process; when a storm topples trees and creates a gap in the forest, fast-growing, light-loving plants and tree seedlings immediately start to fill that gap, aided by seed-dispersing animals. Depending on how large it is, the gap may close in one or two decades; be a relatively healthy forest in one to five centuries; and recover to nearly its original state in a millennium or more. This restoration schedule depends on the presence of intact forest no more than several hundred yards from the gap. Nearby forest is needed to provide seeds, a home to wildlife to disperse the seeds, and fungi.

Throughout their history, human beings have created gaps throughout the world's forests. What appears like virgin forest in the Petén of Guatemala was heavily farmed by ancient Maya civilizations; what stands today is indeed rich in biodiversity, but it is not the same as before the incursion of the Maya farmers. The gaps that today's farmers and ranchers create are often hundreds of acres wide and miles from intact forest; much of this land can never regain anything like its original forest cover, because the soils are exhausted, eroded, and compacted. On thousands of acres of formerly forested, now abandoned cropland in Indonesia, nothing grows but hardy, pioneering grasses and scrubby plants. Throughout the tropics, booming populations compete for limited land, and deforested areas are simply never given a chance to recover.

When nature cannot restore the land, human beings—the same species that degraded it in the first place—can help. Around the world, people are planting trees on deforested land. Reforestation programs are usually spurred by practical needs. After Hurricane Mitch hit Central America in 1998 and caused thousands of deaths from killer landslides and flooding, many communities learned from the tragedy and began replanting the denuded hillsides. Eventually, the new trees will be rooted deeply enough to hold the soil in place, and be big enough to act as sponges, so that future hurricanes will not set into motion such devastating mudslides and floods. Other communities in the tropics are reforesting to protect their watersheds or to provide a future source of fuel wood, timber, and food.

Although most tropical soils are naturally designed for growing trees, reforestation is neither easy nor cheap. Native species grow best, and these are often slow growing. Seeds are expensive and difficult to obtain. Like any new crop, seedlings require care to ensure that they are not overtaken by surrounding vegetation or do not die from disease, predation, or drought. It may be several years before planted trees can make it on their own.

One of the most famous examples of land restoration and reforestation is under way in the dry forest of northwestern Costa Rica. In the mid-1980s, the biologist Daniel Janzen and others began raising funds to buy pastureland outside a 40-square-mile park. Twelve million dollars later, more than 300 square miles of land had been purchased, and the Guanacaste Conservation Area was created. Today the signs of restoration are obvious, thanks to seeds from the original park that the region's high winds and the park's wildlife have deposited, and tree seedlings that people planted. In several centuries or so, a mature, dry forest may indeed be growing where barren pastureland once stretched for miles.

Tree farms are a better use of the tropical soils than cattle pastures, but not much better than any other monoculture, although they generally require fewer agrochemicals than common tropical crops like rice and coffee. Like other crops, trees provide income for their owners and goods that people need: timber and pulp for paper. Tree plantations use up nearly 300 million acres globally; 75 percent are in temperate regions, and more and more farms are being planted in tropical countries. The amount of land in tree farms in Latin America increased by 50 percent between 1985 and 2000, to about 18 million acres. Trees grow faster in the tropics, and labor tends to be cheaper. It costs about $450 to produce a ton of hardwood pulp in the United States, while in Brazil the cost is $350. Northern companies are now buying an increasing amount of land in tropical countries to plant trees for pulp. About 20 percent of pulp production and 25 percent of paper production is now traded internationally, with Japan being the largest importer of wood chips.

By 2050, more than half of the world's industrial wood consumption will be fiber for production of wood-based pulp and paper, so it is likely that tree farming will become more profitable. What conservationists fear is that entrepreneurial lumber companies will cut down natural forests and replace them with tree plantations, and in fact, this is already happening. Plantation owners would obviously prefer to plant a crop on healthy soils near water sources than on degraded land by a dried-up stream. Perhaps 60 percent of the nearly five million acres of plantations in Indonesia is believed to have directly displaced natural forest. Worldwide, it's estimated the planting of

15 percent of all tree farms caused destruction of natural forest.

Foresters believe that it's possible to carefully manage a tree farm, planted on land cleared for another purpose, so that it is environmentally sound. Certification programs like the Rainforest Alliance's SmartWood project inspect and give seals of approval to well-managed tree plantations as well as natural forests. That means that consumers can play a role by selecting pulp and paper products that bear a certification seal. They can also buy recycled paper, recycle the paper they use, and cut back on paper use.

Further Reading:

Allen, "Biocultural Restoration of a Tropical Forest."
Janzen, "Tropical Ecological and Biocultural Restoration."
Matoon, "Paper Forests."
Myers, "Trees by the Billions: A Blueprint for Cooling."
Sedjo and Botkin, "Using Forest Plantations to Spare Natural Forests."
World Commission on Forests and Sustainable Development, *Our Forests, Our Future,* Canada <http://www.iisd.ca/wcfsd>.

SUSTAINABLE DEVELOPMENT

The phrase "sustainable development" was first promoted by the World Commission on Environment and Development, established by the United Nations in 1983, in a 1987 report called *Our Common Future.* The definition of sustainable development offered in the report is: "Sustainable development is development that meets the needs of the present without compromising the ability of future generations to meet their own needs." The definition caught on and was promoted by conservationists, who were often criticized because their initiatives did not seem to consider the need for human economic advancement.

The strong interest in *Our Common Future* prompted the United Nations to hold the UN Conference on Environment and Development, also called the Earth Summit and Rio Conference. The summit, held in Rio de Janeiro in June 1992, was attended by thousands of government delegates, conservationists, and journalists. Delegates from 178 nations discussed how to address such global environmental issues as climate change, ozone depletion, pollution, soil erosion, deforestation, and loss of biodiversity. One result of the Earth Summit was the Rio Declaration on Environment and Development, or Agenda 21, which describes 27 principles to guide international policy. The principles are designed to meet human needs while protecting the environment: sustainable development. Two international treaties that would begin to solve environmental degradation while not impeding development were written at the Earth Summit: the Framework Convention on Climate Change and the Convention on Biodiversity. Hundreds of nations have signed these treaties; by the year 2000, the United States had signed neither.

Countless follow-up meetings have been held since the Earth Summit, including a conference in 1997 called Rio + Five, and thousands of documents have been written as policy makers and conservationists struggle to achieve sustainable development or even understand what it implies. Some believe the term is paradoxical. The World Commission is clear in *Our Common Future* in believing that developing countries must strive for rapid economic growth, since "economic growth and diversification . . . will help developing countries mitigate the strains on the rural environment, raise productivity and consumption standards, and allow nations to move beyond dependence on one or two primary products for the export earnings." But many conservationists observe that with a human population that will double to 12 billion by 2030 and may double again by 2070, unbridled economic growth is impossible—the Earth simply does not hold enough resources to meet the demands of so many people. Without a check in global population growth and high rates of consumption in developed countries, conservationists believe that economic development cannot be sustainable.

How can sustainable development be achieved? Many experts recommend a tectonic shift in the way we think and behave. Here are suggestions of what these crucial changes might entail to attain a truly sustainable life on Earth:

- We must move from societies that satisfy the wants of a few to ones that are committed to meeting the needs of all.

- We should derive self-esteem from our skills, artistry, effort, and integrity, not from wealth or power.

- The amount of goods we harvest from the Earth should be limited to rates of production and not be a response to market demand.

- Market prices and gross national products should reflect costs for environmental services and for ecosystem restoration.

- Economic growth should not be considered as fundamental to social policy. This may require policies for income distribution in the interest of social justice.

- Developed nations need to restructure their economies and retrain workers displaced from industries that are ecologically unsustainable.

Such a major change in human attitude and behavior would greatly increase the possibilities for widespread sustainable development—but that doesn't mean that every corner of the Earth must be developed. To save tropical biodiversity from extinction in particularly vulnerable hotspots on Earth, unrestricted protection may be the only appropriate action.

Further Reading:

Goodland, *Race to Save the Tropics: Ecology and Economics for a Sustainable Future.*

Kramer and Schaik, "Preservation Paradigms and Tropical Rain Forests" in Kramer, van Schaik, and Johnson, *Last Stand: Protected Areas and the Defense of Tropical Biodiversity.*

Milbrath, "Sustainability" in Paehlke, *Conservation and Environmentalism.*

Rees, "The Ecology of Sustainable Development."

Terborgh and van Schaik, "Minimizing Species Loss: The Imperative of Protection" in Kramer, van Schaik, and Johnson, *Last Stand: Protected Areas and the Defense of Tropical Biodiversity.*

World Commission on Environment and Development, *Our Common Future.*

TRADE

On a global level, trade affects the economies of every nation, along with the global environment. Just how significant an impact international trade could have on the environment wasn't apparent until international negotiators established the World Trade Organization (WTO) in 1995. The WTO is responsible for settling disputes about new trade rules among nations and has the power to impose stiff penalties to enforce its rulings. Environmental protection and sustainable development were written in as part of the WTO's governing document, as was a commitment to create a committee charged with analyzing the relationship between trade and the environment, but the committee has taken no concrete action.

WTO rulings, meanwhile, have shocked many environmentalists. In 1991, pre-WTO arbitrators ruled that a U.S. embargo on tuna exported from Mexico violated international trade rules. The U.S. imposed the embargo because many Mexican tuna fishermen were using huge nets at sea that also trap and drown dolphins, a violation of the U.S. Marine Mammal Protection Act. The arbitrators ruled against the U.S. embargo, using a controversial interpretation of the regulations, one that differentiated between characteristics of products and characteristics relating to how products were processed. Some developing countries applauded the decision, as they saw the U.S. embargo as a way for the country to unfairly impose its own environmental values and laws on other nations that are unable to compete economically. But environmentalists, who had fought hard to encourage trade in "dolphin-safe" tuna, saw the ruling as a dangerous precedent, since any international treaty aimed at protecting global resources, whether it be dolphins, forests, or the air, could be deemed illegal by the WTO if the treaty interfered with free trade.

A 1998 WTO ruling added to these fears. The WTO ruled against a U.S. ban on the import of shrimp from countries whose fishermen did not use turtle excluder devices, or TEDs. After conservationists waged a protracted battle—determined to save endangered sea turtles, which are killed more by shrimp fishing than all other human activities combined—the U.S. government required all U.S. shrimp fishermen to use TEDs, simple devices that allow sea turtles to safely escape the nets that capture shrimp. But sea turtles are migratory species, and unless other fishermen in other countries use TEDs, the mortality—an estimated 150,000 sea turtles each year—would continue.

Rather than concede to the U.S.'s TED requirement, India, Malaysia, and Pakistan launched a WTO challenge against the ban. In response to the WTO ruling in favor of the Asian countries, the U.S. now allows the import of individual shrimp shipments that have been certified as "turtle safe," even if the exporting country has not met certification requirements. Environmentalists fear the looser interpretation will allow continued sea-turtle slaughter by countries that do not export to the U.S., as well as "laundering" of shrimp shipments to the United States.

Similarly, the WTO could rule that certification, or eco-labeling, of other products, such as timber that is harvested from well-managed forests or farms, is a violation of trade rules because it jeopardizes products that don't have certification. The WTO is also considering the elimination of tariffs on forest products, such as paper and wood. The result would likely be an increase in logging and consumption of cheaper paper and pulp products, at a time when many conservationists would like to see less consumption and more recycling.

It was the WTO negotiators turn to be shocked when they met in Seattle in November 1999. Thousands of protestors staged massive, nonviolent demonstrations to make clear their objections to the WTO rules that affected the environment, labor, and human rights. The scheduled WTO meeting was not open to the public, and representatives of only a few dozen of the 134 member nations were tapped to participate. Seattle police and demonstrators clashed, and as downtown Seattle filled with tear gas and pepper spray, the meeting was postponed. When negotiators were finally able to meet, talks quickly broke down.

Many conservationists who disagreed with WTO anti-environmental rulings still believe that international trade and investment are needed to attain global environmental solutions, but that trade reforms are essential. WTO meetings and documents should be open to the public, they say. Reforms in international trading regulations should eliminate environmentally damaging subsidies; ensure that actions by multilateral banks promote sustainable development; enforce existing international environmental treaties; and permit legitimate environmental standards adopted by individual countries.

Further Reading:
Durbin, "Trade and the Environment: The North-South Divide."
French, "Challenging the WTO."
Panos Institute, "More Power to the World Trade Organization?"
Scherr and Ward, "Seattle: Act 2."

WILDLIFE RESEARCH AND MANAGEMENT

Wildlife research and management are key to saving endangered tropical fauna. Research is needed so that conservationists and policy-makers know which species are threatened and what their habits and habitats are, so that they can formulate solutions for their salvation. Studying an animal that can fly, sprint, hide, bite, or scratch can be challenging. Wildlife biology requires huge doses of patience and endurance, a willingness to be drenched in sweat or by torrential rains, as the hours pass until the study subject, perhaps a small brownish bird 50 feet up in a dense tangle of vegetation, decides to move. Wildlife science is hard work, requiring no more equipment than binoculars, a notebook, sturdy waterproof boots, and a sharp, investigative mind. But biologists also take advantage of advanced technology.

One technique that researchers use to learn about an animal's habits is radio telemetry. First, the animal must be captured, often in a net. A radio transmitter, big enough to be durable but not so large as to be a burden to the animal, is attached with material that will eventually disintegrate and allow the micro-radio to fall off. Sometimes the animal must be injected with a mild tranquilizer while the transmitter is attached. The animal is released, and the biologists can then track its movements with radio receivers.

In this way, a group of researchers led by the conservation biologists George Powell and Robin Bjork were able to track the movements of the resplendent quetzal (*Pharomachrus mocinno*), considered one of the most beautiful birds in the world, with its bright emerald and ruby plumage and elegant, yard-long, green tail feathers. The scientists attached transmitters to quetzals in Costa Rica, Guatemala, and Mexico and tracked them on foot and via small planes, to learn about the birds' habitat needs. They found that although the quetzals nested in holes in tall, cloud forest trees, they often flew out of the high-altitude forests downslope, to search for their favorite foods—the fruits of the *Lauraceae*, or avocado family of trees. Quetzals are endangered throughout their range from southern Mexico to western Panama. Equipped with the new information, the biologists formulated wildlife-management plans to help save the quetzal. It was not enough to protect the cloud forests in which the birds nest; enough *Lauraceae* trees must remain standing at lower altitudes so the birds have sufficient food year-round.

Governments need information on wildlife habits to help them establish appropriate park and reserve boundaries. Worldwide, many reserves have been established principally to protect a particular species and based on information gathered by wildlife biologists, including the Jaguar Sanctuary in Belize and Kahuzi-Biega National Park in Congo, where 70 mountain gorillas (*Gorilla gorilla beringei*) survive. Unfortunately, rangers have been unable to patrol the latter to check on the gorillas because civil war has made it too dangerous for scientists to monitor the endangered primates.

Officials need to know as much as possible about wildlife populations to impose hunting restrictions or ban hunting of endangered species altogether. Armed with data gathered by biologists, representatives to the international treaty, the Convention on International Trade in Endangered Species of Wild Fauna and Flora (see "Endangered Species"), meet every two years to review which animals should be protected from global trade. At the April 2000 CITES meeting, officials agreed that the ban on trade of elephant ivory should continue, a decision based on the dangerously low populations of African elephants (*Loxodonta africana*).

In addition to guiding the creation of parks to protect wildlife and restrictions on wildlife hunting and trading, wildlife-management specialists are experimenting with farming and ranching. Some tropical animals that have traditionally been hunted for their meat or skins can be successfully farmed, like chickens and pigs, which takes pressure off wild populations.

Conservationists are experimenting with green iguanas (*Iguana iguana*) in Central America, spectacled caiman in Venezuela, and capybaras (*Hydrochaeris hydrochaeris*) in Venezuela to see if farming these animals is practical. Farming might diminish poaching of wild populations; all three species are endangered in the wild from overhunting and deforestation. Wildlife ranching may also be a sustainable way for people to make a living.

Captive breeding and reintroduction are popular, though expensive, wildlife-management techniques. Some species of birds and mammals can be successfully bred in captivity and the resulting offspring released into the wild. In early 2000, the Peregrine Fund in the United States released young harpy eagles (*Harpia harpia*) in Panama. The eagles were born to captive parents and are now flying free in protected rainforest parks. Some conservationists question whether this technique, although it usually attracts a good deal of interest and so serves as a valid public-education tool, makes economic sense. Captive breeding and release is quite costly and perhaps funds would be better spent securing quickly diminishing habitat so existing wild animals survive. Also, reintroductions are extremely complicated and, more often than not, unsuccessful. Wildlife rangers in Borneo and Sumatra had released several thousand orangutans (*Pongo pygmaeus*), most rescued from the illegal pet trade, back into the forest. But biologists recommended that they stop because the few remaining forests were too small to provide sufficient habitat and food for more than the current populations of the primates. Another factor was that the once captive animals could easily introduce diseases to the wild populations.

With increasing human populations resulting in diminishing habitat for animals, the need for more tropical wildlife research and management is greater than ever. When inevitable disputes arise between wildlife and people, it is, of course, the latter species who calls the shots. For people-wildlife conflicts to be resolved intelligently, wildlife managers must play a role in the decision making. The number of qualified wildlife managers is increasing in Latin America, thanks to four graduate schools in wildlife manage-

ment in Argentina, Brazil, Costa Rica, and Venezuela. Along with biology and other science courses, the curriculum at the graduate schools includes resource economics and sociology. As Eduardo Carrillo, a professor at the graduate program in Costa Rica, notes, "Managing wildlife is easy. Managing people is somewhat more difficult."

Further Reading:
Jukofsky, "Mystical Messenger."
Peregrine Fund, USA <http://www.peregrinefund.org>.
Robinson and Redford, *Neotropical Wildlife Use and Conservation.*
Sunquist, "Should We Put Them All Back?"

Mother Nature Wants You!

The conservation community is fighting a brave battle in the tropics, but it is losing and will continue to lose, and thus everyone will lose, unless there is a dramatic change in the number of participants on the conservation side of the war.
—Daniel Janzen

In the 13 years since the biologist and activist Daniel Janzen wrote the above in the journal *Conservation Biology*, many thousands of people worldwide have indeed joined the battle. Yet in spite of tremendous progress, the fight still needs reinforcements—and not only on the front lines but also at the home front. If reading about tropical forests has caused you to hear a call to arms, consider what you can do:

Read

Read more about tropical forests—in books, magazines, and on the World Wide Web.

Visit local museums or zoos that have tropical forest exhibits.

Learn about forests in your own community, whether they are growing in the lot behind you, in a national park, or in a community reserve. Who owns them? How well are they protected? Are any under threat from development?

Teach

Share what you've learned about tropical forests with friends, relatives, and children.

Volunteer your time and skills to a conservation group.

Give conservation-group memberships as birthday, wedding, baby, and holiday presents. Books about the rainforest also make great presents.

Give presentations about tropical biodiversity at schools and community groups. Tell others how to get involved.

Organize a local fund-raising event—a run for the rainforest, a tropical forest walk-a-thon, a rainforest rummage sale.

Choose

Be an environmental consumer: buy energy-efficient appliances, energy-saving windows, and fuel-efficient cars. Look for products with minimal packaging. If you buy a product imported from a tropical country (bananas, coffee, shrimp, or chocolate, for example) or made with a tropical product, such as wood, ask store managers if they know the country of origin. Tell them you would pay more for a product that was certified as eco-friendly. Make the people who depend on your business aware of your concerns. Whenever possible, buy products that have environmental seals of approval. Purchase paper that is recycled, certified, or made from tree-free fibers, and has been processed without chlorine.

Don't buy wildlife or wildlife products such as tortoise shell, coral, or shark teeth.

Recycle as much of everything that you can.

Reuse items as much as possible; avoid buying "disposable" products.

Join a conservation group that is working to save tropical forests.

Write letters to your elected officials to let them know that you care about environmental issues, at home and abroad.

Use cold water when feasible.

Turn down the heat; put on a sweater.

Donate items you no longer want to charitable organizations or thrift shops.

Put up a birdfeeder and birdbath in your yard or a nearby park. Grow native plants in your yard or community park to attract wildlife.

Instead of driving, use public transportation whenever you can. Ride your bike. Walk.

If you are saving up for a big vacation, visit a tropical forest. Tourist dollars directly contribute to rainforest conservation.

Never stop believing you can make a difference.

PART V

Rainforest Resources

Perhaps the previous chapters have awakened a desire to learn more about tropical forests, their flora and fauna, and the people who live in them and work to save them. Whether you want to read more about the binturongs of Borneo, see a video about deforestation in the Amazon, contact a conservation group in Madagascar, or send an email to the United Nations Environment Programme, the information in this section will point you in the right direction.

Reference Publications

The books and magazine and newspaper articles listed in the bibliography below are divided into four categories: tropical forests in general, wildlife, plants, and people. All were used to research this book; also refer to the lists under "Further Reading" that follow each entry in Parts One, Two, Three, and Four. Most of the articles are from popular magazines, leading science journals, or large daily newspapers, and should be available in most large libraries. Addresses for magazines and newsletters can be found in the section "Magazines, Newsletters, and Reports." When an article was published on the World Wide Web, the URL is included.

TROPICAL FORESTS IN GENERAL

Allen, William. "Biocultural Restoration of a Tropical Forest." *BioScience* 38, no. 3 (1988): 156–161.

Baird, Nicola. "Saying 'No' to Asian Loggers." *People & Planet* 5, no. 4 (1996): 26–27.

Becher, Anne. *Biodiversity: A Reference Handbook*. Santa Barbara, CA: ABC-CLIO, 1998.

Belt, Thomas. *The Naturalist in Nicaragua*. Chicago: University of Chicago Press, 1985.

Biodiversity Conservation Network. *Final Stories from the Field*. Washington, DC: Biodiversity Support Program, 1999.

The Biodiversity Project. *Life. Nature. The Public. Making the Connection: A Biodiversity Communications Handbook*. Madison, WI: 1999.

Brandon, Katrina. *Ecotourism and Conservation: A Review of Key Issues*. Environment Department Paper #33. Washington, DC: World Bank, 1996.

Brandon, Katrina, Kent H. Redford, and Steven E. Sanderson, eds. *Parks in Peril: People, Politics, and Protected Areas*. Washington, DC: Island Press, 1998.

Brown, Lester, et al. *State of the World 1999*. New York: W.W. Norton, 1999.

Brown, Michael, and Barbara Wyckoff-Baird. *Designing Integrated Conservation & Development Projects*. Washington, DC: The Biodiversity Support Program, 1992.

Bunch, Roland. *Two Ears of Corn: A Guide to People-Centered Agricultural Improvement*. Oklahoma City: World Neighbors, 1982.

Carr, Thomas, Heather Pedersen, and Sunder Ramaswamy.

"Rain Forest Entrepreneurs." *Environment*, September 1993, 13–15, 33–38.

Caudill, Denise. *Lessons from the Field. Integration of Population and Environment*. Oklahoma City: World Neighbors, 1998.

Cincotta, Richard, and Robert Engelman. *Nature's Place: Human Population and the Future of Biological Diversity*. Washington, DC: Population Action International, 2000.

Clay, Jason W. *Generating Income and Conserving Resources: 20 Lessons from the Field*. Baltimore: WWF Publications, 1996.

Cohen, Sheldon, and Stanley W. Burgiel, eds. *Exploring Biodiversity Indicators and Targets under the Convention on Biological Diversity: A Synthesis Report of the Sixth Session of the Global Biodiversity Forum*. Washington, DC: Biodiversity Action Network and The World Conservation Union, 1998.

Cohn, Jeffrey P. "Culture and Conservation." *Bioscience* 38, no. 7 (1988): 450–453.

Collins, Mark N., Jeffrey A. Sayer, and Timothy C. Whitmore, eds. *The Conservation Atlas of Tropical Forests: Asia and the Pacific*. New York: Simon & Schuster, 1991.

Department of Economic and Social Affairs Statistics Division. United Nations.

Dudley, Nigel, and Sue Stolton. "Management Effectiveness of Protected Areas: An International Workshop." *Arborvitae*, January 2000, 1–12 <http://www.esd.worldbank.org/wwf/new_ind2.htm>.

Durbin, Andrea. "Trade and the Environment: The North-South Divide." *Environment*, September 1995, 16–20, 37–40.

Ecotourism Society. *Ecotourism Guidelines for Nature Operators.* Bennington, VT: Ecotourism Society, 1993.

Engelman, Robert. *Plan & Conserve: A Source Book on Linking Population and Environmental Services in Communities.* Washington, DC: Population Action International, 1998.

Environmental Protection Agency. *Environmental Labeling: Issues, Policies, and Practices Worldwide.* Washington, DC: U.S. Government Printing Office, 1998.

Ericson, Jenny, Mark Freudenberger, and Eckart Boege. *Population Dynamics, Migration, and the Future of the Calakmul Biosphere Reserve.* Occasional Paper No. 1. Washington, DC: American Association for the Advancement of Science, 1999.

Food and Agriculture Organization of the United Nations. *State of the World's Forests 1997.* Rome, Italy: 1997.

Forsyth, Adrian. *Portraits of the Rainforest.* Ontario: Camden House, 1990.

Forsyth, Adrian, and Kenneth Miyata. *Tropical Nature.* New York: Charles Scribner's Sons, 1984.

French, Hilary. "Challenging the WTO." *World Watch,* November–December 1999, 23–27.

Gardner-Outlaw, Tom, and Robert Engelman. *Forest Futures: Population, Consumption and Wood Resources.* Washington, DC: Population Action International, 1999.

Goodland, Robert, ed. *Race to Save the Tropics: Ecology and Economics for a Sustainable Future.* Washington, DC: Island Press, 1999.

Goulding, Michael. *Amazon: The Flooded Forest.* New York: Sterling, 1989.

Goulding, Michael, Nigel J.H. Smith, and Dennis J. Mahar. *Floods of Fortune: Ecology & Economy Along the Amazon.* New York: Columbia University Press, 1996.

Griffith, Daniel M. "Agroforestry: A Refuge for Tropical Biodiversity after Fire." *Conservation Biology* 14, no. 1 (2000): 325–326.

Groombridge, Brian, ed. *Global Biodiversity: Status of the Earth's Living Resources.* New York: Chapman and Hall, 1992.

Guruswamy, Lakshman D., and Jeffrey A. McNeely, eds. *Protection of Global Biodiversity: Converging Strategies.* Durham, NC: Duke University Press, 1998.

Harris, Larry D. *The Fragmented Forest: Island Biogeography Theory and Preservation of Biotic Diversity.* Chicago: University of Chicago Press, 1984.

Hartshorn, Gary, and Nora Bynum. "Tropical Forest Synergies." *Science,* 10 December 1999, 286.

Head, Suzanne, and Robert Heinzman, eds. *Lessons of the Rainforest.* San Francisco: Sierra Club Books, 1990.

Hecht, Susanna, and Alexander Cockburn. *The Fate of the Forest: Developers, Destroyers and Defenders of the Amazon.* New York: Verso, 1989.

Janulewicz, Michael A., et al. *The International Book of the Forest.* London: Mitchell Beazley, 1981.

Janzen, Daniel. "Tropical Ecological and Biocultural Restoration." *Science,* 15 January 1988, 243–244.

Johnson, Nels, and Bruce Cabarle. *Surviving the Cut: Natural Forest Management in the Humid Tropics.* Washington, DC: World Resources Institute, 1993.

Joyce, Christopher. *Earthly Goods: Medicine-Hunting in the Rainforest.* Boston: Little, Brown, 1994.

Jukofsky, Diane. "Path of the Panther." *Wildlife Conservation,* September–October 1992, 18–24.

———. "Can Marketing Save the Rainforest?" *E Magazine,* July–August 1993, 33–39.

———. "After the Pleasure . . . A St. Lucian Hit." *Eco-Exchange,* May–June 1996, 2.

———. "Raptor Rapture in Veracruz." *Eco-Exchange,* November–December 1996, 1.

———. "Same Time, Next Year: More Fires?" *Eco-Exchange,* May–June 1998, 1.

———. "Guides in their Own Backyard." *Nature Conservancy,* July–August 1998, 18–23.

Kates, Robert W. "Population and Consumption: What We Know, What We Need to Know." *Environment,* April 2000, 10–19.

Keohane, Robert O., and Marc A. Levy, eds. *Institutions for Environmental Aid.* Cambridge: MIT Press, 1996.

Kirchner, James W., and Anne Weil. "Delayed Biological Recovery from Extinctions throughout the Fossil Record." *Nature,* 9 March 2000, 177–180.

Kramer, Randall, Carel van Shaik, and Julie Johnson, eds. *Last Stand: Protected Areas and the Defense of Tropical Biodiversity.* Oxford: Oxford University Press, 1997.

Kricher, John C. *A Neotropical Companion.* Princeton, NJ: Princeton University Press, 1989.

Lankford, William F. "Making a Difference with Solar Ovens." *EPA Journal,* July–August 1990, 47–49.

Mabey, Nick, Aimee Gonzalez, and David Schorr. "The WTO: A Comment." *Arborvitae,* January 2000, 5.

MacKerron, Conrad B., and Douglas G. Cogan, eds. *Business in the Rain Forest: Corporations, Deforestation and Sustainability.* Washington, DC: Investor Responsibility Research Center, 1993.

Margolius, Richard, and Nick Salafsky. *Measures of Success: Designing, Managing, and Monitoring Conservation and Development Projects.* Washington DC: Island Press, 1998.

Mattoon, Ashley T. "Paper Forest." *World Watch,* March–April 1998, 20–28.

Miller, Kenton, and Laura Tangley. *Trees of Life: Saving Tropical Forests and Their Biological Wealth.* Boston: Beacon Press, 1991.

Mitchell, Andrew W. *The Enchanted Canopy: Secrets from the Rainforest Roof.* Glasgow: Collins, 1986.

Moffett, Mark W. *The High Frontier: Exploring the Tropical Rainforest Canopy.* Cambridge: Harvard University Press, 1993.

Moore, Peter. "What Makes Rainforests So Special?" *New Scientist,* 21 August 1986, 38–40.

Moore, Peter, Stewart Maginnis, and Bill Jackson. "Firefight." *Arborvitae,* January 2000, 13.

Morell, Virginia. "Restoring Madagascar." *National Geographic,* February 1999, 60–71.

———. "The Variety of Life." *National Geographic*, February 1999, 6–31.

Muñoz, Nefer. "Ambitious Plan to Protect Biodiversity." InterPress Service, 16 September 1999 <http://forests.org/gopher/centamer/ambbiodp.txt>.

Myers, Norman. *A Wealth of Wild Species: Storehouse for Human Welfare*. Boulder, CO: Westview Press, 1983.

———. *The Primary Source: Tropical Forests and Our Future*. New York: W.W. Norton, 1984.

———. *Tackling Mass Extinction of Species: A Great Creative Challenge* (The Horace M. Albright Lectureship in Conservation, XXVI). Berkeley, CA: University of California, College of Natural Resources, 1986.

———. "Trees by the Billions: A Blueprint for Cooling." *International Wildlife*, September–October 1991, 12–15.

Neugebauer, Bernd. *Ecologically Appropriate Agri-Culture: A Manual of Ecological Agri-Culture Development for Smallholders in Development Areas*. Feldafing, Germany: German Foundation for International Development, 1995.

Newman, Arnold. *Tropical Rainforest*. New York: Facts on File, 1990.

Nilsson, Annika. *Greenhouse Earth*. West Sussex, England: John Wiley & Sons, 1992.

Noss, Reed F. "Corridors in Real Landsapes." *Conservation Biology* 1, no. 6: 159–164.

Office of International Affairs, U.S. Fish and Wildlife Service. "A Map Is Worth 1,000 Words." *Wildlife Without Borders*, fall 1999, 3.

Paehlke, Robert, ed. *Conservation and Environmentalism: An Encyclopedia*. New York: Garland, 1995.

Pagiola, Stefano, and John Kellenberg. *Mainstreaming Biodiversity in Agricultural Development: Toward Good Practice*. World Bank Environment Paper Number 15. Washington, DC: The World Bank, 1997.

Panos Briefing. *More Power to the World Trade Organization?* London: Panos Institute, November 1999 <http://www.oneworld.org/panos>.

———. *Economics for Ever: Building Sustainability into Economic Policy*. London: Panos Institute, March 2000.

Patterson, Alan. "Debt for Nature Swaps and the Need for Alternatives." *Environment*, December 1990, 5–13; 31–32.

Peters, Charles M. *Sustainable Harvest of Non-Timber Plant Resources in Tropical Moist Forest: An Ecological Primer*. Washington, DC: Biodiversity Support Program, 1994.

Peters, Charles M., Alwyn Gentry, and Robert Mendelsohn, "Valuation of an Amazonian Rainforest." *Nature*, 29 June 1989, 655–656.

Phillips, Kathryn. "Where Have All the Frogs and Toads Gone?" *BioScience* 40, no. 6 (1990): 422–424.

Plotkin, Mark J. *Tales of Shaman's Apprentice*. New York: Viking, 1993.

Pollack, Andrew. "Biological Products Raise Genetic Ownership Issues." *New York Times*, 26 November 1999, A1.

Rainforest Alliance. *Consensus Statement on Monitoring and Evaluation of Conservation Projects in the Neotropics*. New York: 1998

Raven, Peter, and George Johnson. *Biology*. New York: McGraw-Hill, 1999.

Redford, Kent, and Jane A. Mansour, eds. *Traditional Peoples and Biodiversity Conservation in Large Tropical Landscapes*. Arlington, VA: The Nature Conservancy, 1996.

Rees, William E. "The Ecology of Sustainable Development." *The Ecologist* 20, no. 1 (1990): 18–23.

Reuters News Service. "Honduran Congress Seeks Troops to Protect Forests," 18 April 2000 <http://www.envirolink.org/environews/reuters/articles/>.

Richards, Michael, and Pedro Moura Costa. "Can Tropical Forestry Be Made Profitable by 'Internalizing the Externalities'?" *Natural Resource Perspectives*, October 1999, 1–6.

Roper, John, and Ralph W. Roberts. *Deforestation: Tropical Forests in Decline*. Canada: CIDA Forestry Advisers Network, 1999.

Ross-Larson, Bruce, ed. *World Development Indicators*. Washington, DC: The World Bank, 1998.

Savitsky, Basil, and Thomas Lacher Jr. *GIS Methodologies for Developing Conservation Strategies: Tropical Forest Recovery and Wildlife Management in Costa Rica*. New York: Columbia University Press: 1998.

Sayer, Jeffrey A., Caroline S. Harcourt, and N. Mark Collins, eds. *The Conservation Atlas of Tropical Forests: Africa*. New York: Simon & Schuster, 1992.

Scherr, Jacob, and Justin Ward. "Seattle: Act 2." *Amicus Journal*, spring 2000, 34.

Scott, J. Michael, Blair Csuti, James Jacobi, and John Estes. "Species Richness: A Geographic Approach to Protecting Future Biological Diversity." *BioScience* 37, no. 11 (1987): 782–788.

Sedjo, Roger A., and Daniel Botkin. "Using Forest Plantations to Spare Natural Forest." *Environment*, December 1997, 14–20, 30.

Severin, Timothy. *The Spice Islands Voyage: The Quest for the Man Who Shared Darwin's Discovery of Evolution*. New York: Carroll & Graf, 1998.

Sheehan, Molly. "Gaining Perspective." *World Watch*, March–April 2000: 14–24.

Sheldon, Jennie, Michael J. Balick, and Sarah A. Laird. *Medicinal Plants: Can Utilization and Conservation Coexist?* New York: New York Botanical Garden, 1997.

Skutch, Alexander. *A Naturalist Amid Tropical Splendor*. Iowa City: University of Iowa Press, 1987.

Souder, William. "Frog Decline Linked to Climate Shift." *Washington Post*, 15 April 1999, A3.

Statistical Yearbook. 1995. New York: Department of Economic and Social Affairs Statistics Division. United Nations, 1997.

Swerdlow, Joel L. "Biodiversity: Taking Stock of Life." *National Geographic*, February 1999, 2–5.

Taylor, David. "Saving the Forest for the Trees." *Environment*, January 1997, 6–11; 33–36.

ten Kate, Kerry, and Sarah Laird. *The Commercial Use of Biodiversity: Access to Genetic Resources and Benefit-Sharing.* London: Earthscan, 1999.

Terborgh, John. *Where Have All the Birds Gone?* Princeton, NJ: Princeton University Press, 1989.

———. *Diversity and the Tropical Rain Forest.* New York: Scientific American Library, 1992.

———. *Requiem for Nature.* Washington, DC: Island Press, 1999.

Thrupp, Lorri A, ed. *New Partnerships for Sustainable Agriculture.* Washington, DC: World Resources Institute, 1996.

Turner, Barry, ed. *Statesman's Yearbook: The Essential Political and Economic Guide to All the Countries of the World, 1998–99.* London: MacMillan, 1998.

Tuxill, John. "The Latest News on the Missing Frogs." *World Watch*, May–June 1998, 9–10.

———. *Nature's Cornucopia: Our Stake in Plant Diversity.* Worldwatch Paper 148. Washington, DC: Worldwatch Institute, 1999.

United Nations Environment Programme News Release. "Going Up in Smoke," 24 July 1998.

———. "UNEP Welcomes Tough Stance on Forest Fires." 9 March 2000.

Veevers-Carter, W. *Riches of the Rainforest.* Singapore: Oxford University Press, 1984.

Vogel, Carolyn, and Robert Engelman. *Forging the Link: Emerging Accounts of Population and Environment Work in Communities.* Washington, DC: Population Action International, 1999.

Wells, Michael, and Katrina Brandon. *People and Parks: Linking Protected Area Management with Local Communities.* Washington, DC: The International Bank for Reconstruction and Development/The World Bank, 1992.

Wille, Chris. "The Greening of Debt." *E Magazine*, September–October 1991, 17–19.

Williams, Huntington III. "Banking on the Future." *Nature Conservancy*, May–June 1992, 23–27.

Wilson, Edward O., ed. *Biodiversity.* Washington DC: National Academy Press, 1986.

———. *The Diversity of Life.* Cambridge: Harvard University Press, The Belknap Press, 1992.

Wolfe, Art, and Ghillean T. Prance. *Rainforests of the World: Water, Fire, Earth & Air.* New York: Crown, 1998.

Wood, David S., and Diane Walton Wood. *How to Plan a Conservation Education Program.* Washington, DC: World Resources Institute, 1990.

World Commission on Environment and Development. *Our Common Future.* London: Oxford University Press, 1987.

World Commission on Forests and Sustainable Development. *Our Forests, Our Future.* Cambridge: Cambridge University Press, 1999.

World Conservation Monitoring Centre. *Global Biodiversity: Status of the Earth's Living Resources.* London: Chapman & Hall, 1992.

The World Resources Institute, the United Nations Environment Programme, The United Nations Development Program, The World Bank. *World Resources 1998–99.* New York: Oxford University, 1998.

Zollinger, Peter, and Roger Dower. *Private Financing for Global Environmental Initiatives: Can the Climate Convention's Joint Implementation Pave the Way?* Washington, DC: World Resources Institute, 1996.

WILDLIFE

Anderson, S., and J. K. Jones Jr., eds. *Orders and Families of Recent Mammals of the World.* New York: John Wiley & Sons, 1984.

Angier, Natalie. "Rare Bird Indeed Carries Poison in Bright Feathers." *New York Times,* 30 October 1992, A1.

———. "Chimps Exhibit, er, Humanness, Study Finds." *New York Times*, 15 June 1999.

Astor, Michael. "Song Led to Finding New Bird Discovered in Brazil." Associated Press, 24 April 1998 <http://abcnews.go.com/sections/science/DailyNews/newbird980424.html>.

Badger, David, and John Netherton. *Frogs.* New York: Facts on File, 1987.

Banister, Keith. *Encyclopedia of Aquatic Life.* New York: Facts on File, 1989.

Basler, Barbara. "Vietnam Forest Yields Evidence of New Animal." *New York Times,* 8 June 1993, C1.

Beehler, Bruce M., Thane K. Pratt, and Dale A. Zimmerman. *Birds of New Guinea.* Princeton, NJ: Princeton University Press, 1986.

Beletsky, Les. *Belize and Northern Guatemala: Ecotravellers' Wildlife Guide.* San Diego, CA: Academic Press, 1998.

Bengwayan, Michael. "New Damselfly Found in Threatened Rainforest Could Fight Disease." Environment News Service, 18 October 1999 <http://ens.lycos.com/ens/oct99/1999L-10-18-02.html>.

Berger, Joel, and Carol Cunningham. "Natural Variation in Horn Size and Social Dominance and Their Importance to the Conservation of Black Rhinoceros." *Conservation Biology* 12, no. 3 (1998): 708–711.

Berra, Tim M. *An Atlas of Distribution of the Freshwater Fish Families of the World.* Lincoln and London: University of Nebraska Press, 1981.

Blake, Edgar. "Fighting for a Rare Bird." *International Wildlife*, March–April 1999, 40–45.

Branch, Bill. *Field Guide to Snakes and Other Reptiles of Southern Africa.* Sanibel Island, Florida: Ralph Curtis Books, 1998.

Breining, Greg. "The Ghosts of Way Kambas." *International Wildlife*, September–October 1998, 22–29.

Bruemmer, Fred. "Celestial Visions." *International Wildlife*, March–April 1998, 44–49. Also <http://www.nwf.org/nwf/intlwild/1998/fairy.html>.

Buchsbaum, Ralph, et al. *The Audubon Society Encyclopedia of Animal Life.* New York: Clarkson N. Potter, 1982.

Burns, John F. "Medicinal Potions May Doom Tiger to Extinction." *New York Times,* 15 March 1994, C1.

Burton, Maurice, and Robert Burton. *The Marshal Cavendish International Wildlife Encyclopedia.* New York: Marshal Cavendish, 1989.

Castner, James L. "Please Don't Eat the Katydids." *International Wildlife*, May–June 1998, 22–29.

Chadwick, Douglas H., and Mark W. Moffett. "Planet of the Beetles." *National Geographic*, March 1998, 103–119.

Champlin, Susan. "Searching for Hope in the Family Tree." *National Wildlife,* April–May 1998, 37–41.

Chocran, Doris. *Living Amphibians of the World.* New York: Doubleday, 1961.

Christen, Anita. "The Most Enigmatic Monkey in the Bolivian Rainforest—*Callimico goeldii.*" *Neotropical Primates* 6, no. 2 (1998): 35–37.

Coborn, John. *The Atlas of Snakes of the World.* Neptune City, NJ: T. F. H. Publications, 1991.

Cogger, Harold G. *Reptiles and Amphibians of Australia.* New York: Cornell University Press, 1994.

Cogger, Harold G., and Richard G. Zweifel. *Encyclopedia of Reptiles and Amphibians.* San Diego, CA: Academic Press, 1998.

Cohn, Jeffrey. "Duke Primate Center Fosters Research." *BioScience* 35, no. 11 (1985): 691–695.

Conrow, Joan. "Born to Be Wild." *Audubon*, January–February 1999, 64–71.

Cox, Merel J. *The Snakes of Thailand and Their Husbandry.* Malaber, Florida: Krieger, 1991.

De Carli, Franco. *The World of Fish.* New York: Abbeville, 1975.

De Roy, Tui. "King of the Jungle." *International Wildlife*, November–December 1998, 52–57.

de Schauensee, Rodolphe M., and William H. Phelps, Jr. *A Guide to the Birds of Venezuela.* Princeton, NJ: Princeton University Press, 1978.

Denny, Norman, and Jorge Arias. *Insects of an Amazon Forest.* New York: Columbia University Press, 1982.

Deoras, P. J. *Snakes of India.* New Delhi, India: National Book Trust, 1965.

DeVries, Philip J. *The Butterflies of Costa Rica and Their Natural History: Volume I.* Princeton, NJ: Princeton University Press, 1987.

Ditmars, Raymond L. *Snakes of the World.* New York: Macmillan, 1931.

Dixon, James R., and Pekka Soini. *The Reptiles of the Upper Amazon Basin, Iquitos Region, Peru.* Milwaukee, WI: Milwaukee Public Museum, 1986.

Domingues-Brandao, Leticia, and Pedro Ferreira-Develey. "Distribution and Conservation of the Buffy Tufted-Ear Marmoset, *Callithrix aurita*, in Lowland Coastal Atlantic Forest, Southeast Brazil." *Neotropical Primates* 6, no. 3 (1998): 86–88.

Dublin, Holly, Tom O. McShane, and John Newby. "Conserving Africa's Elephants." World Wildlife Fund, USA <http://www.panda.org/resources/publications/species/elephant/elephant.html>.

Ebersole, Rene S. "Trouble for Monarchs." *International Wildlife*, September–October 1999, 12.

Ehrlich, Paul R., David S. Dobkin, and Darryl Wheye. *The Birder's Handbook: A Field Guide to the Natural History of North American Birds.* New York: Simon & Schuster, 1988.

Elphick, Jonathan, ed. *The Atlas of Bird Migration.* New York: Random House, 1995.

Emmons, Louise H. *Neotropical Rainforest Mammals: A Field Guide.* Chicago: University of Chicago Press, 1990.

Ernst, Carl H., and Roger W. Barbour. *Turtles of the World.* Washington, DC: Smithsonian Institution Press, 1989.

Fay, Michael, and Michael Nichols. "Forest Elephants." *National Geographic,* February 1999, 100–113.

Fjeldsa, Jon, and Niels Krabbe. *Birds of the High Andes.* Copenhagen: Zoological Museum, University of Copenhagen, 1990.

Garel, Tony, and Sharon Matola. *A Field Guide to the Snakes of Belize.* Belize City: The Belize Zoo and Tropical Education Center, 1996.

Gharpurey, Lieut.-Colonel K. G. *The Snakes of India and Pakistan.* Bombay: The Popular Book Depot, 1954.

Glanz, James. "Tracking Gorillas and Rebuilding a Country." *New York Times*, 11 April 2000, F3 <http://www10.nytimes.com/library/national/science/041100sci-animal-gorilla.html>.

Gonzales, P.C., and Rees, C. P. *Birds of the Philippines.* Quezon City, Philippines: Haribon Foundation, 1988.

Graham, Jeffrey. *Air-Breathing Fishes: Evolution, Diversity, and Adaptation.* San Diego, CA: Academic Press, 1997.

Grzimek, Bernhard. *Grzimek's Animal Life Encyclopedia.* New York: Van Nostrand Reinhold, 1972.

Halliday, Tim, and Kraid Adler. *Encyclopedia of Reptiles and Amphibians.* Oxford: Equinox, 1986.

Haltenorth, Theodor, and Helmut Diller. *Collins Field Guide: Mammals of Africa Including Madagascar.* London: HarperCollins, 1994.

Hawkins, Frank, Roger Safford, Will Duckworth, and Mike Evans. "Field Identification and Status of the Sunbird Asities of Madagascar." *Bulletin of the African Bird Club* 4, no. 1 (1997) <http://www.africanbirdclub.org/feature/asities.html>.

Herald, Earl. *Living Fishes of the World.* New York: Doubleday & Company, 1961.

Hill, J. E., and J. D. Smith. *Bats: A Natural History.* Austin: University of Texas Press, 1992.

Hogue, Charles L. *Latin American Insects and Entomology.* Berkeley and Los Angeles: University of California Press, 1993.

Hölldobler, Bert, and Edward O. Wilson. *The Ants.* Cambridge: Harvard University Press, The Belknap Press, 1990.

Holmes, Hannah. "The Lizard Wizard." *Wildlife Conservation*, March–April 1997, 22–29.

Horton, Sydney. "Bat du Jour." *Audubon,* November–December 1999, 26.

Howell, Steve N.G., and Sophie Webb. *A Guide to the Birds of Mexico and Northern Central America.* Oxford: Oxford University Press, 1995.

Hubers, Jos. "The Ultramarine Lory." *Lori Journal International* 4, 1997 <http://www.parrotsociety.org.au>.

Isemonger, R. M. *Snakes of Africa: Southern, Central, and East.* New York: Thomas Nelson and Sons, 1962.

Israel, Samuel, and Toby Sinclair. *Indian Wildlife.* Singapore: Apa Publications, 1994.

Jackson, Robert. "Portia Spider: Mistress of Deception." *National Geographic,* November 1996, 104–115.

Janzen, Daniel, ed. *Costa Rican Natural History.* Chicago: University of Chicago Press, 1983.

Jukofsky, Diane. "Mystical Messenger." *Nature Conservancy,* November–December 1993, 24–29.

———. "Raptor Rapture in Veracruz." *Eco-Exchange,* November–December 1996, 1.

———. "New Chick in Idaho to Soar in Panama." *Eco-Exchange,* March–April 1997, 2.

Juniper, Tony, and M. Parr. *Parrots: A Guide to Parrots of the World.* New Haven, CT: Yale University Press, 1998.

Kannan, Raguapathy, and Douglas A. James. "Fruiting Phenology and the Conservation of the Great Pied Hornbill (*Buceros bicornis*) in the Western Ghats of Southern India." *Biotropica* 31, no. 1 (1999): 167–177.

Kingdon, Jonathan. *Island Africa: The Evolution of Africa's Rare Animals and Plants.* London: William Collins Sons, 1990.

———. *The Kingdon Field Guide to African Mammals.* London: Harcourt Brace, 1997.

Kinnaird, Margaret. "Macaque Island." *Wildlife Conservation,* April 1998 <http://www.wcs.org/news/magazine/199804/feature/index.html>.

———. "Big on Figs." *International Wildlife,* January–February 2000, 12–21.

Kirchner, James W., and Anne Weil. "Delayed Biological Recovery from Extinctions Throughout the Fossil Record." *Nature,* 9 March 2000, 177–180.

Kirkpatrick, David. "The Matamata." *Reptile & Amphibian Magazine,* September–October 1992, 32–39, <http://www.unc.edu/dtkirkpa/stuff/mata.html>.

Knott, Cheryl, and Tim Larman. "Orangutans in the Wild." *National Geographic,* August 1998, 30–57.

Kottelat, Maurice. *Freshwater Biodiversity in Asia.* Washington DC: The World Bank, 1996.

Lahm, Sally. "Devils of the Forest." *Wildlife Conservation,* March–April 1997, 31–35.

Langrand, Oliver. *Guide to the Birds of Madagascar.* New Haven, CT: Yale University Press, 1990.

Larousse Encyclopedia of Animal Life. New York: McGraw Hill, 1967.

Leary, Warren E. "Dashing Across Water's Surface: How One Creature Does It." *New York Times,* 9 April 1996, C4.

Lee, Julian C. *Amphibians and Reptiles of the Yucatan Peninsula.* Ithaca, NY: Cornell University Press, 1996.

Leveque, Christian. *Biodiversity Dynamics and Conservation: The Freshwater Fish of Tropical Africa.* Cambridge: Cambridge University Press, 1997.

Levi, Herbert W. *A Guide to Spiders and Their Kin.* New York: Golden, 1968.

Line, Les. "Silence of the Songbirds." *National Geographic,* June 1993, 68–90.

———. "A Newfound Mammal of Philippine Treetops Gets High-Flown Name." *New York Times,* 20 February 1996, C4.

———. "New Branch of Primate in the Family Tree." *New York Times,* 18 June 1996, C4.

———. "Eating on the Run." *National Wildlife,* October–November 1997, 46–50.

———. "Is This the World's Rarest Bird?" *National Wildlife,* December–January 1998, 46–47.

———. "Put on a Happy Face." *Audubon,* November–December 1999, 152.

Lips, Karen. "Mass Mortality and Population Declines of Anurans at an Upland Site in Western Panama." *Conservation Biology* 13, no. 1 (1999): 117–125.

"A Little Night Musician." *Nature Conservancy,* September–October 1998, 42.

Long, Michael E. "The Shrinking World of Hornbills." *National Geographic,* July 1999, 52–71.

Loveridge, Arthur. *Reptiles of the Pacific World.* New York: Macmillan, 1945.

Lovette, Irby J., Eldredge Bermingham, and Robert E. Ricklefs. "Mitochondrial DNA Phylogeography and the Conservation of Endangered Lesser Antillean *Icterus* Orioles." *Conservation Biology* 13, no. 5 (1999): 1088–1096.

Lowe-McConnell, R.H. *Ecological Studies in Tropical Fish Communities.* Cambridge: Cambridge University Press, 1987.

Mason, Georgia. "How Ants Make the Most of Foraging Trips." *New Scientist,* February 1991, 25.

Matthiessen, Peter. "The Last Wild Tigers." *Audubon,* March–April 1997, 54–63, 122.

Mattison, Chris. *Frogs & Toads of the World.* New York: Facts on File, 1987.

———. *The Encyclopedia of Snakes.* New York: Facts on File, 1995.

McGuire, Tamara L., and Kirk O. Winemiller. "Occurrence Patterns, Habitat Associations, and Potential Prey of the River Dolphin, *Inia geoffrensis,* in the Cinaruco River, Venezuela." *Biotropica* 30, no. 4 (1998): 625–638.

McNeil Jr., Donald G. "The Great Ape Massacre." *New York Times Magazine,* 9 May 1999, 54–57.

Michener, Charles D. *The Social Behavior of the Bees.* Cambridge: Harvard University Press, The Belknap Press, 1974.

Moffett, Mark W. "Poison-dart Frogs: Lurid and Lethal." *National Geographic*, May 1995, 98–105.

———. "Ants and Plants—a Profitable Partnership." *National Geographic,* February 1999, 122–132.

Morell, Virginia. "Looking for Big Pink." *International Wildlife*, November–December 1997, 26–31.

Munn, Charles A. "Winged Rainbows: Macaws." *National Geographic*, January 1994, 118–140.

Nobbe, "Walking on Water." *Wildlife Conservation,* March–April 1997, 14.

Norman, David. *Common Amphibians of Costa Rica*. Heredia, Costa Rica: David Norman, 1998.

Oaks, Tammy. "Lemurs among Madagascar Victims." *MSNBC*, 23 March 2000 <http://www.msnbc.com/news/384591.asp?cp1=1>.

O'Connor, Anahad. "Thought Extinct, a Few Javan Rhinos Are Seen in Vietnam." *New York Times*, 20 July 1999 <http://www.nytimes.com/library/national/science/072099sci-animal-rhinoceros.html>.

Orsak, L. J. "Killing Butterflies to Save Butterflies: A Tool for Tropical Forest Conservation in Papua New Guinea." *News of the Lepidopterists Society* 3 (1993): 71–80 <http://www.aa6g.org/Butterflies/pngletter.html>.

O'Shea, Mark. *A Guide to the Snakes of Papua New Guinea*. Port Moresby, Papua New Guinea: Independent Group Printing, 1996.

O'Toole, Christopher. *Encyclopedia of Insects*. New York: Facts on File, 1986.

Peregrine Fund. "First Madagascar Serpent-Eagle Nest Discovered." Press release, 24 July 1998.

Phelps, Tony. *Poisonous Snakes*. New York: Blandford, 1989.

Phillips, Kathryn. "Where Have All the Frogs and Toads Gone?" *BioScience* 40, no. 6 (1990): 422–424.

Pinhey, Elliot. *Emperor Moths of South and South Central Africa*. Cape Town: C. Struik, 1972.

Pizzey, Graham. *A Field Guide to the Birds of Australia*. Princeton, NJ: Princeton University Press, 1980.

Prasad, "Sri Lanka Caters to Tourist Golfers at Elephants' Expense." Environment News Service, 1 March 2000 <http://ens.lycos.com/ens/jan2000/2000L-01–03–01.html>.

Preston-Mafham, Ken, and Rod Preston-Mafham. *Spiders of the World*. New York: Facts on File, 1985.

———. *The Natural History of Spiders*. Ramsbury, Marlborough: The Crown, 1996.

Pritchard, Peter. *Encyclopedia of Turtles*. Neptune City, NJ: T. F. H. Publications, 1979.

Pyle, Robert M. *Handbook for Butterfly Watchers*. New York: Houghton Mifflin, 1984.

Raffaele, Herbert, et al. *A Guide to the Birds of the West Indies*. Princeton, NJ: Princeton University Press, 1998.

Recer, Paul. "Poison from South American Frog Leads to Discovery of Powerful Painkiller." *Washington Post*, 2 January 1998, A16.

Redford, Kent H., and John F. Eisenberg. *Mammals of the Neotropics: The Southern Cone*. Chicago: University of Chicago Press, 1992.

Reid, Fiona A. *A Field Guide to the Mammals of Central America and Southeast Mexico*. Oxford: Oxford University Press, 1997.

Rettig, Neil. "Remote World of the Harpy Eagle." *National Geographic*, February 1995, 40–49.

Rivas, Jesus. "Tracking the Anaconda." *National Geographic*, January 1999, 62–69.

Robinson, John, and Kent Redford, eds. *Neotropical Wildlife Use and Conservation*. Chicago: University of Chicago Press, 1991.

Robinson, Michael, and Barbara Robinson. *Smithsonian Contributions to Zoology, No. 149*. Washington, DC: Smithsonian Institution, 1973.

Ross, Gary N. "Butterfly Carnivale." *Wildlife Conservation*, May–June 1998, 48–55.

Royte, Elizabeth. "On the Brink: Hawaii's Vanishing Species." *National Geographic*, September 1995, 9–37.

Schmidt, Jeremy. "Soldiers in the Gorilla War." *International Wildlife*, January–February 1999, 13–21.

Serle, William, and Gérard J. Morel. *Birds of West Africa*. Hong Kong: HarperCollins, 1977.

Sick, Helmut. *Birds in Brazil: A Natural History*. Princeton, NJ: Princeton University Press, 1993.

Silvius, Kirsten. "Where Have All the Peccaries Gone?" *Wildlife Conservation*, May–June 1995, 10.

Skaife, S.H. *African Insect Life*. Cape Town: C. Struik, 1979.

Skutch, Alexander. *Birds of Tropical America*. Austin: University of Texas Press, 1983.

Slater, Pat, and Steve Parish. *A First Field Guide to Australian Mammals*. Queensland: Steve Parish Publishing, 1997.

Smith, Hobart M. *Handbook of Lizards: Lizards of the United States and of Canada*. Ithaca, NY: Cornell University Press, 1946.

Souder, William. "Frog Decline Linked to Climate Shift." *Washington Post*, 15 April 1999, A3.

Stewart, Doug. "Prosimians Find a Home Far from Home." *National Wildlife,* February–March 1998, 30–34.

Stewart, Margret. *Amphibians of Malawi*. Albany, NY: State University of New York Press, 1967.

Stiles, F. Gary, and Alexander F. Skutch. *A Guide to the Birds of Costa Rica*. Ithaca, NY: Cornell University Press, 1989.

Stolzenburg, William. "Andean Ambassador." *Nature Conservancy*, July–August 1997, 10–15.

Struhsaker, Tom. "Zanzibar's Endangered Red Colobus Monkeys." *National Geographic*, November 1998, 73–79.

Strum, Shirley C. "Moving the Pumphouse Gang." *International Wildlife*, May–June 1998, 12–21.

Sugal, Cheri. "Elephants of Southern Africa Must Now 'Pay Their Way.' " *World Watch*, September–October 1997, 9.

Sunquist, Fiona. "Blessed Are the Fruit-Eaters." *International Wildlife*, May–June 1992, 4–10.

———. "Should We Put Them All Back?" *International Wildlife,* September–October 1993, 34–40.

Sunquist, Fiona, and Mel Sunquist. "New Look at Cats!" *International Wildlife*, March–April 2000, 12–23.

Sunquist, Mel. "What I've Learned About Tigers." *International Wildlife*, November–December 1997, 12–19.

Swanson, Stephen. *Lizards of Australia.* Sydney: Angus and Robertson, 1976.

Taylor, Caroline. "The Challenge of African Elephant Conservation." *Conservation Issues* 4, no. 2 (1997): 1–11.

Taylor, David. "The Agouti's Nutty Friend." *International Wildlife*, March–April 2000, 32–38.

Tennesen, Michael. "High-Flying Bats." *Wildlife Conservation*, May–June 1998, 17.

Terborgh, John. *Where Have All the Birds Gone?* Princeton, NJ: Princeton University Press, 1989.

Thorbjarnarson, Tom. "The Hunt for Black Caiman." *International Wildlife*, July–August 1999, 12–19.

"Tiny Frog with a Sizable Name." *National Geographic,* May 1997.

Tirira, Diego. *Mamíferos del Ecuador.* Quito: Pontífica Universidad Católica del Ecuador, 1999.

Tucker, Neely. "Poachers Kill 350 Elephants." *Miami Herald,* 12 December 1999, A8.

Tuxill, John. "Death in the Family Tree." *World Watch,* September–October 1997, 13–21.

———. "The Latest News on the Missing Frogs." *World Watch,* May–June 1998, 9–10.

Tweedie, M. W. F. *The Snakes of Malaysia.* Singapore: Government Printing Office, 1953.

Valicenti, Trish. "Paradise Park for Parrots." *Wildlife Conservation,* September–October 1999, 96.

Villa, Jaime. "The Venomous Snakes of Nicaragua." *Contributions in Biology and Geology,* 1 November 1984.

Wade, Nicholas. "For Leaf-Cutter Ants, Farm Life Isn't So Simple." *New York Times,* 3 August 1999, D1.

Ward, Geoffrey C. "Making Room for Wild Tigers." *National Geographic*, December 1997, 2–35.

Webster, Donovan. "The Looting and Smuggling and Fencing and Hoarding of Impossibly Precious, Feathered and Scaly Wild Things." *New York Times Magazine,* 16 February 1997, 26–33, 48–49, 53, 61.

Weintraub, Boris. "Chimps Find a Way to Eat in Comfort." *National Geographic,* September 1997.

Wheeler, Alwyne. *The World Encyclopedia of Fishes.* London: Macdonald & Co. Ltd., 1985.

Whitfield, Philip. *Macmillan Illustrated Animal Encyclopedia.* New York: Macmillan, 1984.

Wildash, Philip. *Birds of South Vietnam.* Tokyo: Charles E. Tuttle, 1968.

Wille, Chris. "Mystery of the Missing Migrants." *Audubon,* May 1990, 81–141.

Williams, John G., and Norman Arlott. *Birds of East Africa.* Hong Kong: HarperCollins, 1980.

Williams, Keith. "Super Snoots." *Wildlife Conservation,* July–August 1991, 70–75.

Wootton, Anthony. *Insects of the World.* New York: Facts on File, 1984.

World Wildlife Fund. "Kenyas' Rhinos Benefit from WWF Donation." *Focus,* September–October 1999, 1; 6.

Yoon, Carol K. "Splendor in the Mud: Unraveling the Lives of Anacondas." *New York Times,* 2 April 1996, C1.

York, Andy. "Termites Are a Rainforest's Best Friend." *New Scientist,* 12 September 1998, 20.

York, Barbara. *Spiders of Australia.* Sidney: Jacaranda, 1967.

Youth, Howard. "The Bear Facts." *ZooGoer* 28, no. 2 (1999) <http://www.fonz.org/zoogoer/zg1999/28(2) spec tacled.htm>.

Zimmerman, Dale A., Donald A. Turner, and David J. Pearson. *Birds of Kenya and Northern Tanzania.* Princeton, NJ: Princeton University Press, 1996.

PLANTS

Acharya, Keya. "Tree Provides Biodiesel for India." Environment News Service, 8 March 1998 <http://www.envirolink.org/archives/enews/0790.html>.

———. "Prostate Demand Strips Curative Trees." Environment News Service, 13 March 1998 <http://ens.lycos.com>.

Armitano, Ernesto, ed. *Garden Plants of the Tropics.* Caracas, Venezuela: Ediciones Armitano, 1986.

Armstrong, W.P. "The Unforgettable Acacias." *Zoonoo,* August 1998, 28–31 <http://daphne.palomar.edu/wayne/plaug99.htm>.

Balick, Michael J., and Hans T. Beck. *Useful Palms of the World: A Synoptic Bibliography.* New York: Columbia University Press, 1990.

Balick, Michael J., and Robert Mendelsohn. "Assessing the Economic Value of Traditional Medicines from Tropical Rain Forests." *Conservation Biology* 6, no. 1 (1992): 128–130.

Balick, Michael J., Elaine Elisabetsky, and Sarah A. Laird, eds. *Medicinal Resources of the Tropical Forest.* New York: Columbia University Press, 1995.

Bannan, Jan. "The Cornucopia Tree." *Wildlife Conservation,* May–June 1993, 8.

Bhattacharyya, Bharati, and B.M. Johri. *Flowering Plants: Taxonomy and Phylogeny.* New Delhi: Narosa Publishing House, 1998.

Bird, Chris. "Medicines from the Rainforest." *New Scientist,* 17 August 1991, 34–39.

Blackwell, Will H. *Poisonous and Medicinal Plants.* Englewood Cliffs, NJ: Prentice-Hall, 1990.

Blombery, Alec, and Tony Rodd. *An Informative, Practical Guide to Palms of the World.* North Ryde, Australia: Angus and Robertson, 1982.

Chin, Wee Yeow. *Ferns of the Tropics.* Singapore: Times Editions, 1997.

Cox, Paul A., and Michael J. Balick. "The Ethnobotanical Approach to Drug Discovery." *Scientific American,* June 1994, 2–7.

Croft, Jim. "Ferns and Man in New Guinea." Centre for

Plant Biodiversity Research, 1982 <http://www.anbg.gov.au/jrc/ferns-man-ng.html>.

Cullen, J., ed. *The Orchid Book*. Cambridge: Cambridge University Press, 1992.

Davies, Frank. "Patent Dispute Germinates Around Plant Used in Amazon Rituals." *Miami Herald*, 31 March 1999, 14A.

Dold, Catherine. "Tropical Forest Found More Valuable for Medicine than Other Uses." *New York Times*, 28 April 1992, C1.

The Economist. "All Bark, No Bite." 15 April 2000, 88.

Graf, Alfred Byrd. *Exotica*. East Rutherford, NJ: Roehrs, 1974.

Hargreaves, Dorothy, and Bob Hargreaves. *Tropical Trees*. Lahaina, Hawaii: Ross-Hargreaves, 1965.

Hartshorn, Gary, and Nora Bynum. "Tropical Forest Synergies." *Science,* 10 December 1999: 2093–2094.

Haynes, Jody L. *The Virtual Palm Encyclopedia*. Palm and Cycad Societies of Florida <http://www.plantapalm.com/>.

Heywood, V.H. *Flowering Plants of the World*. New York: Oxford University Press, 1993.

Irvine, F.R. *Woody Plants of Ghana*. London: Oxford University Press, 1961.

Janzen, Daniel, ed. *Costa Rican Natural History*. Chicago: University of Chicago Press, 1983.

Jones, David L. *Encyclopedia of Ferns*. Portland, OR: Timber Press, 1987.

———. *Cycads of the World*. Washington, DC: Smithsonian Institution Press, 1993.

———. *Palms Throughout the World*. Washington, DC: Smithsonian Institution Press, 1995.

Jukofsky, Diane. "Medicinal Plant Research Leads Scientists to Rainforest." *Drug Topics*, 22 April 1991, 26–28.

Kahn, Francis, and Jean-Jacques de Granville. *Palms in Forest Ecosystems of Amazonia (Ecological Studies 95)*. New York: Springer-Verlag, 1992.

Kingdon, Jonathan. *Island Africa: The Evolution of Africa's Rare Animals and Plants*. London: William Collins Sons, 1990.

Kinnaird, Margaret. "Big on Figs." *International Wildlife*, January–February 2000, 12–21.

Kubitzki, K. *The Families and Genera of Vascular Plants*. New York: Springer-Verlag, 1990.

Laman, Tim. "Borneo's Strangler Fig Trees." *National Geographic,* April 1997, 41–55.

Lennox, G.W., and S.A. Seddon. *Flowers of the Caribbean*. London: Macmillan, 1978.

Line, Les. "The Biggest, Stinkiest Flower." *Audubon*, July–August 1999, 128.

Miller, Susan K. "High Hopes Hanging on a 'Useless' Vine." *New Scientist*, 16 January 1993, 12.

Mueller-Dombois, Dieter, and F. Raymond Fosberg. *Vegetation of the Tropical Pacific Islands (Ecological Studies 132)*. New York: Springer-Verlag, 1998.

Paige, Jeffery M. *Coffee and Power: Revolution and the Rise of Democracy in Central America*. Cambridge: Harvard University Press, 1998.

Peters, Charles M. *Sustainable Harvest of Non-Timber Plant Resources in Tropical Moist Forest: An Ecological Primer*. Washington, DC: Biodiversity Support Program, 1994.

Pierot, Suzanne, and John M. Hall, III. *Easy Guide to Tropical Plants*. San José, Costa Rica: Publicaciones los Olivos, 1996.

Pijl, L. van der, and Calaway H. Dodson. *Orchid Flowers: Their Pollination and Evolution*. Coral Gables, Florida: University of Miami Press, 1996.

Rice, Robert A., and Justin R. Ward. *Coffee, Conservation, and Commerce in the Western Hemisphere*. Washington, DC: Smithsonian Migratory Bird Center and Natural Resources Defense Council, 1996.

Salvesen, David. "The Grind over Sun Coffee." *ZooGoer*, July–August 1996, 5–13.

Sancho, Ellen, and Marcia Baraona. *Fruits of the Tropics*. Heredia, Costa Rica: Programa de Publicaciones e Impresiones de la Universidad Nacional, 1996.

Schultes, Richard Evans, and Robert E. Raffauf. *The Healing Forest: Medicinal and Toxic Plants of the Northwest Amazonia*. Portland, OR: Timber Press, 1990.

Tewari, D.N. *A Monograph on Teak*. New Delhi: Vedams Books International, 1992 <http://www.vedamsbooks.com/no7501.htm>.

Tsuruoka, Doug. "Nature's Laboratory: Sarawak Tree Yields Potential Aids Treatment." *Far Eastern Economic Review*, 24 March 1994, 43–44.

Tuxill, John. *Nature's Cornucopia: Our Stake in Plant Diversity*. Worldwatch Paper 148. Washington, DC: Worldwatch Institute, 1999.

Veevers-Carter, W. *Riches of the Rainforest*. Singapore: Oxford University Press, 1984.

Yoon, Carol K. "Biologists Find Progenitors of Earth's Flowering Plants." *New York Times*, 29 October 1999.

Young, Allen M. *The Chocolate Tree: A Natural History of Cacao*. Washington: Smithsonian Institution Press, 1994.

PEOPLE

Alcorn, Janis B. "Indigenous Peoples and Conservation." *Conservation Biology* 7, no. 2 (1993): 424–426.

Bland, John. "Giving WHO Moral Clout." *People & the Planet* 8, no. 1 (1999): 15–16.

Carpenter, Betsy. "Faces in the Forest." *U.S. News & World Report*, 4 June 1990, 65–69.

Chapin, Mac. "Losing the Way of the Great Father." *New Scientist*, 10 August 1991, 40–44.

Chatterjee, Pratap. "Indigenous Communities Fend Off Gold Miners." Inter Press Service, 25 June 1997 <http://www.oneworld.org/ips2/jun/brazil2.html>.

Chernos, Saul. "Ken Saro-Wiwa's Son Continues Fight Against Big Oil." Environment News Service, 10 November 1999 <http://ens.lycos.com/ens/nov99/1999L-11-09-02.html>.

Cultural Survival. "The Peoples of 'Millennium.'" *Cultural Survival Quarterly*, spring 1992, 60–71.

Cunningham, A. B. "Healthcare, Conservation, and Medicinal Plants in Madagascar." *People & Plants Online*, April 1993 <http://www.rbgkew.org.uk/peopleplants/dp/dp2/issues.htm>.

Daly, Douglas. "Wonder, Astonishment, and Devotion: 13 Explorers in the Americas." *Audubon*, March–April 1997, 118–121.

Dam, Julie. "The Green Team: Defender of the Pearl." *Time*, 29 April 1996 <http://www.pathfinder.com/time/international/1996/960429/environment.html>.

Davis, Wade. "Vanishing Cultures." *National Geographic*, August 1999, 62–89.

Dillon, Sam. "Jailed Mexican Wins Environmental Prize." *New York Times*, 5 April 2000, A4.

Dreifus, Claudia. "Saving the Orangutan, Preserving Paradise." *New York Times*, 21 March 2000, F3.

Duffy, Kevin. *Children of the Forest*. New York: Dodd, Mead, 1984.

Finkel, Mike. "Queen of the Canopy." *Audubon*, September–October 1998, 32–34.

Gilmour, Don. "Rainforest Pioneer." *People & the Planet* 8, no. 4 (1999): 20.

Golden, Frederic. "A Century of Heroes." *Time*, spring 2000, Special Earth Day Edition, 54–57.

Gonen, Amiram, ed. *The Encyclopedia of the Peoples of the World*. New York: The Jerusalem Publishing House, 1993.

Hallowell, Christopher. "Search for the Shamans' Vanishing Wisdom." *Time*, 14 December 1998, 71.

———. "Rainforest Pharmacist." *Audubon*, January-February 1999, 28–29.

Hart, Teresa B., and John A. Hart. *The Ecological Basis of Hunter-Gatherer Subsistence in African Rain Forests: The Mbuti of Eastern Zaire*. In *Case Studies in Human Ecology*, edited by Daniel G. Bates and Susan H. Lees. New York: Plenum Press, 1996.

Hoffman, Eric. "Wonder Woman of Belize." *International Wildlife*, November–December 1992: 14–18.

Honan, William. "Did the Maya Doom Themselves by Felling Trees?" *New York Times*, 11 April 1995.

Iams, David. "Trotting Out Honors and Looking Ahead." *Philadelphia Inquirer*, 27 May 1999 <http://www.phillynews.com/inquirer/99/May/27/magazine/SOCI27.htm>.

Jackson, Nancy B. "Through Politicking for Plants, He Made His Garden Grow." *New York Times*, 4 August 1998, F3.

Jacobson, Daniel. *Great Indian Tribes*. Maplewood, NJ: Hammond, 1970.

Johnson, Tim. "Energy Projects Opposed: Latin Indians Using Sabotage." *Miami Herald*, 18 November 1999, 1C-2C.

Jukofsky, Diane. "Unnatural Disaster: Conservation Lessons from Hurricane Mitch." *Nature Conservancy*, September–October 1999, 18–26.

Kane, Joe. *Savages*. New York: Alfred A. Knopf, 1995.

Krauss, Clifford. "Chilean Indians Divided Over a River, and Their Fate." *New York Times*, 16 August 1998.

Kreisler, Harry. "The Journey of an Environmental Scientist: Conversation with Norman Myers." Institute of International Studies, University of California at Berkeley, 11 November 1998 <http://www.globetrotter.berkeley.edu/people/Myers/myers-con1.html>.

Leach, Monty. "Don Quixotes of the Environment." *San Diego Earth Times*, February 1998 <http://www.sdearthtimes.com/et0298/et0298s6.html>.

Lebar, Frank M., ed. *Ethnic Groups of Insular Southeast Asia*. New Haven, CT: Human Relations Area Files, 1972.

Lentz, David. "Medicinal and Other Economic Plants of the Paya of Honduras." *Economic Botany* 47, no. 4 (1993): 358.

Levinson, David, ed. *Encyclopedia of World Cultures*. Boston: G.K. Hall, 1995.

Levy, Betsy. "Ethnobotanist Paul Cox Helps Save Samoan Rainforest." *Whole Foods*, April 1999 <http://www.wholefoods.com/magazine/articles/0,botantists,00.html>.

Liebhold, David. "Crusader for Indonesia's Enchanted Forests." *Time*, 11 January 1999 <http://www.pathfinder.com/time/reports/environment/heroes/heroesgallery/0,2967,hafild,00.html>.

Linden, Eugene. "Lost Tribes, Lost Knowledge." *Time*, 23 September 1991, 46–56.

Line, Les. "Indiana Jones Meets His Match in Burma Rabinowitz." *New York Times*, 3 August 1999, D5.

Lipske, Michael. "American Heroes: He Wrote the Book on Birds." *National Wildlife*, December–January 1997: <http://nwf.org/natlwild/herodj7.html>.

Luoma, Jon. "The Sociobiologist: Edward O. Wilson." *Audubon*, November–December 1998, 90.

Maslow, Jonathan. *Footsteps in the Jungle: Adventures in the Scientific Exploration of the American Tropics*. Chicago: Ivan R. Dee, 1996.

McClanahan, T.R. "The Rhino Man's Final Rest." *Wildlife Conservation*, November–December 1999, 4–11.

Menchú, Rigoberta. *I, Rigoberta Menchú: An Indian Woman in Guatemala*. London: Verso, 1984.

Minority Rights Group International. *World Directory of Minorities*. London: Minority Rights Group International, 1997.

Mutiso, Clive. "Her Women's Army Defies an Iron Regime." *Time*, 14 December 1998, 73.

Núñez-Farfan, Juan, and Arturo Gómez-Pompa. "Carlos Vázquez-Yanes," *Tropinet* 10, no. 4 (1999): 1.

Okoko, Tervil. "Kenya's Indigenous Honey Hunters Lose Their Forest Home." Environment News Service, 24 March 2000 <http://ens.lycos.com/ens/mar2000/2000L-03-24-01.html>.

Oliveira, Paulo S. "Edward O. Wilson, Doyen of

Biodiversity's Crusade." *Biotropica* 31, no. 4 (1999): 538–539.

Park, Chris C. *Tropical Rainforests*. London: Routledge, 1992.

Plotkin, Mark. "In Search of the Shamans' Vanishing Wisdom." *Time,* 14 December 1998, 71.

Quammen, David. *Song of the Dodo: Island Biogeography in an Age of Extinctions*. New York: Scribner, 1996.

Rawat, Rajiv. "Nature as Laboratory." *Cornell Science and Technology Magazine*, fall 1994 <http://www.englib.cornell.edu/Scitech/f94/eisner.html>.

Redford, Kent H., and Allyn Maclean Stearman. "Forest-Dwelling Native Amazonians and the Conservation of Biodiversity: Interests in Common or in Collision?" *Conservation Biology* 7, no. 2 (1993): 248–255.

Redford, Kent H., and Allyn Maclean Stearman. "On Common Ground?" *Conservation Biology* 7, no. 2 (1993): 427–428.

Reed, Susan. "Sorcerers' Apprentice: What Amazon Shamans Know Could Save Your Life." *People*, June 1996, 143–145.

Ressa, Maria. "Gold and Blood in the Wilderness: Indonesia Mine a Blessing and a Curse." CNN Interactive, 21 February 1996 <http://www.cgi.cnn.com/EARTH/9602/irian_jaya/index.html>.

Revkin, Andrew. *The Burning Season: The Murder of Chico Mendes and the Fight for the Amazon Rain Forest*. Boston: Houghton Mifflin, 1990.

Ridge, Mian. "India's Tiger Threatened by Land Use and Parts Demand." Reuters News Service, 26 December 1999 <http://www.forests.org/archive/asia/tigers26.txt>.

Rosenblatt, Roger. "Earth's Green Gown." *Time*, 14 December 1998, 60–65.

Runyan, Curtis. "Colombia Opts for Oil Over Indigenous Rights." *World Watch*, January–February 2000, 9.

Salmon, Katy. "Mobilising the Mothers." *People & the Planet* 8, no. 4 (1999): 22–23.

Schemo, Diana Jean. "The Last Tribal Battle." *New York Times Magazine,* 31 October 1999.

Schmidlin-Moore, Heidi. "Mapuche Family Beaten Over Forest Land." Environment News Service, 23 October 1998.

Schultes, Richard Evans. *Where the Gods Reign: Plants and Peoples of the Colombian Amazon*. London: Synergetic Press, 1988.

Schwartz, David M. "Drawing the Line in a Vanishing Jungle." *International Wildlife,* July–August 1991, 4–11.

Shoumatoff, Alex. *The World Is Burning: The Tragedy of Chico Mendes*. Boston: Little, Brown, 1990.

Skutch, Alexander. *A Naturalist amid Tropical Splendor*. Iowa City: University of Iowa Press, 1987.

Sonne, Lisa. "Paul Cox." *Time*, 11 January 1999, <http://www.pathfinder.com/time/reports/environment/heroes/tfk/0,2967,tfk_cox,00html>.

Stang, Karla. "Saving Liberia's Rainforests." *San Diego Earth Times*, 20 February 1997 <http://www.sdearthtimes.com/et0894/et0894s4.html>.

Steinmetz, George. "Irian Jaya's People of the Trees." *National Geographic*, February 1996, 35–43.

Stevens, William K. "Biologists' Deaths Set Back Plan to Assess Tropical Forests." *New York Times*, 17 August 17, C1.

Strieker, Gary. "Rainforest Tribe Takes Battle with Developers to Court." Cable News Network, 25 November 1999. <http://cnn.com/>.

Synge, Hugh. "One Man, One River; A Story of Amazon Exploration." *Rainforest Medical Bulletin* 5, no. 2 (1998) <http://www.xs4all.nl/~rainmed/bulletin/amazo2-e.html>.

Tesoro, José Manuel. "Millennium Politics and Power." CNN Asia Now, 5 November 1999 <http://www.cnn.po.com/ASIANOW/asiaweek/magazine/99/1105/leader.indonesia.html>.

Toth, Mike. "Stanford's Nuttiest Tenured Turkey." *Stanford Review*, 10 March 1998 <http://www.junkscience.com/news2/ehrlich.htm>.

Tranberg, Pernille. "Unique Biodiversity Program in Costa Rica." Earth Times News Service, 25 February 1996 <http://www.csf.colorado.edu/elan/96/feb96/0075.html>.

Turnbull, Colin M. *The Mbuti Pygmies: Change and Adaptation*. New York: Holt, Rinehart and Winston, 1983.

Wade, Nicholas. "From Ants to Ethics: A Biologist Dreams of Unity of Knowledge." *New York Times*, 12 May 1998, F1, F6.

Watkins, T.H. "The Greening of the Empire: Sir Joseph Banks." *National Geographic*, November 1996, 28–52.

———. "One Hundred Champions of Conservation." *Audubon,* November–December 1998, 121.

Wilson, Edward O. *Naturalist*. Washington, DC: Island Press, 1994.

Magazines, Newsletters, and Reports

The publications in this section do not carry articles exclusively about tropical biodiversity, but frequently publish articles about tropical forest conservation, flora, fauna, and research. Check the World Wide Web URLs listed for each publication to find out more. Nearly all require subscription fees, but a few are available free of charge on the Web.

Ambio: A Journal of the Human Environment (eight times a year); Royal Swedish Academy of Sciences, Box 50005, S-104 05 Stockholm, Sweden; <http://www.ambio.kva.se/>.

The Amicus Journal (quarterly); Natural Resources Defense Council, 20 West 20th Street, New York, NY 10011, USA; <http://www.nrdc.org>.

Arborvitae (monthly newsletter); IUCN/WWF, Avenue de Mont Blanc, CH-1106, Gland, Switzerland; <http://www.panda.org> or <http://www.iucn.org>.

Audubon (bimonthly magazine); 700 Broadway, New York, NY 10003, USA; <http://magazine.audubon.org>.

Biodiversity and Conservation (monthly journal); Dept. of Biosciences, University of Kent, Canterbury, England; <http://www.wkap.nl/journalhome.htm/0960–3115>.

BioScience (monthly journal); 1444 I Street NW, Suite 200, Washington DC 20005, USA; <http://www.aibs.org>.

Biotropica (quarterly journal); Association for Tropical Biology, Dept. of Biology, University of Missouri-St. Louis, 8001 Natural Bridge Road, St. Louis, MO 63121–4499 USA; <http://atb.botany.ufl.edu>.

Conservation Biology (bimonthly journal); Society for Conservation Biology, Blackwell Science, Inc., Commerce Place, 350 Main Street, Malden, MA 02148, USA; <http://conbio.rice.edu/scb/>.

Cultural Survival Quarterly; 221 Prospect Street, Cambridge, MA 02139, USA; <http://www.cs.org>.

E Magazine (monthly); P.O. Box 5098, Westport, CT 06881, USA; <http://www.emagazine.com>.

The Earth Times (biweekly newspaper; daily on the World Wide Web); P.O. Box 3363, Grand Central Station, New York, NY 10163, USA; <http://www.earthtimes.org/>.

Eco-Exchange (bimonthly newsletter); Rainforest Alliance, 65 Bleecker Street, New York, NY 10012, USA; <http://www.rainforest-alliance.org / programs / cmc / newsletter/ index.html>.

The Ecologist (magazine; 10 times a year); Station Road, Sturminster Newton, Dorset, England DT101BB; <http://www.gn.apc.org/ecologist/>.

Environment (monthly journal); 1319 18th Street NW, Washington, DC 20036-1802, USA; <http://www.heldref.org/html/body_env.html>.

Focus (monthly newsletter); World Wildlife Fund, 1250 24th Street NW, Washington, DC 20037–1175, USA; <http://www.worldwildlife.org>.

International Wildlife (bimonthly magazine); National Wildlife Federation, 8925 Leesburg Pike, Vienna, VA 22184, USA; <http://www.nwf.org/nwf/intlwild/>.

The Living Bird (quarterly magazine); Cornell Laboratory of Ornithology, 159 Sapsucker Woods Road, Ithaca, NY 14850, USA; <http://www.ornith.cornell.edu/Publications/livingbird/>.

Lori Journal International (quarterly journal); Parrot Society of Australia, PO Box 75, Salisbury, Queensland 4107, Australia; <http://www.parrotsociety.org.au>.

National Geographic (monthly magazine); 1145 17th Street NW, Washington, DC 20036, USA; <http://www.nationalgeographic.com/ngm/index.html>.

Natural Resource Perspectives (occasional brief reports); Overseas Development Institute, Portland House, Stag Place, London SW1E 5DP, England; <http://www.oneworld.org/odi/nrp/>.

Nature (weekly journal); Porters South, 4 Crinan Street, London, England N19XW; <http://www.nature.com>.

Nature Conservancy (monthly magazine); The Nature Conservancy, 4245 North Fairfax Drive, Suite 100, Arlington, VA 22203, USA; <http://www.tnc.org>.

New Scientist (weekly magazine); 151 Wardover Street, London W1F 8WE, England; <http://www.newscientist.com.uk/>.

Orion Magazine (quarterly); P.O. Box 195, Berrington, MA 07230, USA; <http://www.orionsociety.org/orion.html>.

People & the Planet (special issues online); Suite 112, Spitfire Studios, 63–71 Collier Street, London N1 9BE, England; <http://www.peopleandplanet.net>.

Science (weekly journal); American Association for the Advancement of Science, 1200 New York Avenue NW, Washington, DC 20005, USA; <http://www.sciencemag.org/>.

Science News (weekly magazine); 1719 N Street NW, Washington, DC 20036, USA; <http://www.sciencenews.org>.

State of the World Series (annual); Worldwatch Institute, 1776 Massachusetts Avenue NW, Washington, DC 20036, USA; <http://www.worldwatch.org>.

Tropinet (quarterly newsletter online); Association for Tropical Biology, National Museum of Natural History, and Organization for Tropical Studies, Duke University, Box 90630, Durham, NC 27708–0630, USA; <http://www.atb.botany.ufl.edu/atb/>.

Wildlife Conservation (monthly magazine); Wildlife Conservation Society, 2300 Southern Boulevard, Bronx, NY 10460, USA <http://www.wcs.org/news/magazine>.

World Resources Report (annual); World Resources Institute, P.O. Box 4852 Hampden Station, Baltimore, MD 21211, USA; <http://www.wri.org>.

World Watch (bimonthly magazine); World Watch Institute, 1776 Massachusetts Avenue NW, Washington, DC 20036, USA; <http://www.worldwatch.org>.

Videos

The videos noted here are among the best available for a reasonable fee. Most are geared toward adult audiences, but 4 of the 38 videos listed (*Jaguar Trax, Songbird Story, A Walk in the Rainforest,* and *You Can't Grow Home Again*) are for schoolchildren. Contact the producers or check the Web site URLs for more information.

The Amazon: A Vanishing Rainforest. 1988; 29 minutes; videotape.
Focuses on the work of the National Institute of Amazon Research and shows how development poses a threat to the Amazon's ecosystem.
Cinema Guild; 212/246–5522; <http://www.cinemaguild.com>.

Amazon Journal. 1995; 58 minutes; videotape.
Begins with the assassination of Chico Mendes in 1988 and ends with a return trip to Yanomami Territory in 1995.
Filmmakers Library; 212/808–4980; <http://www.filmakers.com>.

Amazonia: A Burning Question. 1984; 58 minutes; videotape.
Surveys efforts to understand and conserve the South American Amazon.
State University of New York; 516/444–3132.

Amazonia: The Road to the End of the Forest. 1991; 96 minutes; videotape.
Reveals the failure of a massive resettlement program in the Amazon Basin, which lured millions of colonists to the region with the promise of free farmland.
Filmmakers Library; 212/808–4980; <http://www.filmakers.com>.

Amazonia: Voices from the Rainforest. 1991; 70 minutes; videotape.
Interviews with indigenous people of the Amazon, as well as riverine dwellers, rubber tappers, and small farmers. Also available with a 92-page resource book.
The Video Project; 831/336–0160; E-mail: video@videoproject.net; <http://www.videoproject.org>.

Arrows Against the Wind. 1992; 52 minutes; videotape.
West Papua, also known as Irian Jaya, is the "Amazon of Asia," where numerous indigenous tribes have lived for 25,000 years. Filmed in Irian Jaya, this documentary is the story of two tribes, the Dani and the Asmat, and their social, political and environmental upheaval.

Bullfrog Films; 800/543-FROG; E-mail: info@bullfrogfilms.com; <http://www.bullfrogfilms.com>.

Black Water. 1990; 28 minutes; videotape.
Tells the story of the people of Sao Braz and their struggle to survive the pollution from upstream factories and mills.
First Run/Icarus Films; 800/876–1710; E-mail: info@Irif.com; <http://www.frif.com>.

Blowpipes and Bulldozers. 1988; 60 minutes; videotape.
Recounts the struggle for survival of Borneo's nomadic Penan people, who, after 40,000 years of living in the forests of Sarawak, are being logged out of existence.
Bullfrog Films; 800/543-FROG; E-mail: info@bullfrogfilms.com; <http://www.bullfrogfilms.com>.

Burning Rivers. 1992; 28 minutes; videotape.
Examines the links between environmental and social crisis in Guatemala. Also available with 32-page discussion guide.
The Video Project; 831/336–0160; E-mail: project@videoproject.net; <http://www.videoproject.org>.

Cry of the Forgotten Land. 1995; 26 minutes; videotape.
Reveals the little-known destruction of one of the world's most magnificent rainforests in Papua New Guinea, home to ancient human cultures and more than 100,000 species found nowhere else. A study of the forests and the ancient Moi culture, whose existence hangs in the balance as the rainforest, which has sustained them for over 2500 years, is being rapidly eliminated.
The Video Project; 831/336–0160; E-mail: video@videoproject.net; <http://www.videoproject.org>.

Cry of the Muriqui. 1982; 28 minutes; film and videotape.
Relates the story of endangered monkeys in the largely deforested mountains of southeastern Brazil, north of Rio de Janeiro.
State University of New York; 516/444–3132.

The Decade of Destruction. 1990; series of five videotapes.

Highlights the conflicts that erupt when indigenous tribes encounter outsiders eager to exploit forest resources.

Part 1: "In the Ashes of the Forest, Part 1." 1990; 55 minutes.

Saga of two colonists in the Amazon rainforest and their interaction with the Uru Eu Wau Wau.

Part 2: "In the Ashes of the Forest, Part 2." 1990; 57 minutes.

Concludes the colonists' struggle. In retaliation for the clearing of their land, the Uru Eu Wau Wau indigenous people kidnap a colonist's son. A team with Brazil's Indian Agency searches for the hostage.

Part 3: "The Killing of Chico Mendes." 1990; 55 minutes. The story of Chico Mendes, whose brutal murder on December 22, 1988, provoked international protest and brought worldwide attention to the problem of Amazonian deforestation.

Part 4: "Killing for Land." 1990; 51 minutes.

Squatters face off against gunmen hired by absentee landlords in the Amazon.

Part 5: "Mountains of Gold." 1990; 54 minutes.

The camera follows Jova, a prospector famous among his colleagues for his illegal gold strikes, as he plays hide-and-seek with the security forces of Brazil's largest mining multinational.

Bullfrog Films; 800/543–FROG; E-mail: info@bullfrog films.com; <http://www.bullfrogfilms.com>.

Earth First: The Struggle for the Australian Rainforest. 1989; 58 minutes; videotape.

Documents the fight to save 17,000 acres of Australian rainforest.

The Video Project; 415/655–9050; E-mail: video@ videoproject.net; <http://www.videoproject.org>.

Equatorial River: The Amazon. 1987; 23 minutes; videotape. Explains how the water and nutrient cycles work in the Amazon basin.

Bullfrog Films; 800/543–FROG; E-mail: info@ bullfrog-films.com; <http://www.bullfrogfilms.com>.

A Future for Forests. 1993; 25 minutes; videotape.

Examines the aims of the Forest Stewardship Council to promote responsible forest management by monitoring the international trade of timber. In an effort to prevent the disastrous ecological consequences of deforestation and the increase in illegal logging operations, the council has instituted a sophisticated plan—utilizing handheld computers, laser scanners, and barcode identification for logs—whereby consumers can be assured of the ecological validity of their timber purchase.

Cinema Guild; 212/246–5522; <http://www.cinema guild.com>.

Goodwood. 1998; 45 minutes; videotape.

Looks at four communities that are managing nearby forests, harvesting timber sustainably.

Bullfrog Films; 800/543–FROG; E-mail: info@ bullfrog-films.com; <http://www.bullfrogfilms.com>.

Halting the Fires. 1991; 52 minutes; videotape.

Brazilian director Octavio Bezerra looks beyond the continuing tragedy of deforestation in his country to explore the practical, political, and social problems involved. The film focuses on different groups (ranchers, miners, rubber tappers, loggers, indigenous peoples, and colonists) and how their activities affect the environment.

Filmmakers Library; 212/808–4980; <http://www. filmakers.com>.

In a Time of Headlong Progress. 1994; 45 minutes; videotape. A comprehensive look at the economic and social factors behind deforestation, and the hopeful, innovative work of pioneering Brazilian conservationist Cristina Alves, who is working to establish an indigenous conservation ethic.

The Video Project; 831/336–0160; E-mail: video@ videoproject.net; <http://www.videoproject.org>.

Jaguar Trax. 1998; 26 minutes; videotape.

Featuring a talented cast of kids, this is an entertaining, fact-based tale designed to teach the value of tropical rainforests, biodiversity, and sustainably grown products. For ages 10–15; study guide included.

Video Project; 800/4–PLANET; E-mail: video@ videoproject.net; <http://www.videoproject.org>.

Journey into Amazonia. 2000; 180 minutes; two videotapes.

Captures the spirit of the world's largest tropical wilderness—the rainforest of the Amazon Basin—and offers a closer look at its swollen rivers, flooded forests, dense canopy, and wealth of wildlife.

PBS Films; 800/645–4PBS; <http://www.shop.pbs.org>.

Jungle Pharmacy. 1989; 53 minutes; videotape. Dedicated physicians, scientists, and environmentalists work with indigenous shamans, or healers, to tap the rainforest for vital new drugs and medicines.

Cinema Guild; 212/246–5522; <http://www.cinema guild.com>.

The Last Forest. 1989; 20 minutes; videotape.

Portrays one teenager's awakening to the importance of tropical rainforests and how her life will be different if the destruction is not stopped.

CARE Film Unit; 404/681–2552; E-mail: info@care.org; <http://www.care.org>.

The Living Edens: Borneo, Island in the Clouds. 1999; 60 minutes; videotape.

Explores the magnificent wildlife of Borneo's rainforests and mountains.

PBS Home Video; 800/645–4PBS; <http://shop.pbs. org/>.

The Living Edens: Madagascar, A World Apart. 1998; 60 minutes; videotape.

A view of the sun-scorched deserts, luxurious tropical forests, and amazing wildlife of this island nation.

PBS Home Video; 800/645–4PBS; <http://shop.pbs. org/>.

The Living Edens: Manu, Peru's Hidden Rainforest. 1997; 60 minutes; videotape.

A tour of Manu and its wildlife.

PBS Home Video; 800/645–4PBS; <http://shop.pbs. org/>.

Microchip al Chip. 1991; 18 minutes; videotape.

Examines the destruction of Chilean forest in order to sustain its paper exports to other nations, principally Japan. First Run/Icarus films; 212/727–1711; E-mail: info@frif.com; <http://www.frif.com>.

Rain Forests: Proving Their Worth. 1990; 30 minutes; video tape.

Explores the movement to sustainably collect and market products from tropical forests, including foods and fibers. Video Project; 800/4–PLANET; E-mail: video@videoproject.net; <http://www.videoproject.org>.

Saviors of the Forest. 1994; 90 minutes; videotape.

Tired of filming TV commercials, two well-intentioned Los Angeles "camera guys" decide to do their part for the environment by exposing the villains responsible for destroying the rainforests, so they innocently pack up their video cameras and head to South America looking for the bad guys.

The Video Project; 831/336–0160; E-mail: video@videoproject.net; <http://www.videoproject.org>.

Secrets of the Chocó. 1995; 52 minutes; videotape.

Looks at the choices for development facing the Chocó in Colombia, one of the largest relatively unspoiled rainforests on the planet, which still hides thousands of undiscovered plants and animals, and contains undisturbed ancient forests and shorelines. The future of Chocó and its inhabitants is threatened by large-scale development plans.

Bullfrog Films; 800/543–FROG; E-mail: info@bullfrogfilms.com; <http://www.bullfrogfilms.com>.

Songbird Story. 1994; 13 minutes; videotape.

In an animated dream, two young children fly along with the songbirds on one of their migration paths to the tropical rainforest. While there, they see how quickly the rainforests in Central and South America are being cut down and understand that without the forests, migratory songbirds will have no place to survive through the winter. For grades K–6.

Bullfrog Films; 800/543–FROG; E-mail: info@bullfrogfilms.com; <http://www.bullfrogfilms.com>.

Trekking the Mayan Ruins in the Guatemalan Rainforest. 2000; 30 minutes; videotape.

Kim Batchelder, a forester with the Rainforest Alliance, leads a trek through Guatemala's Tikal National Park, describing the magnificent Maya ruins and the abundance of tropical flora and fauna found there. In addition to discovering many of Tikal's ecological and archeological wonders, viewers can learn helpful dos and don'ts about hiking and camping in the rainforest. Trailside; 800/872–4574; <http://www.trailside.com>.

Trinkets and Beads. 1996; 52 minutes; videotape.

The story of how the U.S. petroleum company Maxus set out to convince the Huaorani, an indigenous group of the Ecuadorian Amazon, to allow drilling on their land. Filmed over two years, it reveals the story of the battle waged by a small band of Amazonian warriors to preserve their way of life.

First Run/Icarus films; 212/727–1711; E-mail: info@frif.com; <http://www.frif.com>.

A Walk in the Rainforest. 1990; 11 minutes; videotape.

Eight-year-old Jason Harding of Belize takes two friends on a guided tour of the rainforests in his Central American country. For grades Pre-K–6.

Bullfrog Films; 800/543–FROG; E-mail: info@bullfrogfilms.com; <http://www.bullfrogfilms.com>.

Yanomami: Keepers of the Flame. 1992; 58 minutes; videotape.

Documents an expedition by a group of journalists, anthropologists, and doctors who journeyed to the Venezuelan rainforest to visit a Yanomami indigenous settlement never contacted by the outside world.

The Video Project; 831/336–0160; E-mail: videovideoproject.net; <http://www.videoproject.org>.

You Can't Grow Home Again. 1995; 58 minutes; videotape.

Provides an excellent introduction for younger children to the rainforests and the concept of biodiversity. This video goes on location to Costa Rica for a friendly view of the rainforest and some of the animals that live there. For grades 2–7.

The Video Project; 831/336-0160; video@videoproject.net; <http://www.videoproject.org>.

Embassies and Government Agencies

If you need specific information about a tropical country, try contacting the country's embassy in Washington or the environmental agency, usually located in the capital city. Not all have e-mail addresses, but they are listed whenever possible. You can also use this information when you want to write a letter of protest—or congratulations—to a country's government. Two Web sites that can steer you to good information about individual countries are: <http://www.embassyworld.com> and <http://www.worldrover.com>.

AFRICA

Angola

Embassy of the Republic of Angola
1615 M Street NW, Suite 900
Washington, DC 20036, USA
Tel: 202/785–1156
Fax: 202/785–1258
E-mail: angola@angola.org

Ministry of Agriculture
Avenida Comandante
Gika 2nd floor
Caixa postal 527
Luanda, Angola
Tel: 244/32–2694
Fax: 244/32–0553

Benin

Embassy of Benin
2737 Cathedral Avenue NW
Washington, DC 20008, USA
Tel: 202/232–6656
Fax: 202/265–1996

Ministry of the Environment and Habitat
PO Box 01–3621
Cotonou, Republic of Benin, West Africa

Tel: 229/31–4661
Fax: 229/31–5081

Burundi

Embassy of Burundi
2233 Wisconsin Avenue NW, Suite 212
Washington, DC 20007, USA
Tel: 202/342–2574

Ministry of Land Use Planning and Environment
Bujumbura, Burundi
Tel: 257/22–0626
Fax: 257/22–8902
E-mail: igebu@cbinf.com

Cameroon

Embassy of Cameroon
2349 Massachusetts Avenue NW
Washington, DC 20008, USA
Tel: 202/265–8790

Ministry in Charge of the Environment and Forest
Yaounde, Cameroon
Tel: 237/22–1454
Fax: 237/22–9489

Central African Republic

Embassy of Central Africa
2002 R Street NW
Washington, DC 20009, USA
Tel: 202/462–4009
Fax: 202/265–1937

Ministry of the Environment and Water
B.P. 80
N'Djamena, Central Africa
Tel: 235/52–6012
Fax: 235/52–3829

Congo (Democratic Republic)

Embassy of Democratic Republic of Congo
1800 New Hampshire Avenue NW
Washington, DC 20009, USA
Tel: 202/234–7690, 202/234–7691
Fax: 202/686–3631

Congo (Republic)

Embassy of Republic of Congo
4891 Colorado Avenue NW
Washington, DC 20011, USA
Tel: 202/726–5500
Fax: 202/726–1860

Ministry of Mining, Industry, and Environment
BP 2124
Brazzaville, Republic of Congo
Tel: 242/81–0291
Fax: 242/81–2611

Côte d'Ivoire

Embassy of Côte d'Ivoire
2412 Massachusetts Avenue NW
Washington, DC 20008, USA
Tel: 202/797–0313

Ministry of Environment
Abidgan, Côte d'Ivoire
Tel: 225/21–1671
Fax: 225/21–4071

Djibouti

Embassy of Djibouti
1156 15th Street NW, Suite 515

Washington, DC 20005, USA
Tel: 202/331–0270
Fax: 202/331–0302

Ministry of Habitat, Urbanism, and the Environment
PO Box 6
Republic of Djibouti, East Africa
Tel: 253/35–0006
Fax: 253/35–3840

Ethiopia

Embassy of Ethiopia
2134 Kalorama Road NW
Washington, DC 20008 USA
Tel: 202/234–2281
Fax: 202/483–8407
E-mail: info@ethiopianembassy.org

Minister of Water Resources
PO Box 1034
Addis Ababa, Ethiopia
Tel: 2511/51–0455
Fax: 2511/51–3042

Gabon

Embassy of Gabon
2034 20th Street NW, Suite 200
Washington, DC 20009, USA
Tel: 202/797–1000
Fax: 202/332–0668

Ministry of Environment and Tourism
B.P. 803
Libreville, Gabon
Tel: 241/74–7196

The Gambia

Embassy of the Gambia
1155 15th Street NW, Suite 1000
Washington, DC 20005, USA
Tel: 202/785–1399
Fax: 202/785–1430

Ministry of Forestry and Environment
Bakau Fajara
The Republic of Gambia
Tel: 220/22–7537
Fax: 220/22–7339

Ghana

Embassy of Ghana
3512 International Drive NW
Washington DC, 20008 USA
Tel: 202/686–4520

Ministry of Lands and Forestry
PO Box M212
Accra, Ghana
Tel: 233/2166–5421

Guinea

Embassy of Guinea
2112 Leroy Place NW
Washington, DC 20008, USA
Tel: 202/483–9420
Fax: 202/483–8688

Ministry of Public Works and Environment
Boulevard of Commerce
Conakry, Republic of Guinea
Tel: 224/44–4024

Guinea-Bissau

Embassy of Guinea-Bissau
918 16th Street NW, Mezzanine Suite
Washington, DC 20006, USA
Tel: 202/872–4222
Fax: 202/872–4226

Ministerio do Desenvolvimentio Rural
does Recursos Naturais e do Ambiente
Bissau, Republic da Guinea-Bissau
Tel: 245/21–1756

Kenya

Embassy of Kenya
2249 R Street NW
Washington, DC 20008 USA
Tel: 202/387–6101
Fax: 202/462–3829

Ministry of Environment & Natural Resources
PO Box 30126
Nairobi, Kenya
Tel: 254/222–9261
Fax: 254/233–8272

Liberia

Embassy of the Republic of Liberia
5201 16th Street NW
Washington, DC 20011, USA
Tel: 202/723–0437
Fax: 202/723–0436
E-mail: info@liberiaemb.org

Ministry of Lands, Mines, and Energy
PO Box 10–9024
Capitol Hill
1000 Monrovia 10, Liberia
Tel: 231/22–6281

Madagascar

Embassy of Madagascar
2374 Massachusetts Avenue NW
Washington, DC 20008, USA
Tel: 202/265–5525
E-mail: malagasy@embassy.org

Ministry of the Environment
B.P. 651 Anosy
Antananarivo 101, Madagascar
Tel: 261/22–4710

Malawi

Embassy of Malawi
2408 Massachusetts Avenue NW
Washington, DC 20008, USA
Tel: 202/797–1007

Ministry of Forestry, Fisheries, and Environmental Affairs
PO Box 30048
Lilongwe 3
Malawi, Central Africa
Tel: 265/78–1000
Fax: 265/78–4268

Mauritius

Embassy of the Republic of Mauritius
4301 Connecticut Avenue NW, Suite 441
Washington DC 20008, USA
Tel: 202/244–1491
Fax: 202/966–0983
E-mail: Mauritius.Embassy@WIX.com

Ministry of Environment and Employment
Treasury Building
Port Louis, Mauritius
Indian Ocean
Tel: 230/208–0281
Fax: 230/201–1266

Mozambique

Embassy of Mozambique
1990 M Street NW, Suite 570
Washington, DC 20036, USA
Tel: 202/293–7146
Fax: 202/835–0245

**Ministry for Co-ordination of Environmental
    Affairs**
Avenida Acordos de Lusaka 2115
C.P. 2020
Maputo, Mozambique
Tel: 258/146–5848
Fax: 258/146–5849
E-mail: MICOA@ambinet.uem.mz

Nigeria

Embassy of Nigeria
1333 16th Street NW
Washington, DC 20036, USA
Tel: 202/986–8400
Fax: 202/775–1385

Federal Environmental Protection Agency
FEPA Office Complex
Independence Way, South
PMB, 265
Garki, Abuja, Nigeria
Tel: 234/9–523–3379

Rwanda

Embassy of the Republic of Rwanda
1714 New Hampshire Avenue NW
Washington, DC 20009, USA
Tel: 202/232–2882
Fax: 202/232–4544
E-mail: rwandemb@rwandemb.org

Senegal

Embassy of Senegal
2112 Wyoming Avenue NW

Washington, DC 20008, USA
Tel: 202/234–0540, 202/234–0541

Ministry of Environment and Natural Protection
Building administratif, 7 ème étage
BP 4055
Dakar, Sénégal
Tel: 221/821–1240
Fax: 221/822–2180

Seychelles

Ministry of the Environment and Transport
Botanical Gardens
PO Box 445
Mont Fleuri, Mahe, Seychelles
Tel: 248/224–4644

**Permanent Mission of Seychelles
to the United Nations**
820 Second Avenue, Suite 900F
New York, NY 10017, USA
Tel: 212/972–1785
Fax: 212/972–1786

Sierra Leone

Embassy of Sierra Leone
1701 19th Street NW
Washington, DC 20009
Tel: 202/939–9261
Fax: 202/483–1793

**Ministry of Agriculture, Forestry, and
    Environment**
Yonyi Building, Brookfields Freetown
Sierra Leone, West Africa
Tel: 232/2222–2907

Tanzania

Embassy of Tanzania
2139 R Street NW
Washington, DC 20008, USA
Tel: 202/939–6125
Fax: 202/797–7408

Ministry of Natural Resources
PO Box 9372
Dar es Salaam, Tanzania
Tel: 255/5111–6681
Fax: 255/5113–2302

Togo

Embassy of Togo
2208 Massachusetts Avenue NW
Washington, DC 20008, USA
Tel: 202/234–4212
Fax: 202/232–3190

Ministry of Environment and Tourism
Lome, Togo
Tel: 228/21–5285

Uganda

Embassy of Uganda
5911 16th Street NW
Washington, DC 20011, USA
Tel: 202/726–7100
Fax: 202/726–1727
E-mail: ugaembassy@rocketmail.com

Ministry of Natural Resources
Attn: Department of Environment Protection
PO Box 7172
Kampala, Uganda
Tel: 25641/23–3331
Fax: 25641/23–0220

Zimbabwe

Embassy of Zimbabwe
1608 New Hampshire Avenue NW
Washington DC 20009, USA
Tel: 202/332–7100
Fax: 202/483–9326

Ministry of Mines, Environment, and Tourism
14th Floor Karigamombe Centre
Private Bag 7753
Causeway, Harare, Zimbabwe
Tel: 263/475–1720
Fax: 263/475–7877

ASIA AND THE PACIFIC

American Samoa

American Samoa Environmental Protection Agency
American Samoa Government
Pago Pago, American Samoa 96799

Tel: 684/633–2304
Fax: 684/633–5801

Australia

Embassy of Australia
1601 Massachusetts Avenue NW
Washington, DC 20036, USA
Tel: 202/797–3000
Fax: 202/797–3168

Environment Australia
GPO Box 787
Canberra Act 2601, Australia
Tel: 612/627–41292
Fax: 612/627–41970
E-mail: ciu@ea.gov.au
<http://www.environment.gov.au>

Bangladesh

Embassy of Bangladesh
2201 Wisconsin Avenue NW, Suite 300
Washington DC 20007, USA
Tel: 202/342–8372
Fax: 202/333–4971
E-mail: BanglaEmb@aol.com
<http://members.aol.com/banglaemb/index.html>

Ministry of the Environment and Forests
Bangladesh Secretariat
Dhaka–1000, Bangladesh
Tel: 880/286–5436
Fax: 880/286–9210
<http://www.bangladeshonline.com/gob/>

Brunei

Embassy of Brunei Darussalam
Watergate Suite 300
2600 Virginia Avenue NW
Washington DC 20037, USA
Tel: 202/342–0159
Fax: 202/342–0158

Ministry of Development
Environmental Unit, 4th Floor
Old Airport Road
Bandar Seri Begawan BB 3510
Negara Brunei Darussalam
Tel: 673/238–3222

Fax: 673/238–3644
E-mail: modenv@brunet.bn

Cambodia

Embassy of Cambodia
4500 16th Street, NW
Washington DC 20011, USA
Tel: 202/726–7742
Fax: 202/726–8381
<http://www.embassy.org/cambodia>

Ministry of Environment
48 Preah Sihanouk
Tonle Bassac, Chamkar Morn
Phnom Penh, Cambodia
Tel: 85523/72–1073
Fax: 85523/72–1073
E-mail: moedncp@forum.org.kh

Federated States of Micronesia (Chuuk, Kosrae, Pohnpei, Yap)

Embassy of the Federated States of Micronesia
1725 N Street NW
Washington DC 20036, USA
Tel: 202/223–4383
Fax: 101/223–4391

Department of Health Services
FSM National Government
PO Box PS70
Palikir, Pohnpei FM 96941
Tel: 691/320–2872
Fax: 691/320–5263

Fiji

Embassy of the Republic of the Fiji Islands
2233 Wisconsin Avenue NW, Suite 240
Washington DC 20007, USA
Tel: 202/337–8320
Fax: 202/337–1996

Ministry of Agriculture, Fisheries, and Forests
Department of Forests
PO Box 2218
Government Buildings
Suva, Fiji Islands
Tel: 679/30–1611
Fax: 679/30–1595

Guam (USA)

Guam Environmental Protection Agency
PO Box 2950
Agana, Guam 96910

India

Embassy of India
2107 Massachusetts Avenue NW
Washington DC 20008, USA
Tel: 202/939–7000
Fax: 202/265–4351
<http://www.indianembassy.org/>

Ministry of Environment and Forests
Government of India
Paryawaran Bhawan
CGO Complex, Lodhi Road
New Delhi-110 003, India
Tel: 9111/436–1509
Fax: 9111/436–3957
E-mail: secy@envfor.delhi.nic.in

Indonesia

Embassy of the Republic of Indonesia
2020 Massachusetts Avenue NW
Washington DC 20036, USA
Tel: 202/775–5200
Fax: 202/775–5365
E-mail: info@kbri.org
<http://www.kbri.org>

Ministry of Forestry and Estate Crops
Gedung Mamggala Wanabakti Blok, 4th floor
Jl. Eatot Subroto
Jakarta 10270, Republic of Indonesia
Tel: 6221/570–4501
Fax: 6221/570–0226
<http://www.dephut.go.id>

Laos

Embassy of Laos
2222 S Street NW
Washington DC 20008, USA
Tel: 202/332–6416
Fax: 612/332–4923
<http://www.laosembassy.com>

Science and Technology and Environment Organization
PO Box 2279
Vientiane, Laos
Tel: 85621/21–3470
Fax: 85621/21–8874
E-mail: somTel@steno.gov.la

Malaysia

Department of Environment Malaysia
Ministry of Science, Technology and the Environment
Tingkat 12 & 13, Wisma Sime Darby
Jalan Raja Laut, 50662 Kuala Lumpur, Malaysia
Tel: 603/294–7844
Fax: 603/293–1480
E-mail: doe@jas.sains.my

Embassy of Malaysia
2401 Massachusetts Avenue NW
Washington DC 20008, USA
Tel: 202/328–2700
Fax: 202/483–7661

Maldives

Ministry of Home Affairs, Housing, and Environment
Republic of Maldives
Tel: 960/32–1752
Fax: 960/32–4739
E-mail: env@environment.go.mv

Marshall Islands

Environmental Protection Agency
Tel: 692/625–3035
Fax: 692/625–5202
E-mail: eparmi@ntamar.com

Republic of Marshall Islands Embassy
2433 Massachusetts Avenue NW
Washington DC 20008, USA
Tel: 202/234–5414
Fax: 202/232–3236
E-mail: info@rmiembassyus.org
<http://rmiembassyus.org>

Northern Mariana Islands

Department of Environmental Quality
Department of Public Works
Lower Base
Saipan MP 96950
Tel: 670/234–1011
Fax: 670/234–1003

Papua New Guinea

Embassy of Papua New Guinea
1615 New Hampshire Avenue NW, 3rd Floor
Washington, DC 20009, USA
Tel: 202/745–3680
Fax: 202/745–3679

Ministry of Environment and Conservation
PO Box 6601
Boroko, NCD Papua New Guinea
Tel: 675/327–1692
Fax: 675/325–0433

Philippines

Department of Environment & Natural Resources
Visayas Avenue, Diliman
Quezon City, Metro-Manila, Philippines
Tel: 632/929–6626
Fax: 632/920–4352

Embassy of the Philippines
1600 Massachusetts Avenue NW
Washington DC 20036, USA
Tel: 202/467–9300
Fax: 202/328–7614

Samoa

Embassy of Samoa
800 Second Avenue, Suite 400D
New York, NY 10017, USA
Tel: 212/599–6196
Fax: 212/599–0797

Ministry of Land Survey and Environment
Government Building Private Bag
Apia, Independent State of Samoa
Tel: 685/22–481
Fax: 685/23–176

Sri Lanka

Embassy of Sri Lanka
2148 Wyoming Avenue NW
Washington DC 20008, USA
Tel: 202/483–4025
Fax: 202/232–7181

Ministry of Forestry and Environment
6th Floor, Unity Plaza Building
Colombo 04, Sri Lanka
Tel: 941/58–8274
Fax: 941/50–2566

Solomon Islands

Embassy of the Solomon Islands
800 Second Avenue, Suite 400L
New York, NY 10017, USA
Tel: 212/599–6192
Fax: 212/661–8925

**Ministry of Forestry, Environment, and
    Conservation**
PO Box G24
Honiara, Solomon Islands
Tel: 677/25–849
Fax: 677/21–245
E-mail: mosesb@welkam.solomon.com.sb

Thailand

Embassy of Thailand
1024 Wisconsin Avenue NW, Suite 202
Washington DC 20007, USA
Tel: 202/467–6790
Fax: 202/429–2949

Ministry of Science, Technology, and Environment
60/1 Soi Phibun Wattana
Rama VI Road
Ratchathewi, Bangkok 10400, Thailand
Tel: 66/2–246–0064 x146
Fax: 66/2–246–5146
E-mail: helpdesk@moste.go.th

Royal Forest Department
61 Thanon Phahon Yothin
Chatuchak, Bangkok 10900, Thailand
Tel: 662/561–4292
Fax: 662/579–2036

Tonga

Ministry of Lands, Survey and Natural Resources
PO Box 5
Nuku'alofa, Tonga
Tel: 676/23–611
Fax: 676/23–888

Vietnam

Embassy of Vietnam
1233 20th Street NW, Suite 400
Washington, DC 20036, USA
Tel: 202/861–0737
Fax: 202/861–0917
E-mail: vietnamembassy@msn.com
<http://www.vietnamembassy-usa.org>

**Ministry of Science, Technology, and
Environment of Viet Nam**
39 Tran Hung Dao St.
Ha No, Vietnam
Tel: 844/825–2731
Fax: 844/825–2733

LATIN AMERICA AND THE CARIBBEAN

Antigua & Barbuda

Embassy of Antigua and Barbuda
3216 New Mexico Avenue NW
Washington, DC 20016, USA
Tel: 202/362–5122
Fax: 202/362–5225

Ministry of Tourism, Culture, and Environment
Queen Elizabeth Highway
St. John's, Antigua
Tel: 268/462–1960
Fax: 268/462–2836
E-mail: minofeco@candw.ag

Argentina

Embassy of Argentina
1600 New Hampshire Avenue NW
Washington, DC 20009, USA
Tel: 202/939–6400
Fax: 202/332–3171

Ministry of Public Health and the Environment
Defensa 192, 4th Piso
1065 Capital Federal
Buenos Aires, Argentina
Tel: 541/41–9721

Aruba

Ministry of Economic Affairs, Tourism, Culture, and Environment
LG Smith Boulevard 76
Oranjestad, Aruba
Tel: 297/839–079
Fax: 297/839–693

Barbados

Barbados Ministry of Health and the Environment
Sir Frank Walcott Building, Fourth Floor
Culloden Road, St. Michael, Barbados
Tel: 246/431–7680
Fax: 246/437–8859
E-mail: envdivn@mail.caribsurf.com

Embassy of Barbados
2144 Wyoming Avenue NW
Washington, DC 20008, USA
Tel: 202/939–9200
Fax: 202/332–7467

Belize

Embassy of Belize
2535 Massachusetts Avenue NW
Washington, DC 20008, USA
Tel: 202/332–9636
Fax: 202/332–6888
E-mail: belize@oas.org

Ministry of Natural Resources
Belmopan, Belize
Tel: 501/882–2630
Fax: 202/882–2333

Bolivia

Embassy of Bolivia
3014 Massachusetts Avenue NW
Washington, DC 20008, USA
Tel: 202/483–4410

Fax: 202/328–3712
E-mail: bolemb2@erols.com

Ministry of Sustainable Development and Planning
PO Box 3116
La Paz, Bolivia
Tel: 5912/39–0630
Fax: 5912/36–9304

Brazil

Embassy of Brazil
3006 Massachusetts Avenue NW
Washington, DC 20008, USA
Tel: 202/238–2700
Fax: 202/ 238–2827

Chile

Embassy of Chile
1732 Massachusetts Avenue NW
Washington, DC 20006, USA
Tel: 202/785–1746
Fax: 202/887–5579

National Planning Office
Ahumeda 48
Santiago, Chile
Tel: 562/72–2033

Colombia

Embassy of Colombia
2118 Leroy Place NW
Washington, DC 20007, USA
Tel: 202/387–8338
Fax: 202/232–8643

National Institute for the Development of Natural Renewable Resources and the Environment
Apartado Aereo 13458
Bogota, Colombia
Tel: 571/285–4417

Costa Rica

Embassy of Costa Rica
2114 S Street NW
Washington, DC 20008, USA
Tel: 202/234–2945
Fax: 202/265–4795

Ministry of Environment and Energy
PO Box 10104–1000
San Jose, Costa Rica
Tel: 506/257–1417
Fax: 506/257–0697
<http://www.minae.go.cr>

Cuba

Cuban Interest Section
Swiss Embassy
2630 16th Street NW
Washington, DC 20009, USA
Tel: 202/797–8518
Fax: 202/797–8521

**National Commission for the Protection of the
    Environment and the Conservation of Natural
    Resources**
Capitolio Nacional Industrial y San Jose
Municipio Habana Vieja, Havana, Cuba
Tel: 537/61–2440

Dominica

Consulate General of Dominica
Second Avenue, Suite 900B
New York, NY 10017, USA
Tel: 212/599–8478
Fax: 212/808–4975
E-mail: dmaun@undp.org

Ministry of Agriculture and the Environment
Government Headquarters
Roseau, Commonwealth of Dominica
Tel: 767/448–2401
Fax: 767/448–7999

Dominican Republic

Department of National Parks
PO Box 2487
Santo Domingo, Dominican Republic
Tel: 809/685–1316

Embassy of the Dominican Republic
1715 22nd Street NW
Washington DC 20008, USA
Tel: 202/332–6280
Fax: 202/265–8057

Ecuador

Embassy of Ecuador
2535 15th Street NW
Washington, DC 20009, USA
Tel: 202/234–7200
Fax: 202/667–3482
E-mail: mecuawaa@pop.erols.com

Ministry of Natural Resources and Energy
Santa Prisca 223
Quito, Ecuador
Tel: 5932/57–0090

El Salvador

Embassy of El Salvador
2308 California Street NW
Washington, DC 20008, USA
Tel: 202/331–4032
Fax: 202/328–0563

Ministry of Agriculture and Livestock
Department of Renewable Resources, Forestry and
    Wildlife Services
National Parks and Wildlife Unit
Apartado Postal 2265
Soyapango, San Salvador, El Salvador
Tel: 503/227–0622

Grenada

Embassy of Grenada
1701 New Hampshire Avenue NW
Washington, DC 20009, USA
Tel: 202/265–2561
Fax: 202/265–2468

Ministry of Health and Environment
V Carenage
St. George's, Grenada
Tel: 473/440–2962
Fax: 473/440–4127

Guatemala

Embassy of Guatemala
2220 R Street NW
Washington, DC 20008, USA
Tel: 202/745–4952
Fax: 202/745–1908

National Protected Areas Council (CONAP)
7 Avenida 709, Zona 13
Guatemala City, Guatemala
Tel: 502/440–7916

Guyana

Embassy of Guyana
2490 Tracy Place NW
Washington, DC 20008, USA
Tel: 202/265–6900
Fax: 202/232–1297
E-mail: maoishmael@aol.com

Ministry of Agriculture
Vlissengen Road
Georgetown, Guyana
Tel: 592/02–7151

Haiti

Directorate General of Natural Resources
Service of Soils Conservation, Forests, and Wildlife
Port-au Prince, Haiti

Embassy of Haiti
2311 Massachusetts Avenue NW
Washington, DC 20008, USA
Tel: 202/332–4090
Fax: 202/745–7215

Honduras

Embassy of Honduras
3007 Tilden Street NW, POD 4M
Washington, DC 20008, USA
Tel: 202/966–7702
Fax: 202/966–9751

Ministry of Natural Resources
Department of Renewable Natural Resources
Apartado 1389
Tegucigalpa, Honduras
Tel: 504/237–5725
Fax: 504/237–5726

Jamaica

Embassy of Jamaica
1520 New Hampshire Avenue NW
Washington, DC 20036, USA

Tel: 202/452–0660
Fax: 202/452–0081
E-mail: emjam@sysnet.net

Ministry of Development, Planning, and Production
Resources Management Division
53⅓ Molynes Road
Kingston 10, Jamaica
Tel: 876/923–5155, 876/923–5070

Mexico

Embassy of Mexico
1911 Pennsylvania Avenue NW
Washington, DC 20006, USA
Tel: 202/728–1600
Fax: 202/728–1698

Ministry of the Environment and Natural Resources
Periferico Sur #4209
Franccionamiento Jardines en la Montana,
Delegacion Tlalpan 14210, Mexico DF
Tel: 525/5628–0600
Fax: 525/5628–0644

Nicaragua

Embassy of Nicaragua
1627 New Hampshire Avenue NW
Washington, DC 20009, USA
Tel: 202/939–6570
Fax: 202/939–6545

Ministry of Natural Resources
Apartado 5123
Managua, Nicaragua
Tel: 505/263–2824

Panama

Embassy of Panama
2863 McGill Terrace NW
Washington, DC 20008, USA
Tel: 202/483–1407
Fax: 202/483–8413
E-mail: eaa@panaemba-dc.ccmail.compuserve.com

National Authority for the Environment
Apartado 2016

Paraiso, Ancon, Panama
Tel: 507/232–5939
Fax: 507/232–6612

Paraguay

Embassy of Paraguay
2400 Massachusetts Avenue NW
Washington, DC 20008, USA
Tel: 202/483–6960
Fax: 202/234–4508
E-mail: embapar@erols.com

Natural Resources Ministry
Presidente Franco #475 y 14 de Mayo
Asunción, Paraguay
Tel: 595/2144–9951
Fax: 595/2149–7965

Peru

Embassy of Peru
1700 Massachusetts Avenue NW
Washington, DC 20036, USA
Tel: 202/833–9860
Fax: 202/659–8124

National Office of Natural Resources Evaluation
Apartado Postal 4992
Lima, Peru
Tel: 511/41–4606

St. Lucia

Embassy of St. Lucia
3216 New Mexico Avenue NW
Washington, DC 20016, USA
Tel: 202/364–6792
Fax: 202/364–6723

Energy, Science, and Technology Section
PO Box 709
Castries, Saint Lucia
Tel: 758/451–8746
Fax: 758/452–2506
E-mail: est_mpde_candw.lc

St. Kitts and Nevis

Embassy of St. Kitts and Nevis
3216 New Mexico Avenue NW
Washington, DC 20016, USA

Tel: 202/686–2636
Fax: 202/686–5740
E-mail: oliburd@aol.com

Ministry of Tourism, Culture, and the Environment
Department of Environment
Pelican Mall, Basseterre, St. Kitts
Tel: 869/465–4040
Fax: 869/465–8794

St. Vincent and the Grenadines

Embassy of St. Vincent and the Grenadines
1717 Massachusetts Avenue NW, #102
Washington, DC 20036, USA
Tel: 202/364–6730
Fax: 202/364–6736

Suriname

Embassy of the Republic of Suriname
4301 Connecticut Avenue NW, #108
Washington, DC 20008, USA
Tel: 202/244–7488
Fax: 202/244–5878
E-mail: embsur@erols.com

Ministry of Natural Resources
Suriname Forest Service
PO Box 436
Paramaribo, Suriname
Tel: 597/474–931
Fax: 597–410–250

Trinidad and Tobago

Embassy of Trinidad and Tobago
1708 Massachusetts Avenue NW
Washington, DC 20036, USA
Tel: 202/467–6490
Fax: 202/785–3130
E-mail: embttgo@erols.com

Ministry of Agriculture, Land, and Marine Resources
Farm Road
St. Joseph, Trinidad and Tobago
Tel: 868/662–5114, 662–220
Fax: 868/645–4288

Venezuela

Embassy of Venezuela
1099 30th Street NW
Washington, DC 20007, USA
Tel: 202/342–2214
E-mail: embavene@dgsys.com

Ministry of the Environment and Renewable Natural Resources
Centro Simon Bolivar, Torre Sur, Piso 19
Caracas 100, Republica de Venezuela
Tel: 582/483–3164
Fax: 582/408–1464

International Conservation Groups and Organizations

If you want to find out more about each tropical country's environmental problems and most promising solutions, contact the leading conservation groups and research organizations below. Many of the Web sites listed for groups in Europe yielded valuable information for this book. Remember that organizations in foreign countries do not necessarily have anyone on staff who can read English and do not always have the resources to provide you with a lot of information. But they will be pleased to learn of your interest and support. Knowing that someone on the other side of the world cares about their work and wants to help means a great deal to grassroots groups.

AFRICA

Angola

Action for Rural Development and Environment
Pruceta Farinba Leitao
No 27, 1 Dto
Luanda 3788
Tel: 244/239–6683
Fax: 244/239–6683
E-mail: adra@angonet.gn.apc.org

Angolan Environment Association
PO Box 2110
Luanda
Republic of Angola
Tel: 244/239–0334

Cameroon

Center for Studies in the Environment and the Development of Cameroon
Boite Postale 410
Maroua
Tel: 237/29–3061
Fax: 237/29–3391

National Coordinator of Natural Resource Management Project
PVO-NGO/NRMS Project
B.P. 422
Yaounde
Tel: 237/20–8622
Fax: 237/22–1873

Congo, Democratic Republic of

Congo Institute for the Conservation of Nature
Ave. des Cliniques N 13
B.P. 868
Kinshasa 1
Tel: 243/123–3250
Fax: 243/122–7547

National Nature Alliance
Dongou, 50 bis
Brazzaville 1
Ouenze
Tel: 242/82–0237
Fax: 242/83–4907

Ethiopia

Ethiopian Wildlife and Natural History Society
PO Box 60074

Addis Ababa
E-mail: Humber@UEL.AC.UK

Ethiopian Wildlife Conservation Organization
PO Box 386
Addis Ababa
Tel: 2511/44–5970

Ghana

Friends of the Earth
PO Box 3794
Accra
Fax: 233/21–22–7993

Ghana Greens Society
C/o Ministry of Local Government
Rural Department and Cooperatives
PO Box M50
Accra
Tel: 233/66–5421
Fax: 233/66–7911

Green Earth Organization
PO Box 16641
Accra-North
Tel: 233/212–32762
Fax: 233/212–30455
E mail: greeneth@ncs.com.gh

Guinea-Bissau

This Land Is Ours (TINIGUENA)
Bairro de Belem
B.P. 667
Bissau
Tel: 245/25–1907
Fax: 245/25–1906

Kenya

African NGOs Environment Network
PO Box 53844
Nairobi
Tel: 254/228–138
Fax: 254/233–518

Environmental Liaison Centre International
PO Box 72461
Nairobi
Tel: 254/256–2015

Global Coalition for Environment and Development
PO Box 72461
Nairobi
Tel: 254/256–2015
Fax: 254/256–2175

International Council for Research in Agroforestry
PO Box 30677
Nairobi
Tel: 254/252–1450
Fax: 254/252–1001

Society for Protection of the Environment in Kenya
Tom Mboya Street
Mwalimu co-op House R. 101
PO Box 60125
Nairobi
Tel: 254/224–6317
Fax: 254/223–0230

Madagascar

Conservation International
II L 1 A, Ankadivato
Antananarivo
Tel: 2612/23–5289
Fax: 2612/22–0422
E-mail: malagasy@embassy.org

World Wide Fund for Nature
B.P. 738
Antananarivo 101
Tel: 2612/34–638
Fax: 2612/34–888

Mauritius

Council for Nature Conservation
6 Edith Cavell
Port Louis
Tel: 230/21–5593
Fax: 230/212–7882

Mozambique

Environmental Working Group
PO Box 2775
Maputo
Tel: 258/149–3102
Fax: 258/149–3049

Nigeria

Nigerian Conservation Foundation
Plot 5, Moseley Road, Ikoyi
PO Box 74638, Victoria Island
Lagos
Tel: 2341/68–4717
Fax: 2341/68–5378

Rwanda

Nyungwe Forest Conservation Project
Wildlife Conservation International
B.P. 363
Cyangugu
Tel: 250/37–193
Fax: 250/72–128

Senegal

Environment 2000
B.P. 5319
Dakar
Tel: 221/820–2985

Rural Foundation for the Development of the Earth
Monde
Km 6 route de Sangalkam
B.P. 184 Rufisque
Tel: 221/936–8831
Fax: 221/924–9246

Senegal Friends of Nature Association
15 bis rue Jules Ferry
Dakar 1810
Tel: 221/825–2966
Fax: 221/826–9363

Sierra Leone

Concerned Environmentalists Action Group
22A Hindowa Street, Kenema
Freetown
Conservation Society of Sierra Leone
4 Sanders Street
PO Box 1292
Freetown
Tel: 2322/222–9716
Fax: 2322/222–4439
E-mail: slango@sl.baobab.com

Council for the Protection of Nature
Daphne Tuboku-Metzger
P.M.B. 2001
Freetown

Friends of the Earth
P.M. Bag 950, 33 Robert Street
Freetown
Tel: 2322/222–5223

Tanzania

National Environment Management Council
Sokoine Drive
PO Box 63154
Dar es Salaam
Tel: 255/51–12–1334
Fax: 255/51–12–1334
E-mail: nemc.crossborder@twiga.com
<http://www.nemc@simbanet.net>

Wildlife Conservation Society of Tanzania
PO Box 70919
Dar es Salaam
Tel: 255/51–11–2518
Fax: 255/51–12–4572
E-mail: wcst@africaonline.co.tz

Uganda

Impenetrable Forest Conservation Project
World Wildlife Fund
PO Box 7487
Kampala
Tel: 25641/24–5597

Wildlife Clubs of Uganda
Plot 31
Kanjokya, Kamwokya
PO Box 4596
Kampala
Tel: 25641/25–6531

Zaire

African Society for the Study of the Environment and Conservation of Nature
B.P. 5698
Kinshasa Gombe 10

Zimbabwe

Association of Zimbabwe Traditional Environmental Conservationists
P.B. 9286
Masvingo
Tel: 263/139–66006
Fax: 263/139–64889

Environment 2000
Box A639
Avondale, Harare
Tel: 263/430–2276
Fax: 263/433–9691
E-mail: e2000@samara.co.zw

National Conservation Trust
PO Box 8575, Causeway
Harare
Tel: 263/44–6105
Fax: 263/44–6105

Zimbabwe Trust
4 Lanark Road, Box 4027
Harare
The Old Lodge, Christchurch Road
Epsom, Surrey KT19 8NE, UK
Tel: 263/472–2957
Fax: 263/479–5150

ASIA AND THE PACIFIC

Australia

Australian Conservation Fund
340 Gore Street
Fitzroy 3065
Victoria
Tel: 6139/416–1166
Fax: 6139/416–0767
<http://www.aefonline.org.au>

Australian Trust for Conservation Volunteers
Box 423
Ballarat 3353
Victoria
Tel: 615/333–1483
Fax: 615/333–2290

Centre for Plant Biodiversity Research
GPO Box 1600
Canberra
ACT 2601
Tel: 02/625–09450
Fax: 02/625–09499
E-mail: anbg-info@anbg.gov.au

Cooperative Research Centre for Tropical Rainforest Ecology and Management
PO Box 6811
Cairns 4870
Queensland
Tel: 61704/042–1246
Fax: 61704/042–1247
<http://www.cimm.jcu.edu.au/rainforestCRC>

Friends of the Earth Australia
Box 222
Fitzroy, VIC, 3065
Tel: 6139/419–8700
Fax: 6139/416–2081
E-mail: foe@foe.org.au
E-mail: foefitzroy@peg.apc.org
<http://www.foe.org.au>

The John Sinclair Trust for Conservation
PO Box 71 (36 Kemp Street)
Gladesville, NSW, 2111
Tel: 02/817–4660
Fax: 02/816–1642

Rainforest Conservation Society
19 Colorado Avenue
Bardon 4065
Queensland
Tel: 6173/368–1318
Fax: 6173/368–3938
E-mail: arcs@gil.com.au

Rainforest Information Center
PO Box 368
Lismore
NSW 2480
Tel: 6126/621–8505
E-mail: rainfaus@mullum.com.au
<http://forests.org/ric/>

Bangladesh

Center for Environmental Studies and Research
68.1 Purana Pattan

Dhaka-1000
Tel: 880/225–4936

Nature Conservation Movement
29-C-1 North Kamalapur
Dhaka 1217
Tel: 880/241–8883
Fax: 880/283–3495

Wildlife and Nature Conservation Society of Bangladesh
House 26, Road 4
Dhanmondi
Dhaka 1205
Tel: 8802/955–3429
Fax: 8802/86–3879
E-mail: salimam@citecho.net

Brunei
Brunei Rainforest Project
Universiti Brunei
Darussalam
Bandar Seri Begawan 3186
Negara
Tel: 010/6732–27001

Fiji

National Trust for Fiji
3 Maafu Street
PO Box 2089
Government Buildings
Suva
Tel: 679/30–1807
Fax: 679/30–5092
E-mail: ntf@pactok.peg.apc.org

South Pacific Action Committee for Human Ecology and the Environment
PO Box 16737
Suva
Tel: 679/31–2371
Fax: 679/30–3053
E-mail: korovulavula_i@iqusp.ac.fj

India

Asian Environmental Society
U-112 (3rd Floor), Vidhata House
Vikas Marg. Shakarpur
New Delhi 110092

Tel: 9111/222–3311
Fax: 9111/331–9435

Bombay Environmental Action Group
9 Street James Court
Marine Drive, Bombay 400020
Tel: 9122/287–3500
Fax: 9122/220–187

Center for Science and Environment
Founder and Director (Int'l)
F-6 Kailash Lolony
New Delhi 110048
Tel: 9111/643–3394
Fax: 9111/644–1711

Development Alternatives
B-32 Tara Crescent
Qutub Institutional Area
New Delhi 110016
Tel: 9111/66–5370
Fax: 9111/68–66031
E-mail: tara@sdalt.ernet.in

Environment Society of India
Karuna Sadan
Sector 11-B
Chandigarh 160011
Tel: 9117/23–2351
Fax: 9117/254–4533

Indian Ecological Society
C/o Bungalow 51 Sector 4
Chandigarh 160001
Tel: 9111/54–0814

Indian Institute of Forest Management
PO Box 357
Nehru Nagar
Bhopal 462 003, MP
Tel: 9175/577–5998
Fax: 9175/577–2878
E-mail: ramprasad@iifm.org
<http://www.iifm.org>

International Institute of Sustainable Development and Management
1982, Subhashnagar, Chandkheda
Ahmedabad 382 424
Tel: 91079/750–2219

Neem Foundation
67-A, Vithalnagar, Road No. 12 NS

Juhu Scheme, Mumbai 400049
Tel: 9122/620–6367
Fax: 9122/620–7508
E-mail: info@neemfoundation.org
<http://neemfoundation.org/>

Indonesia

Friends of the Earth Indonesia
Jl. Penjernihan I, Komp. Keuangan No. 15
Pejompongan, Jakarta 10210
Tel: 622/158–6820
Fax: 622/158–8416

The Indonesian Forum for Environment
Jl. Penjernihan I, Kompleks Keuangan
Jakarta 10210
Tel: 622/158–3975
Fax: 622/158–8416

Malaysia

Environmental Analysis
Kuala Lumpur
Tel: 603/689–3444

Forest Research Institute
Kepong, 52109
Kuala Lumpur
Tel: 603/634–2633
Fax: 603/636–7753
E-mail: webgroup@frim.gov.my
<http://www.frim.gov/my>

Malaysian Nature Society
JKR 641 Jalan Kelantan
Bukit Persekutuan
50480 Kuala Lumpur
Tel: 603/ 287–9422
Fax: 603/287–8773
E-mail: natsoc@po.jaring.my
<http://www.mns.org.my/mns>

World Rainforest Movement
87 Jalan Cantonment
10250 Pulau Pinang
Tel: 04/37–3612

World Wide Fund for Nature
Jalan Sultan PO
46990 Petaling Jaya
Selangor Darul Ehsan

Tel: 603/703–3772
Fax: 603/703–5157
E-mail: wwfmal@wwfnet.org
<http://wwfmal.cjb.net>

New Caledonia

Association Pour La Sauvagarde de la Nature Neo-Caledonia
37 Fue Georges Clemenceau/ FP 1772
Noumea
Tel: 687/28–3275
Fax: 687/28–3275

Papua New Guinea

Conservation Melanesia Inc.
PO Box 735
Boroko, NCD
Tel: 675/323–2758
Fax: 675/323–2773
E-mail: conmelpng2@global.net.pg

Environmental, Cultural, & Spiritual Conservation
PO Box 45, University
Waigani Campus/NCD
Tel: 675/323–0545
Fax: 675/323–5289

Foundation for People and Community Development Inc.
PO Box 1119, Boroko
Port Moresby, NCD
Tel: 675/325–8470
Fax: 675/325–2670
E-mail: fpcd@datec.com.pg

Friends of the Earth
PO Box 4028
Boroko
Tel: 675/21–4673
Fax: 675/25–6797

Philippines

Earth Savers/Population Communications International
Ninoy Aquiro Park, Quezon Blvd.
Manila
Tel: 632/494–9113
Fax: 632/407–637

Ecological Society of the Philippines
C/o 53 Tamarind Road
Forbes Park
Makati, PO Box 2871
Metro Manila
Tel: 632/633–9626
Fax: 632/637–2060

Environmental Network Center Inc.
9th Floor Mondragon Building, 324 Sen. Gil
Puyat Ave., Makati
Metro Manila 1200
Tel: 633/818–1506

Green Coalition Inc.
Suite 421 Equitable Bank II, Ortigas Avenue
Greenhills, San Juan
Metro Manila
Tel: 632/773–419
Fax: 632/493–029

Green Earth Movement
113 B. Gonzales Street, Loyola Heights
Quezon City
Tel: 63/299–7404
Fax: 63/299–7404

World Ecologists Foundation
Gold Building, 15 Annapolis
Greenhills, San Juan
Metro Manila
Tel: 632/722–4016
Fax: 632/721–3356

Samoa

National Environmental Management Strategies
South PAC
PO Box 240
Apia
Tel: 68/52–1929
Fax: 68/52–0231

Solomon Islands

Solomon Islands Development Trust
EcoForestry Unit
PO Box 147
Honiara
Tel: 67/72–2289
Fax: 67/72–1131
E-mail: sidtefu@welkam.solomon.com.sb

Sri Lanka

Environmental Foundation Ltd.
3 Campbell Terrace
Colombo 10
Tel: 941/69–7226
Fax: 941/69–7226
E-mail: efl@ef.is.lk

Organization for Resource Development and Environment
No. 193 Welewewa
Nawagattegama-5815
Kurunegala
North Western Province
Tel: 94/161–2442
Fax: 9410/322–3267

Public Campaign on Environment and Development
50/7C Siripa Road
Colombo 5
Tel: 94/158–2439
Fax: 94/154–6518

The Tree Society of Sri Lanka
50/7C Siripa Road
Colombo 5
Fax: 941/58–8804

Wildlife and Nature Protection Society of Sri Lanka
No. 86 Rajamalwatta Road
Battaramulla
Tel: 94/188–7390
Fax: 94/188–7664
E-mail: advarcht@slt.lk

Thailand

Asian Society for Environmental Protection
C/o Asian Institute of Technology
PO Box 4, Klongluang
Pathumthani 12120
Tel: 662/524–5245, 662/524–6658
Fax: 662/524–5236
E-mail: asep@ait.ac.th

Land Development Department
Phahonyothin Road, Chatuchak
Bangkok 10900

Tel: 662/579–5546
Fax: 662/561–3029
E-mail: Idd@mozart.inet.co.th

Project for Ecological Recovery
1705 Rama 4 Road
Bangkok 10500
Tel: 662/252–5940

Regional Community Forestry Training Center
Kasetsart University
PO Box 1111
Bangkok 10903
Tel: 662/940–5700
Fax: 662/561–4880
E-mail: ftcsss@nontri.ku.ac.th
<http://www.recoftc.org>

Seub Nakhasathien Foundation
Kasetsart University
Alumni Association Building
50 Phaholyothin Road, Ladyao
Jatujak Bangkok 10900
Tel: 662/561–2469
Fax: 662/561–2470
<http://www.seub.ksc.net/aboutus/introseub/his1life-e.htm>

Vietnam

Center for Natural Resources and Environmental Studies
167 Bui Thi Xuan
Hanoi
Tel: 844/976–1080
Fax: 844/821–8934
E-mail: cuc@uplands.ac.vn

EUROPE

England

Friends of the Earth
26–28 Underwood Street
London N17JQ
Tel: 020/7490–1555
Fax: 020/7490–0881
E-mail: tims@foe-couk
<http://www.foe.co.uk>

The Hawk Conservancy
Weyhill, Andover
Hants SP11 8DY
Tel: 44/1264–773850
Fax: 44/1264–773772
E-mail: keith@hawk-conservancy.org
<http://www.hawk-conservancy.org/>

Kielder Water Bird of Prey Centre
Leaplish Waterside Park
Northumberland
Tel: 44/01434–250400
E-mail: eanaylor.jadan@pop3.hiway.co.uk
<http://www.discoverit.co.uk/falconry/>

Living Earth
4 Great James Street
London WC1N 3DA
Tel: 44/0171–2423816
Fax: 44/0171–2423817
E-mail: livearth@gn.apc.org
<http://www.gn.apc.org/LivingEarth/>

The Natural History Museum
Cromwell Road
London SW7 5BD UK
Tel: 44/0207–9389459
Fax: 44/0207–9388937
E-mail: pe@nhm.ac.uk
<http://www.nhm.ac.uk/science/entom/project3/index.html>

Panos Institute
9 White Lion Street
London N1 9PD
Tel: 0171/278–1111
Fax: 0171/278–0345
E-mail: panoslondon@gn.apc.org
<http://www.oneworld.org/panos>

Rainforest Foundation
Suite A5, City Cloisters
196 Old Street
London EC1V 9FR
Tel: 44/020–7251–6345
Fax: 44/020–7251- 4969
E-mail: rainforestuk@rainforestuk.com
<http://www.rainforestfoundationuk.org/rain-home.html>

The Royal Botanic Gardens, Kew
Richmond

Surrey TW9 3AB
Tel: 44/020–8332–5000
Fax: 44/020–8332–5197
E-mail: info@rbgkew
<http://www.rbgkew.org.uk/>

Species 2000
Centre for Plant Diversity and Systematics
School of Plant Sciences
The University of Reading
Reading RG6 6AS, England
Tel: 44/1189–316466
Fax: 44/1189–753676
E-mail: sp2000@sp2000.org
<http://www.species2000.org>

Survival for Tribal Peoples
11–15 Emerald Street
London WC1N 3QL
United Kingdom
Phone: 44/0207–2421441
Fax: 44/0207–2421771
E-mail: info@survival-international.org
<http://www.survival-international.org>

The Whitley Awards Foundation
Royal Geographical Society
1 Kensington Gore
London SW7 2AR
Tel/fax: 44/020–7591
E-mail: info@whitleyaward.org
<http://www.whitleyaward.org>

World Conservation Monitoring Centre
219 Huntingdon Road
Cambridge CB3 0DL
Tel: 44/1223–277314
Fax: 44/1223–277136
E-mail: info@wcmc.org.uk
<http:// www.wcmc.org.uk>

Italy

Food and Agriculture Organization of the United Nations
Viale delle Terme di Caracalla
00100 Rome
Tel: 396/57–0551
Fax: 396/570–53152
E-mail: gII@fao.org
<http://www.fao/org>

The Netherlands

Friends of the Earth International
PO Box 19199
1000 GD Amsterdam
Tel: 3120/622–1369
Fax: 3120/639–2181
E-mail: foei@foei.org
<http://www.foei.org>

Switzerland

Convention on International Trade in Endangered Species of Wild Fauna and Flora (CITES)
15 chemin des Anemones
1219 Chatelaine-Geneva
Tel: 412–2/917–8139/40
Fax: 412–2/797 3417
E-mail: cites@unep.ch
<http://www.cites.org/CITES/eng/index.shtml>

World Conservation Union (IUCN)
28 rue Mauverney
CH-1196 Gland
Tel: 4122/999–0001
Fax: 4122/999–0025
E-mail: forests@hq.iucn.org
<http://iucn.org/themes/>
IUCN Species Survival Commission: <http://iucn.org/themes/ssc/index.htm>

World Wide Fund for Nature (WWF)
Avenue du Mont-Blanc
CH-1196-Gland
Tel: 4122/364–9111
Fax: 4122/364–5358
<http://www.panda.org/>

LATIN AMERICA AND THE CARIBBEAN

Argentina

Foundation for the Defense of the Environment
Casilla de Correo 83, Correo Central
5000 Cordoba
Tel: 54/351–469–0282
Fax: 54/351–452–0260
E-mail: funam@funam.org.ar
<http://www.funam.org.arg/>

Foundation for the Environment and Natural Resources
Monroe 2142, piso 1
Buenos Aires 1428
Tel/Fax: 541/1478–84266
E-mail: info@farn-sustentar.org
<http://www.farn-sustentar.org>

Foundation for Species and Environmental Conservation
Pringels 10, piso 3
Buenos Aires 1183
Tel: 541/981–4792
Fax: 541/983–7949
E-mail: cacheng@fucema.org.ar

Friends of the Earth Foundation
Zapata 343
1426 Capital Federal
1000 Buenos Aires
Tel: 541/553–4318
Fax: 541/331–6720

Wildlife Foundation
Defensa 245–251, 6to. piso
Buenos Aires 1065
Tel: 541/343–3778

Barbados

Barbados Environmental Association
Stephen Boyce
PO Box 132
Bridgetown
Fax: 246/427–0619

Barbados National Trust
Ronald Tree House
Wildey House
Wildey, Street Michael
Tel: 246/426–2421
Fax: 246/429–9055
E-mail: natrust@sunbeach.net
<http://www.trust.funbarbados.com>

Caribbean Conservation Association
Savannah Lodge, The Garrison
Street Michael
Tel: 246/426–5374
Fax: 246/429–8483

Belize

Belize Audubon Society
12 Fort Street
PO Box 1001
Belize City
Tel: 501/23–5004
Fax: 501/23–4985
E-mail: base@btl.net
<http://www.belizeaudubon.org >

Belize Center for Environmental Studies
55 Eve Street PO Box 666
Belize City
Tel: 501/224–5545
E-mail: roses@bces.org.bz

Belize Zoo & Tropical Education Center
PO Box 1787
Belize City
Tel/Fax: 501/281–3004
E-mail: belizezoo@btl.net

Program for Belize
PO Box 749
No. 1 Eyre Street
Belize City
Tel: 501/27–5616
Fax: 501/27–5635
E mail: pfbel@btl.net
<http://www.belizenet.com/pfbel.html>

Wildlife Conservation Society
Gallon Jug
c/o PO Box 37
Belize City
Tel/Fax: 501/21–2002
<http://www.fwie.fw.vt.edu/wcs/>

Bolivia

Bolivian Conservation Association
J.J. Perez N268, piso 1
La Paz 11250
Tel: 591/239–0565
Fax: 591/237–5371
E-mail: tropico@tropico.rds.org.bo

Bolivian Environmental Association
Av. Camacho 1277, 5th piso, Of. 503
PO Box 14174
La Paz
Fax: 5912/093–22007

Eco Bolivia Foundation
Casilla de Correo 8505
La Paz
Tel/Fax: 591/232–5776
E-mail: ecobolivia@mail.megalink.com
<http://www.ecobolivia.com>

Ecological Association of the Orient
Nordenskiold 5
entre Av. Landivar y Alameda Junin
PO Box 4831
Santa Cruz
Tel: 591/332–7877
Fax: 591/332–6367
E-mail: aseo@bibosi.scz.entelnet.bo

Friends of Nature Foundation
Avenida Irala 421
PO Box 2241
Santa Cruz
Tel: 591/333–3806
Fax: 591/334–1327

Brazil

Biodiversity Foundation
Av. do Contorno 9.155, 11oandar
Caixa postal 1462
Belo Horizonte 30110.130-MG
Tel: 5531/292–8235
Fax: 5531/291–7658
E-mail: biodiversitas@biodiversitas.org

Brazilian Ecological Association
Av. Nilo Pecanha, 12 G.
801/803–8vo Andar
Rio de Janeiro 20020–000-RT
Tel: 5521/533–6374
Fax: 5521/27–5011

Brazilian Foundation for the Conservation of Nature
Rua Miranda Valverde 103
Botafago
Rio de Janeiro 22281–000-RJ
Tel: 5521/537–7565
Fax: 5521/537–1343

Brazilian Institute for the Environment and Renewable Natural Resources
Sain, Av. L4 Norte
Ed. Sede-Bloco A-Sala 1

Brasilia 70800–200-DF
Tel: 5561/316–1090
Fax: 5561/322–1058
E-mail: rsoavins@ibama.gov.br

Forest Development Fund
Florestar Sao Paulo
Rua do Horto 931
Sao Paulo 02377–00-SP
Tel: 5511/695–35692
Fax: 5511/695–35331
E-mail: cmaretti@uol.com.br

Foundation SOS Atlantica Forest
Rua Manoel da Nobrega, 456
04001 Sao Paulo- SP
Tel: 5511/887–1195
Fax: 5511/885–1680

Institute of Management and Certification of Forestry and Agriculture
Av. Carlos Botelho
853-conj 2, 13416–145
Piracicaba- SP
Tel/Fax: 5519/422–6258
E-mail: imaflora@merconet.com.br

Pro Nature Foundation
SCLN 107, Bloco B, Salas 201–207
CEP- 70.743–520
Braslia- DF
Tel: 5561/274–5449
Fax: 5561/274–5324
E-mail: funatura@essencial.com.br

Chile

CODDEF
Av. Francisco Bilbao 691
Providencia, CP 6640980
Santiago
Tel/Fax: 562/251–0262
E-mail: codeff@netup.cl

Colombia

Ecological Society
Edificio Camilo Torres
Carrera 50 N27–70
Bl. C Modulo 4, Nivel 5
Bogota 8674
Tel: 571/221–7458
Fax: 571/282–2860

Environmental Protecion Fund
Calle 62A N 468
Bogota 052986
Tel: 571/249–0437
Fax: 571/255–5427
E-mail: efenfund@co11.teleecom.com.co

Fundación Natura
Calle 61 No. 4–26
AA 55402
Bogota
Tel: 571/345–1216
Fax: 571/249–6250
E-mail: enatura@impsat.net.co
<http://www.natura.org.co>

World Wildlife Fund of Colombia
Carrera 35 #4A-25
San Fernando
Cali
Tel: 572/558–2577
Fax: 572/558–2588

Costa Rica

Association for the Preservation of Flora and Fauna
Apartado 917–2150
Moravia
Tel: 506/240–6087
Fax: 506/236–3210
E-mail: preserve@sol.racsa.co.cr
<http://www.preserveplantet.org>

Earth Council
Apdo. 2323–1002
Paseo de los Estudiantes
Tel: 506/256–1611
Fax: 506/255–2197
E-mail: info@terra.ecouncil.ac.cr

Environmental and Natural Resources Law Center
Del Higueron de San Pedro
100 mts al suroeste y 450 mts al este
Apdo. 134–2050
San Pedro, Montes de Oca
Tel: 506/283–7080
Fax: 506/224–1426
E-mail: cedarena@sol.racsa.co.cr

Monteverde Conservation League
Apdo. 10581–1000
San José
Tel: 506/645–5003
Fax: 506/645–5104
E-mail: acmmcl@sol.racsa.co.cr

National Institute of Biodiversity
Apdo. 22–3100
Santo Domingo- Heredia
Tel: 506/244–0690
Fax: 506/244–2816

Neotropica Foundation
Apartado 236–1002
Paseo de los Estudiantes
San José
Tel: 506/253–2130
Fax: 506/253–4210
E-mail: fneotrop@sol.racsa.co.cr

Regional Program for Wildlife Management
Apdo. 1350
Universidad Nacional
Heredia
Tel: 506/237–7039
Fax: 506/237–7036
E-mail: prmvs@irazu.una.ac.cr

Union for the Conservation of Nature and Natural Resources
Apdo. 146–2150
Moravia
Tel: 506/236–2733
Fax: 506/240–9934
E-mail: uicncr@sol.racsa.co.cr

Cuba

Cuban Society for the Conservation of Nature and Natural Resources
Apdo. 7097
Havana 6

Dominica

Archbold Tropical Research Center
PO Box 456
Roseau
Tel: 767/91–401
Fax: 767/449–2160

Dominican Republic

Dominican Society for the Conservation of Natural Resources
Casilla de Correo 174–2
Santo Domingo
Tel: 809/567–6211

ENDA-Caribe
Apdo. 3370
Santo Domingo
Tel: 809/566–8321
Fax: 809/541–3259

Green Caribbean Foundation
David Ben Gurion N2
Esq. Winston Churchill, Plaza
Solangel 3piso, Local 3-F
Santo Domingo
Tel: 809/566–4000
Fax: 809/567–9064

Ecuador

Abya Yala
Casilla 17–12–719
Quito
Tel: 593/250–6247
Fax: 593/250–6255
E-mail: admin-info@abyayala.org
<http://www.abyayala.org>

Conservation and Development
Apdo. 1716–1855
Quito
Tel: 593/246–5845
Fax: 593/255–2614
E-mail: ccd@ccd.org.ec

Ecological Action
Casilla 17–15–246C
Quito
Tel: 593/252–7583
Fax: 593/254–7516
E-mail: verde@hoy.net

Fundación NATURA
Av. America 5653 y Voz Andes
Quito
Tel: 593/420–5152
Fax: 593/420–6777
E-mail: natura@natura.satnet.net

Rainforest Rescue
Apdo. 17–12–105
Quito
Tel: 593/252–6789
E-mail: mlgambo@uio.satnet.net

El Salvador

CESTA
Apdo. 3065
San Salvador
Tel: 503/25–6746

SalvaNATURA
77 Avenida Norte No. 304
San Salvador
Tel: 503/263–1111
Fax: 503/263–3466
E-mail: salnatura@insatelsa.com

French Guiana

ENGREF
Centre de Kourou BP 316
97379
Kourou
Tel: 594/32–2675
Fax: 594/32–4302
E-mail: stephan.a@kourou.cirad.fr

Grenada

Grenada National Trust/Historical Society
Grenada National Museum
Young Street
Street George's
Tel: 473/440–2198
Fax: 473/440–4179

Guatemala

Central America Protected Areas Project/Costas
3ra Av 7–53, Zona 14
Colonia El Campo
Guatemala
Tel: 502/367–5326
Fax: 502/368–3276
E-mail: wwfund@ns.guate.net

Central American Regional Program for the Environment
10a. Calle 6–40

Zona 9
01009 Guatemala
Tel: 502/331–3373
Fax: 502/362–2044

Conservation Endowment Fund
Ruta 7, 6–42, Zona 4
Edificio Torre 6–42, Of. 201
Guatemala
Tel: 502/334–3547
Fax: 502/334–3548
E-mail: fcgua@pronet.net.gt

**Foundation for the Environment and Eco-
Development**
7 Calle A, 20–53 Zona 11
Col. Mirador 01011
Guatemala
Tel: 502/474–3645
Fax: 502/440–4605
E-mail: fundaeco@quetzal.net and
fecoreg@quetzal.net

Interamerican Foundation for Tropical Research
Av. Hincapie 31–31 Zona 13
Km. 10 y 1/2 Mision del Fortín, Of 106
Guatemala
Tel/Fax: 502/233–3555
E-mail: fiit@c.net.gt

Nature Defenders
14 Calle, 6–49, Zona 9
Guatemala 01009
Tel: 502/361–7001
Fax: 502/361–7011
E-mail: defensores@pronet.net.gt

Haiti

Agroforestry Extension Project
Pan American Development Foundation
Project Trees
Delmas 31 #27 Port-au-Prince
B.P. 15574
Petionville
Tel: 509/46–3286
Fax: 509/46–4616

Honduras

Bay Islands Conservation Association
Edif. Cooper, Calle Principal

Coxen Hole, Roatan
Bay Islands
Tel: 504/445–1424

**Committee for the Development and Defense of
the Flora and Fauna of the Gulf of Fonseca**
Apdo. 3663
Tegucigalpa
Tel: 504/238–0415
Fax: 504/238–0415
E-mail: cgolf@sdnhon.org.hn

**Foundation for the Environment and Development
of Honduras**
Apdo. 4252
Tegucigalpa
Tel: 504/239–1462
Fax: 504/239–1645
E-mail: funvida@sdnhon.org.hn

**Hector Rodrigo Pastor Fasquelle Ecological
Foundation**
Apdo. 2479
San Pedro Sula
Tel: 504/552–1014
Fax: 504/557–6220
E-mail: fundeco@netsys.hn

MOPAWI
Apdo. 2175
Tegucigalpa
Tel: 504/235–8659
Fax: 504/239–9234
E-mail: mopaw@optinet.hn

World Neighbors
Apdo. 3385
Tegucigalpa
Tel: 504/230–2002
Fax: 504/230–2006
E-mail: scvm@egroups.com

Jamaica

Birdlife Jamaica
2 Starlight Avenue
Kingston 6
Tel: 1–876/927–8444
Fax: 1–876/978–3243
E-mail: birdlife@yahoo.com

Jamaica Conservation and Development Trust
PO Box 1225
Kingston 8
Tel: 876/960–2848
Fax: 876/960–2850
E-mail: jcdt@kasnet.com

Mexico

Conservation International
Carretera Club Campestre KM 2.2
Rancho El Arenal, Teran 29050
Tuxtla Gutierrez, Chiapas
Tel/Fax: 529/615–1951

Forest Stewardship Council
Ave. Hidalgo 502
OAX 68000
Oaxaca
Tel: 529/514–6905
Fax: 529/516–2110
E-mail: cvallejo@fscoax.org
<http://www.fscoax.org>

Friends of Sian Ka'an A.C.
Apdo. 770
CP 77500 Cancun, Quitana Roo
Tel: 5298/84–9583
Fax: 5298/87–3080
E-mail: sian@cancun.com.mx

Mexican Conservationist Federation
Progreso N.5
Col. Coyoacan 04110, Mexico, D.F.
Tel: 525/658–3112
Fax: 525/273–8718
E-mail: sdelamo@laneta.apc.org

Naturalia, A.C.
Apdo. 21–541
C.P. 04021
Mexico, D.F.
Tel: 525/674–6678
Fax: 525/674–3876
E-mail: naturalia@servidor.unam.mx

PROESTEROS
Av. Ruiz 1687, Esquina calle 17
Zona Centro, Ensenada
22800 Baja California

Tel: 526/178–0162
Fax: 526/178–6050
E-mail: proestr@telnor.net
<http://www.cocese.mx/proester>

Pronatura
Aspergulas #22, Col. San Clemente
Delegacion A. Obregon
01740 Mexico, D.F.
Tel: 525/635–5054
Fax: 525/635–5054
E-mail: pronatura@compuserve.com
<http://www.pronatura.org.mx>

World Wildlife Fund
Av. Mexico 51, Hipodromo
06100 Mexico, D.F.
E-mail: avinawwfmex@compuserve.com.mx

Netherlands Antilles

Netherlands Antilles National Park Foundation
Piscaderabaai z/n
PO Box 2090
Willemstad, Curacao
Tel: 599/962–4242
Fax: 599/962–7680

Nicaragua

Alexander Humboldt Center
Apdo. 768
Managua
Tel: 505/249–2903
E-mail: humboldt@ibw.com.ni

Alistar Foundation
Shell Plaza El Sol
2 cuadras Sur, 1 y 1/2 cuadra arriba No. 4
Managua
Tel: 505/270–4017
Fax: 505/270–4023
E-mail: alistar@ibw.com.ni

Cocibolca Foundation
Apdo. C-212
Managua
Tel: 505/278–3224
Fax: 505/270–0578
E-mail: fcocibol@ibw.com.ni

Community of Young Environmentalists
Apdo. C-101
Managua
Tel: 505/260–0136
Fax: 505/260–0136
E-mail: ja@nicarao.apc.org.ni
<http://www.uca.edu.ni/guia/guia.html>

Friends of the Earth
Apdo. 20A
Managua
Tel: 505/270–5434
Fax: 505/270–5434
E-mail: atenic@sdnnic.org.ni

NICAMBIENTAL
Apdo. 3772
Managua
Tel: 505/267–8267
Fax: 505/267–8267
E-mail: nicam@ns.sdnnic.org.ni

Nicaraguan Environmental Movement
Apdo. A-99
Managua
Tel: 505/277–4835
Fax: 505/278–4863
E-mail: ecoman@sdnnic.org.ni

Panama

Agricultural Frontier Program
Apdo. 87–2733
Zona 7
Panama
Tel: 507/236–8186
Fax: 507/223–3966
E-mail: pfa@sino.net
<http://www.sdnp.org.pa/PFA>

National Association for Conservation
Apdo. 1387
Zona 1, Panama
Tel: 507/263–7950
Fax: 507/264–1836
E-mail: ancon@pty.com and aurena@ancon.org

Nature Foundation
Apdo. 2190
Panama 1

Tel: 507/228–1935
Fax: 507/228–1934
E-mail: naturafi@panama.phoenix.net

Panama Audubon Society
Apdo. 2026
Balboa
Ancon
Tel: 507/224–9371
Fax: 507/224–4740
E-mail: audupan@pananet.com
<http://www.pananet.com/audubon>

PA.NA.MA. Foundation
Apdo. 6–6623
El Dorado
Ciudad de Panam
Tel: 507/225–7325
Fax: 507/225–7314
E-mail: Panabird@sinfo.net

Paraguay

Moises Bertoni Foundation for the Conservation of Nature
PO Box 714
Asuncion
Tel: 59521/44–60–9810
Fax: 59521/44–0239
E-mail: idea@pla.net.py

Peru

Andean Regional Ecological Institute
Jr. Arequipa N 470
Hyancayo
Tel: 51/642–37169
Fax: 51/642–31111

Association for the Conservation and Development of the Environment
Av. Progreso H-15, Urb. R. Castilla
Cajamarca
Tel: 51/449–23356
Fax: 51/449–23356

Ecology and Conservation Association
Calle dos de mayo 527
Miraflores
Tel: 511/42–7369

Peruvian Association for the Conservation of Nature
Parque Jose de Acosta 187
Magdalena
Casilla 621
Lima 17
Tel: 511/264–0970
Fax: 511/264–3027
E-mail: apeco@datos.limaperu.net

Peruvian Society of Environmental Law
Prolongacion Arenales 437
San Isidro
Lima 27
Tel: 511/441–9171
Fax: 511/442–4365
E-mail: caillaux@spda2.org.pe

Pro-Nature-Peruvian Foundation for the Conservation of Nature
Pasaje Parque Blume N106 Con Av
Gnrl. Cordova 518
PO Box 18–1393
Miraflores
Lima 18
Tel: 511/441–3800
Fax: 511/441–2151
E-mail: fpcn@mail.cosapidata.com.pe
<http://www.pronaturaleza.com.pe>

Proterra
Calle Madrid 166, Miraflores
Lima 18
Tel: 511/242–0238
Fax: 511/446–6363
E-mail: proterra@mail.cosapidata.com.pe

Saint Lucia

Organization of Eastern Caribbean States
PO Box 1383
Castries
Tel: 758/453–6208
Fax: 758/452–2194
E-mail: oecsnrmu@candw.lc
<http://www.oecsnrmu.org>

Saint Lucia National Trust
Clarke Avenue
PO Box 595
Castries

Tel: 758/452–5005
Fax: 758/453–2791
E-mail: natrust@candw.lc

Suriname

Conservation International
Burenstraat No. 17
PO Box 2420
Paramaribo
Tel: 597/42–1305
Fax: 597/42–1172

Venezuela

Audubon Society of Venezuela
Apdo. 80450
Caracas 1080-A
Tel: 58/292–2812
Fax: 58/291–0716
E-mail: crodner@conicit.ve

ECODESARROLLO
PO Box 750
Mérida
Tel: 58/747–13814
Fax: 58/747–14576
E-mail: Jcenteno@telcel.net.ve
<http://www.ciens.ula.ve/jcenteno/>

Foundation for the Defense of Nature
Apdo. 70376
Caracas 1071-A
Tel: 58/238–1793
Fax: 58/238–6547
E-mail: fundena@telcel.net.ve
<http://www.fudena.org>

Friends of the Earth
Lucia Antillano Armas
Apdo. 718
Maracaibo
Tel: 58/615–23030

MANFAUNA
UNELLEZ, Mesa de Cavacas 3323
Guanare, Edo. Portuguesa
Tel: 58/576–8130
Fax: 58/576–8130
E-mail: angonfer@dino.conicit.ve

PROVITA
Apdo. 47552
Of 105 y 106
Caracas 1041-A
Tel: 58/25–76–2828
Fax: 58/25–76–1579
E-mail: provita1@telcel.net.ve

Wildlife Foundation
Av. Perimetral
Edif. Mamatica 3, Cumana
Estado de Sucre
Tel: 58/93–33–0265
Fax: 58/93–33–0265
E-mail: fdsv@ven.net

Organizations and Government Agencies Based in the United States and Canada

> The groups listed in this section include conservation groups, research institutions, universities, and government agencies. All of them are directly involved in tropical biodiversity conservation and research and can provide general information as well as information about their work. Many of the Web sites listed provided valuable information for this book.

CANADA

University of Toronto
Urban Entomology Program
33 Willcocks Street
Toronto, Ontario M5S 3B3
Tel: 416/978–5755
Fax: 416/978–3834
E-mail: t.myles@utoronto.ca
<http://www.utoronto.ca/forest/termite/termite.htm>

World Commission on Forests and Sustainable Development
c/o International Institute for Sustainable Development
161 Portage Avenue East, 6th floor
Winnipeg, Manitoba R3B 0Y4
Tel: 204/958–7700
Fax: 204/958–7710
E-mail: wcfsd@iisd.ca
<http://www.iisd.ca/wcfsd>

UNITED STATES

Action for Community & Ecology in the Rainforests of Central America
PO Box 57
Burlington, VT 05402
Tel: 802/863–0571
Fax: 802/864–8203
E-mail: acerca@sover.net
<http://www.acerca.org>

African Wildlife Foundation
1400 16th Street NW, Suite 120
Washington, DC 20036
Tel: 202/939–3333
Fax: 202/939–3332
E-mail: africanwildlife@awf.org
<http://www.awf.org/home.html>

Amanaka'a Amazon Network
PO Box 509
New York, NY 10276
Tel: 212/479–7360
E-mail: info@amanakaa.org>
<http://www.amanakaa.org>

American Bamboo Society
755 Tiziano Avenue
Coral Gables, FL 33143–6263
Tel: 305/662–2315
Fax: 305/669–6842
E-mail: bambuscape@aol.com
<http://www.tropicalbamboo.org/>

American Bird Conservancy
1250 24th Street NW, Suite 400
Washington, DC 20037

Tel: 800/BIRD–MAG
E-mail: abc@abcbirds.org>
<http://www.abcbirds.org>

American Botanical Council
PO Box 144345
Austin, TX 78714–4345
Tel: 512/926–4900
Fax: 512/926–2345
E-mail: abc@herbalgram.org>
<http://www.herbalgram.org>

American Zoo and Aquarium Association
8403 Colesville Road, Suite 710
Silver Spring, MD 20910–3314
Tel: 301/562–0777
Fax: 301/562–0888
http://www.aza.org/about/

Ancient Forest International
PO Box 1850
Redway, CA 95560
Tel/Fax: 707/923–3015
E-mail: afi@ancientforests.org>
<http://www.ancientforests.org>

Bat Conservation International
PO Box 162603
Austin, TX 78716
Tel: 512/327–9721
Fax: 512/327–9724
E-mail: batinfo@batcon.org
<http://www.batcon.org>

Biodiversity Action Network
1630 Connecticut Avenue NW, 3rd Floor
Washington, DC 20009
Tel: 202/238–0550
Fax: 202/238–0579
E-mail: bionet@igc.org
<http://www.bionet–us.org>

Biodiversity Conservation Network
1250 24th Street NW, Suite 600
Washington, DC 20037
Tel: 202/861–8348
Fax: 202/861–8324
E-mail: bcn@wwfus.org
<http://www.BCNet.org>

Biodiversity Project
214 N. Henry Street, Suite 203

Madison, WI 53703
Tel: 608/250–9876
Fax: 608/257–3513
<http://www.biodiversityproject.org>

Caribbean Conservation Corporation
4424 NW 13th Street, Suite #A1
Gainesville, FL 32602–2866
Tel: 800/678–7853
Fax: 352/373–6441
E-mail: ccc@cccturtle.org
<http://www.cccturtle.org>

Center for Biodiversity and Conservation
American Museum of Natural History
Central Park West at 79th Street
New York, NY 10024
Tel: 212/769–5742
Fax: 212/769–5292
E-mail: biodiversity@amnh.org
<http://research.amnh.org/biodiversity/center/
    cmada.html>

**Center for Sustainable Development in the
    Americas**
1700 Connecticut Avenue NW, #403
Washington, DC 20009
Tel: 202/588–0155
Fax: 202/588–0756
E-mail: info@csdanet.org
<http://www.csdanct.org>

**Coalition for Amazonian Peoples & Their
    Environment**
1511 K Street NW, Suite 627
Washington, DC 20005
Tel: 202/785–3334
Fax: 202/785–3335
E-mail: amazoncoal@igc.apc.org
<http://www.amazoncoalition.org>

Conservation International
2501 M Street NW, Suite 200
Washington, DC 20037
Tel: 202/429–5660, 800/429–5660
Fax: 202/887–5187
E-mail: newmember@conservation.org
<http://www.conservation.org>

**Consultative Group on International Agricultural
    Research**
The World Bank

U-3-173
1818 H Street NW
Washington, DC 20433
Tel: 202/473–8918
E-mail: Avonderosten@worldbank.org
<http://www.cgiar.org/>

Cultural Survival
221 Prospect Street
Cambridge, MA 02139
Tel: 617/441–5400
Fax: 617/441–5417
E-mail: csinc@cs.org
<http://www.cs.org/>

The Duke University Primate Center
3705 Erwin Road
Durham, NC 27705
Tel: 919/489–3364
E-mail: primate@acpub.duke.edu
<http://www.duke.edu/web/primate/home.html>

Earthwatch Institute
3 Clock Tower Place, Suite 100
Box 75
Maynard, MA 01754
Tel: 978/461–0081, 800/776–0188
Fax: 978/461–2332
E-mail: info@earthwatch.org
<http://www.earthwatch.org>

Ecotourism Society
PO Box 755
North Bennington, VT 05257
Tel: 802/447–2121
Fax: 802/447–2122
E-mail: ecomail@ecotourism.org
<http://www.ecotourism.org>

Environmental Defense Fund
257 Park Avenue South
New York, NY 10010
Tel: 800/684–3322
E-mail: contact@environmentaldefense.org
<http://www.edf.org>

Environmental Law Institute
1616 P Street NW, Suite 200
Washington, DC 20036
Tel: 202/939–3800
Fax: 202/939–3868
E-mail: widholim@eli.org
<http://www.eli.org>

Forests of the World
PO Box 2693
Durham, NC 27715
Tel: 919/549–7333
Fax: 919/549–8355
E-mail: fowllc@igc.org
<http://www.worldforests.com>

Germplasm Resources Information Network
U.S. Department of Agriculture
Bldg. 003, Rm. 407, BARC–West
10300 Baltimore Avenue
Beltsville, MD 20705–2350
E-mail: dbmu@ars–grin.gov
<http://www.ars–grin.gov/npgs/>

Global Environment Facility
1818 H Street NW
Washington, DC 20433
Tel: 202/473–0508
Fax: 202/522–3240
<http://www.gefweb.org>

Greenpeace USA
1436 U Street NW
Washington, DC 20009
Tel: 202/462–1177
Fax: 202/462–4507
E-mail: greenpeace.usa@wdc.greenpeace.org
<http://www.greenpeace.org>

Harvard University Herbaria
22 Divinity Avenue
Cambridge, MA 02138
Tel: 617/495–2365
Fax: 617/495–9484
<http://www.herbaria.harvard.edu>

Hawaii Biological Survey
Bishop Museum
1525 Bernice Street
Honolulu, HI 96817–2704
Tel: 808/847–3511
E-mail: webmaster@bishopmuseum.org
<http://www.bishop.hawaii.org/>

Healing Forest Conservancy
3521 S Street NW
Washington, DC 20007
Tel/Fax: 202/333–3438
<http://www.shaman.com/Healing_Forest.html>

Inter-American Development Bank
Environmental Division
1300 New York Avenue NW
Washington, DC 20577
Tel: 202/623–1000
Fax: 202/623–1786
E-mail: webmaster@iadb.org
<http://www.iadb.org>

International Aviculturists Society
PO Box 2232
LaBelle, FL 33975
Tel: 941/–674–0321
Fax: 941/–675–8824
E-mail: Richard@funnyfarmexotics.com
<http://www.funnyfarmexotics.com/IAS>

International BioPark Foundation
PO Box 69069
Oro Valley, AZ 85737
Tel/Fax: 520/531–5580
E-mail: admin@biopark.org>
<http://www.biopark.org>

International Primate Protection League
PO Box 766
Summerville, SC 29484
Tel: 843/871–2280
E-mail: ippl@awod.com
< http://www.ippl.org>

International Rivers Network
Latin America Program
1847 Berkeley Way
Berkeley, CA 94703–1576
Tel: 510/848–1155
Fax: 510/848–1008
E-mail: irn@irn.org>
<http://www.irn.org>

LightHawk
Presidio Building 1007, 1st Floor
General Kennedy Avenue
PO Box 29231
San Francisco, CA 94129–0231
Tel: 415/561–6250
Fax: 415/561–6251
E-mail: sfo@lighthawk.org
<http://www.lighthawk.org>

Man and Biosphere Program
U.S. Department of State
OES/ETC/MAB

Washington, DC 20522–4401
Tel: 202/776–8318
Fax: 202/776–8367
<http://www.mabnet.org>

Missouri Botanical Gardens
PO Box 299
Street Louis, MO 63166
Tel: 800/642–8842
<http://www.mobot.org>

National Audubon Society
700 Broadway
New York, NY 10003
Tel: 212/979–3000
Fax: 212/ 979–3188
E-mail: webmaster@audubon.org
<http://www.audubon.org>

National Center for Biotechnology Information
Building 38A, Room 8N805
Bethesda, MD 20894
Tel: 301/496–2475
Fax: 301/480–9241
E-mail: info@ncbi.nlm.nih.gov
<http://www.ncbi.nlm.nih.gov/Taxonomy/tax.html>

National Wild Turkey Federation
PO Box 530
Edgefield, SC 29824
Tel: 803/637–3106
Fax: 803/637–0034
E-mail: nwtf@nwtf.net
<http://www.nwtf.org/>

National Wildlife Federation
8925 Leesburg Pike
Vienna, VA 22184
Tel: 703/790–4000
Fax: 703/827–2585
E-mail: info@nwf.org
<http://www.nwf.org>

Natural Resources Defense Council
1350 New York Avenue NW
Washington, DC 20005
Tel: 202/289–6868
Fax: 202/289–1060
E-mail: proinfo@nrdc.org>
<http://www.nrdc.org>

The Nature Conservancy
4245 North Fairfax Drive, Suite 100
Arlington, VA 22203–1606
Tel: 703/841–5300, 800/628–6860
<http://www.tnc.org>

The New York Botanical Garden
200 Street and Kazimiroff Boulevard
Bronx, NY 10458
Tel: 718/817–8700
<http://www.nyb.org>

Orangutan Foundation International
822 S. Wellesley Ave
Los Angeles, CA 90049
Tel: 310/207–1655
Fax: 310/207–1556
E-mail: ofi@orangutan.org
<http://www.orangutan.org/>

Organization for Tropical Studies
PO Box 90630, Duke Station
Durham, NC 27708–0630
Tel: 919/684–5774
Fax: 919/684–5661
E-mail: nao@acpub.duke.edu
<http://www.ots.duke.edu>

Oxfam America
South–American Program
26 West Street
Boston, MA 02111–1206
Tel: 800/77–OXFAM
Fax: 617/728–2594
E-mail: info@oxfamamerica.org
<http://www.oxfamamerica.org>

Peregrine Fund
566 W. Flying Hawk Lane
Boise, ID 83709
Tel: 208/362–3716
Fax: 208/362–2376
E-mail: tpf@peregrinefund.org
<http://www.peregrinefund.org>

Plant Explorer
6281 NE 6th Street
Fridley, MN 55432–5038
Tel: 612/571–5013
E-mail: j5rson@iversonsoftware.com
<http://www.iversonsoftware.com/>

Population Action International
1300 19th Street NW, 2nd Floor
Washington, DC 20036
Tel: 202/557–3400
Fax: 202/728–4177
E-mail: pai@popact.org
<http://www.populationaction.org>

Population Resource Center
1725 K Street NW, Suite 1102
Washington, DC 20006
Tel: 202/467–5030
Fax: 202/467–5034
E-mail: prc@prcdc.org
<http://www.prcdc.org>

Purdue University
Center for New Crops and Plant Products
1165 Horticulture Building
West Lafayette, IN 47907–1165
Tel: 765/494–6968
Fax: 765/494–0391
E-mail: janick@hort.purdue.edu
<http://www.hort.purdue.edu/newcrop/new-crop.html>

Rainforest Action Network
221 Pine Street, Suite 500
San Francisco, CA 94104
Tel: 415/398–4404
Fax: 415/398–2732
E-mail: rainforest@ran.org
<http://www.ran.org>

Rainforest Alliance
65 Bleecker Street, 6th Floor
New York, NY 10012
Tel: 212/677–1900
Fax: 212/677–2187
E-mail: canopy@ra.org
<http://www.rainforest–alliance.org>

Rainforest Foundation International
270 Lafayette Street, Suite 1107
New York, NY 10012
Tel: 212/431–9008
Fax: 212/431–9197
E-mail: rffny@rffny.org
<http://www.savetherest.org>

The Rainforest Trust
6001 SW 63rd Avenue
Miami, FL 33143
Tel: 305/667–2779
Fax: 305/665 –0691
E-mail: rforest@rainforesttrust.com
<http://www.rainforesttrust.com>

Raintree Nutrition, Inc.
10609 Metric Boulevard, Suite 101
Austin, Texas 78758
Tel: 800/780–5902, 512/633–5006
Fax: 512/833–5414
E-mail: info@rain-tree.com
<http://www.rain-tree.com>

RARE Center for Tropical Conservation
1840 Wilson Boulevard, Suite 402
Arlington, VA 22201–3000
Tel: 703/522–5070
Fax: 703/522–5027
E-mail: rare@rarecenter.org
<http://www.rarecenter.org>

Resources for the Future
1616 P Street NW
Washington, DC 20036–1400
Tel: 202/328–5000
Fax: 202/939–3460
E-mail: webmaster@rff.org
<http://www.rff.org>

Smithsonian Migratory Bird Center
National Zoological Park
Washington, DC 20008
Tel: 202/673–4908
Fax: 202/673–4916
<http://www.si.edu/smbc/start.html>

Tortoise Trust USA
PMB #292
204 North Oak Avenue
Owatonna, MN 55060
E-mail: ttrust@globalnet.co.uk
<http://www.tortoisetrust.org/>

Tropical Conservation and Development Program
Center for Latin American Studies
PO Box 115531
304 Grinter Hall
University of Florida

Gainesville, FL 32611–5531
Tel: 352/392–6548
Fax: 352/392–0085
E-mail: tcd@tcd.ufl.edu
<http://www.tcd.ufl.edu>

Tropical Forest Foundation
225 Reinekers Lane, Suite 770
Alexandria, VA 22314
Tel: 703/518–8834
Fax: 703/518–8974
E-mail: tff@igc.apc.org>

United Nations Development Programme
1 United Plaza, FF 1056
New York, NY 10017
Tel: 212/906–6755
Fax: 212/906–6973
<http://www.undp.org>

United Nations Environment Programme
Room DC2–0803
2 United Plaza
New York, NY 10017
Tel: 212/963–8138
Fax: 212/963–7341
E-mail: ipainfo@unep.org
<http://www.unep.org>

U.S. Agency for International Development Information Center
Ronald Reagan Building
Washington, DC 20523–1000
Tel: 202/712–4810
Fax: 202/216–3524
<http://www.usaid.gov>

U.S. Geological Survey
Biological Resources Division
Pacific Island Ecosystems Research Center
PO Box 44/Bldg. 344
Hawaii National Park, HI 96718
E-mail: biologywebteam@usgs.gov
<http://www.its.nbs.gov>

University of California
Museum of Paleontology
1101 Valley Life Sciences Building
Berkeley, CA 94720–4780
Tel: 510/642–1821
Fax: 510/642–1822
E-mail: webmaster@ucmp1.berkeley.edu
<http://www.ucmp.berkeley.edu/>

University of Connecticut
Ecology and Evolutionary Biology Conservatory
75 North Eagleville Road, Box U–3043
Storrs, CT 06269–3043
Tel: 860/486–0809
Fax: 860/486–6364
E-mail: clinton.morse@uconn.edu
<http://florawww.eeb.uconn.edu/>

University of Florida
Department of Entomology & Nematology
Gainesville, FL 32611–0620
Tel: 352/392–1901 ext. 130
Fax: 352/392–0190
E-mail: Foltz@gnv.ifas.ufl.edu
<http://www.ifas.ufl.edu/>

University of Georgia
Department of Botany
Athens, GA 30602–7271
Tel: 706/542–3732
Fax: 706/542–1805
E-mail: admit@dogwood.botany.uga.edu
<http://www.botany.uga.edu/>

Wildlife Conservation Society
Bronx Zoo
2300 Southern Boulevard
Bronx, NY 10460–1099
Tel: 718/367–1010
E-mail: feedback@wcs.org
<http://www.wcs.org>

Wildlife Preservation Trust International
1520 Locust Street, Suite 704
Philadelphia, PA 19102
Tel: 215/731–9770
Fax: 215/731–9766
E-mail: homeoffice@wpti.org
<http://www.wpti.org>

Winrock International
38 Winrock Drive
Morrilton, AK 72110–9370
Tel: 501/727–5435
Fax: 501/727–5242
E-mail: Information@winrock.org
<http://www.winrock.org/>

The World Bank
1818 H Street NW
Washington, DC 20433
Tel: 202/458–4682
Fax: 202/522–3256
<http://www.worldbank.org>

World Forestry Center
4033 SW Canyon Road
Portland, OR 97221
Tel: 503/228–1367
Fax: 503/228–4608
E-mail: mail@worldforest.org
<http://www.worldforest.org>

World Neighbors International Headquarters
4127 NW 122 Street
Oklahoma City, OK 73120
Tel: 800/242–6387, 405/752–9700
E-mail: info@wn.org
<http://www.wn.org>

World Resources Institute
10 G Street NE, Suite 800
Washington, DC 20002
Tel: 202/729–7600
Fax: 202/729–7610
E-mail: front@wri.org
<http://www.wri.org>

World Wildlife Fund
1250 24th Street, NW
PO Box 97180
Washington, DC 20077–7180
Tel: 202/293–4800, 800/225–5993
Fax: 202/293–9211
E-mail: membership@wwfus.org
<http://www.worldwildlife.org>

Worldwatch Institute
1776 Mass Avenue NW
Washington, DC 20036
Tel: 202/452–1999
Fax: 202/296–7365
E-mail: worldwatch@worldwatch.org
<http://www.worldwatch.org>

Index

Diane Jukofsky is the director of the Rainforest Alliance's Conservation Media Center, which is based in San José, Costa Rica. She edits a bimonthly, bilingual news bulletin that features articles about conservation issues in the Neotropics and manages and edits a Web site, called the Eco-Index (www.eco-index.org), which is a bilingual almanac of conservation initiatives in Mexico and Central America. She also organizes and leads communications skills workshops for conservation groups in Latin America and the Caribbean and environmental-reporting training seminars for journalists in the region. Her articles on tropical conservation and wildlife have appeared in dozens of science, nature, travel, news, and academic publications, and she has authored chapters in two books about tropical conservation.

In 2001, Jukofsky was named an honorary member of Sigma Xi, the Scientific Research Society. She is a member of the board of the Rainforest Alliance and World Teach-Costa Rica.

The Rainforest Alliance works to protect endangered ecosystems and the people, wildlife, and plants that live within them by transforming land use, business practices, and consumer behavior. The Alliance realizes this mission by designing and implementing environmentally and socially responsible management practices for natural resource-based activities such as logging, farming, and tourism. The Alliance focuses its efforts in the tropics, where most of the world's endangered ecosystems and biodiversity are located, and plays a unique role in ensuring environmentally responsible land use around parks and protected areas. The Alliance accomplishes these goals by: (1) creating alliances with farmers, scientists, industry representatives, political leaders, and other conservation and community groups, (2) providing technical and financial assistance and training to those groups and others, (3) developing, monitoring and certifying compliance with conservation standards, (4) raising consumer awareness, and (5) promoting complementary international and national policy change. For more information: Rainforest Alliance, 65 Bleecker Street, 6th floor, New York, NY 10012; tel: 212/677-1900, fax: 212/677-2187; e-mail: canopy@ra.org <www.rainforest-alliance. org>.